Lecture Notes in Bioinformatics 7605

Edited by S. Istrail, P. Pevzner, and M. Waterman

Subseries of Lecture Notes in Computer Science

David Gilbert Monika Heiner (Eds.)

Computational Methods in Systems Biology

10th International Conference, CMSB 2012
London, UK, October 3-5, 2012
Proceedings

Series Editors

Sorin Istrail, Brown University, Providence, RI, USA
Pavel Pevzner, University of California, San Diego, CA, USA
Michael Waterman, University of Southern California, Los Angeles, CA, USA

Volume Editors

David Gilbert
Brunel University
School of Information Systems, Computing and Mathematics
Uxbridge, Middlesex UB8 3PH, UK
E-mail: david.gilbert@brunel.ac.uk

Monika Heiner
Brandenburg University of Technology, Computer Science Institute
03013 Cottbus, Germany
E-mail: monika.heiner@informatik.tu-cottbus.de

ISSN 0302-9743 e-ISSN 1611-3349
ISBN 978-3-642-33635-5 e-ISBN 978-3-642-33636-2
DOI 10.1007/978-3-642-33636-2

Springer Heidelberg Dordrecht London New York

Library of Congress Control Number: 2012947593

CR Subject Classification (1998): F.1.1-2, I.6.3-5, J.3, I.2.3, D.2.2, D.2.4, F.4.3, I.1.3

LNCS Sublibrary: SL 8 – Bioinformatics

Typesetting: Camera-ready by author, data conversion by Scientific Publishing Services, Chennai, India

Printed on acid-free paper

Springer is part of Springer Science+Business Media (www.springer.com)

Preface

These proceedings contain the accepted contributions to the 10th International Conference on Computational Methods in Systems Biology (CMSB 2012), held at the Royal Society's headquarters in central London, UK, 3–5 October 2012.

The conference was an opportunity to hear about the latest research on the analysis of biological systems, networks, and data ranging from intercellular to multiscale. Topics included high-performance computing, and for the first time papers on synthetic biology. The conference brought together computer scientists, biologists, mathematicians, engineers, and physicists interested in a system-level understanding of biological processes. This was the first time that short contributions were accepted in the form of 'flash posters' – four-page papers published in the proceedings, together with a flash presentation advertising a regular poster.

The submissions comprised 46 regular papers and 16 flash posters by a total of 181 authors from 24 countries. All submitted papers were peer-reviewed by a program committee of 40 members coming from 12 countries, supported by 33 external reviewers. There were 152 reviews, with a range of 3 to 4 reviews for regular papers and 2 to 4 reviews for flash posters. The acceptance rate was 37% (regular papers) and 50% (flash posters).

The final program consisted of 3 invited talks, 17 full-length papers, 8 flash poster presentations, and a poster session.

We are delighted to acknowledge substantial support by the EasyChair management system, see http://www.easychair.org, during the reviewing process and the production of these proceedings. We would like to thank Teresa Czachowska for her help with the organization of the meeting, and Crina Grosan for her assistance with the website and publicity.

October 2012

David Gilbert
Monika Heiner

Organization

Program Committee

Paolo Ballarini	École Centrale de Paris, France
Alexander Bockmayr	Freie Universität Berlin, Germany
Kevin Burrage	University of Oxford, UK
Muffy Calder	University of Glasgow, UK
Luca Cardelli	Microsoft Research Cambridge, UK
Claudine Chaouiya	Instituto Gulbenkian de Ciência - IGC, Portugal
Attila Csikasz-Nagy	Microsoft Research – University of Trento, Italy
Vincent Danos	University of Edinburgh, UK
Finn Drabløs	Norwegian University of Science and Technology, Norway
François Fages	INRIA Rocquencourt, France
Jasmin Fisher	Microsoft Research Cambridge, UK
David Gilbert	Brunel University, UK
Stephen Gilmore	University of Edinburgh, UK
Simon Hardy	Université Laval Quebec, Canada
Sampsa Hautaniemi	University of Helsinki, Finland
Monika Heiner	Brandenburg University of Technology, Germany
Thomas Henzinger	IST, Austria
John Heath	Birmingham University, UK
Hidde de Jong	INRIA Rhône Alpes, France
Andzrej Kierzek	University of Surrey, UK
Natalio Krasnogor	Nottingham University, UK
Christopher Langmead	Carnagie Mellon University, USA
Nicolas Le Novere	EMBL - European Bioinformatics Institute, UK
Pietro Liò	University of Cambridge, UK
Molly Maleckar	Simula, Norway
Wolfgang Marwan	University of Magdeburg, Germany
Tommaso Mazza	I.R.C.C.S. Casa Sollievo della Sofferenza, Italy
Pedro Mendes	University of Manchester, UK
Satoru Miyano	University of Tokyo, Japan
Ion Moraru	University of Connecticut Health Center, USA
Joachim Niehren	INRIA Lille, France
Dave Parker	Oxford University, UK
Gordon Plotkin	University of Edinburgh, UK

Corrado Priami	Microsoft Research - University of Trento, Italy
Davide Prandi	CIBIO, Italy
Ovidiu Radulescu	Université de Montpellier 2, France
Nigel Saunders	Brunel University, UK
Koichi Takahashi	RIKEN, Japan
Carolyn Talcott	Stanford Research Institute, USA
P.S. Thiagarajan	National University of Singapore, Singapore
Adelinde Uhrmacher	University of Rostock, Germany
Verena Wolf	Saarland University Saarbrücken, Germany

Additional Reviewers

d'Alché-Buc Florence
Andreychenko, Aleksandr
Batmanov, Kirill
Bing, Liu
Feret, Jerome
Fokkink, Wan
Gao, Qian
Grosan, Crina
Gupta, Ashutosh
Gyori, Benjamin
Herajy, Mostafa
Hoehme, Stefan
John, Mathias
Kirste, Thomas
Klarner, Hannes
Kugler, Hillel
Kuwahara, Hiroyuki
Llamosi, Artemis
Maus, Carsten
Mikeev, Linar
Monteiro, Pedro
Nikolić, Durica
Pedersen, Michael
Pfeuty, Benjamin
Phillips, Andrew
Piterman, Nir
Rohr, Christian
Schwarick, Martin
Sezgin, Ali
Spieler, David
Sunnaker, Mikael
Trybiło, Maciej
Wu, Zujian
Zunino, Roberto

Table of Contents

Invited Talks

Regular Papers

Flash Posters

Differential and Integral Views of Gene-Phenotype Relations: A Systems Biological Insight

Denis Noble

Department of Physiology, Anatomy and Genetics,
Oxford University, UK
denis.noble@dpag.ox.ac.uk

Abstract. This lecture uses an integrative systems biological view of the relationship between genotypes and phenotypes to clarify some conceptual problems in biological debates about causality. The differential (gene-centric) view is incomplete in a sense analogous to using differentiation without integration in mathematics. Differences in genotype are frequently not reflected in significant differences in phenotype as they are buffered by networks of molecular interactions capable of substituting an alternative pathway to achieve a given phenotype characteristic when one pathway is removed. Those networks integrate the influences of many genes on each phenotype so that the effect of a modification in DNA depends on the context in which it occurs. Mathematical modelling of these interactions can help to understand the mechanisms of buffering and the contextual-dependence of phenotypic outcome, and so to represent correctly and quantitatively the relations between genomes and phenotypes. By incorporating all the causal factors in generating a phenotype, this approach also highlights the role of non-DNA forms of inheritance, and of the interactions at multiple levels.

References

1. Noble, D.: Differential and Integral view of genetics in computational systems biology. Interface Focus 1, 7–15 (2011)
2. Noble, D.: The Music of Life; Biology Beyond the Genome. Oxford University Press (2008)

D. Gilbert and M. Heiner (Eds.): CMSB 2012, LNCS 7605, p. 1, 2012.

Resolving the Three-Dimensional Histology of the Heart

Matthew Gibb[1], Rebecca A.B. Burton[2], Christian Bollensdorff[3], Carlos Afonso[4,7], Tahir Mansoori[1], Ulrich Schotten[5], Davig J. Gavaghan[1], Blanca Rodriguez[1], Jurgen E. Schneider[6], Peter Kohl[1,3], and Vicente Grau[4,7]

[1] Department of Computer Science, University of Oxford
[2] Department of Physiology, Anatomy and Genetics, University of Oxford
[3] The Heart Science Centre, National Heart and Lung Institute, Imperial College London
[4] Oxford e-Research Centre, University of Oxford
[5] Department of Physiology, Maastricht University
[6] British Heart Foundation Experimental MR Unit, Department of Cardiovascular Medicine, University of Oxford
[7] Institute of Biomedical Engineering, Department of Engineering Science, University of Oxford

Abstract. Cardiac histo-anatomical structure is a key determinant in all aspects of cardiac function. While some characteristics of micro- and macrostructure can be quantified using non-invasive imaging methods, histology is still the modality that provides the best combination of resolution and identification of cellular/sub-cellular substrate identities. The main limitation of histology is that it does not provide inherently consistent three-dimensional (3D) volume representations. This paper presents methods developed within our group to reconstruct 3D histological datasets. It includes the use of high-resolution MRI and block-face images to provide supporting volumetric datasets to guide spatial reintegration of 2D histological section data, and presents recent developments in sample preparation, data acquisition, and image processing.

Keywords: Cardiac imaging, cardiac microstructure, histology, three-dimensional reconstruction.

1 Introduction

Cardiac structure is a key determinant of all relevant aspects of cardiac function in health and in pathological states. This includes normal electrophysiological activity [1,2,3], as well as the initiation or termination of arrhythmias [4,5,6,7]. Similarly, mechanical activity [8,9] is fundamentally affected by cardiac structure. Importantly, the relation between structure and function is bi-directional: cardiac structure is affected by the mechanical and electrophysiological environments as well. These processes occur at multiple spatial and temporal scales, from acute changes [10,11] to medium-term effects such as cardiac memory [12] and long-term tissue remodeling [13,14]. This structure-function cross-talk involves all cell populations in the heart, from myocytes to cells such as are contained in the cardiac connective tissue, vasculature, neurons, endothelium, etc., whose discrimination requires differential

D. Gilbert and M. Heiner (Eds.): CMSB 2012, LNCS 7605, pp. 2–16, 2012.

approaches that are offered by 'classical' histology techniques. In short, there are multiple and dynamic interactions between cardiac structure and mechano-electrical function, and these are of crucial relevance for normal beat-by-beat activity of the heart, as well as for pathogenesis and therapy. The study of these interactions requires accurate knowledge of cardiac three-dimensional (3D) structure, at multiple scales from sub-cellular levels to whole organ.

Computational models have been proposed, and are increasingly being applied, as a way to link spatio-temporal scales, complementing traditional "wet-lab" approaches and projecting between bench and bed-side [15,16]. State-of-the-art models link protein structures of ion-channels to cell electrophysiology, multi-cellular coupling, and representations of heart anatomy that take into account locally prevailing cell alignment (usually, if inaccurately in terms of the histological substrate, termed fiber orientation), projecting through to clinical relevance, such as for drug actions [17,18,19].

Initial descriptions of cardiac microstructure in general, and myocyte orientation in particular, have arisen from serial histological sectioning, usually of selected locations [20] rather than full hearts. Also important in this context is the calculation of deformation patterns: the analysis of motion can, in principle, provide indirect knowledge about the microstructure. Linking microstructure to motion is, however, a fundamental unresolved issue. Deformation patterns can be obtained by *in vivo* imaging modalities, including echocardiographic strain-rate monitoring [21], or tagged / phase contrast MRI [22,23]. These techniques have limited spatial resolution, precluding analysis of fiber directions. Diffusion Tensor-MRI (DT-MRI) has represented a breakthrough in terms of non-invasive assessment of cell alignment in tissue, but again offers limited resolution, providing an indirect measure of the microstructure through the quantification of water diffusion patterns over volume-averaged locations. Diffusion Spectrum Imaging (DSI), however, resolves minute detail, including fibers crossing within an imaging volume unit (voxel), by increasing the number of directions in which the 3D diffusion function (i.e. q-space) is sampled. From this data, a probability density function of fiber orientation per voxel can be extracted for fiber tract construction. Recent developments have also shown the possibility of quantifying structure in the myocardium with para-cellular resolution, using contrast-enhanced MRI [24], in particular in the *ex vivo* setting. However, in contrast to histological approaches, this does not allow, at present, positive identification of cell-size and -type.

Micro-computed tomography (μCT) is an emerging, non-invasive imaging modality that allows for non-destructive, high-resolution imaging of tissue. It has been shown that μCT is well-suited for imaging of bones and calcified structures, but it provides low soft tissue contrast. Tissue such as heart muscle therefore requires pretreatment with specific stains or contrast agents. While μCT provides superior spatial resolution, compared to MRI (typically $<10\,\mu m$ *vs.* $>20\,\mu m$, respectively), it cannot resolve cellular composition of tissue.

Histology, therefore, is still the only way to comprehensively characterize cardiac microstructure including identification of different cell types within the myocardium. But, it suffers from two fundamental drawbacks: first, it is an inherently 'destructive' technique and can thus be used only on explanted tissue fragments or organs; second, acquired histological section images do not form an inherently consistent 3D data set.

This limits the analysis to in-plane characteristics, a major limitation when dealing with complex, functionally relevant 3D structures, as is the case for myocardium. Methods have been proposed to reconstruct 3D volumes from histology sections, particularly for the brain [25,26,27], with leading examples projecting right through to mapping functional observations to structural substrates [28]. Other examples include skeletal muscle [29] and the atrio-ventricular node of rabbit heart [30]. To our knowledge, however, 3D reconstruction of a whole heart, or even of only the ventricles, from histology sections has not been implemented yet. This paper presents an update on recent methods developed in our group, combining improvements in sample processing, data acquisition, and image analysis. These have reached the point where 3D histology of the heart is becoming a reality.

2 3D Histology from Serial Sections

2.1 Acquisition

All 3D histology reconstruction methods are based on the acquisition of individual 2D histology images, whether as sections [1,33], or from scans of the un-cut surface of the embedded tissue (so-called block-face imaging) [31]. Here we summarize the main steps of our section-based protocol; a more detailed explanation can be found elsewhere [1,33]. The methods described (and the data used for illustration) in this paper are applicable independently of species (images include tissue from New Zealand white rabbit, Wistar rat, and goat hearts).

Excised hearts were perfused in a Langendorff system and fixed by coronary perfusion using Karnovsky's fast-acting fixative (a paraformaldehyde-glutaraldehyde mix). Hearts were left in the fixative overnight, including gadodiamide as a contrast agent for anatomical MRI scanning (more detail is provided in Section 3.1). After MRI, hearts were prepared for paraffin wax embedding. This included the process of sequential dehydration in a series of alcohol concentration steps (20%-100%), clearing in xylene, and impregnation with molten black paraffin wax (subsequently allowed to solidify by cooling). Samples embedded in the wax block were mounted on a Leica SM2400 sledge-microtome, where 10 μm sections were cut. Sections were transferred to a water bath (Leica HI1210), allowing them to relax on the water surface before they were collected onto glass slides and allowed to air-dry. After overnight de-waxing at 60°C, slides were trichrome stained (labeling connective tissue blue-green, myocytes pink-red and nuclei blue-black; see Fig. 1) using an automated stainer (Leica Autostainer XL), mounted in DPX, and left overnight in a fume hood to dry. Sections were then imaged on a Leica DM4000B light microscope, fitted with an automated motorized platform for mosaic imaging. Individual image tiles were taken using a 3.3 megapixel camera (Leica DFC320) in 24-bit color mode, and either a x10 or a x5 objective so that individual (square) voxels had an edge-length of either 0.55 μm or 1.1 μm (respectively). Overall dataset sizes (up to 50 GB per rat heart; exceeding 1.5 TB per rabbit) pose significant challenges for subsequent visualization [32] and computational processing. Fig. 1 shows a sample long-cut section, obtained from a rabbit heart, and a zoomed-in illustration of inherent image resolution.

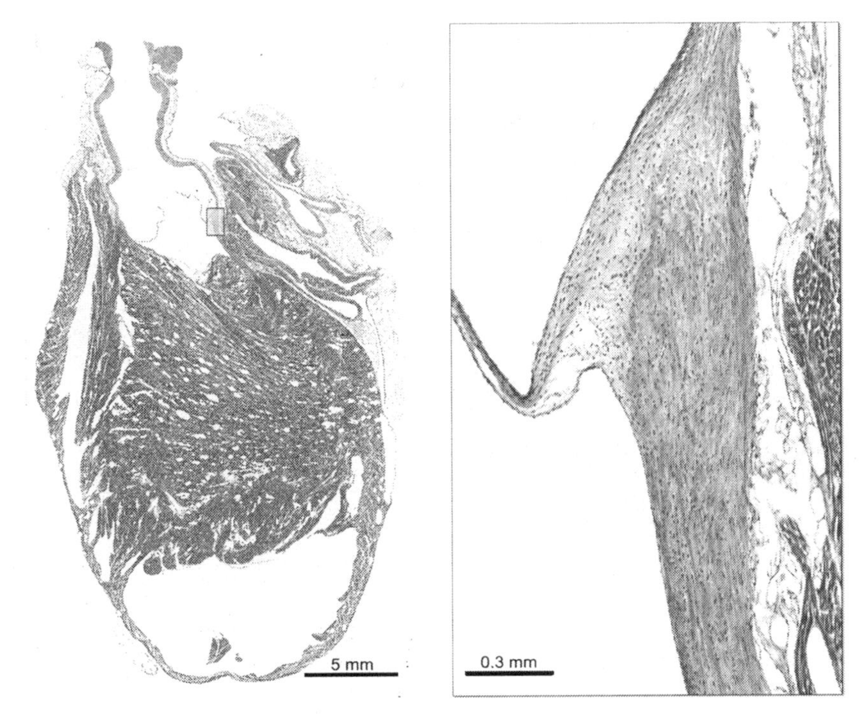

Fig. 1. Illustration of extended-plane high-resolution histological image generation, here from a long-cut of rabbit whole-heart tissue. The image on the left covers an area of 36 mm × 22 mm. It is mosaically assembled from individual microscopic image tiles (pixel size 1.1 μm × 1.1 μm; see enlarged view of the aortic valve region on the right). A 'live zoomable' version of this image is available at `http://gigapan.com/gigapans/c9a83b097c274d94a8b3a59b955ec39b/`

2.2 Section to Section Registration

Even in the absence of 3D reference geometry it is possible, in principle, to produce a volumetric reconstruction by applying sequential, section-to-section registration (image alignment). In this process, one of the sections is chosen as a reference, and the neighboring sections are registered sequentially to the previous one, until the whole volume is covered. Coherence of volume registration manifests itself in smooth contours in the section-normal direction. It is clear, though, that section-to-section alignment will tend to reduce gradual inter-section differences, thus distorting the resulting volumetric shapes. To illustrate this artifact, one can think of the result of reconstructing a cross-sectioned banana-like shape by optimizing the alignment of each section to its neighbor: the reconstructed shape would be straightened, resembling a cylinder.

Fig. 2 illustrates the relevance of this artifact for cardiac tissue reconstruction. A transversal cut through the volume obtained after section-to-section affine registration applied to a whole left ventricular histology data set is shown in Fig. 2 (left). For comparison, Fig. 2 (right) shows the same sections, reconstructed by registration with a corresponding block-face image-based volume, as detailed in Section 4. Structures in the section-to-section registered volume tend to appear artificially straightened in the cross-section direction (see, for example, the septal tissue region in the centre).

When only histological sections are available, this artifact is unavoidable since the original volumetric structure is unknown (beyond certain 'generic' *a-priori* insight into general shape and size of a sample). In principle, this volumetric reference can be obtained by a 3D MRI scan, as explained in Section 3.

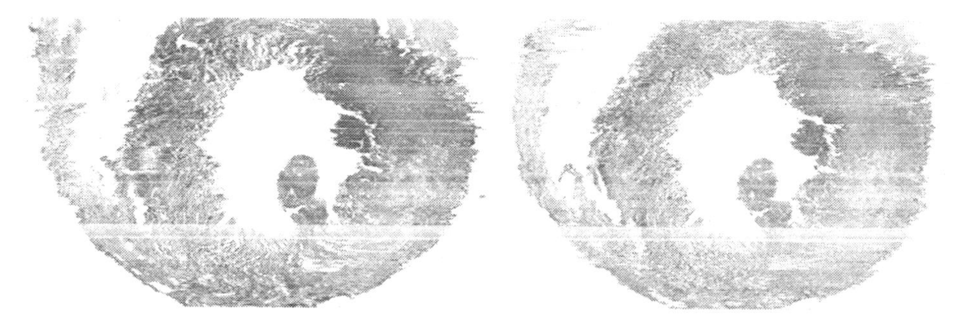

Fig. 2. Comparison of 3D histological section alignment, obtained either by section-to-section affine registration (left), or after prior co-registration with corresponding block-face images (right). The regional "straightening" of structures, apparent on the left, occurs even though the affine transform used here limits excess lateral translocation of structures. This is avoided on the right by reference to volumetric shape information, provided by an additional data set. Note: lateral banding is caused by differences in staining during histological data acquisition.

3 3D Histology Using MRI as Reference

3.1 MRI Acquisition

MRI volumes were acquired on a 9.4 T (400 MHz) MR system (Varian Inc., Palo Alto, CA, USA), using a 3D gradient echo sequence. More detail on MR imaging is available elsewhere [1,33,34]. We obtained images with voxel dimensions of 21.5 μm (isotropic) for rat, and 24.5 μm × 24.5 μm in-plane, 26.5 μm between planes for rabbit.

3.2 Histology / MRI Registration

Initial efforts at obtaining 3D histology volumes within our group focused on the registration of histology sections to corresponding MRI data [35]. Fig. 3 shows the schematic depiction of the procedure. First, histology sections are roughly aligned by section-to-section registration using a rigid transform to compensate for gross misalignments (rotation and translocation). This is followed by independent scaling on each of the three axes to compensate for size differences between histology and MRI data voxels. The volume thus constructed is segmented and aligned rigidly (T1) with the corresponding MRI volume, thus producing an initial, rough correspondence between histology sections and MRI data. The MRI volume is then sliced in a plane corresponding to histology sections, and a 2D registration (T2) between corresponding slices is performed. This produces a stack of 2D MRI images that is roughly aligned with histology sections, and sections in the direction perpendicular to the slices begin to be coherent between MRI and histology data. 3D registration is performed again to improve T1, and the process continues in an iterative loop until convergence is achieved. At this point, each histology section is assigned a corresponding (re-sampled) slice from the MRI volume, and a non-rigid registration (T3) is performed. This involves in-plane (2D) correction of histology sections to identified MRI reference slices. Quantitative validation shows that the procedure is successful in producing a coherent 3D histological volume, as shown in Fig. 4.

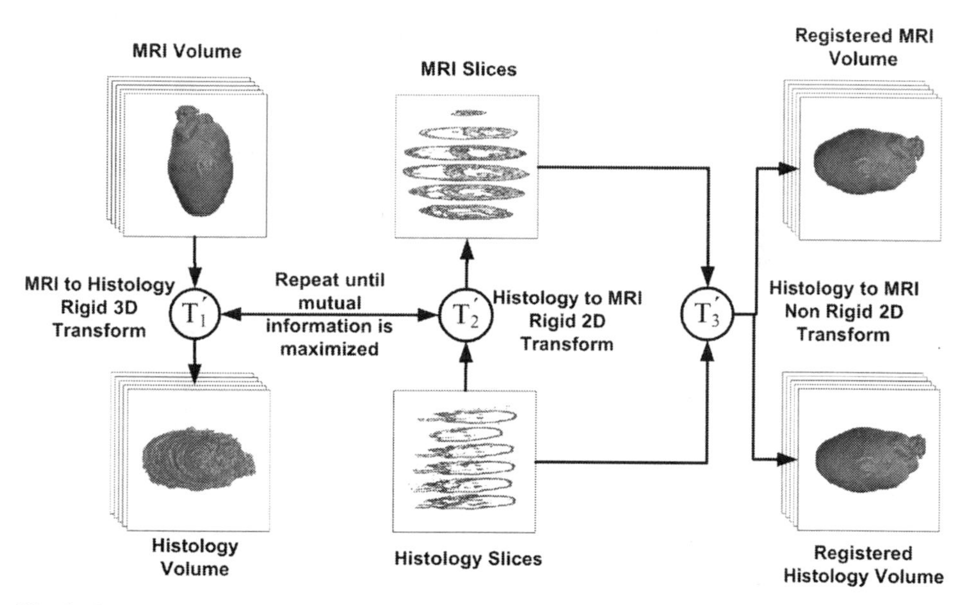

Fig. 3. Iterative algorithm for registration of serial 2D histology sections to a 3D MRI dataset, after [35]; see text for detail

However, by performing non-rigid transformations only in the plane of histological sectioning, we assume that no 3D volume deformations have occurred. This is not a reasonable assumption: in addition to significant non-rigid deformation that occurs during tissue sectioning / relaxation / mounting, it is important to appreciate that the actual shape and size of the dehydrated wax-embedded heart prior to section is different from the same 'wet' heart while it was agar-embedded for 3D MRI data acquisition. This means that even optimal registration of histology sections to MRI slices and in-plane non-rigid correction for deformation will miss any difference present between MRI data and the (un-cut) 3D cardiac anatomy in the section-normal direction.

Fig. 5 shows examples of histology sections, side by side with best-match corresponding MRI slices, resampled in the plane corresponding to the histology section after 3D alignment but prior to a final 2D non-rigid registration step. While the first two rows show good overall correspondence, the third highlights the difficulty associated with out-of-plane deformations.

It is, of course, theoretically possible to extend the algorithm in Fig. 3 to allow non-rigid out-of-plane transformations. This requires true 3D co-registration which is a challenging task and which makes the iterative algorithm highly sensitive to the initial location of sections. Instead, we opted for the inclusion of an intermediate dataset, obtained on the dehydrated wax-embedded tissue before slicing, as explained in Section 4.

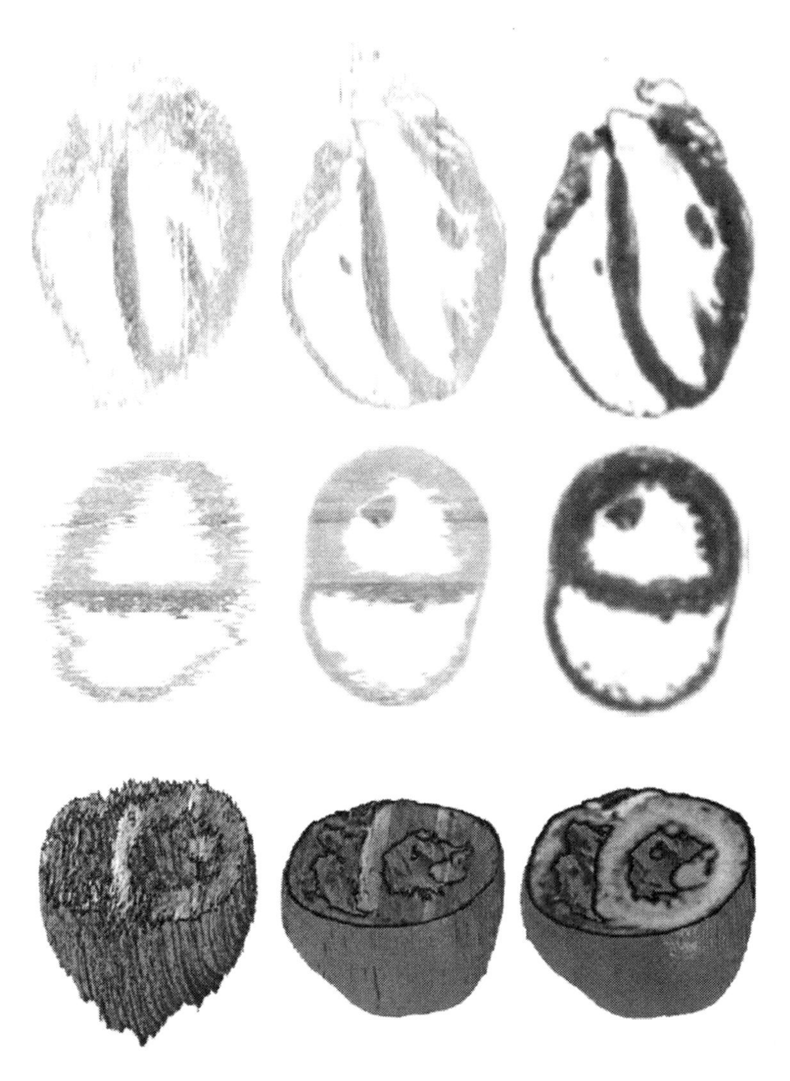

Fig. 4. Results of application of the 3D registration procedure depicted in Fig. 3 to a rabbit heart dataset. The two top rows show alignment in the section-normal direction before (left column) and after (middle column) registration with the corresponding (downsampled) MRI data (right column). The bottom row shows a 3D reconstruction of the volumes.

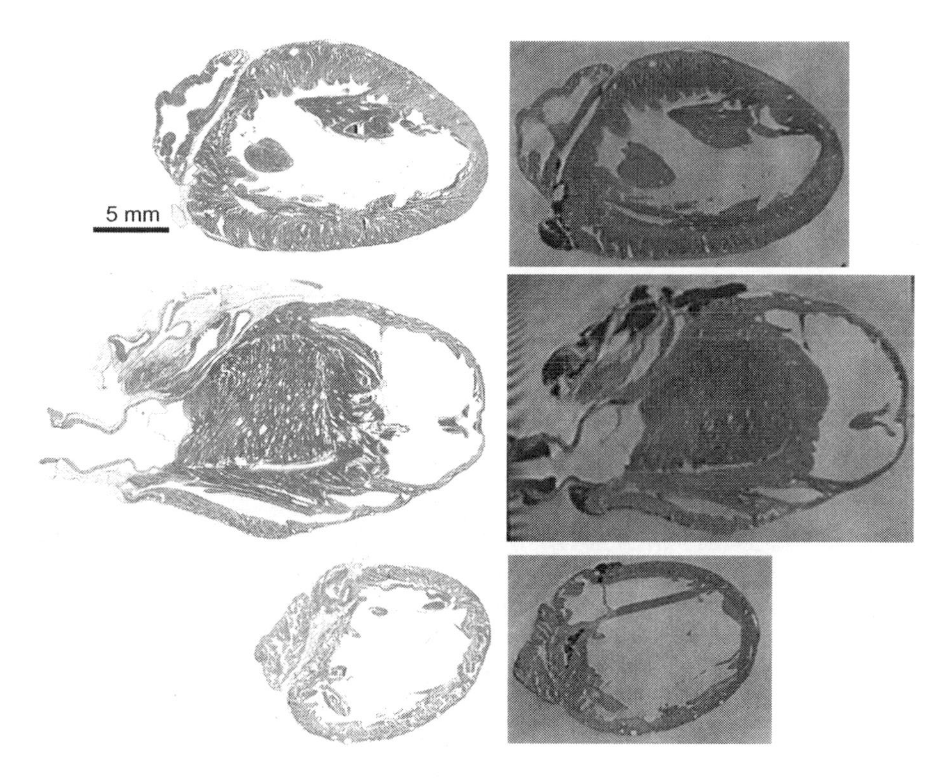

Fig. 5. Examples of histology sections and corresponding slices through the MRI data. While the two top rows show an apparently good match of key structures, the bottom row highlights significant differences present between images: a straight muscle structure, clearly visible in the MRI data, appears as two cross-cuts in the histology section, caused by out-of plane deformations that are not compensated by the registration algorithm shown in Fig. 3.

4 3D Histology Using Block-Face Images for Intermediate Alignment

4.1 Image Acquisition

As highlighted above, the tissue anatomy scanned by MRI is inherently different from the dehydrated wax-embedded sample. In addition, as histology is an inherently destructive technique, it is associated not only with deformation, but also 'irrecoverable' damage. This can give rise to artifacts (folds, rips) which may be minimized by careful processing, but not completely avoided (in particular when hundreds of sections are prepared, e.g. 1,800 per rabbit heart). Tissue deformation is non-linear, and tissue islands that are surrounded by wax (such as are generated when cutting across thin trabaeculae or free-running Purkinje fibers) may dislocate or even detach from the rest of the section. All of this will affect volumetric reconstructions and correspondence between 3D histology and MRI data sets.

To compensate for this, we introduce an additional step, acquiring images of the surface of the wax block containing the tissue, immediately before cutting the next section. Images are acquired using a Canon EOS 450D camera (14.8 mm × 22.2 mm CMOS sensor, 12.2 megapixels) with an EFS 18-55 mm objective including an extension tube EF12 II (all Canon). Block-face imaging provides an in-plane reference with true one-to-one correspondence to the subsequent high-resolution histology image, requiring only 2D correction for sectioning- and processing-related deformations. As the block face images form a consistent 3D dataset (camera and tissue block are in a defined position for image acquisition, and the block surface is always at the same elevation as it is lifted by exactly the amount that will be cut, here 10 μm per step), it can be used to precede non-rigid registration with the MRI scan.

4.2 Image Alignment

As explained in Section 4.1, block-face images provide an intermediate point between 3D MRI and full-resolution serial 2D histology sections (as shown in Fig. 6). On the other hand, block face images are of lower resolution than histology, and are not stained for discrimination of different tissue components. In addition, even using black wax, tissue structures from below the slicing plane will show through. Fig. 7 shows a matching pair of block-face and stained section images, where the differences in resolution and definition of contours are apparent.

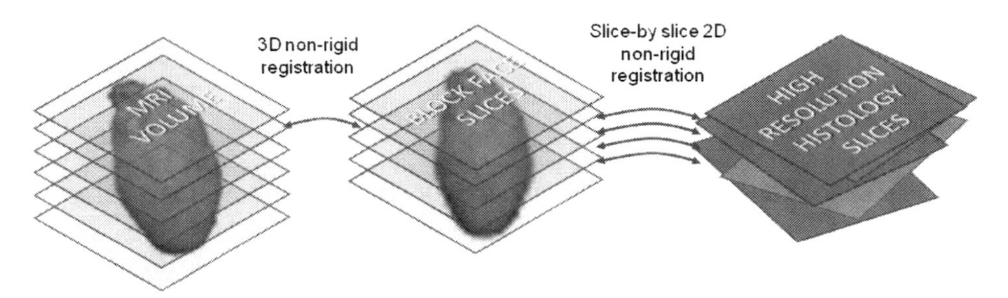

Fig. 6. Diagram showing the use of block face images as an intermediate reference point between the MRI volume and high-resolution 2D histology sections

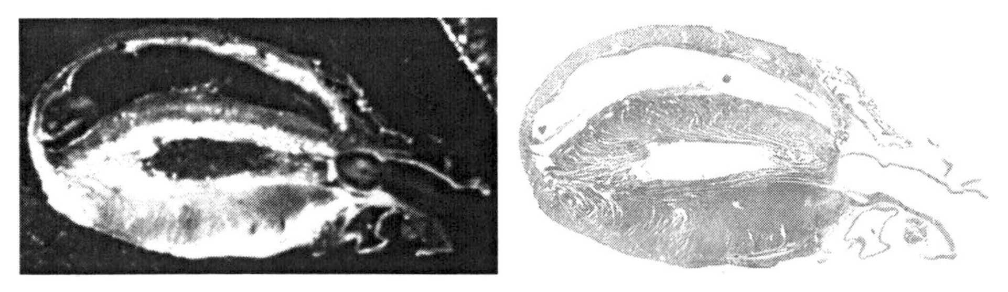

Fig. 7. Sample block face image (left) and corresponding 10 μm section after staining (right)

Section-to-section registration was performed between block-face and histology images. Fig. 8 (left column) shows the result of this registration using an affine transform and a cross-correlation similarity function. As illustrated, the algorithm is successful at aligning sections to their approximate pre-sectioning shape. However, misalignments between sections are still visible. These are due largely to i) the different resolution of block-face images, and ii) the presence of image boundary delineation errors caused by projection of tissue structures from deeper layers of the wax block.

The cross-sections shown on the left hand column of Fig. 8 still show a certain discontinuity of the contours: this is particularly visible on the magnified image in the bottom row. To improve this we have developed a new algorithm for filtering the section-to-section alignment transformations. This technique is based in the idea that the transformation needed to align a section to the one immediately above it is approximately the inverse of the one needed to align it to the one immediately below. Mathematically, the algorithm works by simulating an isotropic diffusion process to guide the transformations applied to sections either side of a slice. The algorithm is applied repeatedly until high frequency transformational noise has been damped and the reconstructed volume is smooth. Results of this new technique, shown in the right hand column of Fig. 8, illustrate the improved smoothness of the contours, while overall tissue geometry is maintained. However, the limited accuracy of tissue boundary outlines in the wax block image remains a constraining factor.

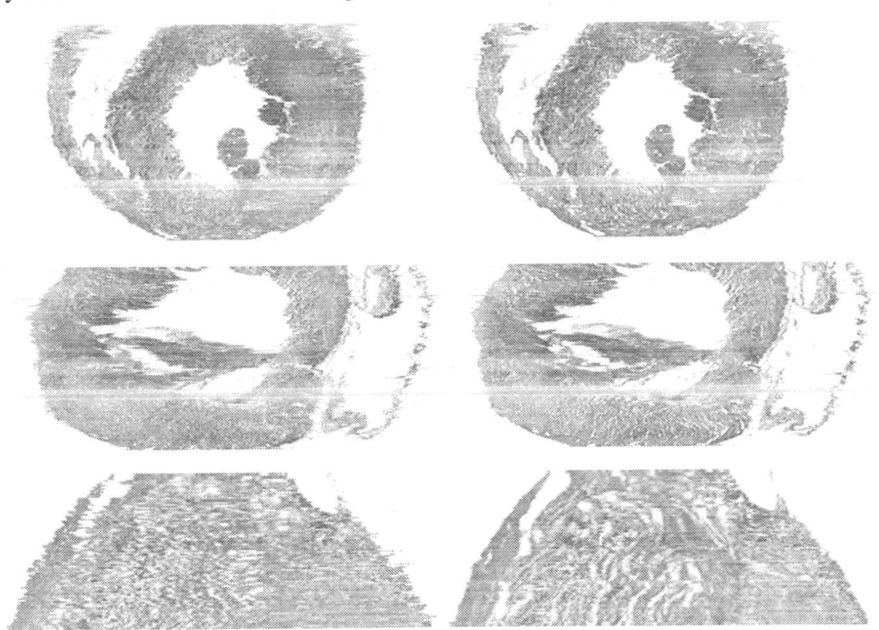

Fig. 8. Results of 2D section to block-face image registration. Top and middle rows show reconstructions orthogonal to the original sections; bottom row shows a close-up view of part of the left ventricular wall. The left column shows results from affine registration (see also Fig. 2). On the right, results after application of our new transformation diffusion step. The improvement of section alignment is particularly visible in the close-up image at the bottom.

4.3 Block Face Images Using Polarized Light

As explained in Section 4.2, the use of black wax is not sufficient to exclude light penetration into the wax block, and reflection from deeper tissue components. However, only the surface is representative of the tissue that will be contained in the next histology section.

To overcome this limitation we introduced the use of blue light to reduce sample penetration (LED CBT90, peak wavelength 459 nm, Luminus Devices), and polarizing filters optimized for collection predominantly of reflected light. A polarizing filter (Canon PLC-B, 58 mm), mounted in front of the camera objective, was rotated to be parallel to the polarization plane of light reflected by the shiny surface of the wax (light reflected from transparent materials at an angle that is different from perpendicular to the surface is partly or fully polarized). In addition, light sources were equipped with additional polarizing filters (Chroma, EM 25 mm), to reduce the fraction of illumination light that penetrates the wax surface.

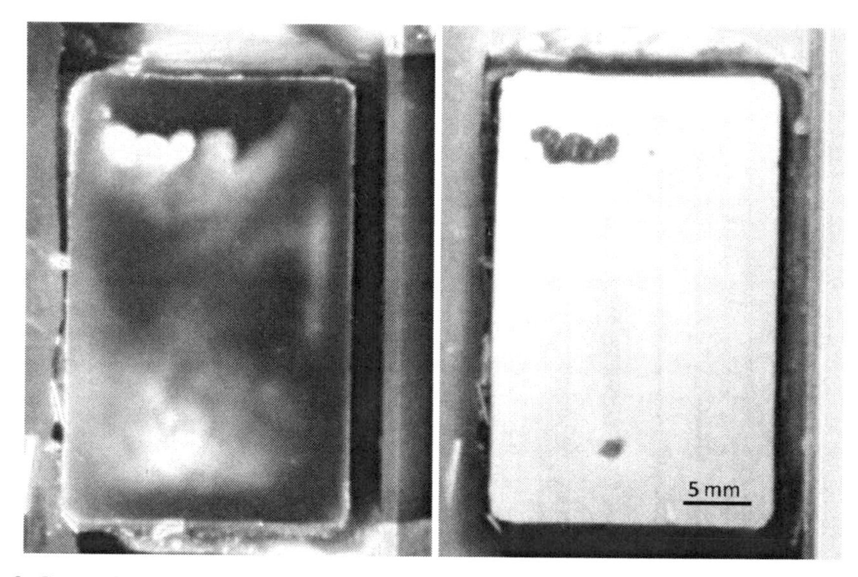

Fig. 9. Comparison between block-face images of goat atrial tissue, obtained using non-polarized (left) and polarized light (right). Normally, light penetrates the wax block and is reflected from deeper tissue structures, forming blurred outlines in the collected image that complicate segmentation and alignment of high-resolution histology sections. Using an approach that reduces light penetration and favors reflected light imaging, a clear contrast between wax and tissue is generated at surface, allowing identification of exactly three discrete sites where tissue penetrates the wax. Their shape and spatial interrelation subsequently allows accurate non-rigid in-plane transformation of the tissue section that will be taken next from this block.

The pronounced difference this makes is apparent in the example shown in Fig. 9, comparing surface images acquired from the same wax block using either non-polarized or polarized light. Serial imaging and processing of whole heart samples, using this new data acquisition approach, is under way.

5 Discussion and Conclusions

While non- or partially-invasive imaging modalities, such as MRI or μCT, are gaining importance for the analysis of cardiac structure at micro scales, histology still offers a unique combination of resolution and discrimination of structures at the cellular and sub-cellular levels. In order to improve our understanding of cardiac structure-function relations, we need to resolve 3D histology of the heart in a manner that maximally preserves original histo-anatomical features. This continues to present substantial challenges for data acquisition and integration.

Reconstruction based on 2D histological sections is burdened by a host of geometrical artifacts, including artificial straightening of structures (as illustrated in Fig. 2). This can be overcome by access to a supporting volumetric imaging modality. MRI scans are an excellent option (Fig. 3 and 4). However, as the actual organ geometry is altered during dehydration and wax embedding, there is no strict correspondence between 3D MRI data and stacked 2D-corrected histology sections. While several methods for histology-MRI alignment have been developed to provide visually satisfactory result, the sequence of registration steps and regularization criteria affect the deformation field. A particularly difficult challenge for post-processing is the presence of out-of-plane deformation. This has significant implications for subsequent use in models of functional activity that span multiple scales of structural complexity (e.g. from cell to ECG) [36], in particular if they are designed to go 'full circle' from live-tissue studies to simulation of observed behavior based on individual histo-anatomy.

This calls for an intermediate reference data-set that is representative of the 3D state of the tissue when it is being sectioned for histology. Block-face images are naturally aligned at acquisition, since the spatial interrelation between sample and camera can be reproducibly defined. The use of high-resolution block-face imaging as a full substitute of single-section histology images is possible, for example by scanning confocal microscopy [31]. However, block-face data gathering is restricted in the extent to which tissue can be histologically stained prior to imaging. We use block-face imaging, therefore, as an intermediate reference point between MRI and full-resolution histology sections. They provide a 3D-registered set of 2D images that have a one-to-one correspondence to 2D images from stained sections. This reduces the complexity of the transformations required to a series of 2D operations.

Block-face imaging still presents inherent imaging challenges. To preserve alignment between adjacent sections, the system (camera, tissue block) needs to be kept aligned, ideally with an accuracy of a fraction of a pixel. In a working laboratory, with imaging spread over several days, this is not always possible. Small image misalignments must be compensated, post-acquisition, by rigid alignment of imaged block-faces. In addition, variations in illumination should also be kept to a minimum to facilitate registration (but this can also be compensated for by post-processing). The most important requirement is to try and restrict imaging to the actual block surface. Here, the challenge is related to transparency of the sample (both of the wax and the tissue), and this can only partially be addressed using deconvolution algorithms. The use of polarized light imaging, essentially taking a snapshot of the wax surface (rather than the tissue), solves this problem to a large extent at the image acquisition stage, significantly facilitating the implementation of truly 3D histology of the heart.

Acknowledgments. The work presented here was supported by BBSRC grants E003443 and BB/I012117. JES and PK are Senior Fellows of the British Heart Foundation. RABB is funded by the Oxford BHF Centre of Research Excellence. CA has a PhD studentship from the Fundação para a Ciência e a Tecnologia, Portugal. The authors thank Dr T. Alexander Quinn for helpful comments on the manuscript.

BR holds a Medical Research Council Career Development Award.

References

1. Plank, G., Burton, R.A.B., Hales, P., Bishop, M., Mansoori, T., Bernabeu, M.O., Garny, A., Prassl, A.J., Bollensdorff, C., Mason, F., Mahmood, F., Rodriguez, B., Grau, V., Schneider, J.E., Gavaghan, D., Kohl, P.: Generation of histo-anatomically respresentative models of the individual heart: tools and applications. Phil Trans R Soc A 367(1896), 2257–2292 (2009)
2. Kanai, A., Salama, G.: Optical mapping reveals that repolarization heart spreads anisotropically and is guided by fiber orientation in guinea pig hearts. Circ Res. 77, 784–802 (1995)
3. Vetter, F.J., Simons, S.B., Mironov, S., Hyatt, C.J., Pertsov, A.M.: Epicardial fiber organization in swine right ventricle and its impact on propagation. Circ Res. 96, 244–251 (2005)
4. Chen, P.S., Cha, Y.M., Peters, B.B., Chen, L.S.: Effects of myocardial fiber orientation on the electrical induction of ventricular fibrillation. Am J Physiol. 264, H1760–H1773 (1993)
5. De Bakker, J.M., Stein, M., van Rijen, H.V.: Three-dimensional anatomic structure as substrate for ventricular tachycardia/ventricular fibrillation. Heart Rhythm 2, 777–779 (2005)
6. Eason, J., Schmidt, J., Dabasinskas, A., Siekas, G., Aguel, F.,, Trayanova, N.: Influence of anisotropy on local and global measures of potential gradient in computer models of defibrillation. Ann. Biomed. Eng. 26, 840–849 (1998)
7. Hooks, D.A., Tomlinson, K.A., Marsden, S.G., LeGrice, I.J., Smaill, B.H., Pullan, A.J., Hunter, P.J.: Cardiac microstructure: implications for electrical propagation and defibrillation in the heart. Circ. Res. 91, 331–338 (2002)
8. Waldman, L.K., Nosan, D., Villarreal, F., Covell, J.W.: Relation between transmural deformation and local myofiber direction in canine left ventricle. Circ. Res. 63, 550–562 (1988)
9. Ashikaga, H., Coppola, B.A., Yamazaki, K.G., Villarreal, F.J., Omens, J.H., Covell, J.W.: Changes in regional myocardial volume during the cardiac cycle: implications for transmural blood flow and cardiac structure. Am. J. Physiol. 295, 610–618 (2008)
10. Iribe, G., Ward, C.W., Camelliti, P., Bollensdorff, C., Mason, F., Burton, R.A.B., Garny, A., Morphew, M., Hoenger, A., Lederer, W.J., Kohl, P.: Axial stretch of rat single ventricular cardiomyocytes causes an acute and transient increase in Ca^{2+} spark rate. Circ. Res. 104, 787–795 (2009)
11. Kohl, P., Bollensdorff, C., Garny, A.: Effects of mechanosensitive ion channels on ventricular electrophysiology: experimental and theoretical models. Exp. Physiol. 91, 307–321 (2006)
12. Ozgen, N., Rosen, M.R.: Cardiac memory: a work in progress. Heart Rhythm 6, 564–570 (2009)

13. Choy, J.S., Kassab, G.S.: Wall thickness of coronary vessels varies transmurally in the LV but not the RV: implications for local stress distribution. Am. J. Physiol. 297, H750–H758 (2009)
14. Cheng, A., Nguyen, T.C., Malinowski, M., Langer, F., Liang, D., Daughters, G.T., Ingels, N.B., Miller Jr., D.C.: Passive ventricular constraint prevents transmural shear strain progression in left ventricle remodeling. Circulation 114, 79–86 (2006)
15. Hunter, P., Coveney, P.V., de Bono, B., Diaz, V., Fenner, J., Frangi, A.F., Harris, P., Hose, R., Kohl, P., Lawford, P., McCormack, K., Mendes, M., Omholt, S., Quarteroni, A., Skår, J., Tegner, J., Randall Thomas, S., Tollis, I., Tsamardinos, I., van Beek, J.H., Viceconti, M.: A vision and strategy for the virtual physiological human in 2010 and beyond. Philos. Transact. A Math. Phys. Eng. Sci. 368(1920), 2595–2614 (2010)
16. Kohl, P., Crampin, E.J., Quinn, T.A., Noble, D.: Systems biology: an approach. Clin. Pharmacol. Ther. 88, 25–33 (2010)
17. Moreno, J.D., Zhu, Z.I., Yang, P.C., Bankston, J.R., Jeng, M.T., Kang, C., Wang, L., Bayer, J.D., Christini, D.J., Trayanova, N.A., Ripplinger, C.M., Kass, R.S., Clancy, C.E.: A computational model to predict the effects of class I anti-arrhythmic drugs on ventricular rhythms. Sci. Transl. Med. 3, 83–98 (2011)
18. Rodriguez, B., Burrage, K., Gavaghan, D., Grau, V., Kohl, P., Noble, D.: The systems biology approach to drug development: application to toxicity assessment of cardiac drugs. Clin. Pharmacol. Ther. 88, 130–134 (2010)
19. Mirams, G.R., Davies, M.R., Cui, Y., Kohl, P., Noble, D.: Application of cardiac electrophysiology simulations to pro-arrhythmic safety testing. Br. J. Pharmacol. (2012) (Epub ahead of print; doi: 10.1111/j.1476-5381.2012.02020)
20. Streeter Jr., D.D., Bassett, D.L.: An engineering analysis of myocardial fiber orientation in pig's left ventricle in systole. Anat. Rec. 155, 503–511 (1966)
21. Dandel, M., Lehmkuhl, H., Knosalla, C., Suramelashvili, N., Hetzer, R.: Strain and strain rate imaging by echocardiography - basic concepts and clinical applicability. Curr. Cardiol. Rev. 5, 133–148 (2009)
22. Zerhouni, E.A., Parish, D.M., Rogers, W.J., Yang, A., Shapiro, E.P.: Human heart: tagging with MR imaging - a method for noninvasive assessment of myocardial motion. Radiology 169, 59–63 (1988)
23. Pelc, N.J., Drangova, M., Pelc, L.R., Zhu, Y., Noll, D.C., Bowman, B.S., Herfkens, R.J.: Tracking of cyclic motion with phase-contrast cine MR velocity data. J. Magn. Reson. Imaging 5, 339–345 (1995)
24. Gilbert, S.H., Benoist, D., Benson, A.P., White, E., Tanner, S.F., Holden, A.V., Dobrzynski, H., Bernus, O., Radjenovic, A.: Visualization and quantification of whole rat heart laminar structure using high-spatial resolution contrast-enhanced MRI. Am. J. Physiol. 302, H287–H298 (2012)
25. Palm, C., Penney, G.P., Crum, W.R., Schnabel, J.A., Pietrzyk, U., Hawkes, D.J.: Fusion of rat brain histology and MRI using weighted multi-image mutual information. In: Medical Imaging 2008: Image Processing, Pts. 1-3, vol. 6914, pp. M9140–M9140 (2008)
26. Yelnik, J., Bardinet, E., Dormont, D., Malandain, G., Ourselin, S., Tandé, D., Karachi, C., Ayache, N., Cornu, P., Agid, Y.: A three-dimensional, histological and deformable atlas of the human basal ganglia. I. Atlas construction based on immunohistochemical and MRI data. NeuroImage 34, 618–638 (2007)
27. Schormann, T., Zilles, K.: Three-dimensional Linear and Nonlinear Transformations: An Integration of Light Microscopical and MRI Data. Human Brain Mapping, 339 (1998)

28. Eickhoff, S.B., Heim, S., Zilles, K., Amunts, K.: A systems perspective on the effective connectivity of overt speech production. Philos. Transact. A Math. Phys. Eng. Sci. 367(1896), 2399–2421 (2009)
29. Breen, M.S., Lazebnik, R.S., Wilson, D.L.: Three-dimensional registration of magnetic resonance image data to histological sections with model-based evaluation. Ann. Biomed. Eng. 33, 1100–1112 (2005)
30. Ko, Y.-S., Yeh, H.-I., Ko, Y.-L., Hsu, Y.-C., Chen, C.-F., Wu, S., Lee, Y.-S., Severs, N.J.: Three-dimensional reconstruction of the rabbit atrioventricular conduction axis by combining histological, desmin, and connexin mapping data. Circulation 109, 1172–1179 (2004)
31. Rutherford, S.L., Trew, M.L., Sands, G.B., Legrice, I.J., Smaill, B.H.: High-resolution 3-dimensional reconstruction of the infarct border zone. Circ. Res. (2012) (Epub ahead of print: doi: 10.1161/CIRCRESAHA.111.260943)
32. Goodyer, C., Hodrien, J., Wood, J., Kohl, P., Brodlie, K.: Using high-resolution displays for high-resolution cardiac data. Philos. Transact. A Math. Phys. Eng. Sci. 367(1898), 2667–2677 (2009)
33. Burton, R.A.B., Plank, G., Schneider, J.E., Grau, V., Ahamer, H., Keeling, S.L., Lee, J., Smith, N.P., Gavaghan, D., Trayanova, N., Kohl, P.: Three-dimensional models of individual cardiac histo-anatomy: tools and challenges. Ann. NY. Acad. Sci. 1080, 301–319 (2006)
34. Schneider, J.E., Bose, J., Bamforth, S., Gruber, A.D., Broadbent, C., Clarke, K., Neubauer, S., Lengeling, A., Bhattacharya, S.: Identification of cardiac malformations in mice lacking PTDSR using a novel high-throughput magnetic resonance imaging technique. BMC. Dev. Biol. 4, 16 (2004)
35. Mansoori, T., Plank, G., Burton, R.A.B., Schneider, J.E., Kohl, P., Gavaghan, D., Grau, V.: Building detailed cardiac models by combination of histoanatomical and high-resolution MRI images. In: IEEE International Symposium on Biomedical Imaging (ISBI), pp. 572–575 (2007)
36. Brennan, T., Fink, M., Rodriguez, B.: Multiscale modelling of drug-induced effects on cardiac electrophysiological activity. Eur. J. Pharm. Sci. 36, 62–77 (2009)

Bimodal Protein Distributions in Heterogeneous Oscillating Systems

Maciej Dobrzyński, Dirk Fey, Lan K. Nguyen, and Boris N. Kholodenko

Systems Biology Ireland,
University College Dublin, Belfield, Dublin 4, Ireland
{maciej.dobrzynski,dirk.fey,lan.nguyen,boris.kholodenko}@ucd.ie

Abstract. Bimodal distributions of protein activities in signaling systems are often interpreted as indicators of underlying switch-like responses and bistable dynamics. We investigate the emergence of bimodal protein distributions by analyzing a less appreciated mechanism: oscillating signaling systems with varying amplitude, phase and frequency due to cell-to-cell variability. We support our analysis by analytical derivations for basic oscillators and numerical simulations of a signaling cascade, which displays sustained oscillations in protein activities. Importantly, we show that the time to reach the bimodal distribution depends on the magnitude of cell-to-cell variability. We quantify this time using the Kullback-Leibler divergence. The implications of our findings for single-cell experiments are discussed.

Keywords: signaling networks, oscillations, bimodality, stochasticity, protein distributions.

1 Introduction

Protein levels in cellular systems undergo constant changes due to varying extra- and intracellular cues that are dynamically processed by cellular machinery as well as due to thermal noise – an inevitable factor affecting all biochemical reactions. It is because of this variability that cells within a population, be it a bacterial colony or tumor cells, at any given point in time exhibit a distribution of values rather than a precise value of concentrations of its biochemical components, such as proteins or mRNA. Such distributions can be assessed as population snapshots in fluorescence-activated assays using flow cytometry or cell imaging. In both cases the measurement of fluorescence intensity in individual cells correlates with protein abundance. This starkly contrasts to bulk measurements such as Western blots where proteins are detected in cell lysates, which only estimates the average (per-cell) concentration of the entire population.

Of particular interest are bimodal protein distributions that indicate a temporal or steady-state phenotypic division of an isogenic cellular population. Bimodality often reflects the existence of two subpopulations, each capable of performing a different task [2] or having an altered survival rate to stress [3] and

D. Gilbert and M. Heiner (Eds.): CMSB 2012, LNCS 7605, pp. 17–28, 2012.

drug treatment [15]. Bimodal distributions may arise in a number of situations: a purely stochastic genetic switch [1], a bistable system with stochastically induced transitions [11], or noisy networks with sigmoidal response function [8,9]. In this paper we address a much less appreciated mechanism: heterogeneous oscillations. We show that cell-to-cell variability in protein abundances can result in bimodal distributions of concentrations of active (e.g. phosphorylated) protein forms, although individual cells display solely deterministic oscillatory dynamics. We examine analytically and numerically conditions under which this phenomenon occurs.

2 Results

A *single* oscillating cell visits all intermediate levels between the low and the high protein concentrations. A histogram, or a distribution, of concentrations assumed over time can be constructed in the following manner. The range of concentrations between oscillation extrema is divided into infinitesimally small bins and the time the system spends in each of the bins is recorded. For deterministic oscillations, a single period suffices to obtain such a distribution. Depending on the shape of these oscillations, a bimodal single-cell time-averaged histogram of concentrations may arise (Fig. 1). The key question, however, is whether in the presence of cell-to-cell variability which affects the amplitude, phase and frequency of oscillations in individual cells, the described mechanism can also evoke bimodality at the level of a cellular population? The question is equivalent to asking about the ergodicity of such a system: does the distribution of a population coincide with the distribution of an individual measured over time? The disparity of the two has been recently demonstrated experimentally for noisy cellular systems [14]. Protein fluctuations that are high in amplitude and slow compared to cells lifetime may drive a number of cells to a range of concentrations that is only a fraction of the entire concentration spectrum. This condition may persist well over a cells generation thus rendering snapshots of the population incapable of reflecting the underlying network dynamics.

Similar phenomenon may affect a population of oscillating cells. Even though our analysis focuses on oscillations that are deterministic in individual cells, biochemical noise manifests itself in cell-to-cell variability. As a result, oscillations across the population differ in the amplitude, phase and frequency. If this variability is not large enough, a population might not cover the entire concentration spectrum at a given point in time, and a bimodal distribution fails to emerge. An additional condition is required to facilitate this emergence and relates to a so-called mixing time – the time after which all individuals within the population of cells assume all states of the asymptotic (stationary) protein distribution. We therefore set out to answer following questions: (1) under what biochemical circumstances can a heterogeneous population of cells exhibiting oscillatory dynamics give rise to bimodal protein distributions? (2) What is the time after stimulus required to reach a time-independent bimodal distribution?

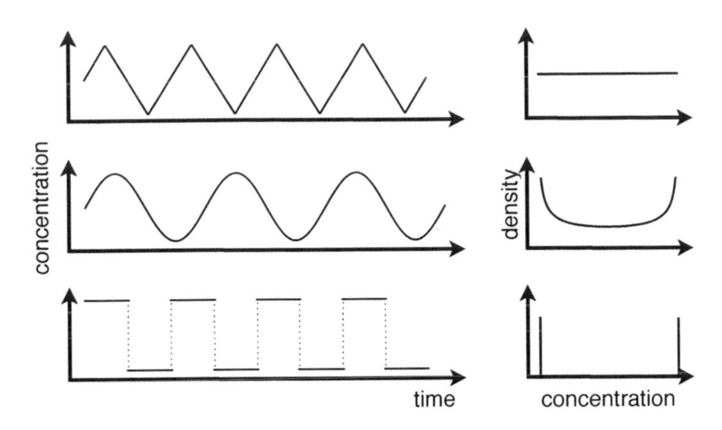

Fig. 1. Protein distributions depend on the functional form of oscillations. Sample time-courses of triangle wave, sinusoidal and step oscillations (left column) along with corresponding time-averaged probability densities ("normalized histograms") of protein concentrations (right).

2.1 Sinusoidal Oscillations Give Rise to Bimodality

We consider an ensemble of cells, each displaying an oscillating level of active protein concentration governed by the intracellular biochemical network dynamics. Cell-to-cell heterogeneity that emerges due to varying gene expression levels induces randomness in the concentration of network components. In an ensemble of oscillating cells, this randomness translates to a distribution of amplitude (A), phase shift (φ), and frequency (ω) of protein activity (y). In order to illustrate the concept of mixing times, we first consider sinusoidal oscillations, $y = A \sin(\omega t + \varphi)$. The three random influences (phase, frequency and amplitude variability) cause qualitatively different behavior with respect to the convergence of the y-distribution in a heterogeneous ensemble of oscillators.

Phase shift variability reflects desynchronization of independent cells within the population and can be quantified in a standard manner, for instance by measuring variance. Narrow distribution of phase shifts compared to the oscillation period, or small desynchronization, implicates that at any point in time the protein levels assumed in the population do not cover the entire range of concentrations. This restricted concentration diversity persists during the time evolution (Fig. 2A, left panel). Stationary distribution emerges instantaneously only when the range of variability uniformly spans the entire oscillation period (Fig. 2A, right panel). In this regime, the probability density function (*pdf*) can be obtained by considering a sine transformation of a uniformly distributed random variable φ restricted to a single oscillation period (cf. Appendix) [6],

$$f(y) = \frac{1}{\pi\sqrt{A^2 - y^2}} \,. \tag{1}$$

The *pdf* is the arcsine distribution (Fig. 2, solid line in *pdf* plots). Notably, it is independent of the time at which the measurement takes place, as well as independent of the frequency of the underlying oscillations.

The variability of frequencies across cellular population stems from intrinsic biochemical noise that affects protein concentrations across the population. Contrary to phase shift variability, an ensemble of sinusoidal oscillators with distributed frequencies reaches the asymptotic stationary distribution regardless of the distribution width; the variance affects only the time to reach it and greater variability accelerates the convergence (Fig. 2B). The asymptotic distribution for uniformly distributed frequencies can be calculated analytically and equals the (previous) result concerning phase-variability (Eq. 1). An intuitive explanation follows from the functional form of the sinusoidal oscillation. The value of random frequency ω is multiplied by time, t. Therefore, regardless of the ω distribution shape, ω is scaled by the increasing time, which accordingly results in the increasing range of frequencies. For large enough t, this range becomes sufficient to facilitate mixing analogous to phase shifts that cover the entire oscillation period.

If cells within the population oscillate with random amplitude only, no stationary distribution can emerge. Since no nonlinear transformation of the random variable takes place, the *pdf* is merely a distribution of the random amplitude A modulated by the sinusoidal wave. The resulting distribution of concentrations cycles over the oscillation period (Fig. 2C).

2.2 Quantification of the Mixing Time

A real biological oscillatory network exhibits a combination of all three types of variability discussed in the previous section. In a typical experimental scenario, the measurement of oscillations is performed on a population of cells and is preceded by a period of starvation followed by addition of a stimulating agent that evokes oscillations. The procedure corresponds to synchronization of cells such that oscillations begin approximately at the same time. Variability among cells still exists, albeit diminished. The emergence of a stationary population-wide bimodal distribution such as the one depicted in Fig. 2D is therefore delayed. The time to approach it, which we shall call the mixing time, depends on the magnitude of contributions to oscillation variability between cells.

A mixing time larger than zero demonstrates a simple fact that the ensemble of oscillators with small variability of frequencies, amplitudes and phase shifts does not immediately reflect time-averaged statistics. As shown in the section above, the system can reach the stationary distribution, provided variability of frequencies exists.

As a quantification of the mixing time we use the Kullback-Leibler divergence (KL), which, in simple terms, measures the divergence of two distributions [13,7]. Let $P(y,t)$ be the probability density of the y concentration at time point t and let further $Q(y)$ be the asymptotic probability density of y for $t \to \infty$, then the KL(t) is defined as

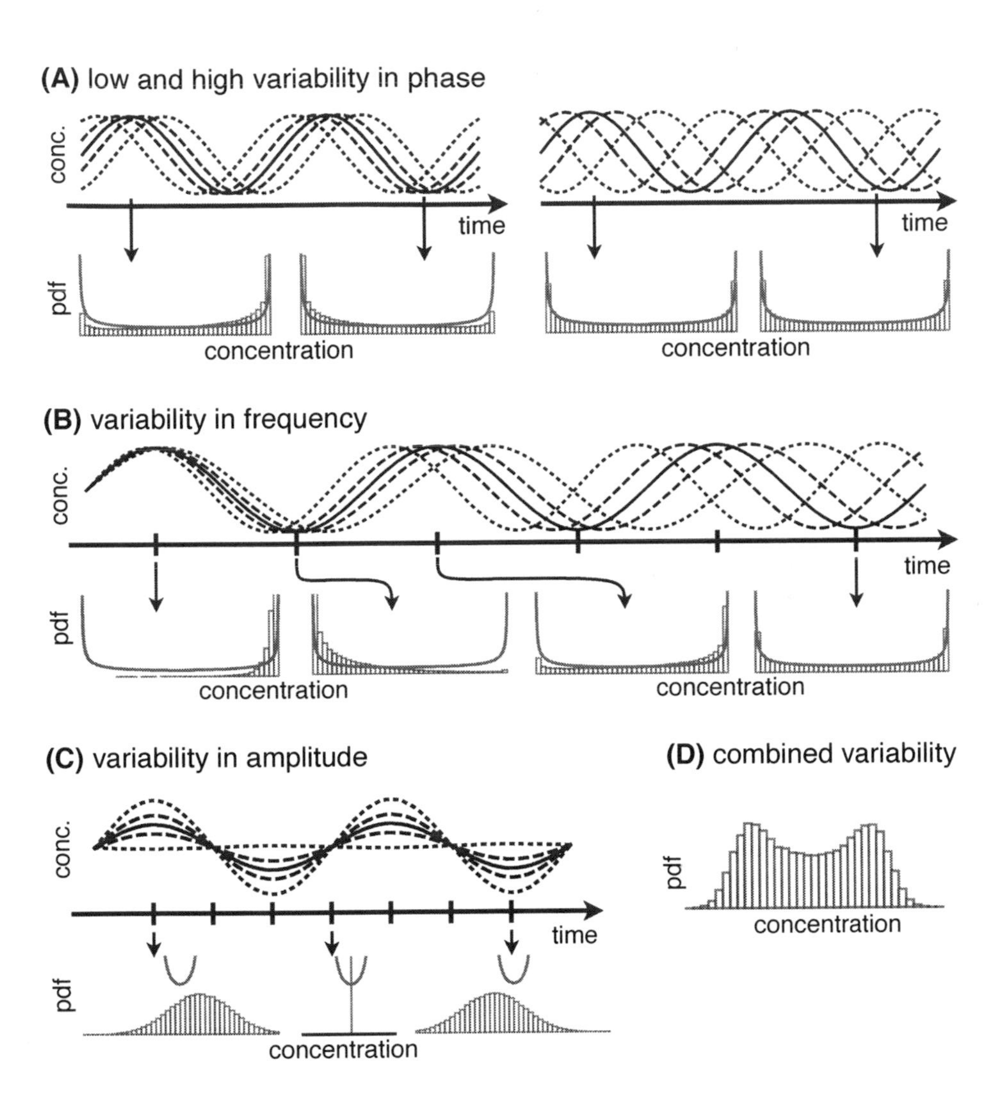

Fig. 2. Approach to an asymptotic distribution. Time courses of oscillations mark the 25th and 75th percentile (dotted), 40th and 60th percentile (dashed), and 50th percentile (solid) of the corresponding parameter distribution. Protein probability density functions (*pdf*) are evaluated numerically at points indicated by arrows. Asymptotic solution, Eq. 1, is marked by the solid line. (A) Phase shifts follow Gaussian distribution with zero mean and standard deviation $\sigma = \pi/2$ (left) and π (right). For large σ, the *pdf* is time-independent and equals the asymptotic *pdf*. (B) Frequency follows log-normal distribution with median 1 and $\sigma = 0.2$. (C) Amplitude follows Gaussian distribution with mean 1 and $\sigma = 1$. The distribution cycles over the oscillation period. (D) Sample stationary protein distribution when all three variability influences are combined.

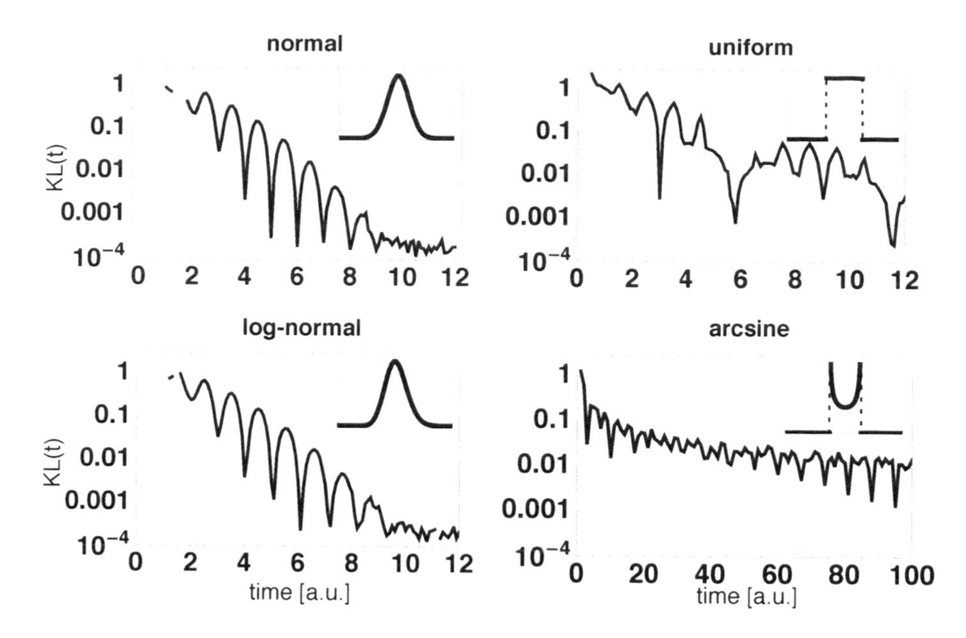

Fig. 3. Kullback-Leibler divergence for a population of sinusoidal oscillators with random frequencies drawn from normal, log-normal, uniform and arcsine distributions (insets). All frequency distributions have the same mean, $\mu = \pi$, and standard deviation, $\sigma = \pi/10$. We calculate KL between numerically sampled protein distributions (based on 100'000 points) at time t and the asymptotic distribution from Eq. 1.

$$KL(t) = \int_{-\infty}^{+\infty} P(y,t) \ln \frac{P(y,t)}{Q(y)} dy \, . \tag{2}$$

Here, $KL(t)$ measures the divergence rather than distance of the snapshot at time t from the asymptotic true snapshot distribution. It is worth emphasizing that KL is always non-negative but it is not a metric in the mathematical sense for it is asymmetric and it does not satisfy triangle inequality.

Temporal behavior of $KL(t)$ is shown in Fig. 3 where we measure the divergence between the protein distribution in an ensemble of oscillators with random frequencies and the asymptotic arcsine distribution. Regardless of the type of frequency distribution, the KL divergence decays at an exponential rate as the oscillating ensemble evolves in time.

2.3 Oscillations and Population Snapshots in a Two-Layer GTPase System

To analyze how oscillations mix in a biologically realistic scenario, we numerically study a model of a generic two-layered GTPase system. Small GTPases can cycle between an inactive GDP-bound state (G) and an active GTP-bound state

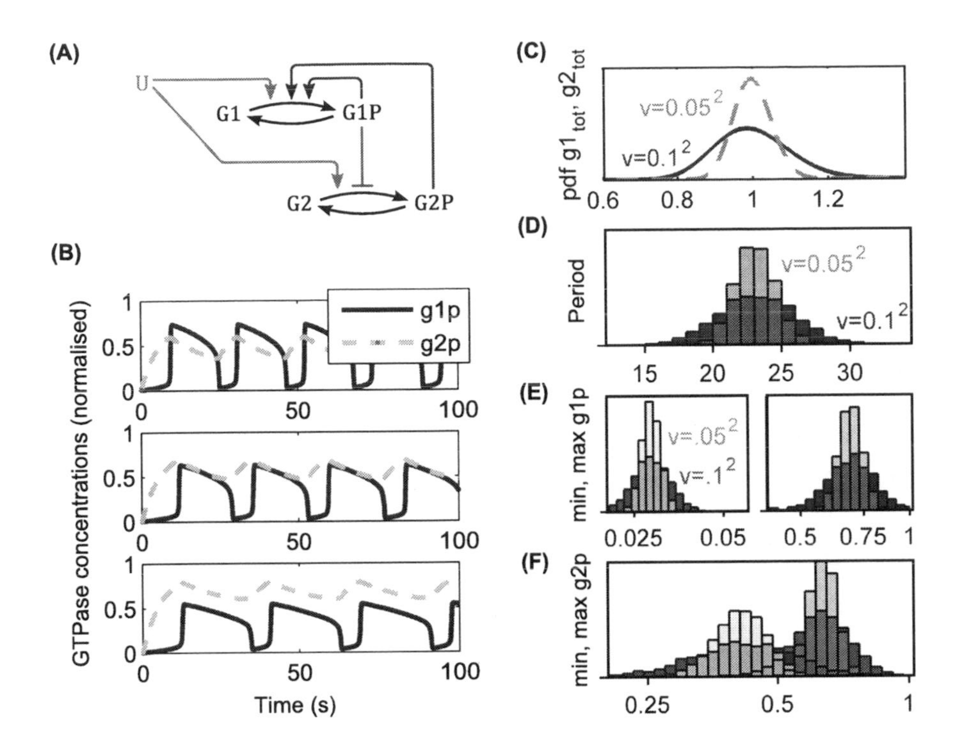

Fig. 4. Dynamic model of GTPase cascade. (A) Interaction scheme. (B) Simulated trajectories of three random cells. (C) Distribution of total GTPase concentrations used for ensemble simulations. The resulting distributions of periods (D), extrema for concentrations of g_1p (E) and g_2p (F). Note the logarithmic x-axis in panels E and F. Parameters used in the simulation: maximal rates $r_1 = 10$, $r_2 = 6.5$, $r_3 = 1$, $r_4 = 0.55$; half-activation constants $m_1 = 25$, $m_2 = 0.09$, $m_3 = 5$, $m_4 = 14$; positive interaction $G1 \rightarrow G1$, $a_{11} = 200$, $m_{11} = 10$; negative interaction $G1 \dashv G2$, $a_{12} = 0.005$, $m_{13} = 0.05$; positive interaction $G2 \rightarrow G1$, $a_{21} = 80$, $m_{21} = 20$.

(GP). They are important transducers of cell signaling that regulate a wide range of biological processes, for instance cell proliferation, cell morphology as well as nuclear and vesicle transport. Individual GTPase are often interlinked, thereby generating positive and negative feedback systems that are theoretically capable of exhibiting rich dynamics including oscillations. Indeed oscillations have been observed experimentally for several GTPases. For example, the small Rho-GTPase cdc42 regulates polarized growth in fission yeast using oscillating activity arising from both positive and negative feedback [4].

We consider the GTPase cascade depicted in Fig. 4A and let G1, G2 and G1P, G2P denote the inactive and active form of the corresponding GTPase, respectively. The system features a positive auto-regulatory loop in which G1P enhances its own activation and a negative feedback loop in which G1P inhibits the activation of G2 and in turn G2P activates G1. The following ordinary differential equations in normalized coordinates represent the system [16],

$$\frac{d}{dt}g_1p = \alpha_{11}\alpha_{21}\frac{r_1\,g_1}{m_1+g_1} - \frac{r_2\,g_1p}{m_2+g_1p}, \qquad \alpha_{11} = \frac{m_{11}+a_{11}\,g_1p}{m_{11}+g_1p},$$

$$\alpha_{21} = \frac{m_{21}+a_{21}\,g_2p}{m_{21}+g_2p},$$

$$\frac{d}{dt}g_2p = \alpha_{13}\frac{r_3\,g_2}{m_3+g_2} - \frac{r_4\,g_2p}{m_4+g_2p}, \qquad \alpha_{13} = \frac{m_{13}+a_{13}\,g_1p}{m_{13}+g_1p}, \tag{3}$$

with $g_1 = g_1^{tot} - g_1p$, $g_2 = g_2^{tot} - g_2p$, and where g_1, g_2 and g_1p, g_2p denote the concentrations of inactive and active GTPases, respectively, and r_i, m_i, a_{ij}, m_{ij} are kinetic parameters. The factors α_{11}, α_{21}, α_{13} model the described interactions with the parameters $a_{11} > 1$, $a_{21} > 1$ (positive interactions) and $0 < a_{13} < 1$ (negative interaction).

In accordance with the literature [12], we model a population of cells as an ensemble [10,5] of single cells in which the total concentrations of both GTPases are log-normally distributed with mean one and standard deviations consistent with experimentally reported values ranging from 0.12 to 0.28 in human cells [14] (Fig. 4C).

Fig. 4B illustrates the dynamics of the model with representative responses of three random cells to a step input. The model exhibits switch-like G1P oscillations and triangle wave-like G2P oscillations, thus providing a convenient tool to investigate how differentially shaped oscillations manifest in the distribution of a population snapshots taken at a particular time point. Figs. 4D-F demonstrate how the periods and the minima and maxima of the oscillations are distributed in the population. Decreasing the variability of the total GTPase distribution (from $\sigma = 0.1$ to 0.05) yields more narrowly distributed periods and extrema while their means remain unchanged.

The distribution of $g_1p(t)$ and $g_2p(t)$ concentrations, in the following referred to as a distribution snapshot, changes over time. For $t < 0$ the entire population is synchronized; the phase of all oscillations is zero and the first peak occurs roughly at the same time; at the initial time all cells exhibit zero GTPase activity, while after 15 seconds most cells have progressed to the first peak. Over time, the cell-to-cell variability has an increasing effect on the population snapshots; different periods shift the phases of subsequent peaks until the phases are uniformly distributed. During this transition period, the snapshot distribution dynamically changes (Fig. 5). The evolution of the distribution crucially depends on the shape of the underlying oscillations. For example, switch-like G1P oscillations result in uni-modal ($t = 5\,$s), bi-modal ($t = 9\,$s) and even tri-modal ($t = 78\,$s) distributions. In contrast, the triangle wave G2P oscillations yield uni-modal snapshot distributions at all times (Fig. 5B).

Next we sought to assess how quickly the snapshot distribution converges to the asymptotic one using Kullback-Leibler divergence and asked whether it is possible to find the time point at which the oscillations are well mixed. The results are shown in Fig. 5C and D. The snapshot distribution converges exponentially

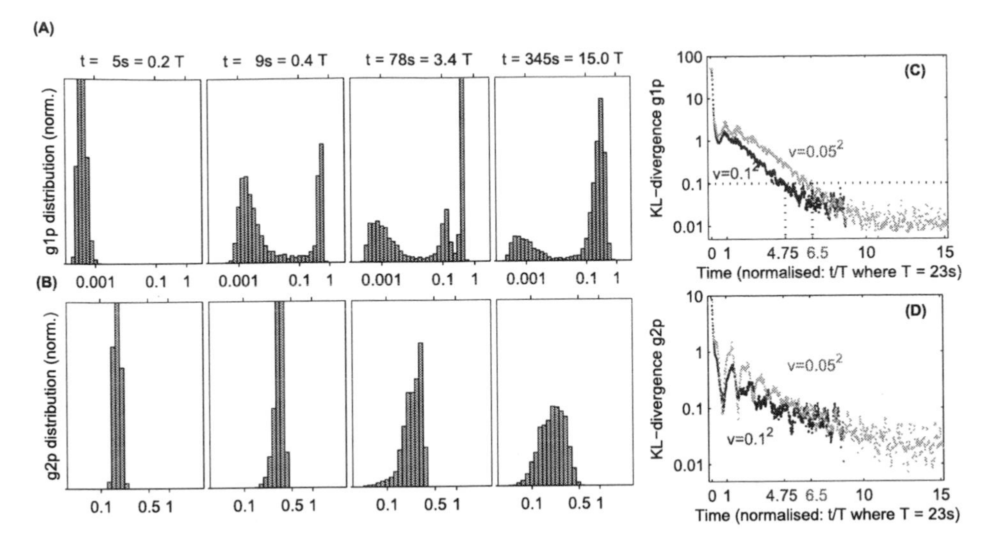

Fig. 5. Numerical simulations of time evolution of g_1p (A) and g_2p (B) protein distribution for indicated time points. The distribution of total G1 and G2 concentrations is normal with mean 1 and standard deviation 0.05. Panels C and D show a comparison of Kullback-Leibler divergence for total G1 and G2 distributions with standard deviation 0.1 and 0.05. Simulation parameters same as described in caption of Fig. 4.

towards its asymptotic distribution. Further, the rate of convergence depends on the cell-to-cell variability in the population; lower variability of the total GTPase concentrations causes lower variability of periods and results in longer mixing times.

3 Discussion

A stationary bimodal protein distribution may arise in a heterogeneous population of independent cells with sustained deterministic oscillations of active protein levels. The emergence of population-level bimodality is inevitable in the presence of cell-to-cell variability that affects oscillation frequency. Importantly, the type of the frequency distribution across the population is irrelevant for the emergence of bimodality; only the time of convergence is affected (Fig. 3).

Detecting the oscillatory nature of signaling networks with bulk measurements (e.g. immunoblotting) is only possible for synchronized cells. This synchronization, for instance, occurs at the point of stimulation preceded by a period of starvation. If this condition is not satisfied, cell-to-cell variability introduces phase desynchronization and diversity in frequencies, which averages out the oscillations. On top of that, sampling frequency in the experiment should be sufficiently higher than frequency of the oscillations. Otherwise the measurement captures only the population mean, which does not oscillate. In this regime, single-cell measurement methods can give an additional insight, for they record the amount of protein in individual cells.

A population snapshot obtained with flow cytometry or imaging is equivalent to the probability density functions discussed throughout the paper. The time dependence of such a distribution may become very intricate for realistic systems, which we demonstrated in Fig. 5. Nonetheless, there exists a time scale – the mixing time, which we estimated using Kullback-Leibler divergence – when the stationary distribution emerges. If amplitude variability is smaller than the oscillation amplitude itself (cf. g_1p and g_2p oscillations in Fig. 4B), the distribution may become bimodal. The time to converge to this distribution is independent of the population size: as long as mixing of frequencies and phases in a population has not commenced, the protein distribution remains different from the asymptotic one. This behavior contrasts with other mechanisms that generate bimodality where increasing the number of independent individuals results in a better indication of the population-wide asymptotic distribution. Experimental verification of sources of bimodality might benefit from this feature.

The second implication of our finding relates to mechanisms that preserve synchronized population-wide oscillations generated by biochemical networks. In the presence of frequency heterogeneity the mixing and eventual convergence to the asymptotic distribution is only a matter of time. If oscillations are a physiologically relevant trait, as is the case of circadian rhythms, the convergence is undesirable because it would indicate that oscillations are out of sync and each cell within an organ, for instance, has its own day and night pattern. This could explain why biochemical coupling is present in such systems in order to facilitate spontaneous synchronization across the population.

Acknowledgments. MD, LKN and BNK were supported by Science Foundation Ireland under Grant No. 06/CE/B1129. DF received funding from the European Union Seventh Framework Programme (FP7/2007-2013) ASSET project under grant agreement number FP7-HEALTH-2010-259348.

References

1. Acar, M., Mettetal, J.T., van Oudenaarden, A.: Stochastic switching as a survival strategy in fluctuating environments. Nat. Genet. 40(4), 471–475 (2008)
2. Ackermann, M., Stecher, B., Freed, N.E., Songhet, P., Hardt, W.D., Doebeli, M.: Self-destructive cooperation mediated by phenotypic noise. Nature 454(7207), 987–990 (2008)
3. Blake, W.J., Balázsi, G., Kohanski, M.A., Isaacs, F.J., Murphy, K.F., Kuang, Y., Cantor, C.R., Walt, D.R., Collins, J.J.: Phenotypic consequences of promoter-mediated transcriptional noise. Molecular Cell 24(6), 853–865 (2006)
4. Das, M., Drake, T., Wiley, D.J., Buchwald, P., Vavylonis, D., Verde, F.: Oscillatory dynamics of Cdc42 GTPase in the control of polarized growth. Science 337(6091), 239–243 (2012)
5. Kuepfer, L., Peter, M., Sauer, U., Stelling, J.: Ensemble modeling for analysis of cell signaling dynamics. Nat. Biotechnol. 25(9), 1001–1006 (2007)
6. Miller, S., Childers, D.: Probability and Random Processes. With Applications to Signal Processing and Communications. Academic Press (January 2012)

7. Mirsky, H.P., Taylor, S.R., Harvey, R.A., Stelling, J., Doyle, F.J.: Distribution-based sensitivity metric for highly variable biochemical systems. IET Syst. Biol. 5(1), 50 (2011)
8. Niepel, M., Spencer, S.L., Sorger, P.K.: Non-genetic cell-to-cell variability and the consequences for pharmacology. Curr. Opin. Chem. Biol. 13(5-6), 556–561 (2009)
9. Ochab-Marcinek, A., Tabaka, M.: Bimodal gene expression in noncooperative regulatory systems. PNAS 107(51), 22096–22101 (2010)
10. Ogunnaike, B.A.: Elucidating the digital control mechanism for DNA damage repair with the p53-Mdm2 system: single cell data analysis and ensemble modelling. J. R. Soc. Interface 3(6), 175–184 (2006)
11. Samoilov, M., Plyasunov, S., Arkin, A.P.: Stochastic amplification and signaling in enzymatic futile cycles through noise-induced bistability with oscillations. PNAS 102(7), 2310–2315 (2005)
12. Schliemann, M., Bullinger, E., Borchers, S., Allgöwer, F., Findeisen, R., Scheurich, P.: Heterogeneity reduces sensitivity of cell death for TNF-stimuli. BMC Syst. Biol. 5, 204 (2011)
13. Shahrezaei, V., Swain, P.S.: Analytical distributions for stochastic gene expression. PNAS 105(45), 17256–17261 (2008)
14. Sigal, A., Milo, R., Cohen, A., Geva-Zatorsky, N., Klein, Y., Liron, Y., Rosenfeld, N., Danon, T., Perzov, N., Alon, U.: Variability and memory of protein levels in human cells. Nature 444(7119), 643–646 (2006)
15. Spencer, S.L., Gaudet, S., Albeck, J.G., Burke, J.M., Sorger, P.K.: Non-genetic origins of cell-to-cell variability in TRAIL-induced apoptosis. Nature 459(7245), 428–432 (2009)
16. Tsyganov, M.A., Kolch, W., Kholodenko, B.N.: The topology design principles that determine the spatiotemporal dynamics of G-protein cascades. Mol. Biosyst. 8(3), 730–743 (2012)

Appendix

Asymptotic Protein Distribution

Consider an ensemble of cells with sinusoidal oscillations of protein levels. Variability of phases φ is accounted for by a random variable Φ uniformly distributed on the range of an oscillation period, 2π. We set out to obtain a distribution of protein levels y denoted by a random variable Y, which is the result of a nonlinear transformation $Y = A\sin(\omega t + \Phi)$. Since the transformation is periodic, without loss of generality we first set $t = 0$ and focus on the shorter range, $-\pi/2 < \Phi < \pi/2$, where sine function is monotonically increasing.

The cumulative distribution function (CDF) of Y is simply the probability that the random variable Y takes on a value less than or equal y, $F_Y(y) = Pr(Y \leq y)$. From this we obtain,

$$F_Y(y) = Pr\left[A\sin(\Phi) \leq y\right] = Pr\left[\Phi \leq \arcsin\left(\frac{y}{A}\right)\right], \text{ and } |y| \leq A\,. \tag{4}$$

The CDF of Y can be therefore expressed in terms of the CDF of Φ. The probability density function (*pdf*), denoted by $f_Y(y)$, is CDF's first derivative,

$$\begin{aligned} F_Y(y) &= F_\Phi\left(\arcsin\left(\frac{y}{A}\right)\right), \\ \frac{d}{dy}F_Y(y) &= f_Y(y) = f_\Phi\left(\arcsin\left(\frac{y}{A}\right)\right)\frac{1}{\sqrt{A^2+y^2}} \\ &= \frac{1}{\pi}\frac{1}{\sqrt{A^2+y^2}}\,. \end{aligned} \tag{5}$$

The procedure can be repeated to yield the same result, the arcsine distribution, for every range of length π, where the transformation is monotonic.

Expressive Statistical Model Checking of Genetic Networks with Delayed Stochastic Dynamics

Paolo Ballarini[1], Jarno Mäkelä[2], and Andre S. Ribeiro[2]

[1] Ecole Centrale Paris, France
paolo.ballarini@ecp.fr
[2] Tampere University of Technology, Finland
{andre.ribeiro,jarno.makela}@tut.fi

Abstract. The recently introduced Hybrid Automata Stochastic Logic (HASL) establishes a powerful framework for the analysis of a broad class of stochastic processes, namely Discrete Event Stochastic Processes (DESPs). Here we demonstrate the potential of HASL based verification in the context of genetic circuits. To this aim we consider the analysis of a model of gene expression with delayed stochastic dynamics, a class of systems whose dynamics includes both Markovian and non-Markovian events. We identify a number of relevant properties related to this model, formally express them in HASL terms and, assess them with COSMOS, a statistical model checker for HASL model checking. We demonstrate that this allows assessing the "performances" of a biological system beyond the capability of other stochastic logics.

Keywords: Statistical model checking, genetic networks, stochastic dynamics, stochastic petri nets.

1 Introduction

Biological systems are regulated by complex information processing mechanisms which are at the basis of their survival and adaptation to environmental changes. Despite the continuous advancements in experimental methods many of those mechanisms remain little understood. The end goal of computational systems biology [26] is to help filling in such knowledge gap by developing formal methods for rigorously representing and effectively analysing biological systems. Understanding what cells actually compute, how they perform computations and, eventually, how such computations can be modified/engineered are essential tasks which computational modelling aims to. In this context, the ability to "interrogate" a model by posing relevant "questions", referred to as *model checking*, is critical. Model checking approaches have proved effective means to the analysis of biological systems, both in the framework of non-probabilistic models [19,12] and in that of stochastic models [28,23].

Our Contribution. We consider the application of a recently introduced stochastic logic, namely the Hybrid Automata Stochastic Logic (HASL), to the verification of biological systems. Our contribution is twofold. In the first part

D. Gilbert and M. Heiner (Eds.): CMSB 2012, LNCS 7605, pp. 29–48, 2012.

we demonstrate the effectiveness of HASL verification by developing a full case study of gene expression, a relevant biological mechanism represented by means of non-Markovian models, and which, therefore, cannot be analysed by means of classical (Markovian) stochastic model checking. In the second part we introduce preliminary results illustrating the effectiveness of HASL verification in dealing with a rather relevant aspect of many biological mechanisms, namely: the analysis of oscillatory trends in stochastic models of biological systems.

Paper Organization. In Section 2 we provide some backgrounds which put into context the proposed approach. In Section 3 we introduce the gene expression mechanism with stochastic delays we refer to in the remainder of the paper. In Section 4 we recall the basics of the HASL formalism. In Section 5 we present the formal analysis (by means of HASL) of the previously introduced single-gene model. In section 6 we illustrate the application of HASL to measurements of oscillations. Conclusive remarks are given in Section 7.

2 Background

Model Checking and Systems Biology. Model checking is a technique addressing the formal verification of discrete-event systems. Its success is mainly due to the following points: (1) the ability to express specific properties by formulas of an appropriate logic, (2) the firm mathematical foundations based on automata theory and (3) the simplicity of the verification algorithms which has led to the development of numerous tools. Initially [17] targeted to the verification of functional *qualitative* properties of non-probabilistic models by means of "classical" temporal logics (i.e. LTL, CTL), model checking has progressively been extended toward the performance and dependability analysis realm (i.e. *quantitative* verification) by adaptation of classical temporal logics to express properties of Markov chains [21],[6]. In systems biology [26] two modelling alternatives are typical: (1) the continuous-deterministic framework, whereby dynamics of biological agents are expressed in terms of (a system of) differential equations (e.g. ODE, PLDE) and (2) the discrete-stochastic framework, whereby dynamics are expressed in terms of a stochastic process (most often a continuous-time Markov chain). The application of model checking to systems biology has targeted both modeling frameworks. BIOCHAM [18], GNA [15], BioDIVINE [10] are examples of tools providing LTL/CTL model-checking functionalities for the verification of *qualitative* properties of biological models represented by means of differential equations. Conversely, PRISM[31], MARCIE [35] are examples of tools featuring Continuos Stochastic Logic (CSL) [6] model-checking for the verification of *quantitative* properties of continuous-time Markov chains (CTMC) models of biological systems. Recently *linear-time* reasoning (as opposed to CSL *branching-time* reasoning) has been extended to the probabilistic framework as well. Examples are: the addition of LTL properties specifications in PRISM; the introduction of the bounded LTL, i.e. BLTL [24].

3 Genetic Networks with Delayed Stochastic Dynamics

Gene expression is the process by which proteins are synthesized from a sequence in the DNA. It consists of two main phases: *transcription* and *translation*. Transcription is the copying of a sequence in the DNA strand by an RNA polymerase (RNAp) into an RNA molecule. This process takes place in three main stages: initiation, elongation and termination. Initiation consists of the binding of the RNAp to a promoter (Pro) region, unwinding the DNA and promoter escape. Afterwards, elongation takes place, during which the RNA sequence is formed, following the DNA code. Once the termination sequence is reached, both the RNAp and the RNA are released. In prokaryotes, translation, the process by which proteins are synthesized from the (transcribed) RNA sequence, can start as soon as the Ribosome Binding Site (RBS) region of the RNA is formed.

The rate of expression of a gene is usually regulated at the stage of transcription, by activator/repressor molecules that can bind to the operator sites (generally located at the promoter region of the gene) and then promote/inhibit transcription initiation. Evidence suggests that this is a highly stochastic process (see, e.g. [4]), since usually, the number of molecules involved, e.g. transcription factors and promoter regions, is very small, ranging from one to a few at a given moment [37]. Due to that, stochastic modeling approaches were found to be more appropriate than other strategies (e.g. ODE models or Boolean logic).

Stochastic Models of Gene Expression with Delayed Dynamics. The first stochastic models of gene expression assumed that the process of gene expression, once initialized, is instantaneous [4]. Namely, each step was modeled as a uni- or bi-molecular reaction and its kinetics was driven by the stochastic simulation algorithm (SSA) [20]. These models do not account for one important aspect of the kinetics of gene expression. Namely, that it consists, as mentioned, of a sequential process whose intermediate steps take considerable time to be completed once initiated (see e.g. [25]). This feature can be accounted for by introducing 'time delays' in the appearance of the products modeling the process [13,34,32].

Biochemical reaction with stochastic delays can be generally denoted as:

$$\sum n_i R_i \xrightarrow{k} \sum m_j P_j^{nd} + \sum m'_k P_k^d(dist_k)$$

where R_i, P_j^{nd} and P_k^d denote, respectively, the i-th reactant, the j-th non-delayed product and the k-th delayed product (n_i, m_j and m'_k being the stoichiometric coefficients) and $dist_k$ denotes the distribution for the delayed introduction of k-th delayed product. For example, reaction $A+B \xrightarrow{k} A+C(\delta(\tau))$ represents a reaction between molecules A and B, that produces molecule C from B by a process that takes τ seconds to occur once initiated (i.e. $\delta(t)$ denotes a delta dirac distribution centred in t). When this reaction occurs, the number of molecules A is kept constant, a molecule B is immediately removed from the system and a molecule C is introduced in the system τ seconds after the reaction takes place.

To deal with delayed reactions different adaptations of Gillespie's SSA algorithm (referred to as "delayed SSA") have been introduced. Initially two methods [13,11] were proposed for implementing reactions with delays. Then [34] introduced, a generalization of the method proposed in [13], in that it allows multiple time delays in a single reacting event. This algorithm allows implementing a generalized modeling strategy of gene networks [32] and is the one which the SGNSim [33] tool is based on.

3.1 Single Gene Expression Model

We consider a model of single gene expression that follows the approach proposed in [32]. Our model differs in that transcription is modeled as a 2-step process so as to accurately account for the open complex formation and promoter escape [25]. Each of these processes duration follows an exponential distribution. The gene expression system we refer to consists of the following reactions[1]:

$$R_1 : \; Pro + *RNAp \xrightarrow{k_t} Prox \tag{1}$$

$$R_2 : \; Prox \xrightarrow{\lambda_1} Pro + RBS + R(\Gamma(G_{len}, 0.09)) \tag{2}$$

$$R_3 : \; *Rib + RBS \xrightarrow{k_{tr}} RBS(\delta(\tau_1)) + Rib(\Gamma(G_{len}, 0.06)) + P(\Gamma(\tau_{5_{sh}}, \tau_{5_{sc}})) \tag{3}$$

$$R_4 : \; RBS \xrightarrow{rbsd} \emptyset \tag{4}$$

$$R_5 : \; Pro + Rep \xrightarrow{k_r} ProRep \tag{5}$$

$$R_6 : \; ProRep \xrightarrow{k_{unr}} Pro + Rep \tag{6}$$

Reactions (1) and (2) model transcription. In (1), an RNAp binds to a promoter (*Pro*), which remains unavailable for more reactions until reaction (2) occurs. Following reaction (2), which models the promoter escape, both the promoter and the RBS become available for reactions. Also from reaction (2), once transcription is completed, at $\Gamma(G_{len}, 0.09)$, a complete RNA (represented by R) is released in the system. R will not be substrate to any reaction, and is only modeled as a means to count the number of RNA molecules produced over a certain period of time. In our model, according to the SSA, the time necessary for any reaction to occur follows an exponential distribution whose mean is determined by the product between the rate constant of the reaction with the number of each of the reacting molecules present in the system at that moment. For simplicity, we assume that the number of RNAPs is constant. In the case of reactions (1) and (2), both k_t and λ_1 are set to 1/400 s [25], following measurements for the lar promoter. Meanwhile, G_{len} is determined by the length of the gene, here set to 1000 nucleotides, and the time spent by the RNAp at each nucleotide, which follows an exponential distribution with a mean of 0.09 s [29].

[1] Note that symbol $*$ prefixing a species name in the above reactions means that the reactant is not consumed in the reaction. This is applied for simplicity to those reactants such as ribosomes, which exists in large amounts, and thus fluctuations in their numbers wont be significant in the propensity of reactions.

In Prokaryotes, translation can begin as soon as the ribosome binding site (RBS) region of the RNA is completed. In reaction (3), a ribosome (*Rib*) binds to the RBS and translates the RNA. The RBS becomes available for more reactions after τ_1 s. The ribosome is released after $\Gamma(G_{len}, 0.06)$ seconds. The initiation rate, k_{tr} is set to 0.00042 s^{-1} [38]. Following measurements from *E. coli*, we have set $\tau_3 = 2$ s, and $\Gamma(G_{len}, 0.06)$ to follow a gamma distribution dependent on the gene's length, where each codon is added following an exponential distribution with a mean of 0.06 s [29]. Finally, $\Gamma(\tau_{5_{sh}}, \tau_{5_{sc}})$ is such that it accounts for the time that translation elongation takes, as well as the time it takes for a protein to fold and become active. In this case, we used the parameter values measured from GFP mutants commonly used to measure gene expression in *E. coli* [30]. Finally we consider also three additional reactions representing, respectively: RBS decay (equation 4) and promoter repression (equation (5)). Initially, the system has 1 promoter and 100 ribosomes. In the remainder of the paper we illustrate a thorough formal analysis of the above described single gene model by means of HASL model checking.

4 HASL Model Checking

The Hybrid Automata Stochastic Logic (HASL) [8] is a novel formalism widening the family of model checking approaches for stochastic models. Its main characteristics are as follows: first the class of models it addresses are the so-called Discrete Event Stochastic Processes (DESPs), a broad class of stochastic processes which includes, but (unlike most stochastic logics) is not limited to, CTMCs. Second the HASL logic turns out to be a powerful language through which temporal reasoning is naturally blended with elaborate reward-based analysis. In that respect HASL unifies the expressiveness of CSL[6] and its action-based [5], timed-automata [16,14] and reward-based [22] extensions, in a single powerful formalism. Third HASL model checking belongs to the family of statistical model checking approaches (i.e. those that employ stochastic simulation as a means to estimate a model's property). More specifically HASL statistical model checking employs confidence-interval methods to estimate the expected value of random variables which may represent either a measure of probability or a generic real-valued measure. In the following we recall the basics of the HASL formalism i.e. the characterization of DESP and of HASL formula. We also quickly outline Cosmos [7] the HASL model checker we employed for analysing the models considered in this paper. For a comprehensive and more formal treatment of HASL we refer the reader to [8].

4.1 DESP

A DESP is a stochastic process consisting of a (possibly infinite) set S of states and whose dynamic is triggered by a (finite) set E of (time-consuming) discrete events. No restrictions are considered on the nature of the delay distribution associated with events, thus any distribution with non-negative support may be considered. For the sake of space in this paper we omit the formal definition of DESP and give an informal description of Generalised Stochastic Petri Nets (GSPNs) [2] the high-level language adopted to characterise DESP in the context of HASL model checking.

DESP in Terms of Generalised Stochastic Petri Nets. Accordind to its definition the characterization of a DESP is a rather unpractical one, requiring an explicit listing of all of its elements (i.e. states, transitions, delay distributions, probability distribution governing concurrent events). However several high-level formalisms commonly used for representing Markov chain models (e.g. Stochastic Petri Nets, Stochastic Process Algebras), can straightforwardly be adapted to represent DESPs. In the context of HASL model checking we consider GSPNs as high level formalism for representing DESPs. The choice of GSPNs is due to two factors: (1) they allow a flexible modeling w.r.t. the policies defining the process (choice, service and memory) and (2) allow for efficient path generation (due the simplicity of the *firing rule* which drives their dynamics). We quickly recall the basics about GSPN models pointing out the correspondence with the various parts of a DESP. A GSPN model (e.g. Figure 1) is a bi-partite graph consisting of two classes of nodes, *places* (represented by circles) and *transitions* (represented by bars). Places may contain *tokens* (e.g. representing the number of molecules of a given species) while transitions (i.e. representing to the events) indicate how tokens "flow" within the net. The state of a GSPN consists of a *marking* indicating the distribution of tokens throughout the places. A transition is enabled whenever all of its *input places* contains a number of tokens greater than or equal to the multiplicity of the corresponding (input) arc. An enabled transition may *fire* consuming tokens (in a number indicated by the multiplicity of the corresponding input arcs) from all of its input places and producing tokens (in a number indicated by the multiplicity of the corresponding output arcs) in all of its output places. Transitions can be either *timed* (denoted by empty bars, if exponential, or gray bars if non-exponential) or *immediate* (denoted by black filled-in bars). Generally speaking transitions are characterized by: (1) a distribution which randomly determines the delay before firing it (corresponding to the DESP $delay()$ function); (2) a priority which *deterministically* selects among the transitions scheduled the soonest, the one to be fired; (3) a weight, that is used in the random choice between transitions scheduled the soonest with the same highest priority (corresponding to the DESP $choice()$ function). With the GSPN formalism [2] the delay of timed transitions is assumed *exponentially* distributed, whereas with GSPN-DESP it can be given by any distribution. Thus whether a GSPN timed-transition is characterized simply by its weight $t \equiv w$ ($w \in \mathbb{R}^+$ indicating an $Exp(w)$ distributed delay), a GSPN-DESP timed-transition is characterized by a triple: $t \equiv (\text{Dist-t}, \text{Dist-p}, w)$, where Dist-t indicates the type of distribution (e.g. Unif), dist-p indicates the parameters of the distribution (e.g $[\alpha, \beta]$) and $w \in \mathbb{R}^+$ is used to probabilistically choose between transitions occurring with equal delay.[2]

[2] A possible condition in case of non-continuous distributions.

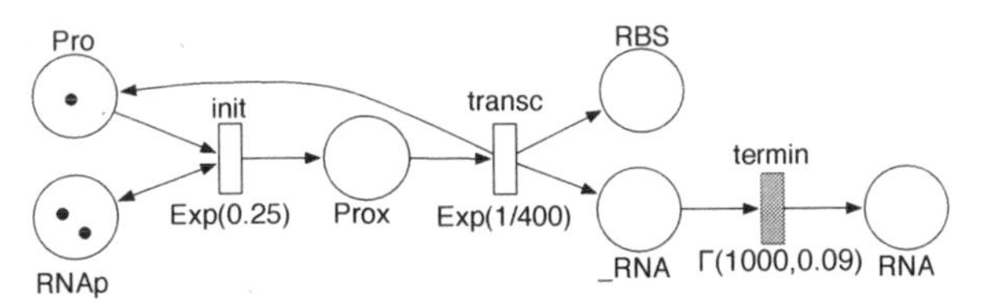

Fig. 1. Example of GSPN-DESP: model of reaction R_1 and R_2 of single-gene system

Example. The GSPN in Figure 1 encodes the transcription phase of the single-gene model (i.e. reaction R_1 and R_2 in Section 3.1). The net has: a place for each species involved in reactions R_1 and R_2 (i.e. Pro, RNAp, Prox, RBS, RNA) plus an extra place (i.e. _RNA) for capturing the intermediate delayed phase of RNA formation; three timed-transition corresponding to the delayed phases of reactions R_1 and R_2. Transitions *init* and *transc* are Exponentially distributed with rate $k_t = 0.25$, respectively $\lambda_1 = 1/400$. Transition *termin* is Gamma distributed (with parameters, *shape* $= 1000$ and *scale* $= 0.09$ as from experimental data) and represent the delayed termination of RNA formation as per R_2. In the initial marking $M_0 = (1, 2, 0, 0, 0, 0)$ we assume a molecule of Pro and two molecules of RNAp are available. Thus in state M_1 *init* is the only reaction enabled, and when it fires it will remove one token from both *Pro* and *RNAp* and add a token in each of its output places (i.e. RNAp and Prox), moving the state of the system in marking $M_1 = (0, 2, 1, 0, 0, 0)$ whereby the only enabled transition is *transc*, and so on.

4.2 Hybrid Automata Stochastic Logic

HASL is a logic designed to analyse properties of a DESP $\mathcal{D}$. A HASL formula is a pair $(\mathcal{A}, Z)$ where $\mathcal{A}$ is Linear Hybrid Automaton (i.e. a restriction of hybrid automata [3]) and Z is an expression involving *data variables* of $\mathcal{A}$. The goal of HASL model checking is to estimate the value of Z by synchronisation of the process $\mathcal{D}$ with the automaton $\mathcal{A}$. This is achieved through stochastic simulation of the synchronised process $(\mathcal{D} \times \mathcal{A})$, a procedure by means of which, infinite timed executions of process $\mathcal{D}$ are selected through automaton $\mathcal{A}$ until some final state is reached or the synchronisation fails. During such synchronisation, data variables evolve and the values they assume condition the evolution of the synchronisation. The synchronisation stops as soon as either: a final location of $\mathcal{A}$ is reached (in which case the values of the variables are considered in the estimate of Z), or the considered trace of $\mathcal{D}$ is rejected by $\mathcal{A}$ (in which case variables' values are discarded).

Synchronised Linear Hybrid Automata. The first component of an HASL formula is an LHA, which is formally defined as follows:

Definition 1. *A* synch. LHA *is a tuple* $\mathcal{A} = \langle E, L, \Lambda, Init, Final, X, flow, \rightarrow \rangle$ *where:*

- $\boldsymbol{E}$, *a finite alphabet of events;*
- $\boldsymbol{L}$, *a finite set of locations;*
- $\boldsymbol{\Lambda : L \rightarrow Prop}$, *a location labelling function;*
- $\boldsymbol{Init}$, *a subset of L called the initial locations;*
- $\boldsymbol{Final}$, *a subset of L called the final locations;*
- $\boldsymbol{X = (x_1, ...x_n)}$ *a n-tuple of data variables;*
- $\mathit{flow} : \boldsymbol{L \mapsto Ind^n}$ *a n-tuple of indicators representing the rate of evolution of each data variable in a location.*
- $\rightarrow \subseteq \boldsymbol{L} \times ((\mathsf{Const} \times 2^{\boldsymbol{E}}) \uplus (\mathsf{lConst} \times \{\sharp\})) \times \mathsf{Up} \times \boldsymbol{L}$, *a set of edges*

where an edge $(l, \gamma, E', U, l') \in \rightarrow$ (also denoted $l \xrightarrow{\gamma, E', U} l'$), consists of: a constraint γ (i.e. a boolean combination of inequalities of the form $\sum_{1 \leq i \leq n} \alpha_i x_i + c \prec 0$ where $\alpha_i, c \in Ind$ are DESP indicators, $\prec \in \{=, <, >, \leq, \geq\}$ and $x_i \in X$; we denote Const the set of such constraints and $\mathsf{lConst} \subset \mathsf{Const}$ the set of left closed constraints, i.e. constraints giving rise to left-closed enabling intervals); a set E' of labels of synchronising events (including the extra label $\sharp$ denoting autonomous edges); a set U of *updates* (i.e. an n-tuple of functions $u_1, ..., u_n$ where each u_k is of the form $x_k = \sum_{1 \leq i \leq n} \alpha_i x_i + c$ where the $\alpha_i, c \in Ind$ are DESP indicators; we denote Up the set of updates). by means of which new values are assigned to variables of X on traversing of the edge).

Edges labelled with a set of events in 2^E are called *synchronized* whereas those labelled with $\sharp$ are called *autonomous*. Furthermore we impose the following (informally described[3]) constraints for an automaton $\mathcal{A}$: (c1) only one initial location can be enabled; (c2) the same event cannot lead to different simultaneous synchronisations; (c3) two autonomous transition cannot be fireable simultaneously (c4) infinite loops without synchronisation are not possible. Informally the synchronisation between ($\mathcal{D}$ and $\mathcal{A}$) works as follows: a *synchronised* transition of the product process ($\mathcal{D} \times \mathcal{A}$) is triggered by the occurrence of a corresponding (time-consuming) event of the DESP, whereas an *autonomous* transition occurs (without synchronisation with a DESP event) as soon as the corresponding constraint is enabled (i.e. the variables of the LHA assume values fulfilling the constraint). Note that *autonomous* transitions may not consume time.

Example: Figure 2 depicts two variants of a simple two locations LHA defining measures of the gene-transcription toy model of Figure 1. Location l_0 is the initial location while l_1 the final location. The automaton employs two data-variables: t registering the simulation-time (hence with flow $\dot{t} = 1$ in every location) and n_1, an event counter (hence with flow $\dot{n}_1 = 0$ in every location), counting the occurrences of transition *transcr*. The automaton has two synchronising edges (the self-loops on l_0) and one autonomous edge (from l_0 to l_1). The top synchronising edge allows to increment the counter n_1 each time a transition *transcr* occurs whereas the bottom synchronising edge, simply reads in all other transitions occurrence without performing any update. The autonomous edge instead leads to acceptance location as soon as its constraint is fulfilled. The LHA in Figure 2(a) represents time-bounded measures as the constraints on the edges (and notably

[3] For the formal characterisation see [8].

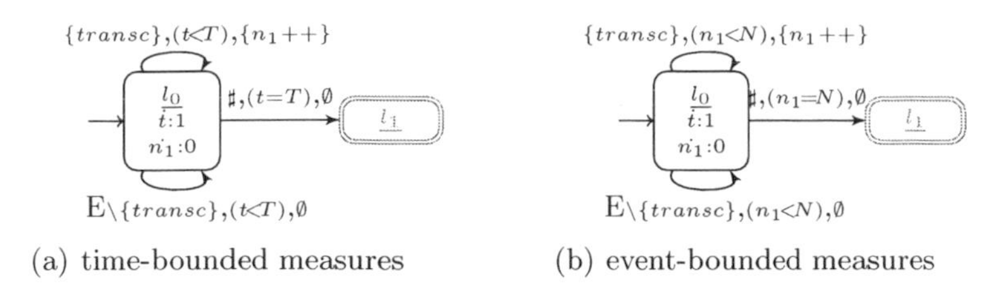

(a) time-bounded measures (b) event-bounded measures

Fig. 2. Example of LHA for simple properties of the GSPN-DESP model of Figure 1

on the edge leading to the acceptance location) refers to the simulation time t, thus: as soon as $t = T$ the read in path is accepted. On the other hand the LHA in Figure 2(b) represents event-bounded measures accepting paths as soon as transition *transc* have occurred $n_1 = N$ times. In the following we provide few simple examples of relevant measures referred to the LHA in Figure 2 in terms HASL expressions.

HASL Expressions. The second component of an HASL formula is an expression, denoted Z and defined by the grammar:

$$\begin{aligned} Z &::= E(Y) \mid Z + Z \mid Z \times Z \\ Y &::= c \mid Y + Y \mid Y \times Y \mid Y/Y \mid last(y) \mid min(y) \mid max(y) \mid int(y) \mid avg(y) \\ y &::= c \mid x \mid y + y \mid y \times y \mid y/y \end{aligned} \tag{7}$$

y is an arithmetic expression built on top of LHA data variables (x) and constants (c). Y is a path dependent expression built on top of basic path random variables such as $last(y)$ (i.e. the last value of y along a synchronizing path), $min(y)(max(y))$ the minimum (maximum), value of y along a synchronizing path), $int(y)$ (i.e. the integral over time along a path) and $avg(y)$ (the average value of y along a path). Finally Z, the target measure of an HASL experiment, is an arithmetic expression built on top of the first moment of Y ($E[Y]$), and thus allowing to consider more complex measures including, e.g. $Var(Y) \equiv E[Y^2] - E[Y]^2$, $Covar(Y_1, Y_2) \equiv E[Y_1 \cdot Y_2] - E[Y_1] \cdot E[Y_2]$.

The COSMOS Tool. Assessment of HASL formulae against a DESP model is performed by means of the COSMOS [1,7] model checker. COSMOS employs a confidence interval method to estimate the target expression Z. The desired accuracy of the target estimation is then set by the end user in terms of *confidence level* and *interval width.*

5 Analysis of SGN Models through HASL

We illustrate the application of HASL model checking to the analysis of a model of gene expression with stochastic delayed dynamics. We (1) present the GSPN-

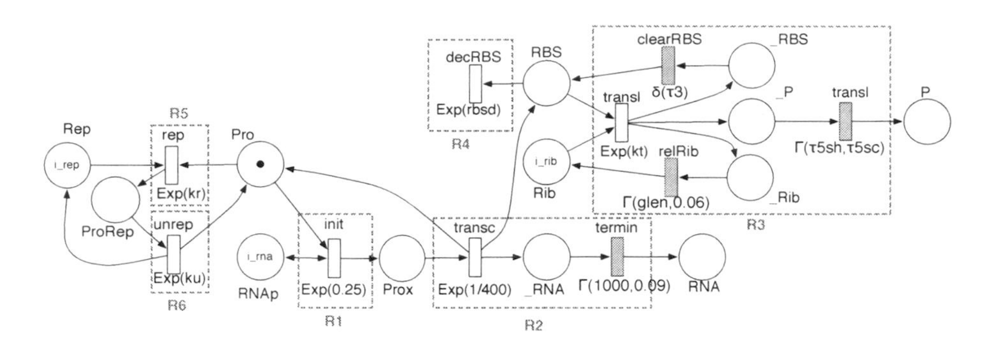

Fig. 3. GSPN model of Single Gene system with delayed stochastic dynamics

DESP codification of the considered model; (2) introduce a number of relevant properties/measures of a model first describing them informally and then providing their encoding in HASL terms; (3) discuss results obtained by evaluation of the presented properties/measures by means of the COSMOS model checker.

5.1 Single Gene Model

The single-gene model described by equations (1) to (6) (Section 3.1) is encoded in GSPN-DESP terms by the net depicted in Figure 3. The net includes a place for each species of the model (i.e. Pro, RNAp, Prox, RNA, RBS, Rib P, Rep and ProRep) plus a number of auxiliary places representing intermediate stages of delayed reactions (i.e. _RNA, _RBS, _P, relRib). Initial marking of the net is set by means of parameters *i_rep*, *i_rnap*, *i_rib*, which correspond to the chosen initial population of the model (note that the promoter place, *Pro*, is initialized with one token, as each gene has one promoter region).

Reactions $\{R_1, \ldots, R_6\}$ of the single gene model correspond to subnets (enclosed in red-dashed rectangles) in Figure 3. Each such subnet contains either a single exponentially-distributed transition (in case of reactions with non-delayed products i.e. R1, R4, R5, R6) or a combination of exponential and non-exponential transitions (in case of reactions with delayed products, i.e. R1 and R2). For example subnet R3 in Figure 3 represents the encoding of the *translation* reaction. It consists of: the translation-start event (i.e. exponentially distributed transition labeled *s_transl*; the RBS release event (i.e. deterministically distributed transition *clearRBS*); the ribosome releasing event (gamma distributed transition *relRib*); the protein production event (i.e. gamma distributed transition *prodP*). Observe that the effect of repressed gene-expression can be promptly analysed by setting of the repressor initial population (parameter *i_rep*= l_r): unrepressed configurations corresponds to $l_r = 0$, whereas $l_r > 0$ settings correspond to repressed model where the level of repression is proportional to $l_r > 0$.

Table 1. Properties of the Single Gene model

performance of TRANSCRIPTION and TRANSLATION mechanisms	
ID	**description**
ϕ_{1_a}	average num. of completed-transcriptions (within T)
ϕ_{1_b}	average num. of completed-translations (within T)
ϕ_{2_a}	prob. density of the number of completed-transcriptions (within T)
ϕ_{2_b}	prob. density of the number of completed-translations (within T)
ϕ_{3_a}	cumulative prob. of the number of completed-transcriptions (within T)
ϕ_{3_b}	cumulative prob. of the number of completed-translations (within T)
efficiency of TRANSLATION wrt TRANSCRIPTION	
ID	**description**
ϕ_4	avg. num. of completed translations between two consecutive transcriptions
ϕ_5	prob. of at least N completed translations between two consecutive transcriptions
REPRESSION related measures	
ϕ_6	percentage of time gene is repressed
ϕ_7	how long does it take for translation to stop once a repression starts (i.e. sustainment of translation under repression)

Properties of Single Gene Model. Table 1 depicts an excerpt of (informally stated) relevant measures of the single-gene model. They are grouped according to different aspects of gene-expression performance. The corresponding HASL encoding is given in Table 2. We briefly illustrate the automata of Table 2 and the associated HASL expressions:

$\mathcal{A}_1$ **:** it is designed for measures concerning the occurrences of *transc* and *transl* events. It accepts all paths of duration T and uses variables, n_1 and n_2 to maintain the number of *transc* and *transl* transitions occurred along a path. Different measures can be assessed through different HASL expressions referred to $\mathcal{A}_1$ including: $\phi_{1_a} = (\mathcal{A}_1, E[last(n_1)])$; $\phi_{1_b} = (\mathcal{A}_1, E[last(n_2)])$ and $\phi_4 = (\mathcal{A}_1, E[last(n_2)/last(n_1)])$ (see Table 1).

$\mathcal{A}_2$ **:** it measures the probability that the number of transcriptions is $n_1 = C$ (within time T). On acceptance (i.e. duration $t = T$), it distinguishes between paths such that $n_1 = C$ (in which case the bernoulli variable OK is set to 1), and paths such that $n_1 \neq C$ (i.e. OK is set to 0). The probability density of n_1 is assessed by re-iterated evaluations of formula $\phi_{2_a} = (\mathcal{A}_2, last(OK))$ corresponding to different values of C[4].

$\mathcal{A}_3$ **:** for measures concerning the amount of time gene is repressed. Apart from the usual global clock t it uses a timer t_r registering the time gene is repressed, hence it grows ($\dot{t}_r = 1$) only in location l_1 (i.e. repression is ON, corresponding to a marking of place $ProRep > 0$), while it is unchanged ($\dot{t}_r = 0$) in location l_0 (i.e. marking of place $ProRep = 0$). Also note that both l_0 and l_1 are initial locations, which is perfectly legal as their constraints make them mutually exclusive (this way $\mathcal{A}_3$ can be used to analyse both *repressed* and *unrepressed* configurations of the model).

[4] Note that simple variants of $\mathcal{A}_2$ can be used to assess the PDF of translations and the CDFs of both transcription and translation.

Table 2. LHA for various measures of the Single Gene model

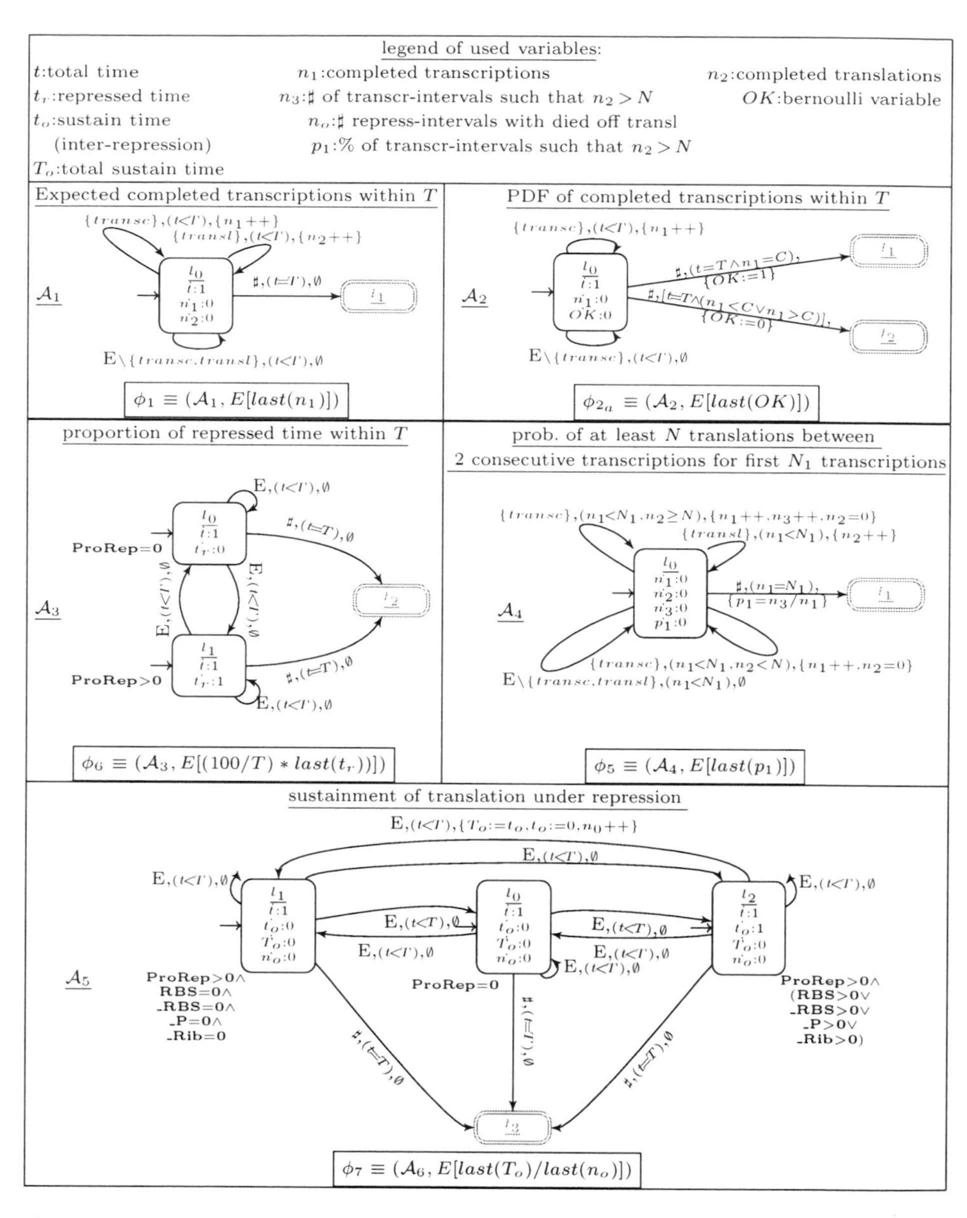

$\mathcal{A}_4$: it measures *"how likely it is that within a transcription interval (i.e. the interval between two occurrences of the* transc *event) at least N translations have been completed"*. It uses variables n_1 and n_2 (as above) and n_3 to count how many transcription intervals (along a path) contain $n_2 \geq N$ translations. The result is stored in $p_1 = n_3/n_1$ on acceptance. Note that, in this case, we consider an event-bounded observation window consisting of $n1 = N_1$ transcription events. Measure ϕ_5 (Table 1) in HASL terms is $\phi_5 = (\mathcal{A}_4, last(p_1))$.

$\mathcal{A}_5$: it is designed for measures of *sustainment of translation activity under repression* (i.e. ϕ_7 in Table 1). It uses the following variables: n_o counting the number of repression intervals (interval between two repression events) in which translation arrested; t_o: measuring the *translation time-to-arrest* in a repression interval (given that translation arrested); T_o timer measuring the cumulated t_o. Note that translation arrest corresponds to the absence of tokens in all translation related places of the GSPN model (Figure 3), corresponding to condition: $(RBS{=}0 \wedge\_RBS{=}0 \wedge\_P{=}0 \wedge\_Rib{=}0)$. Locations l_0, l_1 and l_2 are then associated to the following state conditions of the model: *repression is off* (l_0), *repression is on and translation off* (l_1) and *repression is on and translation ongoing* (l_2). All paths of duration $t{=}T$ are accepted and the target measure[5] is be obtained through expression $Z = E[last(T_o)/Last(n_o)]$.

Remark. A formal assessment of HASL expressiveness is beyond the scope of this paper however we make some considerations in that respect. The peculiarity of HASL based reasoning is that any combination of state and/or transition and/or reward conditions may be employed to characterise the paths of interests. This is the main difference with other stochastic logics, ranging from those limited to state-based reasoning (e.g. CSL, BLTL), to those featuring state/action based reasoning but not supporting rewards (e.g. asCSL [5]), up to the timed-automata ones which mix state/action-based reasoning with (multiple) time-bounding [16,14]. So far reward-based analysis has been added to logics featuring state-based reasoning. For example the rewards enriched version of CSL supported by PRISM [27] allows for considering multiple (state and transition) reward structures and to assess reward measures wrt paths of a given CTMC model. However, even with the addition of rewards, CSL remains a language limited to state-based temporal reasoning, thus, differently from HASL, reward values do not play an active role in characterising relevant paths. As a consequence several measures that can easily be expressed with HASL, do not always have an equivalent in CSL terms (or if they do they require hard wiring of extra information in the original CTMC). For example, measuring the PDF (and CDF) of an event occurrences, is easily done with HASL (e.g. LHA $\mathcal{A}_2$, of Table 2), whereas cannot be naturally achieved with CSL, unless states of the original CTMC are enriched with variables counting the occurrences of relevant events. Similarly, more complex measures involving combination of elaborate conditions on rewards values (i.e. LHA variables) as the factors characterising the selected paths (e.g. those corresponding to automata $\mathcal{A}_4$ and $\mathcal{A}_5$, in Table 2) seem not to be expressible through CSL rewards.

Experiments. We assessed the previously described HASL measures through experiments executed with the COSMOS model checker. For time-bounded

[5] Note that with a time-bounded measurement, as with $\mathcal{A}_5$, measuring may stop in any instant (not necessarily at the end) of a repression interval: this is not a problem as T_o and n_o are updated only when translation arrests, thus if bound T is reached before translation arrests, measure T_o/n_o will correctly refer to the duration of translation sustainment over all completed repression cycles.

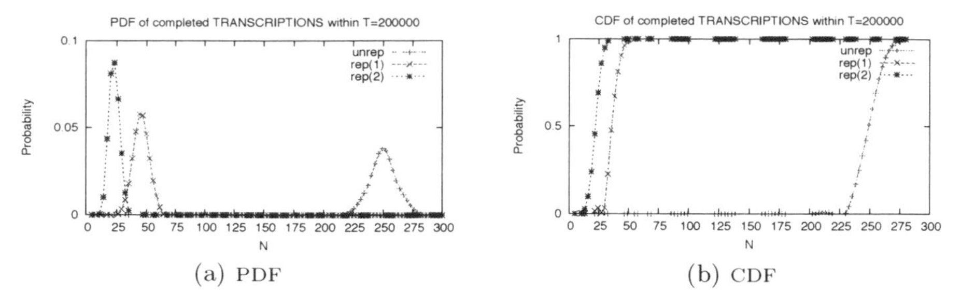

(a) PDF (b) CDF

Fig. 4. PDF and CDF of completed transcriptions within T

measures we have considered (following [33]) $T = 2 \cdot 10^5$ as time horizon which roughly corresponds to 60 cell cycles, considering an average a cell cycle period of about 55 minutes (i.e. 3300s) in the case of E. coli. All experiments have been run with the following setting concerning confidence interval estimation: confidence-level: 99.99%; interval-width: 0.01.

Experiment 1. Figure 4 compares plots of the PDF (Figure 4(a)) and CDF (Figure 4(b)) of random variable n_1: *num. of completed transcription within T* (query ϕ_2 and ϕ_3) of *unrepressed* vs *repressed* configurations (i.e. *rep(1)*, corresponding to initial marking $i\_rep = 1$ and *rep(2)*, corresponding to initial marking $i\_rep = 2$). The effect of repression is evident as the bell-shaped probability density of n_1 is shifted toward lower values for increasing level of repression.

Experiment 2. Figure 5(a) compares the *expected number of completed* transcriptions *vs.* translations *within T* in function of time for unrepressed and repressed (*rep(1)*) configurations. Observe that the throughput of *translation* is roughly twice as much as that of *transcriptions*, (both in unrepressed condition, as well as, in presence of repression). This is due to the rates of RNA degradation and translation initiation.

Experiment 3. Figure 5(b), plots two measures of timing: *the percentage of time gene is repressed* ($\mathcal{A}_3$) and the *percentage of time no translation activity is going on* (variant of $\mathcal{A}_3$) when system is observed for duration T and in function of

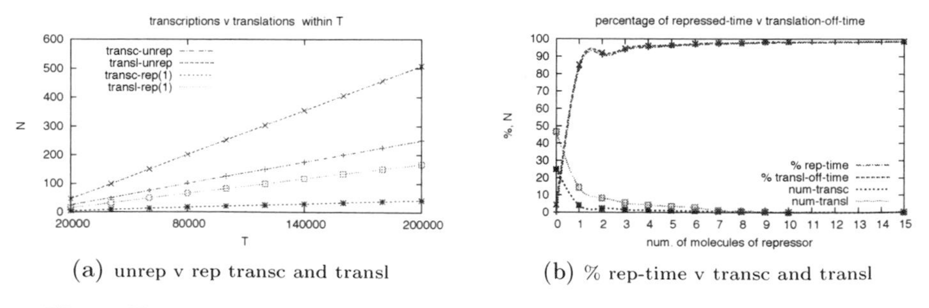

(a) unrep v rep transc and transl (b) % rep-time v transc and transl

Fig. 5. Exp. transcriptions and translations and percentage of repressed time

the level of repression (num. of repressor molecules on the x-axis). To observe also the trend of transcription and translation activity in function of repression level Figure 5(b) also includes two curves referred to the expected number of *transcriptions*, respectively *translations*, within T. Observe that the presence of a single repressor is sufficient for the gene to remain repressed for 83% of the time and, likewise, for translation to be non-existing for 85% of the observation time (whereas in absence of repressor, translation activity is only non-existing for about 4% of the time).

Experiment 4. Figure 6(a) compares the PDFs of random variable n_2: *num. of completed transcription within a transcription interval* (i.e. within two consecutive transcription completions) (query $\phi_4 : (\mathcal{A}_4, Last(p_1))$. This is computed for the unrepressed model and for two configurations of the repressed model (*rep(1)* and (*rep(2)*). Outcomes indicate that in presence of repression the probability density is more "distributed", than the bell shaped one corresponding to the unrepressed configuration. Furthermore increasing the level of repression seems to have no effect on the probability density (plots *rep(1)* and (*rep(2)* are essentially identical).

Experiment 5. Figure 6(b) refers to measurement of the *translation sustainment within a repression-interval* (query $\phi_5 : (\mathcal{A}_5, Last(n_{off})/Last(n_{rep}))$ in function of the RBS decay rate (*rbsd*). We varied *rbsd* in the interval [0.001, 4] which includes *rbsd* = 0.01 i.e. the value complying with experimental evidence used in the"standard" model's configuration. Obtained results indicate, quite sensibly, that translation sustainment is inversely proportional to RBS decay. It should be noted that with *rbsd* < 0.004 the translation sustainment is actually increasing with *rbsd* (not very evident in plot of Figure 6(b)). This is because, by definition, query $\phi_5 : (\mathcal{A}_5, Last(n_{off})/Last(n_{rep}))$ measures the sustainment of translation *on condition that sustainment lasts lesser than repression*. With *rbsd* < 0.004, however, decay is so slow that with high probability sustainment lasts longer than repression, while with low probability it lasts less. In this case (*rbsd* < 0.004) it is sensible that the average value of (low-likely) *translation sustainment not exceeding repression duration* increases with *rbsd*.

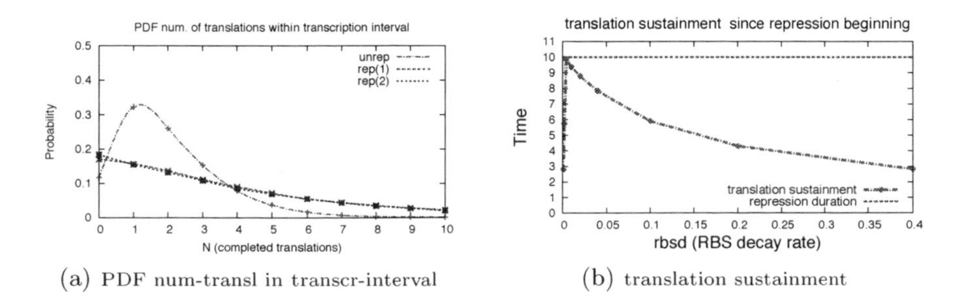

(a) PDF num-transl in transcr-interval (b) translation sustainment

Fig. 6. Measure related to translation activity

6 Measuring Oscillations with HASL

Oscillatory trends are fundamental aspects of the dynamics of many biological mechanisms, therefore the ability to detect/measure oscillations in biological models is crucial. CSL-based characterisation of oscillations in CTMC models of biochemical reactions have been considered in [9], with limited success, and more comprehensively in [36]. Here we show preliminary results concerning the application of HASL to the analysis of oscillations. Let σ be a (infinitely long) simulation trace of an n-dimensional DESP model whose states' form is $s = (s_1, \dots s_n) \in \mathbb{N}^n$, with s_i being the value along the ith dimension (e.g the number of molecules of species i). Let $\sigma_i(t)$ denote the i-projection of $\sigma(t)$ the state σ is at time t. Following the characterisation given in [36] σ_i can either be: *convergent* (i.e. tending to a finite value), *divergent* (i.e. tending to infinity) or *oscillating* (i.e. the lack of the previous two). Furthermore σ_i is *periodic* with period δ iff $\forall t, \sigma_i(t) = \sigma_i(t + \delta)$. Thus σ is periodic oscillatory along the i-th dimension iff σ_i is both oscillating and periodic. Here, instead, we focus on a less restrictive characterisation of oscillatory trends namely that of *noisy periodicity* [36]. Given an upper and a lower bound $b_h, b_l \in \mathbb{N}$ ($b_h > b_l$), inducing intervals $low = (-\infty, b_l]$, $mid = (b_l, b_h)$ and $high = [b_h - \infty)$ (i.e. l,m and h), trace σ_i is said *noisy periodic* iff it perpetually switches from low to $high$ (passing through mid) and returning to low.Note that such trends corresponds to the following regular expression: $e_{np} = l(l)^*m(ll^*m)^*h(mm^*h)^*m(mh^*m)^*l$.

We illustrate preliminary results about application of HASL to oscillations analysis by means of a simple example. Reactions (8) represent, the so called, 3-way doped oscillator, a systems consisting of three species A, B and C which oscillate perpetually. Note that A, B and C form a loop of of dependency whereby A is converted into B, which, in turns, is converted into C, which, in turns, is converted into A. D_A, D_B and D_C, are auxiliary species representing doping substances which guarantees the *liveness* of the conversion loop. It can be easily shown, (e.g. by application of stochastic simulation), that species A, B and C oscillate (Figure 7(a)) with amplitude, period and "noisiness" dependent on the initial population (a_0, b_0, c_0) (by default we assume the population of doping species to be 1).

$$\begin{array}{lll} A + B \xrightarrow{r_A} 2B & B + C \xrightarrow{r_B} 2C & C + A \xrightarrow{r_C} 2A \\ D_A + C \xrightarrow{r_C} A + D_A & D_B + A \xrightarrow{r_A} B + D_B & D_C + B \xrightarrow{r_B} C + D_C \end{array} \quad (8)$$

An LHA to Measure Noisy Periodic Traces: the LHA $\mathcal{A}_{np}$ in Figure 8 is designed to measure the number of noisy periods of amplitude $a \geq (b_h - b_l)$, where b_l and b_h are the bounds inducing the low, mid and $high$ (as above). It consists of three locations l_0, l_1 and l_2, associated to the low (i.e. $A \leq b_l$), respectively the mid (i.e. $b_l < A < b_h$) and the $high$ (i.e. $A \geq b_h$) interval.

It uses three variables: t (total time), n to count the completed noisy periods (i.e. corresponding to regular expression $e_{np} = l(l)^*m(ll^*m)^*h(mm^*h)^*m(mh^*m)^*l$), and *top*, a boolean flag used to condition the increase of n on completion of a noisy period (i.e. *top* is set to 1 entering the $high$ interval and n is incremented on entering the low interval only if $top = 1$,

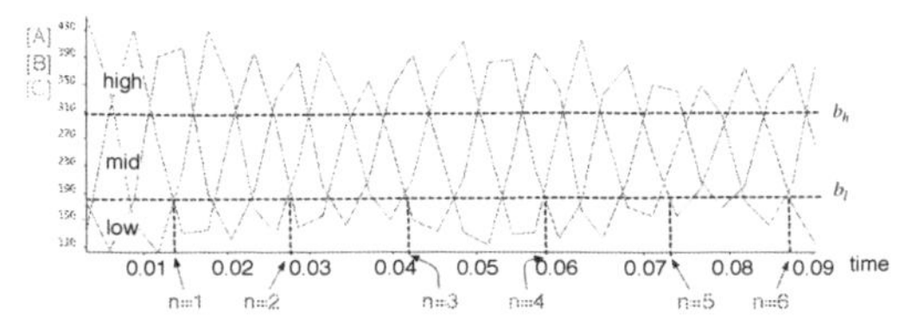

(a) simulation trace of doped oscillator init-population:(100, 350, 310)

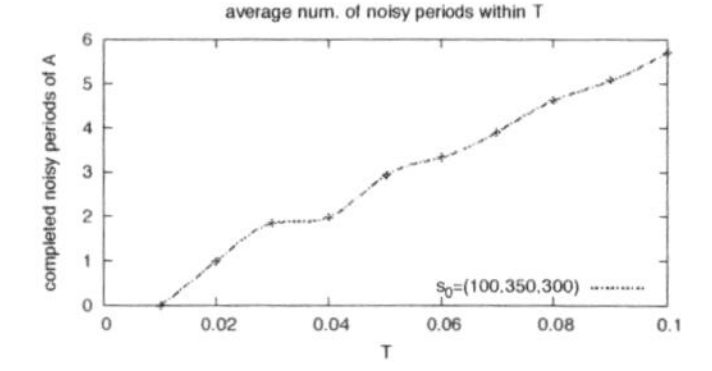

(b) number of periods in doped oscillator

Fig. 7. Number of periods in a simulated trace of the 3-ways doped oscillator (left) and corresponding average number of noisy periods calculated (in function of time) through the HASL query $\phi_{np} = (\mathcal{A}_{np}, E(last(n))$ (right) with bounds set to $b_l = 180$ and $b_h = 300$

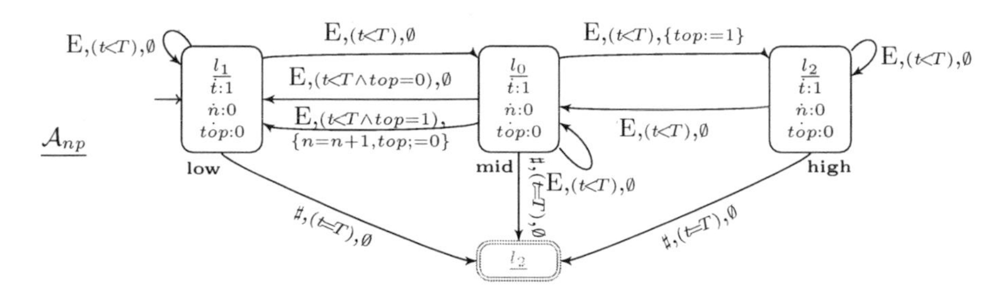

Fig. 8. An LHA for measuring the number of periods of a *periodic noisy* oscillatory trace of the 3-way oscillator

in which case *top* is reset). The average number of completed noisy periods with time T can be assessed by means of HASL formula $\phi_{np} = (\mathcal{A}_{np}, E(last(n)))$. Figure 7(b) depicts the outcome of assessment of ϕ_{np} in function of time bound T. Such results are in good agreement with simulated traces, a sample of which is shown in Figure 7(a).

7 Conclusion

We presented some insights on the application of HASL statistical model checking to the analysis of (non-necessarily Markovian) stochastic models of biological systems. The most important feature of HASL model checking lays in its expressive power: by employing LHA as machinery to characterise relevant trajectories of a model it is possible to identify/assess elaborate measures which may not be naturally accounted for with more popular stochastic logic, i.e. those featuring state-based temporal reasoning, such as, for example, CSL and BLTL. We demonstrated (part of) the potential of the HASL language by developing and assessing a number of properties of a model of single-gene network with delayed non-Markovian dynamic. Although such model is a rather simple, it served well to our aim, which was to show the potential of HASL based analysis. Furthermore we presented preliminary insights on the application of HASL to the analysis of oscillatory trends. Future developments include: 1) the application of HASL approach to more complex systems, such as, for example a model of the P53-Mdm2 feedback loop with stochastic dynamics previously analysed with the GNSim simulator but not yet formally model checked. 2) further development of HASL based oscillation analysis (i.e. measurements of frequency, amplitude of oscillatory trends). The main difficulty of the HASL approach is the technicality of the formalism itself. In particular, specifying an HASL property boils down to specifying an automata, something which may be far from intuitive for non-expert users. Thus, a further direction of development regards working on the definition of a more intuitive property specification language with an associated translator to LHA specifications.

References

1. Cosmos home page, `http://www.lsv.ens-cachan.fr/software/cosmos/`
2. Ajmone Marsan, M., Balbo, G., Conte, G., Donatelli, S., Franceschinis, G.: Modelling with Generalized Stochastic Petri Nets. John Wiley & Sons (1995)
3. Alur, R., Courcoubetis, C., Henzinger, T.A., Ho, P.-H.: Hybrid Automata: An Algorithmic Approach to the Specification and Verification of Hybrid Systems. In: Grossman, R.L., Ravn, A.P., Rischel, H., Nerode, A. (eds.) HS 1991 and HS 1992. LNCS, vol. 736, pp. 209–229. Springer, Heidelberg (1993)
4. Arkin, A.P., Ross, J., McAdams, H.H.: Stochastic Kinetic Analysis of a Developmental Pathway Bifurcation in Phage-l Escherichia coli. Genetics 149(4), 1633–1648 (1998)
5. Baier, C., Cloth, L., Haverkort, B., Kuntz, M., Siegle, M.: Model checking action- and state-labelled Markov chains. IEEE Trans. on Software Eng. 33(4) (2007)
6. Baier, C., Haverkort, B., Hermanns, H., Katoen, J.-P.: Model-checking algorithms for CTMCs. IEEE Trans. on Software Eng. 29(6) (2003)
7. Ballarini, P., Djafri, H., Duflot, M., Haddad, S., Pekergin, N.: COSMOS: a statistical model checker for the hybrid automata stochastic logic. In: Proceedings of the 8th International Conference on Quantitative Evaluation of Systems (QEST 2011), pp. 143–144. IEEE Computer Society Press (September 2011)
8. Ballarini, P., Djafri, H., Duflot, M., Haddad, S., Pekergin, N.: HASL: an expressive language for statistical verification of stochastic models. In: Proc. Valuetools (2011)
9. Ballarini, P., Guerriero, M.L.: Query-based verification of qualitative trends and oscillations in biochemical systems. Theoretical Computer Science 411(20), 2019–2036 (2010)
10. Barnat, J., Brim, L., Cerná, I., Drazan, S., Fabriková, J., Láník, J., Safránek, D., Ma, H.: Biodivine: A framework for parallel analysis of biological models. In: COMPMOD. EPTCS, vol. 6, pp. 31–45 (2009)
11. Barrio, M., Burrage, K., Leier, A., Tian, T.: Oscillatory regulation of hes1: Discrete stochastic delay modelling and simulation. PLoS Computational Biology 2(9), 1017–1030 (2006)
12. Bernot, G., Comet, J.-P., Richard, A., Guespin, J.: Application of formal methods to biological regulatory networks: extending Thomas' asynchronous logical

approach with temporal logic. Journal of Theoretical Biology 229(3), 339–347 (2004)
13. Bratsun, D., Volfson, D., Tsimring, L.S., Hasty, J.: Delay-induced stochastic oscillations in gene regulation. Proc. Natl. Acad. Sci. USA. 102(41), 14593–14598 (2005)
14. Chen, T., Han, T., Katoen, J.-P., Mereacre, A.: Quantitative model checking of CTMC against timed automata specifications. In: Proc. LICS 2009 (2009)
15. De Jong, H., Geiselmann, J., Hernandez, C., Page, M.: Genetic network analyzer: qualitative simulation of genetic regulatory networks. Bioinformatics 19(3), 244–336 (2003)
16. Donatelli, S., Haddad, S., Sproston, J.: Model checking timed and stochastic properties with CSL^{TA}. IEEE Trans. on Software Eng. 35 (2009)
17. Emerson, E.A., Clarke, E.M.: Characterizing Correctness Properties of Parallel Programs Using Fixpoints. In: de Bakker, J.W., van Leeuwen, J. (eds.) ICALP 1980. LNCS, vol. 85, pp. 169–181. Springer, Heidelberg (1980)
18. Fages, F., Soliman, S.: Formal Cell Biology in Biocham. In: Bernardo, M., Degano, P., Zavattaro, G. (eds.) SFM 2008. LNCS, vol. 5016, pp. 54–80. Springer, Heidelberg (2008)
19. Fages, F., Soliman, S., Rivier, C.N.: Modelling and querying interaction networks in the biochemical abstract machine BIOCHAM. Journal of Biological Physics and Chemistry 4(2), 64–73 (2004)
20. Gillespie, D.T.: Exact stochastic simulation of coupled chemical reactions. Journal of Physical Chemistry 81(25), 2340–2361 (1977)
21. Hansson, H.A., Jonsson, B.: A framework for reasoning about time and reliability. In: Proc. 10th IEEE Real -Time Systems Symposium, Santa Monica, Ca, pp. 102–111. IEEE Computer Society Press (1989)
22. Haverkort, B.R., Cloth, L., Hermanns, H., Katoen, J.-P., Baier, C.: Model checking performability properties. In: Proc. DSN 2002 (2002)
23. Heath, J., Kwiatkowska, M., Norman, G., Parker, D., Tymchyshyn, O.: Probabilistic model checking of complex biological pathways. Theoretical Computer Science 319(3), 239–257 (2008)
24. Jha, S.K., Clarke, E.M., Langmead, C.J., Legay, A., Platzer, A., Zuliani, P.: A Bayesian Approach to Model Checking Biological Systems. In: Degano, P., Gorrieri, R. (eds.) CMSB 2009. LNCS, vol. 5688, pp. 218–234. Springer, Heidelberg (2009)
25. Kandhavelu, M., Hakkinen, A., Yli-Harja, O., Ribeiro, A.S.: Single-molecule dynamics of transcription of the lar promoter. Phys. Biol. 9(2) (2012)
26. Kitano, H.: Foundations of Systems Biology. MIT Press (2002)
27. Kwiatkowska, M., Norman, G., Parker, D.: Stochastic Model Checking. In: Bernardo, M., Hillston, J. (eds.) SFM 2007. LNCS, vol. 4486, pp. 220–270. Springer, Heidelberg (2007)
28. Lakin, M., Parker, D., Cardelli, L., Kwiatkowska, M., Phillips, A.: Design and analysis of DNA strand displacement devices using probabilistic model checking. Journal of the Royal Society Interface (to appear, 2011)
29. Makela, J., Lloyd-Price, J., Yli-Harja, O., Ribeiro, A.S.: Stochastic sequence-level model of coupled transcription and translation in prokaryotes. BMC Bioinformatics 12(1), 121 (2011)
30. Megerle, J.A., Fritz, G., Gerland, U., Jung, K., Rädler, J.O.: Timing and Dynamics of Single Cell Gene Expression in the Arabinose Utilization System. Biophysical Journal 95, 2103–2115 (2008)

31. Prism home page, `http://www.prismmodelchecker.org`
32. Ribeiro, A., Zhu, R., Kauffman, S.A.: A general modeling strategy for gene regulatory networks with stochastic dynamics. Journal of Computational Biology: a Journal of Computational Molecular Cell Biology 13(9), 1630–1639 (2006)
33. Ribeiro, A.S., Lloyd-Price, J.: SGN Sim, a stochastic genetic networks simulator. Bioinformatics 23(6), 777–779 (2007)
34. Roussel, M.R., Zhu, R.: Validation of an algorithm for delay stochastic simulation of transcription and translation in prokaryotic gene expression. Physical Biology 3(4), 274–284 (2006)
35. Schwarick, M., Rohr, C., Heiner, M.: Marcie - model checking and reachability analysis done efficiently. In: Proc. 8th International Conference on Quantitative Evaluation of SysTems (QEST 2011), pp. 91–100. IEEE CS Press (2011)
36. Spieler, D.: Model checking of oscillatory and noisy periodic behavior in markovian population models. Master's thesis, Saarland University (2009)
37. Taniguchi, Y., Choi, P.J., Li, G.-W., Chen, H., Babu, M., Hearn, J., Emili, A., Xie, X.S.: Quantifying E. coli Proteome and Transcriptome with Single-Molecule Sensitivity in Single Cells. Science 329(5991), 533–538 (2010)
38. Zhu, R., Ribeiro, A.S., Salahub, D., Kauffman, S.A.: Studying genetic regulatory networks at the molecular level: Delayed reaction stochastic models. Journal of Theoretical Biology 246(4), 725–745 (2007)

Symmetry-Based Model Reduction for Approximate Stochastic Analysis

Kirill Batmanov[1,⋆], Celine Kuttler[1], Francois Lemaire[1], Cédric Lhoussaine[1], and Cristian Versari[1,2]

[1] Lifl (Cnrs Umr 8022), University of Lille 1, France
kirill.batmanov@lifl.fr
[2] Inria, Lille, France

Abstract. For models of cell-to-cell communication, with many reactions and species per cell, the computational cost of stochastic simulation soon becomes intractable. Deterministic methods, while computationally more efficient, may fail to contribute reliable approximations for those models. In this paper, we suggest a reduction for models of cell-to-cell communication, based on symmetries of the underlying reaction network. To carry out a stochastic analysis that otherwise comes at an excessive computational cost, we apply a moment closure (MC) approach. We illustrate with a community effect, that allows synchronization of a group of cells in animal development. Comparing the results of stochastic simulation with deterministic and MC approximation, we show the benefits of our approach. The reduction presented here is potentially applicable to a broad range of highly regular systems.

Keywords: model reduction, stochastic analysis, moment closure, model symmetry, cell-to-cell communication, community effect.

1 Introduction

The dynamics of biochemical reaction systems are traditionally formalized as systems of ordinary differential equations (ODEs), whose variables represent concentrations of molecular species in a well-mixed solution. This assumes that the inherent stochastic fluctuations are negligible. However, this assumption is invalid for certain systems, such as gene regulatory networks [7]. Those systems must be analyzed stochastically, accounting for randomness of biomolecular interactions.

Exact solutions for the dynamics of most non-linear chemical systems are practically impossible to obtain. Various methods for *approximate stochastic analysis* have been suggested: Monte Carlo sampling of probability density functions of species' counts over time, known as Gillespie's algorithm [11]; approximations of this sampling [4,20]; explicit treatment of fluctuations with stochastic differential equations [12]; consideration of subspace of system states with highest probability mass [6,25]; aggregation of states [17].

⋆ Corresponding author.

D. Gilbert and M. Heiner (Eds.): CMSB 2012, LNCS 7605, pp. 49–68, 2012.

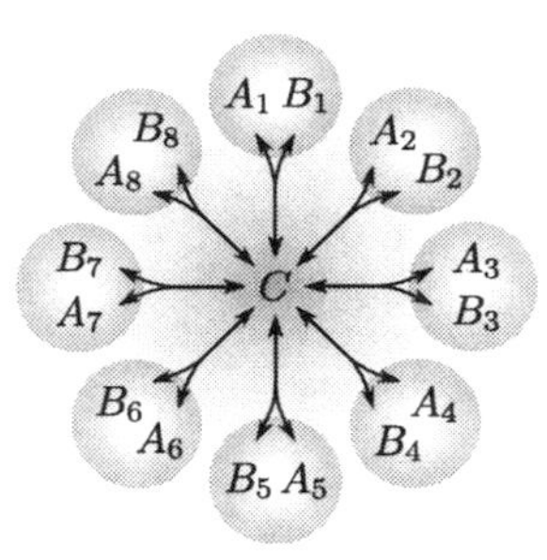

Fig. 1. Symmetry in cell-to-cell communication: n cells with equal intracellular reaction network, involving molecules A and B, interact through the exchange of the extracellular molecule C

Moment closure. (MC) is a promising method for approximate analysis of the behavior of stochastic systems. It allows efficient calculation of approximate dynamics of *moments* of random variables associated with the system under study. MCs were successfully used in the fields of ecology [33], demographics [14], epidemiology [15] and statistical physics [22]. Traditionally they were derived manually. Recently, methods for automatic MC derivation were proposed [10,18,32] for biochemical reaction systems.

Because MC provides a system of ordinary differential equations, it allows computing the approximate dynamics more efficiently than any of the previously cited approaches to stochastic approximation. In particular, it relatively quickly yields solutions for different parameter values, such as reaction rates and initial conditions. This property has been exploited for efficient parameter estimation [23], and can possibly be used for other tasks, such as real-time control [31]. A disadvantage of the MC method is that the number of generated ODEs quickly grows with the system size, making it potentially difficult to scale to larger systems. *Reduction techniques* are needed to keep the analysis tractable.

In this work, we exploit *symmetries in cell signaling* to perform model reduction. Consider a pool of identical cells communicating over a short distance through the exchange of molecules, which are released by one cell, then diffuse and make contact with another cell. The reaction set

$$r_i : \qquad A_i + B_i \rightleftarrows C \qquad , \text{for } i \in \{1 \dots n\} \tag{1}$$

describes such a system with n cells. Its symmetry is illustrated in Figure 1. A reaction between a pair of A and B, within the ith cell, results in a C, which is expelled to the extracellular medium. Note that the extracellular C lacks a positional index, unlike the other molecules. C can migrate back from the extracellular medium to *any* of the n cells. The symmetry of this minimal system clearly appears, with C as the center, around which the n equal cells gather, and through which they communicate. Our model reduction strategy uses a notion of symmetry based on *invariance under certain changes of the chemical reaction network*. Intuitively, we observe that the global dynamics of the system remains invariant as we swap cell indices, because all cells are equal.

Symmetries in cell-to-cell communication are widespread, going far beyond the illustrative example of Figure 1. In bipartite bacterial communities, interspecies exchange of metabolites can enable important metabolic functions, that are not reached by either of the isolated species [29]. One can distinguish between different types of metabolic interaction, the mechanisms of which remain under debate [34]. However, symmetries in the reaction network for cell-to-cell communication are a common feature to most of them. A last prominent example is the one considered by A. Turing in [30]: a ring of identical cells with rotational symmetries.

The community effect is thought to be a widespread phenomenon in animal development [2,5,16]. It allows a cell population within an embryo to forge a common identity, that is, to express a common set of genes. This synchronization is based on cell-to-cell communication, in which cells produce and exchange a diffusible molecule, resembling C in (1). Only when the cell population exceeds a critical number n_c, the common gene expression is maintained over extended periods of time. With fewer cells, after an initial induction, gene expression soon ceases.

A recent model for the community effect in *Xenopus laevis* [27], detailing on intracellular cascade of gene expression, mediated to intercellular communication, requires 17 reactions per cell. The community effect threshold n_c is about a hundred cells. Analysis of this model requires stochastic simulation over a wide range of parameters, cell numbers, and over extended periods of time, in order to yield realistic results, and namely to determine the critical number. The computational cost for the stochastic simulation becomes intractable, due to the multiplication of numbers of reactions per cell and numbers of cells. Deterministic approximations, on the other hand, do not provide precise predictions. The model reduction technique in combination with MC, as presented in this paper, provides a more convincing stochastic approximation of this model. Comparison of the solution of the truncated moment ODEs with stochastic simulations shows that MC is significantly closer to the stochastic dynamics than the deterministic solution.

Related work. Various symmetries have been exploited previously to facilitate finding the solutions of the biochemical models and to infer their properties. In [26], results from group theory are applied to the analytical solution of a simple model of gene expression, demonstrating how interesting properties of the model follow from continuous symmetries. In [13], symmetrically connected cell networks are considered, and some properties of them are shown in the deterministic regime. In [3], a model reduction technique is presented, which is based on particular kinds of symmetries expressed in Kappa language. That model reduction is applicable either to the deterministic approximation or, in fewer cases, to the stochastic semantics; the stochastic version of that reduction is not applicable to the community effect model. Here we present a different method, which provides a way to reduce higher order approximations (MC), under a different set of assumptions.

Paper outline. Section 2 reviews MC, Section 3 shows symmetry-based reduction, Section 4 applies our approach to a community effect model, and Section 5 concludes.

2 Approximate Stochastic Analysis

In this section we assume that the chemical reactions follow the *mass action* law, the first discovered and widely used kinetic law. Therefore the kinetics of a reaction can be described by a single parameter r, called *rate constant*, as well as stoichiometric coefficients.

For a system of n chemical species $A_1, \ldots, A_n$, define tuples

$$(r, \boldsymbol{\alpha}, \boldsymbol{\beta}) \in \mathit{Reacts} = \mathbb{R}_{>0} \times \mathbb{N}^n \times \mathbb{N}^n$$

such that $(r, \boldsymbol{\alpha}, \boldsymbol{\beta})$ represents the reaction

$$\alpha_1 A_1 + \ldots + \alpha_n A_n \xrightarrow{r} \beta_1 A_1 + \ldots + \beta_n A_n$$

In the following, we index chemical species with i and reactions with j. A system of k chemical reactions is a set $\mathcal{R} = \{U_1, \ldots, U_k\}$ of k tuples $(r, \boldsymbol{\alpha}, \boldsymbol{\beta})$. For this set, we define the reactant stoichiometric matrix R, the product stoichiometric matrix P, and the vector of rate constants $\mathbf{r}$ such that:

$$\begin{array}{c} \forall U_j \in \mathcal{R} : \text{denote } (r, \boldsymbol{\alpha}, \boldsymbol{\beta}) = U_j : \\ r_j = r, \forall i : R_{ij} = \alpha_i, P_{ij} = \beta_i \end{array}$$

so that a chemical system is fully determined by the tuple $(\mathbf{r}, R, P)$.

Example 1. Given the 3 molecular species A, B and C, the tuple

$$\left((0.1\ 0.3), \begin{pmatrix} 1 & 0 \\ 1 & 0 \\ 0 & 1 \end{pmatrix}, \begin{pmatrix} 0 & 1 \\ 0 & 1 \\ 1 & 0 \end{pmatrix} \right) \quad \text{represents the reactions} \quad \left\{ \begin{array}{r} A + B \xrightarrow{0.1} C \\ C \xrightarrow{0.3} A + B \end{array} \right.$$

2.1 Deterministic Dynamics: The Coarser Approximation

The computational analysis of chemical reaction systems often assumes deterministic behavior. Following the mass action law, at any given time t a system's dynamics is driven by the *concentrations* of the reacting chemical species. For example, given some initial condition, the dynamics of Example 1 would be described by a set of ordinary differential equations

$$\frac{d[A]}{dt} = \frac{d[B]}{dt} = -0.1\ [A][B] + 0.3\ [C] \qquad \frac{d[C]}{dt} = 0.1\ [A][B] - 0.3\ [C] \quad (2)$$

where $[X]$ denotes the concentration of the species X. Despite the fact that concentrations of species can only assume discrete values, the deterministic approach is often justified by arguing that the high number of molecules usually present in a solution make these discrete values so close that continuous domains for concentrations constitute a safe approximation. The corresponding mathematical argument is that the limit behavior of the system of Example 1 is exactly described by (2) when the volume of the system *tends to infinity* while the concentrations remain the same.

Formally, given a system of chemical reactions $(\mathbf{r}, R, P)$, its state space under the deterministic assumption is the space of the concentrations of the chemical species $A_1, \ldots, A_n$ in the system. Concentration of every species is assumed to be a non-negative real number. Its dynamics is defined, for a given initial state $\boldsymbol{\nu}_0$, by the function of time

$$\mathbf{x}(t) : \mathbb{R}_{\geq 0} \to \mathbb{R}^n_{\geq 0}, \quad \mathbf{x} \text{ is the solution of the ODE } \begin{cases} \dot{\mathbf{x}} &= (P - R)\boldsymbol{\lambda} \\ \mathbf{x}(0) &= \boldsymbol{\nu}_0 \end{cases}$$

where $\boldsymbol{\lambda}$ is the vector of reaction rate laws in state $\mathbf{x}(t)$, that is, for reaction number $j \in \{1, \ldots, k\}$, the rate law is $\lambda_j = r_j \prod_i {x_i}^{R_{ij}}$.

Example 2. For the system of Example 1, we have

$$\boldsymbol{\lambda} = \begin{pmatrix} 0.1x_1x_2 \\ 0.3x_3 \end{pmatrix} \text{ and } P - R = \begin{pmatrix} -1 & 1 \\ -1 & 1 \\ 1 & -1 \end{pmatrix}$$

leading to the system of ODEs

$$\begin{cases} \dot{x}_1 = 0.3\, x_3 - 0.1\, x_1x_2 \\ \dot{x}_2 = 0.3\, x_3 - 0.1\, x_1x_2 \\ \dot{x}_3 = 0.1\, x_1x_2 - 0.3\, x_3 \end{cases} \tag{3}$$

2.2 Stochastic Semantics

The deterministic assumption is often invalid at the cellular scale, where the number of molecules per species and cell can be low – one for genes, or a few for mRNA. In this case, it is preferable to formalize the stochastic behavior of the system in terms of a continuous time Markov chain (CTMC), whose states represent the different possible configurations of the system: each state is determined by the number of molecules per chemical species. Formally, we can define a CTMC as a collection $\{\mathbf{x}(t) \mid t \in \mathbb{R}_{\geq 0}\}$ of n time-dependent *random variables* with state space in $\mathbb{N}^n$.

The system's evolution is then interpreted stochastically: at a given time t, a probability is assigned to each state and its variation in time is governed by a differential equation. The set of all such equations, one per state of the system, constitutes the *chemical master equation (CME)*, whose solution gives the complete information about the system's kinetics at any time. If $\pi_{\boldsymbol{\nu}}(t) = Pr(\mathbf{x}(t) = \boldsymbol{\nu})$ is the probability of being in the state $\boldsymbol{\nu}$ at time t, the CME becomes [32]:

$$\dot{\pi_{\boldsymbol{\nu}}} = \sum_{(r,\boldsymbol{\alpha},\boldsymbol{\beta})\in\mathcal{R}} r\binom{\boldsymbol{\nu} + \boldsymbol{\alpha} - \boldsymbol{\beta}}{\boldsymbol{\alpha}}\pi_{\boldsymbol{\nu}-\boldsymbol{\alpha}+\boldsymbol{\beta}} - r\binom{\boldsymbol{\nu}}{\boldsymbol{\alpha}}\pi_{\boldsymbol{\nu}} \tag{4}$$

where $\binom{\mathbf{a}}{\mathbf{b}}$ denotes the product $\prod_i \binom{a_i}{b_i}$.

The number of states (and of corresponding differential equations in the master equation) scales exponentially with the number of species and possible number of molecules per species. When some reaction creates unbounded numbers of new molecules, it even becomes infinite (but countable). In this case, while it is still possible to find solutions of the CME for some particular systems, the general, automatic numeric solution of the CME becomes intractable. As a consequence, one can only obtain *partial or approximated information* on the CME's solution.

2.3 Moments and Moments Calculation

A radically different perspective on a system is, instead of computing state probabilities, to directly consider the time evolution of the *moments* of its variables. Given a vector $\mathbf{m} \in \mathbb{N}^n_{\geq 0}$, the mixed moment $\mu^{(\mathbf{m})}$ about zero (i.e. *uncentred*) of a subset of variables defined by non-zero elements of $\mathbf{m}$ is

$$\mu^{(\mathbf{m})} : \begin{cases} \mathbb{R}_{\geq 0} \to \mathbb{R}_{\geq 0} \\ \quad t \mapsto E[\mathbf{x}^{\mathbf{m}}(t)] \end{cases}$$

where E denotes the expectation, and $\mathbf{x}^{\mathbf{m}}$ denotes a product $\prod_i {x_i}^{m_i}$. Therefore, $\mu^{(1,0,\ldots,0)}$ denotes, for instance, the dynamics of the expectation of the random variable x_1. The *order* of moment $\mu^{(\mathbf{m})}$ is the sum of the indices $\sum_i m_i$.

This change of perspective arises when general characterizations of a model matter more than the probability of each single state. For example, one may only need to extract the average concentrations of species in time and their stochastic noise. Those correspond to the first two (central) moments: mean and variance.

The moment-based analysis of a chemical system requires a preliminary step, which is to replace the state-centric description of the dynamics given by the CME with a new description directly focused on the evolution of the value of moments in time. In practice, this corresponds to building a new set of differential equations, one for each moment, and can be performed in several ways. For example, when the rate functions associated with chemical reactions are polynomial, moment equations may be calculated using the moment-generating function [10], or equivalently adopting a generator operator [18] or by a probability generating function [32]. Extensions to more general cases may require more sophisticated methods, e.g. when rational rate functions are considered [24].

Although the model reduction presented in this paper is valid for models with kinetic laws of any kind, here we follow the method of moment generation presented in [32], which is reasonably simple and covers those kinetic laws needed for Section 4 (the mass-action family). According to this method, moment equations can be calculated thanks to the probability-generating function

$$\phi(\mathbf{z}, t) = E[\mathbf{z}^{\mathbf{x}(t)}] = \sum_{\boldsymbol{\nu}} \pi_{\boldsymbol{\nu}}(t)\mathbf{z}^{\boldsymbol{\nu}} \tag{5}$$

where $\mathbf{z}$ is a formal parameter consisting of a vector of variables $(z_1, \ldots, z_n)$. The introduction of $\mathbf{z}$ allows the calculation of moments of any order by applying properly the operation of partial differentiation to ϕ with respect to the variables $z_1, \ldots, z_n$, and then by setting their value to 1. For example, the expectation of x_1 of Example 1 can be calculated by differentiating ϕ once with respect to z_1, and then by setting $\mathbf{z} = \mathbf{1} = (1, 1, 1)$:

$$\begin{aligned} \phi_{z_1}(\mathbf{z}, t)_{|\mathbf{z}=\mathbf{1}} &= \Big(\sum_{\boldsymbol{\nu}} \nu_1 \pi_{\boldsymbol{\nu}}(t) \mathbf{z}^{(\boldsymbol{\nu}-(1,0,0))}\Big)_{|\mathbf{z}=\mathbf{1}} \\ &= \sum_{\boldsymbol{\nu}} \nu_1 \pi_{\boldsymbol{\nu}}(t) \quad = \quad E[x_1(t)] = \mu^{(1,0,0)} \end{aligned}$$

where ϕ_{z_1} denotes the partial derivative $\frac{\partial \phi}{\partial z_1}$. The same procedure applies to calculate the expectation of x_2 (respectively x_3), where one has to differentiate with respect to z_2 (respectively z_3). Second order factorial moments can be calculated by differentiating twice with respect to the corresponding variables, from which the general (uncentred) moments can be obtained, for example:

$$\left(\frac{\partial^k \phi}{\partial z_1^k}\right)_{|\mathbf{z}=\mathbf{1}} = E[x_1(t)(x_1(t)-1)\ldots(x_1(t)-k+1)] = \mu_F^{(k,0,0)} = \sum_{i=1}^{k} s(k,i)\mu^{(i,0,0)}$$

where μ_F is a factorial moment and $s(k,i) = (-1)^{k-i}\binom{k}{i}$ is the Stirling's number of the first kind. This procedure generalizes to any order, so that univariate and multivariate higher order factorial moments are calculated by differentiating the proper number of times with respect to each variable in $\mathbf{z}$. Joint factorial moments can be expressed in terms of general moments as follows:

$$\frac{\partial^{\mathbf{m}} \phi}{\partial \mathbf{z}^{\mathbf{m}}}_{|\mathbf{z}=\mathbf{1}} = \frac{\partial^{m_1}}{\partial z_1^{m_1}} \cdots \frac{\partial^{m_n} \phi}{\partial z_n^{m_n}}_{|\mathbf{z}=\mathbf{1}} = \mu_F^{(\mathbf{m})} = \sum_{i_1=1}^{m_1} \cdots \sum_{i_n=1}^{m_n} \mu^{(\mathbf{m})} \prod_{j=1}^{n} s(m_j, i_j) \quad (6)$$

From (4) and (5) it is possible to derive (see [8], (5.60) modified for Kurtz-type combinatorial mass action model used here) the following partial differential equation:

$$\phi_t(\mathbf{z}, t) = H\phi(\mathbf{z}, t) \quad (7)$$

where H is the so-called Hamiltonian operator, obtained by the stoichiometric coefficients and the reaction rates of the chemical system:

$$H = \sum_{(r,\boldsymbol{\alpha},\boldsymbol{\beta})\in\mathcal{R}} \frac{r}{\boldsymbol{\alpha}!}(\mathbf{z}^{\boldsymbol{\beta}} - \mathbf{z}^{\boldsymbol{\alpha}})\left(\frac{\partial}{\partial \mathbf{z}}\right)^{\boldsymbol{\alpha}} \quad (8)$$

where $\boldsymbol{\alpha}!$ denotes the product $\prod_i \alpha_i!$.

The procedure for calculating moment equations summarizes as follows:

1. Determine the Hamiltonian operator for the reaction system by (8).
2. Calculate the moment equation for any desired moment $\mu^{(\mathbf{m})} = \mu^{(m_1,\ldots,m_n)}$ by applying the partial derivative $\frac{\partial^{\mathbf{m}}}{\partial \mathbf{z}^{\mathbf{m}}}$ to both sides of equation (7) and setting $\mathbf{z}$ to 1.

3. Convert higher order factorial moments in the equation to general moments using (6) (the first order factorial moments are equal to the general moments).

Example 3. The Hamiltonian for Example 1 is

$$H = 0.1\ (z_3 - z_1 z_2)\frac{\partial^2}{\partial z_1 \partial z_2} + 0.3\ (z_1 z_2 - z_3)\frac{\partial}{\partial z_3}$$

To get the differential equation for $\mu^{(1,0,0)}$, we first apply to (7) partial differentiation with respect to z_1 and then set $\mathbf{z} = \mathbf{1}$:

$$\phi_{t z_1}(\mathbf{z}, t)_{|\mathbf{z}=\mathbf{1}} = \phi_{z_1 t}(\mathbf{z}, t)_{|\mathbf{z}=\mathbf{1}} = \dot{\mu}^{(1,0,0)}$$

The same procedure is applied to the r.h.s.:

$$\begin{aligned}
\left(\frac{\partial}{\partial z_1} H\phi(\mathbf{z}, t)\right)_{|\mathbf{z}=\mathbf{1}} &= \left(\frac{\partial}{\partial z_1}\left(0.1\ (z_3 - z_1 z_2)\phi_{z_1 z_2}(\mathbf{z}, t)\right)\right)_{|\mathbf{z}=\mathbf{1}} + \\
&\quad \left(\frac{\partial}{\partial z_1}\left(0.3\ (z_1 z_2 - z_3)\phi_{z_3}(\mathbf{z}, t)\right)\right)_{|\mathbf{z}=\mathbf{1}} \\
&= \left(-0.1 z_2 \phi_{z_1 z_2}(\mathbf{z}, t) + 0.1\ (z_3 - z_1 z_2)\phi_{z_1^2 z_2}(\mathbf{z}, t)\right)_{|\mathbf{z}=\mathbf{1}} + \\
&\quad \left(0.3\ z_2 \phi_{z_3}(\mathbf{z}, t)\ +\ 0.3\ (z_1 z_2 - z_3)\phi_{z_1 z_3}(\mathbf{z}, t)\right)_{|\mathbf{z}=\mathbf{1}}
\end{aligned}$$

By applying the property of the probability-generating function, we obtain:

$$\dot{\mu}^{(1,0,0)} = \left(\frac{\partial}{\partial z_1} H\phi(\mathbf{z}, t)\right)_{|\mathbf{z}=\mathbf{1}} = -0.1\ \mu^{(1,1,0)} + 0.3\ \mu^{(0,0,1)}$$

Similarly, one finds that

$$\dot{\mu}^{(0,1,0)} = -\ 0.1\ \mu^{(1,1,0)} + 0.3\ \mu^{(0,0,1)} \tag{9}$$

$$\dot{\mu}^{(0,0,1)} = +\ 0.1\ \mu^{(1,1,0)} - 0.3\ \mu^{(0,0,1)} \tag{10}$$

$$\begin{aligned}\dot{\mu}^{(1,1,0)} = &-\ 0.1\ \mu^{(2,1,0)} + 0.3\ \mu^{(1,0,1)} - 0.1\ \mu^{(1,2,0)} + \\ &0.3\ \mu^{(0,1,1)} + 0.1\ \mu^{(1,1,0)} + 0.3\ \mu^{(0,0,1)}\end{aligned} \tag{11}$$

It is important to note that the moment equations are always linear, like those of the CME (4).

The implementation of analysis tools based on moments usually relies on libraries for symbolic computation that help to automate the calculation of moment equations explained above.

2.4 Moment Closure

Switching from a state-based description to a moment-based one seemingly happens without any particular gain (or loss). Indeed, given a system with a finite number n of states, the number of (independent) equations in the CME is $n-1$. Its characterization in terms of moment equations gives a different set of ODEs, but with the same number $n-1$ of (independent) equations. If the CME is defined by an infinite (countable) number of equations, the same holds for the corresponding system of moment equations. Mathematically, both descriptions contain the same information: switching back and forth between them is fully reversible. Practically, the high number of equations renders both systems equally intractable. However, information is differently distributed across the two ODE systems, such that different approximation techniques can be applied.

Let us point out a dependency in the structure of equation systems associated to moments: an mth order moment generally depends on moments of order at most $m+h$, where h is a constant whose value depends on the stoichiometry matrix. In Example 3 we have $h=1$, so that each moment of order m depends – besides lower and equal order moments – on some moment of order $m+1$: for example, the expectation values of the number of molecules per chemical species $\mu^{(1,0,0)}, \mu^{(0,1,0)}, \mu^{(0,0,1)}$ depend on the moment $\mu^{(1,1,0)}$ of order two, which in turn depends on some moments of order three, and so on. In cases when we were only interested in the mean number per chemical species, we could confine our attention to first order moments. If we also wanted information on the stochastic noise, we could also consider second order moments (and so on: the higher the order considered, the more complete the information about the probability distribution of chemical species). The problem is then how to break the previously described *infinite cascade* of dependencies.

Moment closure denotes a wide set of techniques allowing to effectively break infinite cascades. The closure usually follows from some assumption on the probability distribution of chemical species: the closure of *order* m, for example, allows rewriting m^{th} order equations, such that they no longer depend on higher order moments. Breaking the cascade comes at the cost of introducing an approximation error: the more faithful is the assumption allowing the closure, the closer to the precise solution is the resulting set of equations.

Example 4. Assume the correlation between species A and B in Example 1, given time, is negligible. Then the covariance of x_1 and x_2 is zero at any time t. Thus,

$$\begin{aligned} 0 = \mathrm{Cov}(x_1, x_2) &= E[(x_1 - \mu^{(1,0,0)})(x_2 - \mu^{(0,1,0)})] \\ &= E[x_1 x_2] - \mu^{(1,0,0)}\mu^{(0,1,0)} \end{aligned}$$

and from this it follows that

$$\mu^{(1,1,0)} = E[x_1 x_2] = \mu^{(1,0,0)}\mu^{(0,1,0)} \tag{12}$$

By applying (12) in (9)-(10), one obtains

$$\begin{aligned} \dot{\mu}^{(1,0,0)} &= -\,0.1\,\mu^{(1,0,0)}\mu^{(0,1,0)} + 0.3\,\mu^{(0,0,1)} \\ \dot{\mu}^{(0,1,0)} &= -\,0.1\,\mu^{(1,0,0)}\mu^{(0,1,0)} + 0.3\,\mu^{(0,0,1)} \\ \dot{\mu}^{(0,0,1)} &= +\,0.1\,\mu^{(1,0,0)}\mu^{(0,1,0)} - 0.3\,\mu^{(0,0,1)} \end{aligned}$$

so that equation (11) is no longer considered. Remarkably, the above set of closed moment equations corresponds exactly to the ODE system (3). This simple result supports a different interpretation of the deterministic approximation introduced with equation (2). Under this interpretation, the continuity of the domain of species concentrations is fully justified without introducing any limit behavior, because the "concentration" is instead thought as the expected value of the number of molecules of chemical species. Moreover, the set of ODEs describing the (approximate) evolution of expectations is not derived by the application of a limit involving the number of molecules: it follows as a direct consequence of the assumption of zero correlation between the species participating in higher-order reactions. Although the concentration limit and the zero correlation assumptions are inherently related, they lead to the same result through different mathematical procedures.

More interesting applications of moment closure are those where the deterministic approximation fails to capture the real behavior of the system. Here, higher order closures may be applied in order to get better quantitative approximations, as well as information about the stochasticity of the system under analysis. Important results in this direction are presented in [21] for the class of *zero central moment closures*, where the closure of order m is obtained by setting to zero all $(m+1)^{\text{th}}$ central moments. For this class, the approximation error has been proven to decrease as the order of the closure increases. Similar results have been shown in [28] for the first orders of another class of closures, obtained by a procedure called derivative matching.

Remarkably, the first in the class of zero central moment closures is the deterministic approximation applied in Example 4, therefore it is usually regarded as the coarsest among moment closure approximations.

Normal closure. The second closure of this class, one of the first to be applied and still widely used, is the so called *normal closure*, consistent with the assumption that the counts of all chemical species, at any time point, is jointly normally distributed. This assumption is obviously wrong: first, the support of a Gaussian distribution is continuous, while the probability distributions associated with chemical systems are discrete (in fact their support is given by the set of reachable states). Moreover, the support of normal distributions also includes negative values: in the context of chemical systems, this would correspond to allowing states with a negative number of molecules, which is clearly impossible. However, this assumption is one of the easiest to apply and works very well in many practical cases, including the study of the community effect model presented in Section 4.

Formally, the normal closure is obtained by setting to zero each central moment of order three. Given three random variables x_1, x_2, x_3, the closure follows by the equation

$$
\begin{aligned}
0 &= E\big[(x_1 - E[x_1])(x_2 - E[x_2])(x_3 - E[x_3])\big] \\
&= E[x_1x_2x_3] + 2E[x_1]E[x_2]E[x_3] \\
&\quad - E[x_3]E[x_1x_2] - E[x_2]E[x_1x_3] - E[x_1]E[x_2x_3]
\end{aligned}
$$

from which we get

$$
\begin{aligned}
E[x_1x_2x_3] =& E[x_3]E[x_1x_2] + E[x_2]E[x_1x_3] + \\
& E[x_1]E[x_2x_3] - 2E[x_1]E[x_2]E[x_3]
\end{aligned} \tag{13}
$$

Example 5. In order to apply (13) to (11), we must calculate (13) for the moments $\mu^{(2,1,0)}$ and $\mu^{(1,2,0)}$:

$$
\begin{aligned}
\mu^{(1,2,0)} = E[x_1x_2^2] =& 2\, E[x_2]E[x_1x_2] + E[x_1]E[x_2^2] - 2E[x_1]E[x_2]^2 \\
=& 2\, \mu^{(0,1,0)}\mu^{(1,1,0)} + \mu^{(1,0,0)}\mu^{(0,2,0)} - 2\, \mu^{(1,0,0)}\mu^{(0,1,0)^2} \\
\mu^{(2,1,0)} = E[x_1^2x_2] =& 2\, \mu^{(1,0,0)}\mu^{(1,1,0)} + \mu^{(0,1,0)}\mu^{(2,0,0)} - 2\, \mu^{(0,1,0)}\mu^{(1,0,0)^2}
\end{aligned}
$$

By substituting in (11) we get

$$
\begin{aligned}
\dot{\mu}^{(1,1,0)} = \quad & 0.3\, \mu^{(1,0,1)} + 0.3\, \mu^{(0,1,1)} + 0.1\, \mu^{(1,1,0)} + 0.3\, \mu^{(0,0,1)} \\
& + 0.1\, (2\, \mu^{(0,1,0)}\mu^{(1,0,0)^2} - 2\, \mu^{(1,0,0)}\mu^{(1,1,0)} - \mu^{(0,1,0)}\mu^{(2,0,0)}) \\
& + 0.1\, (2\, \mu^{(1,0,0)}\mu^{(0,1,0)^2} - 2\, \mu^{(0,1,0)}\mu^{(1,1,0)} - \mu^{(1,0,0)}\mu^{(0,2,0)})
\end{aligned}
$$

so that $\dot{\mu}^{(1,1,0)}$ no longer depends on third order moments. In order to eliminate their dependencies on third order moments, the same steps apply to the equations of any further moment of order two – including those of $\mu^{(2,0,0)}, \mu^{(0,2,0)}, \mu^{(1,0,1)}$ and $\mu^{(0,1,1)}$ in the above equation.

The resulting system of ODEs, typically non-linear after the closure, can be solved numerically. The initial values for the moments are usually given under the zero-variance assumption, that is $\mu^{(\mathbf{m})}(0) = \mathbf{x}^{\mathbf{m}}(0)$.

3 Model Reduction Based on Symmetries

Moment closure of order m for a system with n species generates $O(n^m)$ ODEs, because moments for all combinations of n species may be included. The equation system may thus become difficult to handle. Attempts were made to simplify models based on various properties, e.g. conservation laws and bounds on numbers of species [32]. Here we present a model reduction method based on *symmetries in the reaction set*, which can reduce the model dramatically in some cases.

3.1 Reduction by Example

We demonstrate the idea by a simple example. Consider the set of $p = 2n + 1$ chemical species $A_1, \ldots, A_n, B_1, \ldots, B_n, C$ associated with state random variables $\mathbf{x}(t) = (A_1(t), \ldots, A_n(t), B_1(t), \ldots, B_n(t), C(t))$ and the following set of reactions

$$\mathcal{R} = \{A_i + B_i \underset{\kappa_2}{\overset{\kappa_1}{\rightleftarrows}} C \mid i \in \{1, \ldots, n\}\}$$

We assume that for any fixed time t, $\mathcal{R}$ and an initial state $\mathbf{x}(0) = \boldsymbol{\nu}_0 \in \mathbb{N}^p$ defines a probability distribution over $\mathbf{x}(t)$ with probability mass function $\pi_{\boldsymbol{\nu}}(t)$. For instance, considering combinatorial mass action kinetics, $\pi_{\boldsymbol{\nu}}(t)$ will be the solution of the CME (4).

We want to identify moment equalities from simple *symmetries* in the reaction system. Since a moment is fully defined by the *marginal distributions* of the variables composing it, we actually identify equal marginal distributions from symmetries. The marginal distribution of the variables A_1 and B_1 is the probability distribution of this set of variables, ignoring the others. By symmetry of the reaction set, we mean that the reaction set remains *invariant under permutation of the chemical species*. An obvious permutation of this kind for $\mathcal{R}$ is swapping A_1 with A_2 and B_1 with B_2. In that case, $\mathcal{R}$ remains unchanged. Suppose we further assume that the initial state is invariant with respect to the same permutation. That is, initial numbers of A_1 and A_2 are the same, as well as those of B_1 and B_2. Because we consider probability distributions that are completely defined by the reactions $\mathcal{R}$ and the initial state, the stochastic dynamics of the variable set $\{A_1, B_1\}$ cannot be distinguished from that of $\{A_2, B_2\}$. As proved below, this means that their marginal distributions are equal

$$Pr(A_1(t) = a, B_1(t) = b) = Pr(A_2(t) = a, B_2(t) = b), \forall t \in \mathbb{R}_{\geq 0}, a, b \in \mathbb{N}$$

Permuting of C with itself, we get

$$Pr(A_1(t) = a, B_1(t) = b, C(t) = c) = Pr(A_2(t) = a, B_2(t) = b, C(t) = c),$$
$$\forall t \in \mathbb{R}_{\geq 0}, a, b, c \in \mathbb{N} \tag{14}$$

Importantly, this entails the moment equalities $E[A_1^i B_1^j C^k] = E[A_2^i B_2^j C^k]$ for any $i, j, k \geq 0$. As another example of symmetry, assuming $n \geq 4$ one can swap A_1 with A_3, A_2 with A_4, B_1 with B_3, B_2 with B_4, C with itself. Again, also assuming invariance of the initial state by this permutation, we have

$$Pr(A_1(t) = a_1, A_2(t) = a_2, B_1(t) = b_1, B_2(t) = b_2, C(t) = c) =$$
$$Pr(A_3(t) = a_1, A_4(t) = a_2, B_3(t) = b_1, B_4(t) = b_2, C(t) = c),$$
$$\forall t \in \mathbb{R}_{\geq 0}, a_1, a_2, b_1, b_2, c \in \mathbb{N} \tag{15}$$

It is straightforward to use equalities of the form (14)-(15) to reduce a set of moment equations of the system. For example, the system considered above generates, among others, the following moment equations for second order moments:

$$\frac{dE[A_iC]}{dt} = \kappa_1(\sum_{j=1}^{n} E[A_iA_jB_j]) - \kappa_1(E[A_iB_i] + E[A_iB_iC]) - \\ \kappa_2(E[C] - E[C^2] + n \cdot E[A_iC]) \qquad i = 1 \dots n \tag{16}$$

Using relations as (14), we can infer the following moment equalities: $E[A_iC] = E[A_1C]$, $E[A_iB_i] = E[A_1B_1]$, $E[A_iB_iC] = E[A_1B_1C]$, $E[A_i^2B_i] = E[A_1^2B_1]$, and using equalities as (15) we have $E[A_iA_jB_j] = E[A_1A_2B_2], i = 1 \dots n, j = 1 \dots n, i \neq j$. Therefore we can equivalently rewrite all n equations in (16) into one:

$$\begin{aligned}\frac{dE[A_1C]}{dt} &= \kappa_1((n-1)E[A_1A_2B_2] + E[A_1^2B_1]) - \\ &\kappa_1(E[A_1B_1] + E[A_1B_1C]) - \\ &\kappa_2(E[C] - E[C^2] + n \cdot E[A_1C])\end{aligned} \tag{17}$$

We can't exchange moments for A_i and B_i because they may have different initial conditions in general. Using this approach, the system of moment equations up to order two is reduced from $2n^2 + 5n + 2$ to 11 ODEs for any $n \geq 2$. The rest of the equations are redundant and can be safely excluded. The transformation is exact, and we can recover the dynamics of the original system from the reduced one. In order to compute the moment dynamics, it is necessary to perform a closure of the reduced system as described in Section 2.4.

3.2 Formal Reduction

We now formally define the previous notions. We however won't make use of marginal distributions, since equivalence of the full joint probability distribution entails equivalence of its marginal distributions.

We consider permutations σ over the set of species indices $\{1, \dots, n\}$. Permutations of vectors and reaction sets are defined as

$$\begin{aligned}\mathbf{a}_\sigma &= (a_{\sigma(1)}, \dots, a_{\sigma(n)}) \\ \mathcal{R}_\sigma &= \{(\kappa, \boldsymbol{\alpha}_\sigma, \boldsymbol{\beta}_\sigma) \mid (\kappa, \boldsymbol{\alpha}, \boldsymbol{\beta}) \in \mathcal{R}\}\end{aligned}$$

We say that a vector $\mathbf{a}$, resp. a reaction set $\mathcal{R}$, is σ-*invariant*, iff $\mathbf{a} = \mathbf{a}_\sigma$, resp. $\mathcal{R} = \mathcal{R}_\sigma$. We denote $\mathcal{P}$ the function that, for a given set $\mathcal{R}$ of reactions, initial state $\boldsymbol{\nu}_0$ and time t, gives a probability distribution over the counts of the species with probability mass function $\pi_{\boldsymbol{\nu}}(t)$. Somehow $\mathcal{P}$ provides the stochastic semantics of the reactions. For example, $\mathcal{P}$ could be the stochastic semantics of the reactions (Section 2.2), or an approximation of it. We make the following assumption about $\mathcal{P}$.

Assumption 1. *Let $\pi_{\boldsymbol{\nu}}(t) = \mathcal{P}(\mathcal{R}, \boldsymbol{\nu}_0, t)$ and $\pi'_{\boldsymbol{\nu}}(t) = \mathcal{P}(\mathcal{R}_\sigma, \boldsymbol{\nu}_{0\sigma}, t)$, for some reaction set $\mathcal{R}$, initial state $\boldsymbol{\nu}_0$, time t, and permutation σ of the species indices. For any $\boldsymbol{\nu} \in \mathbb{N}^n$, we have $\pi_{\boldsymbol{\nu}}(t) = \pi'_{\boldsymbol{\nu}}(t)$.*

This assumption relates permutations at the level of reactions to permutation at the level of its stochastic semantics. It just states that the stochastic dynamics of a species A provided by $\mathcal{P}$ does not depend on its position in the state vector. Saying it differently, we assume that the stochastic semantics is insensitive to species renaming, provided that this renaming doesn't create name conflicts. This is a reasonable assumption that is, for instance, satisfied by the master equation.

Theorem 1. *Let $\mathcal{R}$ be a set of k reactions of n species, $\boldsymbol{\nu}_0 \in \mathbb{N}^n$ be an initial state, and σ be a permutation over $\{1, \dots, n\}$. Let $\pi_{\boldsymbol{\nu}}(t) = \mathcal{P}(\mathcal{R}, \boldsymbol{\nu}_0, t)$, if $\mathcal{R}$ and $\boldsymbol{\nu}_0$ are σ-invariant, then, for any $\boldsymbol{\nu} \in \mathbb{N}^n$, $\pi_{\boldsymbol{\nu}}(t) = \pi_{\boldsymbol{\nu}_\sigma}(t)$.*

Proof. This theorem is a straightforward consequence of the above assumption. Indeed, let $\pi_{\boldsymbol{\nu}}(t) = \mathcal{P}(\mathcal{R}, \boldsymbol{\nu}_0, t)$ and $\pi'_{\boldsymbol{\nu}}(t) = \mathcal{P}(\mathcal{R}_\sigma, \boldsymbol{\nu}_{0\sigma}, t)$, since $\mathcal{R} = \mathcal{R}\sigma$ and $\boldsymbol{\nu}_0 = \boldsymbol{\nu}_{0\sigma}$, we have $\pi_{\boldsymbol{\nu}}(t) = \pi'_{\boldsymbol{\nu}}(t)$. By the Assumption 1 it follows that $\pi_{\boldsymbol{\nu}}(t) = \pi'_{\boldsymbol{\nu}_\sigma}(t) = \pi_{\boldsymbol{\nu}_\sigma}(t)$. □

Corollary 1. *Let $\mathcal{R}$ be a set of k reactions of n species, $\boldsymbol{\nu}_0 \in \mathbb{N}^n$ be an initial state and σ a permutation of the species indices. If $\mathcal{R}$ and $\boldsymbol{\nu}_0$ are σ-invariant, then $\mu^{(\mathbf{m})} = \mu^{(\mathbf{m}_\sigma)}$.*

Proof. At any time t we have

$$
\begin{aligned}
\mu^{(\mathbf{m})}(t) &= E[\mathbf{x}^{\mathbf{m}}(t)] \\
&= \textstyle\sum_{\boldsymbol{\nu}} \boldsymbol{\nu}^{\mathbf{m}} \pi_{\boldsymbol{\nu}}(t) \\
&= \textstyle\sum_{\boldsymbol{\nu}} \boldsymbol{\nu}_\sigma^{\mathbf{m}_\sigma} \pi_{\boldsymbol{\nu}}(t) \text{ by commutativity of multiplication} \\
&= \textstyle\sum_{\boldsymbol{\nu}} \boldsymbol{\nu}_\sigma^{\mathbf{m}_\sigma} \pi_{\boldsymbol{\nu}_\sigma}(t) \text{ by Theorem 1} \\
&= E[\mathbf{x}^{\mathbf{m}_\sigma}(t)] = \mu^{(\mathbf{m}_\sigma)}(t)
\end{aligned}
$$

□

We denote by $\Sigma(\mathcal{R}, \boldsymbol{\nu}_0)$ the set of permutations σ such that $\mathcal{R}$ and $\boldsymbol{\nu}_0$ are σ-invariant. By Corollary 1, this set defines equivalence classes $[\mu^{(\mathbf{m})}]_\Sigma$ of moments, i.e. the set of moments $\mu^{(\mathbf{m}')}$ such that $\mu^{(\mathbf{m}')} = \mu^{(\mathbf{m}_\sigma)}$ for some $\sigma \in \Sigma = \Sigma(\mathcal{R}, \boldsymbol{\nu}_0)$. As usual, we also write $[\mu^{(\mathbf{m})}]_\Sigma$ for the representative moment of this set that is, for instance, the smallest of those moments for the lexicographical order on $\mathbb{N}^n$. We denote $\boldsymbol{\mu}_k$ a vector of all M moments up to order k. Let

$$\mathcal{M}(\mathcal{R}, k) = \{\dot{\mu}^{(\mathbf{m})} = L \cdot \boldsymbol{\mu}_h \mid order(\mathbf{m}) \leq k, L \in \mathbb{R}^M\}$$

be a set of moment equations obtained by some moment generation method, with moments up to order k (recall that moment equations are always linear). h is the maximum order of the moments in the equations, it can be greater than k for systems of moment equations with an unclosed cascade of dependencies.

The *reduced set of moment equations* is defined by

$$\mathcal{M}_{red}(\mathcal{R}, \boldsymbol{\nu}_0, k) = \{\rho(\dot{\mu}^{(\mathbf{m})}) = L \cdot \rho(\boldsymbol{\mu}_h) \mid (\dot{\mu}^{(\mathbf{m})} = L \cdot \boldsymbol{\mu}_h) \in \mathcal{M}(\mathcal{R}, k)\}$$

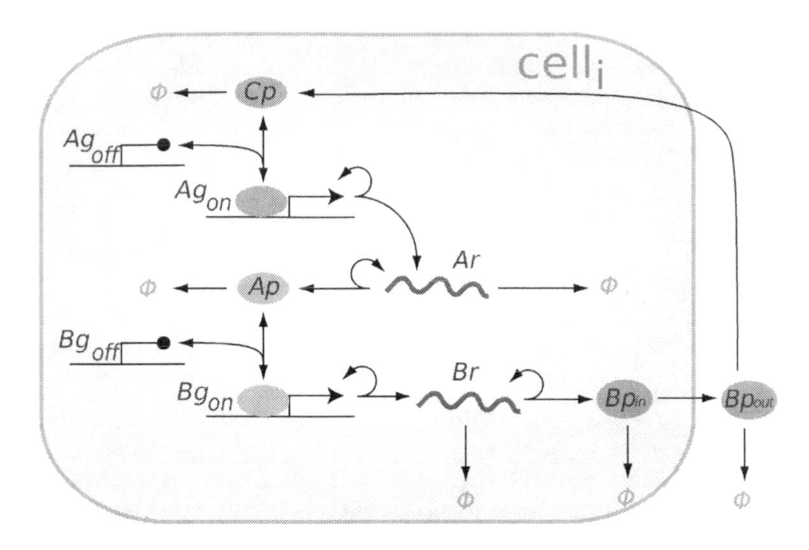

Fig. 2. Model of a community effect in Xenopus [27]. The extracellular molecule Bp_{out} mediates communication between n cells with identical intracellular reaction network.

where ρ is the substitution of moments for their representative

$$\rho = \{\mu^{(\mathbf{m})} \text{ is substituted by } [\mu^{(\mathbf{m})}]_\Sigma \mid order(\mathbf{m}) \leq k \text{ and } \Sigma = \Sigma(\mathcal{R}, \boldsymbol{\nu}_0)\}$$

This transformation just excludes from $\mathcal{M}(\mathcal{R}, k)$ repeated equations for the variables which are provably equal, and therefore is exact.

4 Application to a Community Effect

We applied the analytical tools described in this paper to a model of a community effect in Xenopus [27]. We derived a second order MC using normal approximation for simplicity. In order to do that efficiently, we reduced the moment equations as described in Section 3. Comparison of the solution of the truncated moment ODEs with stochastic simulations shows that the MC is significantly closer to the stochastic dynamics than the deterministic solution.

4.1 Community Effect Model

The model of a community effect in Xenopus [27] is summarized in Fig. 2. It features the species Bp_{out} for communication between n cells, each having the same intracellular network: within a cell, Bp_{out} triggers a cascade of two genes, and results in more Bp_{out} for cell-to-cell communication.

The intracellular details are: In the receiving cell, a signaling mechanism transforms Bp_{out} into the C_p protein, which binds to the first gene, and activates its transcription into mRNA Ar, which in turn translates into the Ap protein. Ap

activates the second gene, yielding mRNA Br, which translates into the Bp_{in} protein. The model distinguishes passive modes of genes (Ag_{off}, Bg_{off}) from active (Ag_{on}, Bg_{on}) – where genes are bound by their respective activator proteins, and constantly produce mRNA.

Leaving its original cell, the protein Bp_{in} becomes Bp_{out} when joining the common pool for cell-to-cell communication. From there, it can reach any cell in the system, and activate the gene cascade there. This closes the positive feedback loop of the community effect. Because diffusion is assumed infinitely fast, Bp_{out} equally likely reaches any of the n cells.

Finally, all species except the genes can degrade, yielding the pseudo species ϕ. The complete model for n cells, with $17n + 1$ reactions and the associated rates, is provided in a supplementary file.

Studies of the deterministic approximation of this model have shown that its behavior changes if the number of cells, n, exceeds a threshold n_c, that we call the *critical number*. If $n > n_c$, all cells continuously express their genes, otherwise all activity ceases after a short time.

Since gene and mRNA concentrations are always low in this system, stochastic effects may play a significant role in its dynamics. Indeed, stochastic simulations indicate that $n_c = 97$ derived from the deterministic approximation in [27] may be imprecise: at $n = 100$ we observed that in all 1000 simulations the gene expression stopped early. We studied the stochastic behavior of this system with the MC method.

4.2 Reduction of the Community Effect Model

The community effect model's structure resembles a star, just as Figure 1 on page 50 does. It is easy to see that the community effect model exhibits the symmetries required by Theorem 1, which allows to reduce its moment equations for any number of cells to a system of constant size, similar to the example in Section 3.1.

The model contains 9 species per cell, and the procedure described in Section 2.3 generates $40.5n^2 + 22.5n + 2$ moment equations[1] up to order two for n cells. For 120 cells, which is near this system's true n_c, it would generate 585902 equations. This by far exceeds the processing capabilities of the software we used. The reduced model contains 146 equations for any n. The deterministic approximation, which is equivalent to the first order MC, can also be reduced using the same method. This kind of reduction, among others, has been done in [27], where the deterministic approximation consisted of only 8 ODEs for any n. Our Maple[2] implementation of the MC method and the reduction for this model are available online[3].

---

[1] This is always an integer.

[2] `http://www.maplesoft.com/`

[3] `http://www.lifl.fr/~batmanov/cmsb2012-files/`

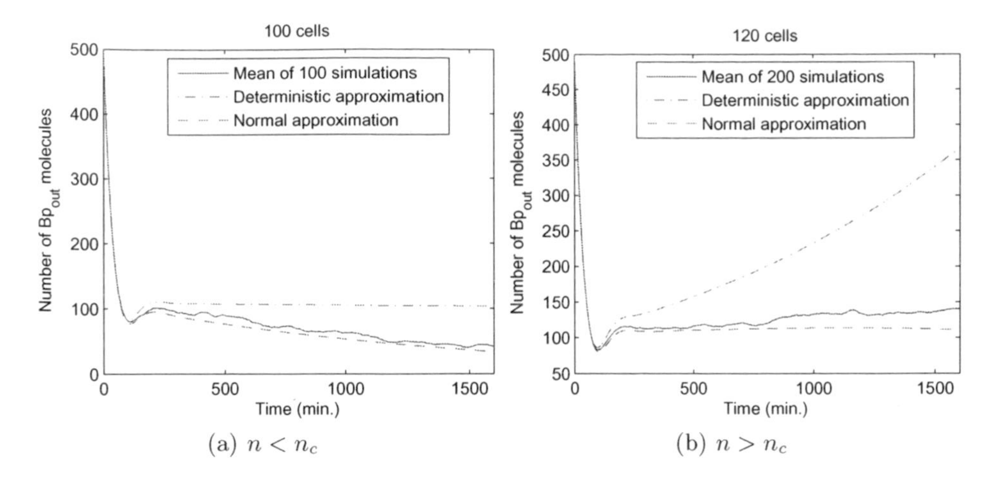

(a) $n < n_c$

(b) $n > n_c$

Fig. 3. Traces of Bp_{out} over time for systems of sizes below and above n_c, computed using stochastic simulations, deterministic approximation and second order MC using normal approximation.

4.3 Comparison of the Approximations

Figure 3 plots the Bp_{out} dynamics, computed by three different approximations. First is a *mean of many stochastic simulations.* As the number of simulations increases, this converges to the true mean - however it is noisy and the simulations take very long time. The simulations were done using COPASI software [19]. Second is a (usual) *deterministic approximation* of the system, with one ODE per species, which is the fastest in terms of computation. However, it tends to diverge from the stochastic estimates. This indicates the presence of significant stochastic effects in the model. Third is a *reduced MC of order two,* using the normal approximation for truncation. The normal approximation is not the best choice for chemical reaction systems generally, but it is simple to implement and it gives good results in this case.

Due to the complexity of the resulting system of ODEs in the MC, we couldn't derive an analytical solution for n_c. By examining the numerical solutions for different values of n, we found that the MC gives $n_c = 117$, the same as derived from statistical analysis of stochastic simulations.

The deterministic approximation, on the other hand, predicts $n_c = 97$, and therefore miscalculates the qualitative behavior of the system for a range of n. In addition, the deterministic estimate of Bp_{out} strongly diverges from the stochastic one, especially when the cell number n is close to n_c.

5 Conclusion

The moment closure method reviewed here is a flexible tool for approximate stochastic analysis. It allows manipulations of moment equations similar to those that can be done with deterministic ODEs, but including, approximately, the stochastic effects.

Model reduction is one kind of such manipulations. It aims to eliminate redundant variables from the system of ODEs, making it easier to solve. For MC, which tend to generate a large number of ODEs, reductions are especially important. We have described a model reduction method based on symmetries, which in case of MC is more complicated than what is used with deterministic approximation.

Currently, the only way to exploit such symmetries while performing stochastic analysis of a system is through MC: reduced models are not amenable to Gillespie simulation. If the corresponding species in the cells are "lumped" together, in the same way as in the deterministic approximation, the results diverge quickly from the non-reduced system. Also note that, as the order of the closure grows, the symmetric reduction can eliminate a smaller fraction of moments, suggesting that for the limit case of the exact solution the gain from the reduction will be negligible. We believe that, for the community effect model, the approach presented here provides the only tractable analysis.

Symmetry-based reduction is potentially applicable to many highly regular systems. For example, in [3] a model reduction method for deterministic approximation is applied to a system that contains a protein with symmetric activation sites. That system is also symmetric in the sense described here, w.r.t. exchanging activation states for different sites. Thus, its MC could be reduced with our method as well. Another example is a discrete ring of identical cells, considered in [30], which is symmetric under rotation of all cells. Many higher order moments could be eliminated with our method using this property.

The approach presented here can be extended in a number of ways. Checking and finding the required symmetric properties of a reaction set can be automated rather easily. The symmetries considered here are just the automorphisms of the reaction graph with the additional constraints that the initial conditions of the corresponding species must be equal and the rates of the corresponding reactions must be equal. The problem of finding all automorphisms belongs to the NP class of complexity, however for real systems the requirement of having the same rates and initial conditions restricts the number of possible symmetries. Verification of a specified symmetry can be done in polynomial time.

One can also directly derive a reduced MC from a rule-based representation, without expanding it to the full system. This becomes interesting if the expanded system's size is huge, but the system is highly symmetric and can be described by a manageable set of moment equations. It resembles what is done in [3].

However, the current method is not applicable to spatial systems with borders. By borders, we mean the outermost cells in a one dimensional row of cells, or in a two-dimensional grid, those cells that frame the grid. In such system, a distinct distance from the border(s) uniquely identifies each cell. For example, one-dimensional spatial models of the community effect [1] can be reduced in half by central symmetry. But, for a second order MC, the quadratic dependency of number of equations on the system size remains. To deal with this, [1] constructed systems of partial differential equations (PDEs), as limit cases when the number of cells tends to infinity and simultaneously their size goes to zero. While this allows efficient treatment in the deterministic regime, stochastic analysis is still required to run many long simulations.

Moment closures for spatial models have been previously derived in ecology [9] and statistical physics [22], and it may be possible to infer them automatically for chemical reaction systems as well.

Acknowledgements. We wish to thank Michel Petitot for discussions, and the Agence Nationale de Recherche for funding us (ANR BioSpace 2009-12, ANR Iceberg 2012-17).

References

1. Batmanov, K., Kuttler, C., Lhoussaine, C., Saka, Y.: Self-organized patterning by diffusible factors: roles of a community effect. Fundamenta Informaticae (2012)
2. Bolouri, H., Davidson, E.H.: The gene regulatory network basis of the "community effect," and analysis of a sea urchin embryo example. Dev. Biol. 340(2), 170–178 (2010)
3. Camporesi, F., Feret, J.: Formal reduction of rule-based models. In: Math Foundations Programming Semantics. ENTCS, vol. 276C, pp. 31–61 (2011)
4. Cao, Y., Gillespie, D.T., Petzold, L.R.: Efficient step size selection for the tau-leaping simulation method. The Journal of Chemical Physics 124, 044109 (2006)
5. Davidson, E.H.: The Regulatory Genome: Gene Regulatory Networks In Development And Evolution. Academic Press (2006)
6. Didier, F., Henzinger, T.A., Mateescu, M., Wolf, V.: Approximation of event probabilities in noisy cellular processes. TCS 412(21), 2128–2141 (2011)
7. Elowitz, M.B., Levine, A.J., Siggia, E.D., Swain, P.S.: Stochastic gene expression in a single cell. Science 297, 1183–1186 (2002)
8. Erdi, P., Toth, J.: Mathematical Models of Chemical Reactions - Theory & Applications of Deterministic & Stochastic Models. In: Nonlinear Science: Theory and Applications. John Wiley & Sons (1992)
9. Gandhi, A., Levin, S., Orszag, S.: Moment expansions in spatial ecological models and moment closure through gaussian approximation. Bull. Math. Biol. 62, 595–632 (2000), 10.1006/bulm.1999.0119
10. Gillespie, C.S.: Moment-closure approximations for mass-action models. IET. Sys. Biol. 3(1), 52 (2009)
11. Gillespie, D.T.: A general method for numerically simulating the stochastic time evolution of coupled chemical reactions. J. Comp. Physics 22, 403–434 (1976)
12. Gillespie, D.T.: Chemical Langevin equation. J. Chem. Physics 113, 297–306 (2000)
13. Golubitsky, M., Pivato, M., Stewart, I.: Interior symmetry and local bifurcation in coupled cell networks. Dynamical Systems 19(4), 389–407 (2004)
14. Goodman, L.A.: Population growth of the sexes. Biometrics 9(2), 212–225 (1953)
15. Grenfell, B.T., Wilson, K., Isham, V.S., Boyd, H.E., Dietz, K.: Modelling patterns of parasite aggregation in natural populations. Parasitology 111 (January 1995)
16. Gurdon, J.B.: A community effect in animal development. Nature 336(6201), 772–774 (1988)
17. Hegland, M.: Approximating the solution of the chemical master equation by aggregation. In: 14th Computational Techniques and Applications Conference. ANZIAM J., vol. 50, pp. C371–C384 (2008)

18. Hespanha, J.P., Singh, A.: Stochastic models for chemically reacting systems using polynomial stochastic hybrid systems. J. Robust Control 15, 669–689 (2005)
19. Hoops, S., Sahle, S., Gauges, R., Lee, C., Pahle, J., Simus, N., Singhal, M., Xu, L., Mendes, P., Kummer, U.: Copasi – a complex pathway simulator. Bioinformatics 22(24), 3067–3074 (2006)
20. Kiehl, T.R., Mattheyses, R.M., Simmons, M.K.: Hybrid simulation of cellular behavior. Bioinformatics 20(3), 316–322 (2004)
21. Lee, C.H., Kim, K.H., Kim, P.: A moment closure method for stochastic reaction networks. The Journal of Chemical Physics 130, 134107 (2009)
22. Levermore, C.: Moment closure hierarchies for kinetic theories. Journal of Statistical Physics 83, 1021–1065 (1996), 10.1007/BF02179552
23. Milner, P., Gillespie, C., Wilkinson, D.: Moment closure based parameter inference of stochastic kinetic models. Statistics and Computing, 1–9 (2012)
24. Milner, P., Gillespie, C.S., Wilkinson, D.J.: Moment closure approximations for stochastic kinetic models with rational rate laws. Mathematical biosciences 231(2), 99–104 (2011)
25. Munsky, B., Khammash, M.: The finite state projection algorithm for the solution of the chemical master equation. Journal chemical physics 124(4), 044104 (2006)
26. Ramos, A., Innocentini, G., Forger, F., Hornos, J.: Symmetry in biology: from genetic code to stochastic gene regulation. IET systems biology 4(5), 311–329 (2010)
27. Saka, Y., Lhoussaine, C., Kuttler, C., Ullner, E., Thiel, M.: Theoretical basis of the community effect in development. BMC Systems Biology 5, 54 (2011)
28. Singh, A., Hespanha, J.P.: Lognormal moment closures for biochemical reactions. In: Proc. of the 45th Conf. on Decision and Contr. (December 2006)
29. Summers, Z.M., Fogarty, H.E., Leang, C., Franks, A.E., Malvankar, N.S., LovleyDirect, D.R.: exchange of electrons within aggregates of an evolved syntrophic coculture of anaerobic bacteria. Science 330(6009), 1413–1415 (2010)
30. Turing, A.M.: The chemical basis of morphogenesis. Philosophical Transactions of the Royal Society of London. Series B, Biological Sciences 237(641), 37–72 (1952)
31. Uhlendorf, J., Hersen, P., Batt, G.: Towards real-time control of gene expression: in silico analysis. IFAC 18, 14844–14850 (2011)
32. Vidal, S., Petitot, M., Boulier, F., Lemaire, F., Kuttler, C.: Models of Stochastic Gene Expression and Weyl Algebra. In: Horimoto, K., Nakatsui, M., Popov, N. (eds.) ANB 2011. LNCS, vol. 6479, pp. 76–97. Springer, Heidelberg (2012)
33. Whittle, P.: On the use of the normal approximation in the treatment of stochastic processes. Journal Royal Statistical Society. Series B 19(2), 268–281 (1957)
34. Zengler, K., Palsson, B.O.: A road map for the development of community systems (CoSy) biology. Nature Reviews Microbiology 10(5), 366–372 (2012)

Detection of Multi-clustered Genes and Community Structure for the Plant Pathogenic Fungus *Fusarium graminearum*

Laura Bennett[1], Artem Lysenko[2], Lazaros G. Papageorgiou[3], Martin Urban[4], Kim Hammond-Kosack[4], Chris Rawlings[2], Mansoor Saqi[2,*], and Sophia Tsoka[1,*]

[1] Department of Informatics, School of Natural and Mathematical Sciences, King's College London, Strand, London, WC2R 2LS, UK
{laura.bennett,sophia.tsoka}@kcl.ac.uk

[2] Department of Computational and Systems Biology, Rothamsted Research, Harpenden, Herts, AL5 2JQ, UK
{chris.rawlings,mansoor.saqi,artem.lysenko}@rothamsted.ac.uk

[3] Centre for Process Systems Engineering, Department of Chemical Engineering, University College London, Torrington Place, London, WC1E 7JE, UK
l.papageorgiou@ucl.ac.uk

[4] Department of Plant Biology and Crop Sciences, Rothamsted Research, Harpenden, Herts, AL5 2JQ, UK
{martin.urban,kim.hammond-kosack}@rothamsted.ac.uk

Abstract. Exploring the community structure of biological networks can reveal the roles of individual genes in the context of the entire biological system, so as to understand the underlying mechanism of interaction. In this study we explore the disjoint and overlapping community structure of an integrated network for a major fungal pathogen of many cereal crops, *Fusarium graminearum*. The network was generated by combining sequence, protein interaction and co-expression data. We examine the functional characteristics of communities, the connectivity and multi-functionality of genes and explore the contribution of known virulence genes in community structure. Disjoint community structure is detected using a greedy agglomerative method based on modularity optimisation. The disjoint partition is then converted to a set of overlapping communities, where genes are allowed to belong to more than one community, through the application of a mathematical programming method. We show that genes that lie at the intersection of communities tend to be highly connected and multifunctional. Overall, we consider the topological and functional properties of proteins in the context of the community structure and try to make a connection between virulence genes and features of community structure. Such studies may have the potential to identify functionally important nodes and help to gain a better understanding of phenotypic features of a system.

Keywords: Community structure, overlapping communities, integrated networks, multi-functional genes, phytopathogenic fungi.

* Corresponding authors.

D. Gilbert and M. Heiner (Eds.): CMSB 2012, LNCS 7605, pp. 69–86, 2012.

1 Introduction

In addition to genome sequence data, a large amount of multiple complimentary types of biological data is available for many organisms, such as gene expression, protein interactions and phenotypic information. Integration of data from various sources gives rise to complex networks where nodes are proteins (or other gene products) and edges capture intricate associations between them [1, 2]. The analysis of topological features in these networks can not only uncover information about the underlying functional properties of individual nodes and relevant gene products, but it can also reveal the principles of how genes group to assemble entire cellular systems.

Network analysis is therefore a popular means of investigating the link between topological and functional features in biological systems. Community structure detection in biological networks is widely employed to derive an understanding of molecular interactions. The standard community structure detection problem involves the identification of a partition of a complex network into disjoint communities (also sometimes known as modules or clusters) such that interactions within a community are maximised and interactions between communities are minimised. Many approaches exist, including divisive [3], agglomerative [4], spectral [5] and mathematical programming methods [6-8]. The communities detected represent semi-independent functional units of an entire system, where members are likely to share some common characteristic. The identification of such communities allows information on members with unknown functional properties to be inferred.

However, the constraint of disjoint communities, where a node can only belong to one community, may not offer the most realistic abstraction of a system. For example, some proteins can carry out more than one task or belong to more than one protein complex [9, 10]. The identification of overlapping communities can reveal the multi-functionality of nodes and determines which nodes act as bridges between different functional groups or co-ordinate multiple tasks so as to hold the system together. For example, such roles are important in social networks modelling the spread of disease, where potential immunisation targets are individuals that bridge communities [11]. Moreover, in a biological system, a multi-clustered gene may act as a communicator, transferring biological information between functional units [12]. It is currently not well explored whether genes/proteins that have multiple community membership also possess particular functional and topological properties. Here we test such hypotheses in the case study of the plant pathogenic fungus *Fusarium graminearum*.

Although a less well-covered area of research than standard community structure detection, several methods for the detection of overlapping communities have been proposed. One of the first methods was the Clique Percolation method [13]. Subsequently, a wide range of approaches followed including spectral methods [14], non-negative matrix factorisation of various feature matrices [15, 16], local optimisation of a fitness function [17, 18], greedy agglomerative algorithms [19] and mathematical programming approaches [8].

Integrated functional networks can provide a framework to begin to explore genotype-phenotype relationships. For example if a gene disruption experiment of a given gene leads to a certain outcome (a given phenotype) the network may provide clues to suggest the underlying mechanisms that are affected and may aid thereby hypothesis generation. Clustering such integrated networks into disjoint and overlapping communities can identify functional communities and give insight into higher levels of biological organisation [20, 21]. As proteins take part in multiple processes a better description of the underlying biological themes may be provided by consideration of the overlaps between communities.

The Ascomycete fungus *Fusarium graminearum* is a major pathogen of wheat and other cereal crops. The complete genome sequence (with about 13,718 protein coding genes) of *Fusarium graminearum* has been determined [22] and additional data on gene expression [23] and predicted protein interactions [24] also exist. Floral infections by *Fusarium* can have a significant impact on grain yield and quality. In addition, infection by the fungus leads to contamination of the grain by various mycotoxins including deoxynivalenol (DON), which makes the grain harmful for human consumption and also for animal feed. In this study we explore the modular properties of an integrated network for *Fusarium graminearum*. Detection of disjoint and overlapping community structure is employed as a means of elucidating topological-functional relationships in the pathogen. Additionally, we relate the modular organisation of the network to virulence genes known to be required for pathogen infection and disease formation. Such an analysis may lead to better understanding of phenotypic features of the system, for example, potential insights into infection-related pathways.

2 Methods

An integrated network for *Fusarium graminearum* was constructed using information from sequence similarity, co-expression and predicted protein interactions (PPI). The sequence similarity network was constructed from all-versus-all sequence matching of the proteins in version 3.2 of the *Fusarium graminearum* annotation (at ftp://ftpmips.gsf.de/FGDB/v32) implemented on a TimeLogic® Tera-BLAST™ (Active Motif Inc., Carlsbad, CA) system with a threshold E-value for bidirectional best hits of 10^{-6}. Co-expression information was obtained from the publicly available set of Fusarium expression studies from PLEXdb [23] that used Fusarium Affymetrix GeneChip arrays. The similarity of expression profiles was measured using weighted Pearson correlation coefficient, according to the method in [25]. The PPI information was taken from the predicted core PPI of [26]. Two nodes are linked if any of the following properties is satisfied: (i) a bi-directional sequence similarity BLAST hit comprised of unidirectional hits with an expected value of less than 10^{-6}, (ii) correlation of gene expression with an absolute value of Pearson correction greater than 0.88, or (iii) PPI link from the dataset from [26]. Integration of the various data sources was carried out using the Ondex data integration platform [27, 28].

The community structure of this network was detected using the greedy agglomerative method known as Louvain [4], where nodes are allocated communities based on the maximum increase in the Newman modularity measure [29]. This results in a *hard partition* of the network into disjoint communities. The hard partition can then be converted into a *soft partition* where communities are allowed to overlap, using the method described in [8]. This mathematical programming approach fixes the community membership of all nodes that only interact with nodes in their own community in the hard partition (*isolated nodes*), whereas nodes that form interactions across communities (*border nodes*) can belong to more than one community. A mixed integer non linear programming (MINLP) model, known as OverWeiMod, is formulated to optimise the sum of the *community strength* (CS) [18] across all communities according to node-community assignments.

The inclusion of an overlapping parameter r allows the user to control the extent of overlapping, where the larger the value of r, the smaller the overlap. The choice of r is user-defined and we show here that certain topological and functional criteria can indicate a range of values. The output of OverWeiMod is a partition of the network nodes belonging to more than one community, which we define as *multi-clustered nodes* (as opposed to *mono-clustered* nodes). The *belonging coefficient* (BC) gives a measure of strength of membership of a node to a community according to the community's gain in CS with the presence of the node. For example, a node belonging to two communities with BC equal to 0.5 in both cases belongs equally to the two communities. However, if the node belongs to one community with a BC equal to 0.7 and to the other with a BC equal to 0.3, this indicates a stronger attachment to the first community over the second.

3 Results

3.1 Disjoint Community Structure Detection

The integrated *Fusarium graminearum* network comprises 9521 nodes (proteins), 80997 links and is made up of 439 disconnected components. Table 1 shows the distribution of sizes of the connected components. This analysis focuses on the largest connected component of 8364 nodes and 79931 links, as community structure of smaller components is of limited scope.

Table 1. Connected components in the integrated network

No. of nodes	2	3	4	5	6	7	9	10	11	16	8364
No. of components	288	101	23	10	5	3	3	2	2	1	1

The main component of the network is partitioned by Louvain [4], which detects a partition of 91 disjoint communities with modularity equal to 0.7973. The resultant community structure has an 'uneven' community size distribution, with 89 communities of size <500 and 2 large communities with 1007 and 1951 nodes. The output of the Louvain method was compared with another well-known community structure method, QCUT [5], based on the spectral properties of the network Laplacian. QCUT finds a partition with 53 communities (modularity equal to 0.7665), 51 of which have <500 nodes and two larger communities with 1198 and 2968 nodes.

This is in agreement with the 'uneven' community structure found by the Louvain method. Based on Louvain finding the slightly larger value of modularity, we use this hard partition in the following analysis.

The above disjoint community structure is illustrated by a 'meta-view' of the partition in Figure 1, with (i) the size of communities, (ii) the number of shared nodes across communities in the overlapping community structure (discussed in section 3.2) and (iii) their functional content. The functional coherence of a community was described by the Average Information Content of the Most Informative Common Ancestor set (AIC-MICA) a metric defined in [28], which can be used to gauge the degree of commonality of gene annotations in a particular set. This method works by identifying a set of representative Most Informative Common Ancestor (MICA) terms, where the information content (IC) is calculated based on how frequently a particular annotation is found in an annotation set for a given species. The MICA term is defined as a term in a hierarchically-organised ontology graph, which has the highest possible IC value whilst also acting as a subsumer for all terms in a particular set. The AIC-MICA approach takes as input a set of annotated entities and returns a non-redundant set of MICA terms that are applicable to at least a certain fraction of these entities, as specified by the user. The AIC-MICA statistic itself is an average of their IC values, which can serve as an indicator of annotation commonality within a set of entities. A higher value would indicate that most of the MICAs for the set are found lower in the ontology and therefore commonality in annotation is at a level with higher specificity. Here we have used the Gene Ontology (GO) [30] in which the functional role of a gene product is described at three levels: biological process (BP), molecular function (MF) and cellular component (CC).

We looked at the annotation for all three aspects of GO for the communities in the Louvain partition with at least 5 annotated nodes and used the AIC-MICA approach to find the most specific terms applicable to at least 60% of the nodes. We find that 43 communities are assigned a term from the BP aspect of the Gene Ontology, 52 are assigned a term from the MF aspect of GO and 35 are assigned a term from the CC aspect of GO. Figure 1 shows the corresponding MICA BP terms and their percentage of coverage for the largest communities. Some highly functionally coherent communities detected were "transport", "blood vessel morphogenesis" and "carbohydrate metabolic process" (communities 3, 31 and 88 respectively, 100% coverage) and "oxidation-reduction process", "transport" and "regulation of transcription, DNA-dependent" (communities 28, 60 and 76 respectively, coverage >90%). Other communities with a strong functional coherence point to 'vitamin transport' (community 78), 'nucleotide biosynthetic process' and 'serine family amino acid metabolic process'. Expectedly, larger communities show less homogeneous functional content and therefore a broader GO term is assigned, e.g. community 79, the largest community is assigned "cellular process". Overall, the hard partition detected by the Louvain method appears to find some biologically coherent communities.

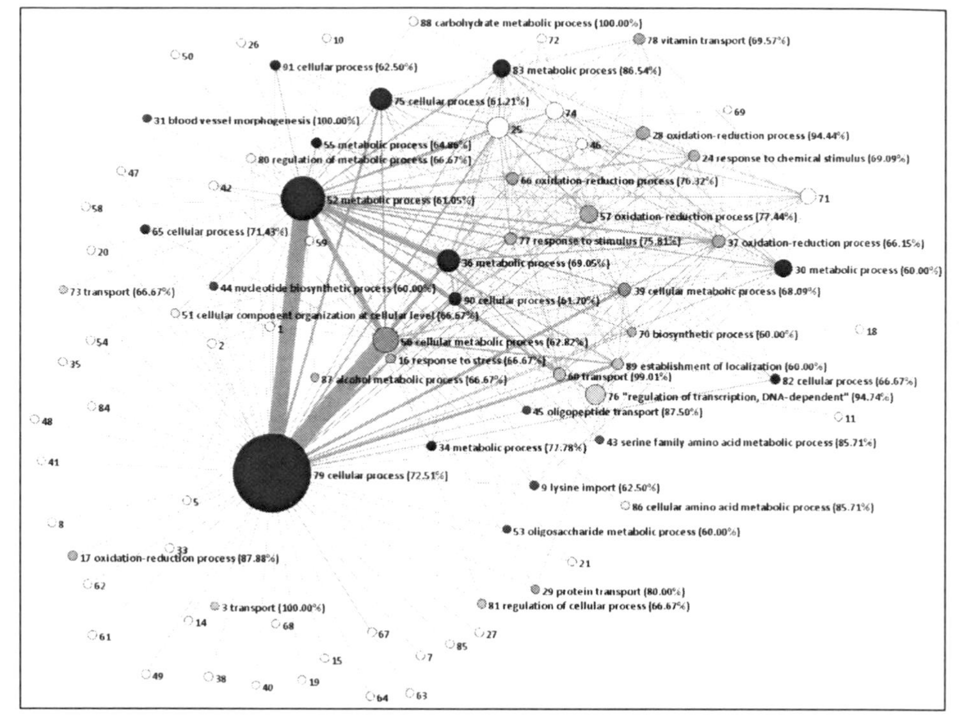

Fig. 1. The meta-view of the hard partition of the main component detected by the Louvain method, where nodes represent communities. The thickness of the links between the communities corresponds to the number of genes that are shared between communities in the overlapping community structure discussed in Section 3.2. For the larger communities, the MICA (BP) term is shown next to the corresponding community and the corresponding percentage of coverage (visualisation generated in Ondex [27, 28]).

3.2 Overlapping Community Structure

The hard partition of the main connected component is converted to a soft partition with overlapping communities using the mathematical programming method, OverWeiMod [8]. The hard partition results in 3877 border nodes, which are the potential multi-clustered nodes. As mentioned earlier, the community membership of 4487 isolated nodes that are only associated to intra-community edges, are fixed and do not change in the course of the conversion procedure. In other words, the MINLP is solved with only border nodes allowed to be assigned to multiple communities. Figure 2 shows the results for r ranging from 0.4 to 1.1. The range of values of r is chosen to some extent arbitrarily and we discuss the suitability of the range in forthcoming sections. Table 2 shows how the number of communities that a multi-clustered node belongs to changes with r. We find that for $0.4 \le r \le 0.5$ the multi-clustered nodes belong to up to 6 communities, but as r increases, and the extent of overlap decreases, this range also decreases. When $r = 1.1$ multi-clustered nodes only belong to two communities maximum.

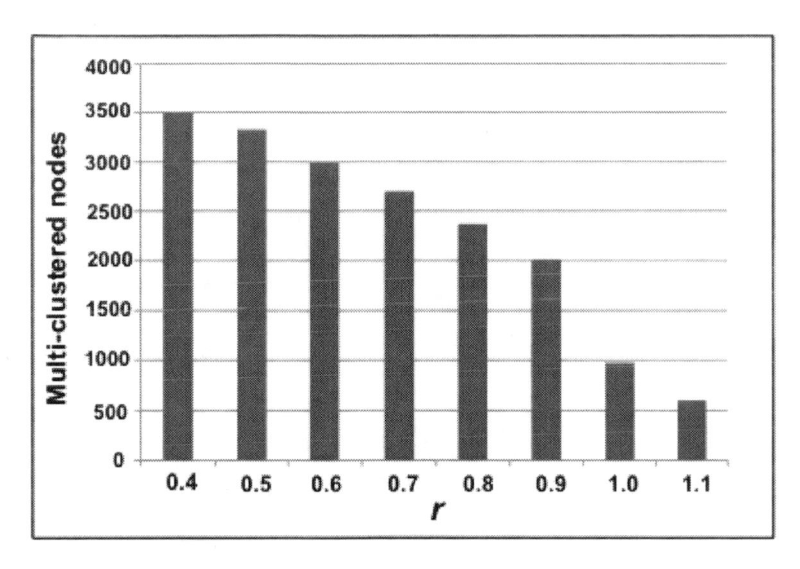

Fig. 2. The number of multi-clustered nodes detected by OverWeiMod for $0.4 \leq r \leq 1.1$

As previously mentioned, the overlapping community structure problem may be subject to multiple interpretations according to the underlying problem statement and user requirements. Our approach is to consider a hard partition of a network and examine the nodes that form interactions across community borders, to assess their associations with communities other than their own according to the hard partition employed. As described in Section 1, many different approaches to the overlapping community structure detection problem exist and the variation in methodology can affect results considerably [8]. Consequently a direct comparison between methods may not be a fair evaluation of performance. In any case, for real life networks the 'real' cover is not known and so therefore the aim is to show that according to the user's interpretation of the problem, the chosen method finds biologically relevant solutions.

In order to explore the robustness of the approach implemented in OverWeiMod we consider the greedy agglomerative method, known as the Overlapping Cluster Generator (OCG) method which has been shown to identify multifunctional proteins in PPI networks [19]. OCG is based on an adapted modularity measure applicable to overlapping communities and bears a similar methodological framework to OverWeiMod. An initial partition of 'centred cliques' is generated. Then, the elements are joined together iteratively in order of increasing average modularity gain, where modularity in this case is an overlapping equivalent of the Newman modularity [29].

OCG detects a soft partition of the main component of the integrated network with 808 communities with 3877 multi-clustered nodes. Of the 808 modules, 201 comprise only 2 nodes, 47 have 3 nodes and 33 have 4 nodes and the remaining modules range from between 5 and 211 nodes. Of the 3877 nodes, 1628 nodes belong to 2 communities, 692 belong to 3, 440 belong to 4 and the remaining multi-clustered nodes belong to between 5 and 58 communities. Figure 3 shows the breakdown of multi-clustered nodes according to the method they were detected by. In each case a considerable level of agreement can be seen between the two methods.

Table 2. Number of communities the multi-clustered nodes belong to ($0.4 \leq r \leq 1.1$)

	Number of communities				
r	**2**	**3**	**4**	**5**	**6**
0.4	2297	804	263	96	29
0.5	2330	718	208	66	4
0.6	2354	512	106	22	0
0.7	2305	319	59	14	0
0.8	2135	217	14	0	0
0.9	1868	138	1	0	0
1	960	16	0	0	0
1.1	601	0	0	0	0

Variation in results is due to fundamental differences in methodology. In particular, OCG starts its agglomerative procedure with an initial cover of the network comprising a large number of modules, which are subsequently fused until one of three stopping criteria are achieved. Consequently, if the stopping criteria are met after relatively few iterations, the resulting number of modules is high. The number of overlapping modules in the final cover of the network detected by OverWeiMod on the other hand depends on the method used to find the hard partition. In this study we use a method based on modularity optimisation, a well-recognised approach to community structure detection, employed by many methods and that has been shown to find relevant solutions in bioinformatics applications [20, 31]. However, the debate about which is the most realistic partition of the network, is beyond the scope of this study. Our main aim being to show that our method assigns structurally and functionally important nodes with biological significance to multiple modules. Here we put less importance on directly comparing methods in terms of the nodes they find to be multi-clustered and more on what 'type' of node is multi-clustered. Such properties are discussed in the next section.

Overall the results of OverWeiMod and OCG vary greatly in terms of (i) number of modules in each of the partitions of the network and (ii) the number of communities that the multi-clustered nodes can belong to. Despite these differences, there are still a considerable number of proteins that are multi-clustered by both methods.

3.3 Evaluation of Multi-clustered Nodes

If we consider genes that belong to more than one community as bridges, connectors between functional units, or communicators that spread information in a system, one would imagine them to exhibit properties that reflect such capabilities. In this section, we consider features that distinguish multi-clustered from mono-clustered genes and additionally show how these features can indicate an appropriate range of values for the overlapping parameter, r.

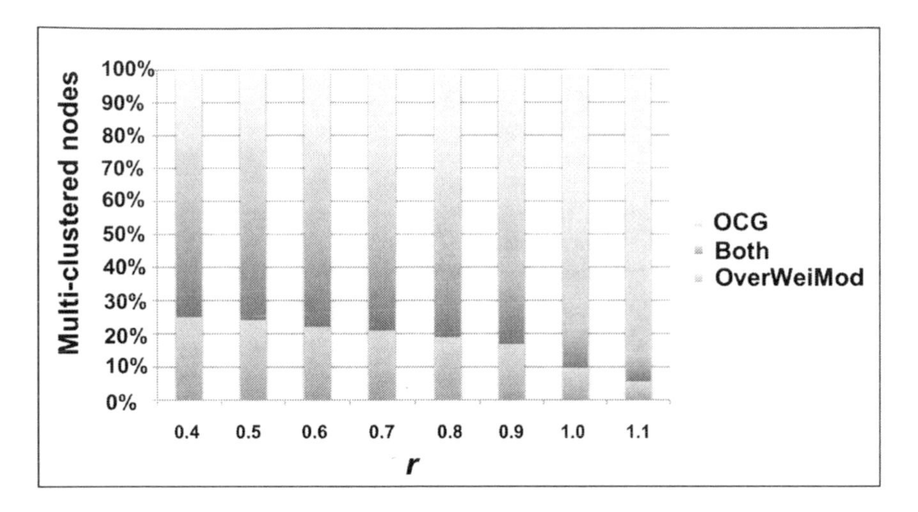

Fig. 3. Breakdown of multi-clustered proteins according to the methods they were found by. For each value of parameter r in the figure, the set of multi-clustered nodes detected by OCG remains constant.

3.3.1 Node Degree

We compare the average degree of the nodes with multiple community membership with the equivalent values for nodes that belong to only one community. For $0.4 \leq r \leq 1.1$, the range of values of parameter r tested in section 3.2, multi-clustered nodes have a higher average degree than the mono-clustered nodes (Table 3). We determine if the population means are statistically significantly different using the Mann–Whitney–Wilcoxon U test as implemented in the R statistical computing environment [32], where a p-value < 0.01 is significant. For all values of r tested, the average node degree of the multi-clustered nodes is significantly larger than the average degree of the mono-clustered nodes (Table 3). This result indicates that multi-clustered genes tend to have a higher number of interactions than those that belong to only one community. This result is intuitive, since if proteins lying in the overlapping sections play a connector role in the system, interacting with two or more communities, then it reasonable that they are more likely to interact with more partners compared to isolated nodes. This result indicates that multi-clustered nodes are topologically significant, however more important perhaps is to establish the likely functional roles of such connector nodes, as described in section 3.3.2.

It is also worth mentioning that, although the inclusion of parameter r is advantageous as it offers greater flexibility to the user, it is also necessary to determine a reasonable range of values for each network. Node degree can be used as an indicator of such a range if we assume that nodes belonging to more than one community should have more interactions than those that do not. Therefore, in terms of node degree, these results indicate our range of values for r is reasonable for this network. In the next section we show that by looking at the functionality of the multi-clustered proteins can be reduced the range of values.

Table 3. The average degree of multi- and mono-clustered nodes detected by OverWeiMod

r	Multi-clustered	Mono-clustered	p-value
0.4	26.98	13.46	< 2.2e-16
0.5	27.37	13.66	< 2.2e-16
0.6	28.13	14.09	< 2.2e-16
0.7	28.33	14.73	< 2.2e-16
0.8	29.34	15.08	< 2.2e-16
0.9	28.91	16.02	< 2.2e-16
1	26.67	18.11	< 2.2e-16
1.1	31.95	18.12	< 2.2e-16

3.3.2 Gene Ontology Term Analysis

Where node degree offers a topological measure for distinguishing between multi-clustered and mono-clustered genes, GO annotations can offer a distinction based on functional features. As seen in [19], and in line with our interpretation of multi-clustered nodes as bridges between multiple functions, one would expect multi-clustered genes to be associated with a higher number of GO annotations than those belonging to only one community.

To test this hypothesis, we compare the number of Gene Ontology terms annotated to multi-clustered and mono-clustered genes to determine which group has a significantly higher number of GO annotations. The annotations are taken from the MIPS *Fusarium graminearum* database [33]. The *F. graminearum* genome has 4915 genes annotated with 13,883 GO terms from all three aspects of GO (molecular function (MF), biological process (BP) and cellular component (CC)). As the complete genome sequence comprises 13,718 protein coding genes, only roughly a third of the genome is annotated. In the integrated network, 4251 proteins have no annotations, 4311 are annotated with at least one GO term and when considering each GO category, there are more proteins unannotated than annotated in the network.

Due to the high number of genes without GO annotations, we first consider all three aspects of GO terms together (ALL GO). Each gene in the main component of the *Fusarium* network is mapped to its GO terms where possible, and the average number of GO terms for all four categories (ALL GO, MF, BP and CC) is calculated. For ALL GO, BP and CC the multi-clustered proteins have a statistically significant higher average number of GO terms than the mono-clustered proteins, for $0.4 \leq r \leq 0.9$. For MF, the average number of GO terms for the multi-clustered proteins is not significantly higher than mono-clustered nodes for all values of *r*. Overall, it seems that multi-functionality of multi-clustered nodes is better endorsed at the BP level (i.e. in terms of participation in multiple biochemical pathways), rather than the MF (i.e. the individual biochemical tasks) of the corresponding gene product. Future work to consolidate such observations by accounting for lack of comprehensive annotations for this genome would be recommended.

Similar to the node degree analysis, we can use the number of GO terms assigned to genes to suggest a potential rage values for *r* where OverWeiMod detects multi-clustered genes with desirable properties. If we assume that multi-clustered nodes are multifunctional, we can use the ALL GO count as an indicator that considering values of *r* between 0.4 and 0.9 inclusive is reasonable for this network.

Table 4. The significance value of the difference between average number of GO annotations for multi- and mono-clustered nodes. Significant p-values are shown in bold (<0.01).

r	ALL GO	MF	BP	CC
0.4	**1.62E-04**	2.31E-01	**4.40E-06**	**1.06E-03**
0.5	**1.95E-03**	1.23E-01	**2.09E-05**	**7.37E-04**
0.6	**7.23E-03**	5.35E-01	**1.29E-03**	**5.33E-04**
0.7	**7.74E-04**	9.28E-01	**5.96E-04**	**9.05E-04**
0.8	**3.48E-04**	3.95E-01	**5.72E-05**	**1.84E-03**
0.9	**4.92E-03**	4.94E-01	**4.91E-07**	**1.03E-04**

3.3.3 Functional Cartography of Multi-clustered Genes

Throughout this study we hypothesise that multi-clustered nodes play an important role topologically and functionally in the network, considering them as bridges or communicators between functional units, helping to maintain the structure of the system. This idea has been reinforced by showing that (i) they are more connected than mono-clustered nodes and (ii) for some aspects of the Gene Ontology, they have more functional annotations that mono-clustered nodes. We relate the overlapping community structure to a node role classification scheme proposed in [20]. Each node is assigned a role based on its position in the hard partition of the network. A node's role is characterised according to two measures: within-community degree z-score and participation coefficient (see [20] for details). The within-community degree z-score measures how well a node is connected with nodes in its own community and the participation coefficient measures how uniformly the nodes' links are distributed among the other communities in the partition.

The node classification scheme in [20] can be summarised as follows. Based on the within-community degree z-score, nodes are classified as hubs and non-hubs, where hubs have a higher number of links with nodes in their own communities. Non-hubs are then classified into 4 roles: R1, ultra-peripheral nodes, R2, peripheral nodes, R3, non-hub connector nodes and R4, non-hub kinless nodes. Hubs are also classified into 3 roles: R5, provincial hubs, R6, connector hubs and R7, global kinless hubs. Both R3 and R6 nodes are labelled 'connector' nodes according to the classification scheme as they have by definition a large participation coefficient, indicating a high distribution of links with communities other than their own. Consequently, the removal of these nodes may result in poorly connected communities or even the disconnection of communities and therefore having an impact on the global structure. On the application of the classification scheme to metabolic networks it is found that R3 and R6 nodes are the most preserved across the species tested, suggesting that their role is more structurally relevant and similar results are predicted for other systems, including protein interaction and gene regulation networks [20].

We assign node roles to the *Fusarium graminearum* network. The distribution of node role types is shown in Table 5. We determine whether the proportion of R3 and R6 nodes is significantly higher in multi-clustered than mono-clustered nodes indicating that the multi-clustered nodes do indeed have a bridge/connector role in the system. For $r = 0.4$, all 165 R3 nodes and all 50 R6 nodes belong to the set of multi-clustered nodes. For $0.4 \le r \le 1$, there is a higher proportion of R3 nodes in the multi-clustered set than the mono-clustered set, and for R6 nodes, the range is $0.4 \le r \le 1.1$. We use the Fisher's exact test to decide if any difference in proportions is significant. We find that there is a significantly higher proportion of R3 nodes in the multi-clustered nodes for $0.4 \le r \le 0.8$ and similarly for R6 nodes for $0.6 \le r \le 0.9$ (results in Table 6).

Table 5. Node role type distribution

Role Types	R1	R2	R3	R4	R5	R6	R7
No. of nodes	4669	3323	165	0	157	50	0

The node classification scheme is employed to help describe the type of node that lies within the intersections of communities, offering a more sophisticated topological description than node degree alone. We find that nodes described as connectors are significantly enriched in the multi-clustered nodes. These are node roles that have previously been shown to be more structurally relevant than other node roles in biological networks [20]. This goes some way to supporting our claims that multi-clustered nodes play an important part in anchoring the communities of the network and thus contributing to the global functioning of the system. Furthermore, OverWeiMod is successfully detecting such nodes.

Table 6. The FDR-adjusted p-values for difference in proportion of R3 and R6 nodes in multi- and mono-clustered sets (significant values shown in bold)

r	**0.4**	**0.5**	**0.6**	**0.7**	**0.8**	**0.9**	**1**	**1.1**
R3	**1.31E-07**	**1.61E-09**	**1.11E-07**	**1.08E-05**	**6.90E-05**	3.96E-02	5.22E-01	1.38E-02
R6	2.21E-02	1.19E-02	**9.36E-05**	**1.08E-05**	**9.28E-07**	**3.74E-07**	1.65E-01	1.41E-01

4 Verified Virulence Genes

The analysis described next is to try to link the modular structure of the integrated network to known virulence genes. The experimentally verified *Fusarium graminearum* virulence genes known to be required for different aspects of the infection and disease formation process, were extracted from the Pathogen-Host Interaction database [30-32] and others were manually obtained from the scientific literature. We found that 79 out of 98 of the verified virulence proteins map to the integrated network of which 75 are in the main component. We also refer to the set of 75 genes as the verified virulence (VV) nodes.

First, the distribution of the VV nodes in the communities of the integrated network is considered. These nodes appear in only 15 of the 91 communities (Table 7), with the largest community (community 79, Figure 1) containing over half of the VV nodes (39 out of 75). For each of the 15 communities, we test if the community has a statistically significant higher proportion of VV nodes than the rest of the network using Fisher's exact test. Only the largest community and another of size 48 (community 80 in Figure 1), with 7 VV nodes, encompass a statistically significant high proportion of the proteins (FDR adjusted p-values 8.38E-07 and 1.35E-06 respectively). Although community 79 does contain a significant number of VV nodes, the corresponding BP MICA term, "cellular process", has an AIC of only 1.29, indicating that this is highly functionally diverse. Community 80 however is more coherent with an AIC of 4.88, and its BP MICA term is "regulation of metabolic process". The 7 VV genes in community 80 are all predicted to be transcription factors of the Zinc finger (Cys_2His_2) type [34].

We observe here that the verified virulence nodes are concentrated in two communities, suggesting that pathogenicity processes may be linked to specific parts of the network. Even at this preliminary stage, such effects indicate that there is a clear link between functional features and related interactions at network level. Similar analyses of community structure can therefore support further reverse molecular genetics and biochemistry experiments and thereby determine how these communities relate to underlying biochemical pathways.

We further partition community 79 and look at the distribution of the VV nodes in this new hard partition. The Louvain method detects a partition with 19 communities. The 39 VV nodes that are in community 79 in the hard partition of the main component are found in 8 of the communities of the re-partitioned community 79 (Table 8). Again, checking for overrepresentation of VV nodes in the individual communities shows that no community is significantly enriched. This may reflect the fact that proteins involved in a wide range of processes have a role in virulence because of the overall complexity of the infection process. In PHI-base [35, 36], there are many proteins with a "general" functional role such in basic metabolism, signal transduction and transcription and far fewer with a specific function such as toxin biosynthesis or infection structure formation.

Table 7. Distribution of the verified virulence (VV) nodes among communities and corresponding biological process average information content (BP AIC)

Comm. no.	7	16	28	39	51	52	56	57	64	71	75	76	79	80	82
No. VVs	1	1	5	1	1	3	2	1	1	2	4	6	39	7	1
BP AIC	-	27.5	3.62	2.07	4.48	1.2	2.07	3.62	-	-	1.29	5.76	1.29	4.88	1.29

Table 8. Distribution of verified virulence (VV) proteins in the hard partition of community 79

Community number	**1**	**5**	**7**	**9**	**14**	**15**	**16**	**17**
No. of VV proteins	9	3	2	2	5	12	5	1

We next consider the connection between multiple community membership and the VV genes. For $0.4 \leq r \leq 1.1$, we show the number of VV genes that are found by OverWeiMod to belong to more than one community in Table 9. The VV genes were not seen to be significantly overrepresented in either the multi-clustered genes or the mono-clustered genes for all values of r. However we do note that nearly half (49.3%) of the VV genes belong to more than one community for $r = 0.4$, and these may still play an important role in the system.

As we have shown previously, the number of multi-clustered nodes detected by OverWeiMod decreases as r increases. Nodes that remain multi-clustered for higher values of r may indicate cases that are more inclined to belong to multiple communities. We find 4 VV genes that are multi-clustered for r equal to 1 and 1.1. This may indicate that these proteins are more robustly multi-clustered than other pathogenicity-associated genes and therefore are more strongly inclined to belong to multiple communities. These four genes are: FGSG_04104 (probable guanine nucleotide-binding protein beta subunit), FGSG_08028 (conserved hypothetical protein), FGSG_10142 (related to transcription factor atf1+) and FGSG_03747 (NPS6 related to AM-toxin synthetase (AMT)). Probable guanine nucleotide-binding protein beta subunits are known to be part of a signalling process upstream of a range of biochemical pathways and therefore likely to be part of several communities. FGSG_03747 is a large protein with 7 predicted InterPro domains. It codes for a non-ribosomal peptide synthetase (NRPS). In this case the product and its function is already known. It is an extracellular siderophore that is used by Fusarium to bring the essential nutrient iron into the fungal hyphae and to protect against the cellular damage caused by various reactive oxygen species [37]. This link to a transfer process from outside the hyphae cell to inside may indicate likely bridging functional roles, therefore justifying why this gene may belong to more than one community. Such results suggest that OverWeiMod has the potential to detect appropriate multi-clustered nodes and therefore shows promise in predicting candidate multi-functional genes.

It should be noted that there is currently a small number of experimentally verified virulence genes, and within the set of known genes there may be a bias in reflecting particular classes of proteins that have been investigated experimentally, for example intracellular signalling and transcription-associated proteins. Therefore, network analyses and systems biology strategies such as the one presented here, offer good potential to plan future experiments using a more rational basis.

Table 9. The number of verified virulence (VV) seeds that belong to more than one community for $0.4 \le r \le 1.1$

r	**0.4**	**0.5**	**0.6**	**0.7**	**0.8**	**0.9**	**1**	**1.1**
No. of VV proteins	37	35	32	26	25	24	3	1

5 Discussion and Conclusions

In this study we have explored the disjoint and overlapping community structure of an integrated network for the globally important plant pathogenic fungus, *Fusarium graminearum*. We have investigated the topological and functional properties of proteins that belong to more than one community compared with those that belong to only one. It is shown that proteins in the intersection between two or more communities tend to be more highly connected than mono-clustered proteins. Furthermore, topological description through node role classification scheme illustrates that multi-clustered nodes tend to be enriched with connector roles (hub and non-hub connectors, roles R6 and R3 respectively) that are structurally important in biological networks [20]. Additionally, we consider the functional properties of the multi-clustered proteins in terms of number of GO annotations. For all three aspects of the GO combined (ALL GO) and for BP and CC multi-clustered proteins tend to have a higher number of annotations, than proteins belonging to only one community, although the same trend is not seen for MF. These results corroborate to some extent the idea that multi-clustered proteins are bridges between communities, which allow the semi-independent functional units to interact and regulate all functions required by the system.

As mentioned previously, the problem of detecting overlapping communities is not as well defined as the standard community structure detection problem, for example due to difficulties in conceptualising a uniform definition of overlapping properties and consequently methodological disparities. For this reason, methods and parameters used vary greatly and comparisons across different methods and benchmark examples are challenging. Therefore, the purpose of comparison of the multi-clustered nodes found by OverWeiMod with those found by OCG is not to assess prediction accuracy, but to demonstrate that OverWeiMod is in line with other methods in this context. In addition, the following characteristics render OverWeiMod a competitive method. First, OverWeiMod has the capability to define the strength of belonging of a node to a community, giving another level of understanding of the system. Future work includes (i) analysing how the belonging coefficients of a multi-clustered gene are distributed between communities and (ii) identifying genes that are more equally spread among functional communities than others. The authors of [20] propose that nodes with the same role should have similar topological properties, therefore an expansion of the functional cartography analysis, where we include all 7 node role types may offer insight into important properties of the nodes.

Additionally, OverWeiMod is applicable to weighted networks making it conducive to further work in assessing the suitability of data sources in the integrated network. The integrated *Fusarium* network contains information from multiple heterogeneous data sources and some of the data sources may be of better quality than others. The inclusion of weighted edges in the network might provide a more accurate and informative description of the community structure of the organism. A simple approach might be to weight the edges heuristically depending on the number of data sources that suggest an association subject to the various thresholds chosen, this can be regarded as an indication of reliability of an interaction. Another approach would be to estimate the likelihood of a functional association between two proteins given evidence from each data source. This procedure requires a benchmark such as proteins known to be functionally associated as determined from experiment or suggested by some measure (such as belonging to the same pathway) and is complicated if the different data sources are not independent (see for example [2]). As we have shown before, community structure detection of a weighted network may result in a different partition as compared to the equivalent binary network [8]. Such effects can be addressed in future work.

Finally, the flexible nature of mathematical programming framework allows for the easy implementation of additional constraints and parameters, again leading to more accurate and detailed network representations. For example, we can use prior knowledge such that nodes with similar functional annotations could be constrained to be in the same community. Furthermore, in terms of methodology, introducing symmetry constraints to as we have done in previous models [7] may improve the efficiency of OverWeiMod.

The motivation behind this study was to gain a better understanding of the fungal pathogen, *Fusarium graminearum*. We used network analysis tools to investigate the underlying mechanisms of the fungus from a community detection perspective. In particular we looked closely at the proteins taking part in more than one functional community in an attempt to identify those that may play a role in maintaining a structurally cohesive system. As the number of verified virulence proteins increases, analytical methods featured in this study could prove promising in the detection of relationships between the topological description and functional properties, potentially leading towards a better understanding of the pathogenicity process.

Acknowledgements. LB acknowledges financial support from the School of Natural and Mathematical Sciences, King's College London. ST acknowledges support from the EU and the Leverhulme Trust (RPG-2012-686).

References

1. Lee, I., Date, S.V., Adai, A.T., Marcotte, E.M.: A probabilistic functional network of yeast genes. Science 306(5701), 1555–1558 (2004)
2. Lee, I., Marcotte, E.M.: Integrating functional genomics data. Methods in Molecular Biology 453, 267–278 (2008)
3. Girvan, M., Newman, M.E.J.: Community structure in social and biological networks. Proc. Natl. Acad. Sci. U.S.A. 99, 7821–7826 (2002)
4. Blondel, V., Guillaume, J.-L., Lambiotte, R., Lefebvre, E.: Fast unfolding of communities in large networks. Journal of Statistical Mechanics: Theory and Experiment 2008(10), P10008 (2008)
5. Ruan, J., Zhang, W.: Identifying network communities with a high resolution. Phys. Rev. E, 77, 016104 (2008)
6. Xu, G., Bennett, L., Papageorgiou, L.G., Tsoka, S.: Module detection in complex networks using integer optimisation. Algorithms for Molecular Biology 5, 36 (2010)
7. Xu, G., Tsoka, S., Papageorgiou, L.G.: Finding community structures in complex networks using mixed integer optimisation. Eur. Phys. J. B 60, 231–239 (2007)
8. Bennett, L., Liu, S., Papageorgiou, L.G., Tsoka, S.: Detection of disjoint and overlapping modules in weighted complex networks. Advances in Complex Systems, 15, 11500 (2012)
9. Kuhner, S., van Noort, V., Betts, M.J., Leo-Macias, A., Batisse, C., Rode, M., Yamada, T., Maier, T., Bader, S., Beltran-Alvarez, P., Castaño-Diez, D., Chen, W.-H., Devos, D., Güell, M., Norambuena, T., Racke, I., Rybin, V., Schmidt, A., Yus, E., Aebersold, R., Herrmann, R., Böttcher, B., Frangakis, A.S., Russell, R.B., Serrano, L., Bork, P., Gavin, A.-C.: Proteome Organization in a Genome-Reduced Bacterium. Science 326(5957), 1235–1240 (2009)
10. Gavin, A.-C., Bosche, M., Krause, R., Grandi, P., Marzioch, M., Bauer, A., Schultz, J., Rick, J.M., Michon, A.-M., Cruciat, C.-M., Remor, M., Hofert, C., Schelder, M., Brajenovic, M., Ruffner, H., Merino, A., Klein, K., Hudak, M., Dickson, D., Rudi, T., Gnau, V., Bauch, A., Bastuck, S., Huhse, B., Leutwein, C., Heurtier, M.-A., Copley, R.R., Edelmann, A., Querfurth, E., Rybin, V., Drewes, G., Raida, M., Bouwmeester, T., Bork, P., Seraphin, B., Kuster, B., Neubauer, G., Superti-Furga, G.: Functional organization of the yeast proteome by systematic analysis of protein complexes. Nature 415(6868), 141–147 (2002)

11. Salathé, M., Jones, J.H.: Dynamics and Control of Diseases in Networks with Community Structure. PLoS Computational Biology 6(4), e1000736 (2010)
12. Zhang, S., Wang, R.-S., Zhang, X.-S.: Identification of overlapping community structure in complex networks using fuzzy c-means clustering. Physica A: Statistical Mechanics and its Applications 374(1), 483–490 (2007)
13. Palla, G., Derényi, I., Farkas, I., Vicsek, T.: Uncovering the overlapping structure of complex networks in nature and society. Nature 435, 814–818 (2005)
14. Ma, X., Gao, L., Yong, X.: Eigenspaces of networks reveal the overlapping and hierarchical community structure more precisely. J. Stat. Mech., P08012 (2010)
15. Zhang, S., Wang, R.S., Zhang, X.S.: Uncovering fuzzy community structure in complex networks. Physical Review E 76(046103) (2007)
16. Zarei, M., Izadi, D., Samani, K.A.: Detecting overlapping community structure of networks based on vertex-vertex correlations. Journal of Statistical Mechanics (P11013) (2009)
17. Lancichinetti, A., Fortunato, S., Kertész, J.: Detecting the overlapping and hierarchical community structure in complex networks. New Journal of Physics 11(033015) (2009)
18. Wang, X., Jiao, L., Wu, J.: Adjusting from disjoint to overlapping community detection of complex networks. Physica A 388, 5045–5056 (2009)
19. Becker, E., Robisson, B., Chapple, C.E., Guénoche, A., Brun, C.: Multifunctional Proteins Revealed by Overlapping Clustering in Protein Interaction Network. Bioinformatics 28(1), 84–90 (2012)
20. Guimera, R., Amaral, L.A.N.: Functional Cartography of Complex Metabolic Networks. Nature 433, 895–900 (2005)
21. Liu, G., Wong, L., Chua, H.N.: Complex discovery from weighted PPI networks. Bioinformatics 25(15), 1891–1897 (2009)
22. Cuomo, C.A., Guldener, U., Xu, J.R., Trail, F., Turgeon, B.G., Di Pietro, A., Walton, J.D., Ma, L.J., Baker, S.E., Rep, M., Adam, G., Antoniw, J., Baldwin, T., Calvo, S., Chang, Y.L., Decaprio, D., Gale, L.R., Gnerre, S., Goswami, R.S., Hammond-Kosack, K., Harris, L.J., Hilburn, K., Kennell, J.C., Kroken, S., Magnuson, J.K., Mannhaupt, G., Mauceli, E., Mewes, H.W., Mitterbauer, R., Muehlbauer, G., Munsterkotter, M., Nelson, D., O'Donnell, K., Ouellet, T., Qi, W., Quesneville, H., Roncero, M.I., Seong, K.Y., Tetko, I.V., Urban, M., Waalwijk, C., Ward, T.J., Yao, J., Birren, B.W., Kistler, H.C.: The Fusarium graminearum genome reveals a link between localized polymorphism and pathogen specialization. Science 317(5843), 1400–1402 (2007)
23. Wise, R.P., Caldo, R.A., Hong, L., Shen, L., Cannon, E., Dickerson, J.A.: BarleyBase/PLEXdb. Methods in Molecular Biology 406, 347–363 (2007)
24. Zhao, X.M., Zhang, X.W., Tang, W.H., Chen, L.: FPPI: Fusarium graminearum protein-protein interaction database. J. Proteome Res. 8(10), 4714–4721 (2009)
25. Obayashi, T., Kinoshita, K., Nakai, K., Shibaoka, M., Hayashi, S., Saeki, M., Shibata, D., Saito, K., Ohta, H.: ATTED-II: a database of co-expressed genes and cis elements for identifying co-regulated gene groups in Arabidopsis. Nucleic Acids Research 35(suppl. 1), D863–D869 (2007)
26. Zhao, X.-M., Zhang, X.-W., Tang, W.-H., Chen, L.: FPPI: Fusarium graminearum Protein-Protein Interaction Database. Journal of Proteome Research 8(10), 4714–4721 (2009)
27. Kohler, J., Baumbach, J., Taubert, J., Specht, M., Skusa, A., Ruegg, A., Rawlings, C., Verrier, P., Philippi, S.: Graph-based analysis and visualization of experimental results with ONDEX. Bioinformatics 22(11), 1383–1390 (2006)

28. Lysenko, A., Defoin-Platel, M., Hassani-Pak, K., Taubert, J., Hodgman, C., Rawlings, C., Saqi, M.: Assessing the functional coherence of modules found in multiple-evidence networks from Arabidopsis. BMC Bioinformatics 12(1), 203
29. Newman, M., Girvan, M.: Finding and evaluating community structure in networks. Phys. Rev. E 69, 026113 (2004)
30. Ashburner, M., Ball, C.A., Blake, J.A., Botstein, D., Butler, H., Cherry, J.M., Davis, A.P., Dolinski, K., Dwight, S.S., Eppig, J.T., Harris, M.A., Hill, D.P., Issel-Tarver, L., Kasarskis, A., Lewis, S., Matese, J.C., Richardson, J.E., Ringwald, M., Rubin, G.M., Sherlock, G.: Gene Ontology: tool for the unification of biology. Nat. Genet. 25(1), 25–29 (2000)
31. Chen, J., Yuan, B.: Detecting functional modules in the yeast protein-protein interaction network. Bioinformatics 22(18), 2283–2290 (2006)
32. R Development Core Team, R: A language and environment for statistical computing. R Foundation for Statistical Computing, Vienna, Austria (2010)
33. http://mips.helmholtz-muenchen.de/genre/proj/FGDB/
34. Son, H., Seo, Y., Min, K., Park, A., Lee, J., Jin, J., Lin, Y., Cao, P., Hong, S., Kim, E., Lee, S., Cho, A., Lee, S., Kim, M., Kim, Y., Kim, J., Kim, J., Choi, G., Yun, S., Lim, J., Kim, M., Lee, Y., Choi, Y., Lee, Y.: A phenome-based functional analysis of transcription factors in the cereal head blight fungus, Fusarium graminearum. PLoS Pathogens 7, e1002310 (2011)
35. Winnenburg, R., Baldwin, T.K., Urban, M., Rawlings, C., Kohler, J., Hammond-Kosack, K.E.: PHI-base: a new database for pathogen host interactions. Nucleic Acids Res. 64(Database issue), D459–D464 (2006)
36. Winnenburg, R., Urban, M., Beacham, A., Baldwin, T.K., Holland, S., Lindeberg, M., Hansen, H., Rawlings, C., Hammond-Kosack, K.E., Kohler, J.: PHI-base update: additions to the pathogen host interaction database. Nucleic Acids Res. 6(Database issue), D572–D576 (2008)
37. Oidea, S., Moederb, W., Krasnoffc, S., Gibsonc, D., Haasd, H., Yoshiokab, K., Turgeona, B.G.: NPS6, Encoding a Nonribosomal Peptide Synthetase Involved in Siderophore-Mediated Iron Metabolism, Is a Conserved Virulence Determinant of Plant Pathogenic Ascomycetes. The Plant Cell 18, 2836–2853 (2006)

Predicting Phenotype from Genotype through Automatically Composed Petri Nets

Mary Ann Blätke[1], Monika Heiner[2], and Wolfgang Marwan[1]

[1] Magdeburg Centre for Systems Biology and Lehrstuhl für Regulationsbiologie, Otto-von-Guericke-Universität, Magdeburg, Germany
[2] Chair of Data Structures and Software Dependability, Brandenburg Technical University, Cottbus, Germany
mary-ann.blaetke@ovgu.de

Abstract. We describe a modular modelling approach permitting curation, updating, and distributed development of modules through joined community effort overcoming the problem of keeping a combinatorially exploding number of monolithic models up to date. For this purpose, the effects of genes and their mutated alleles on downstream components are modeled by composable, metadata-containing Petri net models organized in a database with version control, accessible through a web interface (www.biomodelkit.org). Gene modules can be coupled to protein modules through mRNA modules by specific interfaces designed for the automatic, database-assisted composition. Automatically assembled executable models may then consider cell type-specific gene expression patterns and the resulting protein concentrations. Gene modules and allelic interference modules may represent effects of gene mutation and predict their pleiotropic consequences or uncover complex genotype/phenotype relationships. Forward and reverse engineered modules are fully compatible.

Keywords: Biomodel engineering, formal language, data integration, high-throughput, quantitative trait loci.

1 Introduction

Systems biology witnesses the evolution of experimental high-throughput methods with steadily increasing power regarding the quantification of nucleic acids, proteins, covalent modifications, and metabolites. Soon, these methods will broadly allow *omics* scale analyses of molecules involved in cellular regulatory control that capture the time-resolved response to stimulation or (genetic) network perturbation [22]. These advances challenge the development of integrative and modular modelling frameworks that support the combination of findings obtained through different, qualitative and quantitative experimental approaches. To be useful, such models will need to be multi-level in terms of integrating multiple levels of abstraction in causally connected and experimentally well established processes at the molecular level with adjustable resolution of details for higher level phenomena. These may include cell fate decisions for

D. Gilbert and M. Heiner (Eds.): CMSB 2012, LNCS 7605, pp. 87–106, 2012.

simulating the intrinsic heterogeneity of clonal populations of cells following individual trajectories during development. Petri nets are an ideal formalism for the formal description of processes at multiple levels of abstraction for systems biology purposes [13,17,18,28].

For the sake of creating realistic scenarios, it will presumably become indispensable to compare, on a regular basis, simulation results on one and the same network topology as obtained by employing continuous, stochastic, and hybrid paradigms. In the software tool Snoopy, a given Petri net graph can be interpreted and simulated as continuous (ODE), stochastic, hybrid, or simply as qualitative model with export option to SBML [29]. Interpretation as coloured Petri net, again in Snoopy, provides advanced options for biomodel engineering [19] as coloured Petri nets combine the formalism of Petri nets with the expressive power of a programming language. For this reason, and because of the intuitively accessible graphical representation, we have chosen Petri nets as framework for modelling and simulation.

Repeated iterations of experimental data acquisition, modelling, and simulation can evaluate the consistency of the interpretation of experimental results. This is especially true when high-throughput data come into play. However, conventional monolithic models usually represent certain aspects of a phenomenon at a certain resolution in detail and are restricted to a certain mathematical modelling paradigm. Such models can neither be easily combined with other models nor can they be easily updated by persons other than the author of the particular model. One solution to this limitation is to create a collection of Petri net modules that can be automatically linked in order to obtain and to update coherent models covering all or selected aspects of a biological process. Conceptually, such modules may be contributed, curated, and updated by individuals of the community with special expertise in certain aspects. Being organized within a database with version control, modules obtained by reverse engineering of experimental data can also be integrated and, as we will show below, help to import complex data sets into models that have been automatically generated by composition of pre-existing modules.

This paper makes three major contributions which fundamentally enhance the versatility of modular Petri net modelling by (1) linking regulated gene expression to protein concentration, (2) allowing the fully automated generation of models for the application in genome-wide (*omics*) approaches, and (3) linking gene mutation to complex phenotypic consequences in generating predictable models. Let us briefly elaborate on these claims.

1. The introduction of gene modules and mRNA modules allows to model regulated gene expression and protein biosynthesis. As the gene expression pattern of a cell is not constant and can drastically change dependent on cell type, physiological state, or experimental conditions, cells are definitely equipped with specific sets of proteins of variable relative abundance. By introducing gene modules into the model, differentially regulated gene activity and the resulting gene expression patterns translate into the marking of places of the protein modules. As the rates of biochemical reactions always

depend on both, the kinetic rate constants and the concentrations of the reactants, changed gene expression will also change the rates of biochemical reactions, which in turn may drastically alter the dynamic behaviour of a regulatory network. Moreover, changed concentrations in regulatory proteins (e.g. transcription factors) may feed back in a complex manner onto the gene level by changing gene expression profiles. This circuitry of interwoven regulatory control becomes systematically accessible through the model.
2. Gene and mRNA module prototypes permit the fully automated generation of modules by simply uploading lists of gene names. This allows the automatic creation of models representing hundreds or even thousands of genes, their mRNAs and the proteins they form. By importing transcriptomic or proteomic data sets obtained in high throughput experiments [30], one can infer rate constants and reverse engineer regulatory mechanisms with the help of the model and predict changes in the proteome in response to differential gene regulation. Such models will also support the interpretation of phenomena observed in systematic RNAi screens where individual genes are knocked down [25,10,9].
3. Introduction of allelic influence modules extends the modelling of gene activity to the modelling of the regulatory consequences, which gene mutations have on cellular processes. This sets the formal framework to reverse engineer biomodels from complex phenotype data sets resulting from genotypic variation e.g. by employing Petri net compatible algorithms [16,12,11]. It is obvious that such type of models have a high potential for the application to various areas from basic research to synthetic biology or personalized medicine.

We are not aware of any modelling framework providing a comparable versatility and integrative power in terms of combining forward and reverse biomodel engineering.

This paper is organised as follows. In the next section we briefly summarise relevant own previous work on modular Petri net modelling and the automatic composition of modules with the help of a module database. Then we will provide the rules for the generation of modules and modular models and introduce gene and mRNA modules in Section 3, the application of which is demonstrated by the first case study in Section 4. Section 5 explains the specific features of gene and mRNA modules required for modelling gene expression and its differential regulation in eukaryotes. We continue in Section 6 with a second case study on cell differentiation and eukaryotic gene regulation, define allelic influence modules, and explain how these work together with gene, mRNA, and protein modules in generating models that integrate forward and reverse biomodel engineering approaches. In Section 7 we conclude with discussing the versatility of our approach and provide future perspectives regarding the application to synthetic biology and *omics* approaches.

2 Previous Work

Initially, we developed our modular modelling approach to represent biochemical reaction networks made of protein-protein interactions. The core idea is to take an object-oriented approach where the objects correspond to the natural modular building blocks of life. We represent individual proteins as independent and self-contained hierarchically structured Petri nets, called modules. Thus, modules correspond to natural units, each of which comprises intramolecular regulatory mechanisms of the respective protein and of all its interactions with other molecules. Modules of interacting proteins can be coupled by logical nodes of identical shared subnets describing their interaction with each other. The assembly of models from a set of modules needs no further modifications at the module level. An essential advantage of this approach is the reusability of all constructed modules in arbitrary combination to obtain models representing specific pathways [4,5].

A crucial point of our modular modelling approach is that each module is self-contained and can be evaluated on its own. Composed networks which in terms of their behaviour correspond to the conventional, monolithic networks need never to be explicitly shown. Their validity is entirely assessable by understanding the individual modules and the connection rules.

Our modular modelling approach has been deployed to construct modular Petri nets for two non-trivial case studies: (1) the JAK/STAT pathway in IL-6 signalling [4] and (2) nociceptive network in pain signalling [5]. The JAK/STAT pathway is one of the major signalling pathways in multicellular organisms controlling cell development, growth and homeostasis by regulating the gene expression. The modular network of the JAK/STAT pathway in IL-6 signalling comprises 7 protein modules (IL6, IL6-R, gp130, JAK1, STAT3, SOCS3, and SHP2). Overall, the model consists of 92 places, 102 transitions spread over 58 pages with a nesting depth of 4. The nociceptive network in pain signalling consists of several crucial signalling pathways, which are hitherto not completely revealed and understood. The latest version of the nociceptive network consists of 38 modules, among them are several membrane receptors, kinases, phosphatases and ion-channels. So far, the model is made up by 713 places and 775 transitions spread over 325 pages, again with a nesting depth of 4.

To support convenient module selection and network composition, we developed a database prototype, which is accessible by a web-interface [4,6], see Figure 1. The database holds the qualitative Petri net structure of each module, as well as the kinetic information assigned to each transition. In addition, the database contains also meta-information about the corresponding proteins (extracted from UniProt) and information about the literature used to construct the modules (extracted from PubMed), which can be associated with each module. The organization of the modules in such a database enables the user to (1) search for individual modules, places, transitions, etc., (2) store modules in collections, and (3) assemble a modular network from a chosen collection.

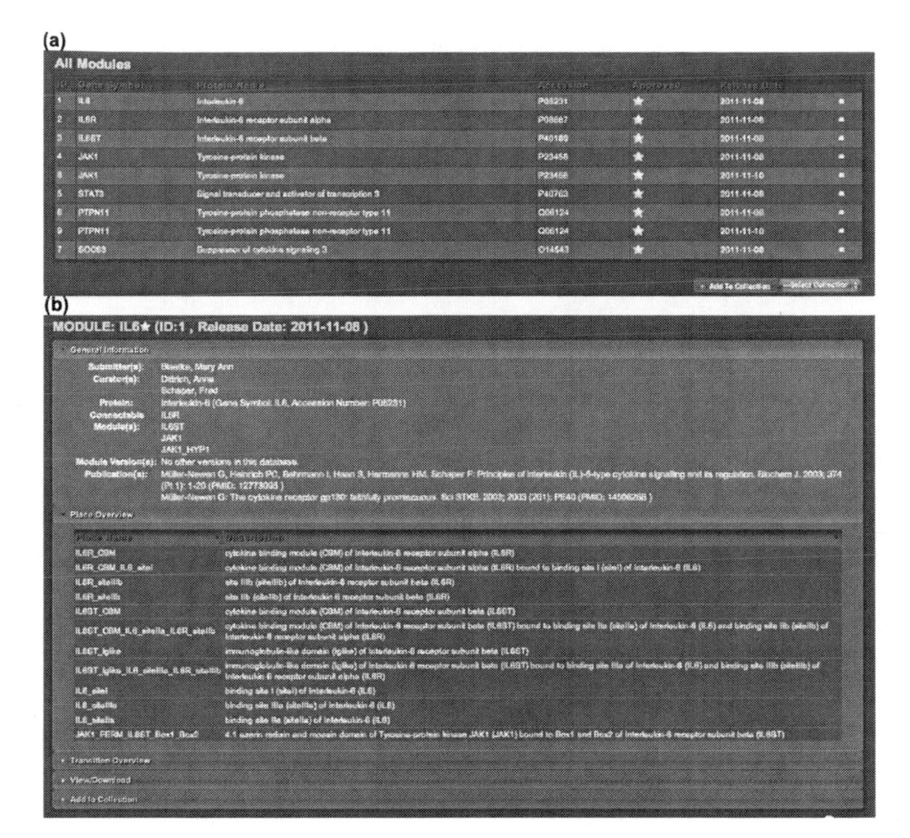

Fig. 1. Selected screenshots of the prototype database [6]**.** (a) The web-interface enables the user to browse and/or search for modules of specific proteins. Modules can be stored in collections by selecting them. (b) Detailed information about each module can be shown on a separate page.

3 Petri Net Modules

3.1 How Modules Are Built and Composed

We use Petri nets structured in the form of modules that allow the automatic composition of executable models [4,5] based on Snoopy [29]. The modules and their associated metadata are organized in a database accessible through a web interface designed to manage multiple versions of each module and supporting the automatic composition of models from modules (interactively) selected from a potentially rich collection [4]. Initially, the approach was designed to model protein-protein interactions (see Section 2) in the context of signal transduction networks [4,5]. In this paper we go one step further by defining gene modules, mRNA modules and allelic influence modules which considerably enhance power and versatility of our modelling approach.

A module in general is centred around one entity describing all its interactions with other components to which the module can be linked. The entity can be a protein, an mRNA, a gene, or the specific allele of a gene. Conceptually, it is possible to extend this definition to admit other entities as well. Before going into detail, let us first briefly explain how the modular modelling approach technically works and how modules are connected to each other.

The basic principle is explained taking the reaction of an autophosphorylating kinase with its substrate as example (Figure 2). The kinase X autophosphorylates to give XP. XP transfers its phosphate group to the substrate Y. The phosphate group of YP then hydrolyses spontaneously. The Petri net describing these reactions (Figure 2a) can be decomposed into two modules each representing the reactions of one of the two proteins involved, X and Y, respectively (Figure 2b, 2c). Places and transitions that occur in more than one module are defined as so-called logical nodes. Logical nodes appear as multiple graphical copies of a given place or transition. In this paper, logical nodes are shaded in grey. Braking a Petri net up into modules introduces redundancy in terms of the graphical display of nodes which might appear unnecessarily complicated at first. For complex modules or when many modules are involved, the benefits are indeed tremendous as we have previously shown ([4,5], and see Discussion). The biosynthesis and the degradation of a protein or a nucleic acid are modelled by separate biosynthesis or degradation modules, respectively. Accordingly, the user can choose for each protein or mRNA whether or not its biosynthesis and degradation should be considered in the assembled model.

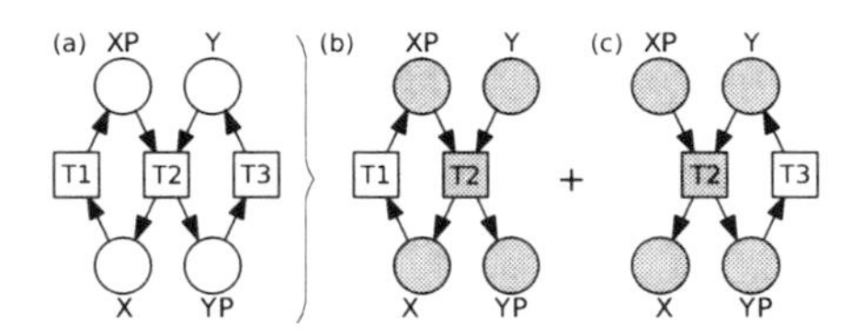

Fig. 2. The principle of modular modelling based on the use of logical nodes (connectors). (a) A Petri net model of the phosphorylation of protein Y by the autophosphorylating kinase X is split into modules for (b) protein X and (c) protein Y, respectively. Places and transitions that are shared by the two modules are implemented as logical nodes shaded in grey.

3.2 Definition of Modules Representing the Function of Genes

We now extend the initial approach constrained to protein modules by defining gene modules. A gene module considers the mechanisms of a gene being regulated through the reversible transition between its on and off state. Assigned metadata information includes the *Genbank* database accession number providing the DNA sequence information as a cross-reference. Logical nodes of a gene module are used to link the gene to other modules.

In general, multiple forms of each gene, so-called alleles, do exist that differ in one or more base pairs. These differences are due to mutations. Mutations can be silent in not altering the amino acid sequence of the encoded protein. Mutations can also be neutral in not changing the properties of the encoded protein although its amino acid sequence is changed due to mutation. These sequence polymorphisms are commonly found in wild-type populations. Alternatively, a mutation can change the properties of the encoded protein due to its altered amino acid sequence or it may even prevent the formation of the protein at all, e.g. by introducing a stop codon. Following strictly the modularity principle in designing Petri nets, a separate gene module is created for each allele of a gene to represent mutations that change relevant properties of the gene products (RNAs and proteins) as compared to the wild-type. Entering a query for a gene, the module database will list all modules corresponding to alleles of that gene.

We will now use two case studies to demonstrate how gene modules work. The first case study (see Section 4) concerns metabolic regulation in bacteria. It is taken to explain the principle of gene modules with the help of a simple example. In Section 5, we explain the specific features of gene and mRNA modules required for modelling gene expression and its differential regulation in eukaryotes, which is the topic of our second case study (see Section 6). Here, we show a more complex scenario involving multiple layers of regulatory control. In addition, the second case study will provide an example of how modules can be obtained through reverse engineering of experimental data, it will introduce allelic influence modules and it will reveal the scalability of the approach.

4 Case Study: The Phosphate Regulatory Network

In the first case study, we consider the response of enteric bacteria like *Escherichia coli* to the limitation in inorganic phosphate which is required for the biosynthesis of nucleic acids and other cellular components (compare Figure 2).

When inorganic phosphate (P_i) becomes low in the environment, it may turn into a growth-limiting factor even if sufficient nutrients are available [26]. When present, inorganic phosphate is taken up from outside of the cell through an ABC transporter system, the PstSCAB transmembrane complex (Figure 3 [21]). With sufficient P_i outside, the PstSCAB complex actively pumps P_i across the cell membrane into the cytoplasm. Under this condition, the PhoU protein forms, according to the proposed mechanistic model [21], a complex with the pstSCAB transporter system and the PhoR histidine kinase. Complex formation prevents the kinase to autophosphorylate caused by the interaction with PhoU. PhoU is a chaperone-like PhoR/PhoB inhibitory protein. When external P_i is low and the PstSCAB complex is inactive, PhoU dissociates and allows the autophosphorylation of PhoR. PhoR then phosphorylates and thereby activates the transcription factor PhoB. The phosphorylated form of PhoB, namely PhoBP, then activates the transcription of at least 31 genes organised into 9 transcriptional units (*eda*, *phnCDEFGHIJKLMNOP*, *phoA*, *phoBR*, *phoE*, *phoH*, *psiE*, *pstSCAB-phoU*, and *ugpBAECQ*) [21]. One of the activated genes, *phoA*, encodes the PhoA protein which is a bacterial alkaline phosphatase. PhoA is exported across the membrane into the periplasm where it degrades organic phosphorous compounds to liberate P_i which is then taken up into the cell to overcome the limitation.

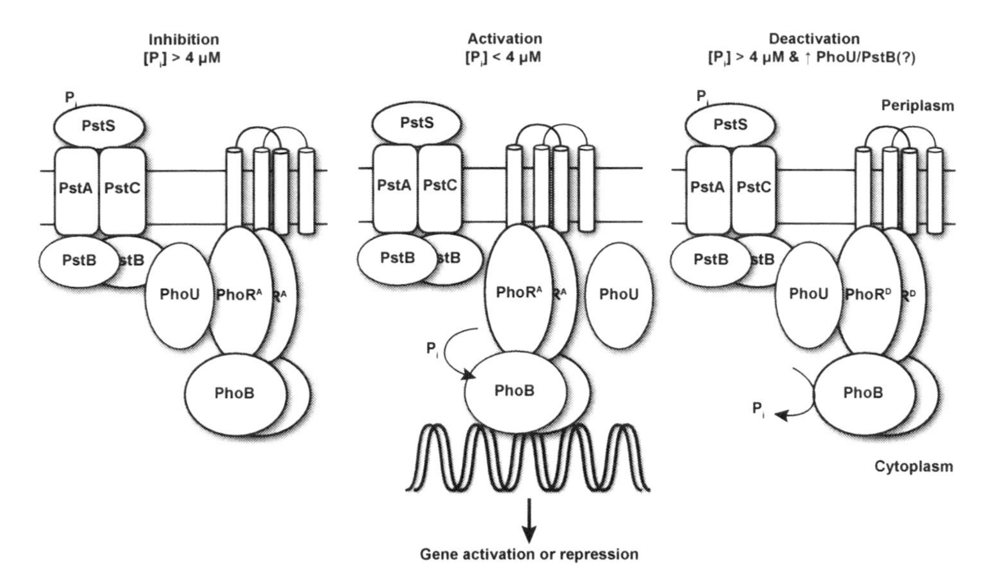

Fig. 3. Biochemical model for sensing extracellular inorganic phosphate (P_i) and transduction of the signal to control gene expression. The PstSCAB transmembrane complex serves as an ABC transporter for the uptake of environmental P_i. At high extracellular P_i concentration, the binding protein PstS is fully saturated and this signal is relayed to the cytoplasmic part of the receptor that forms an inhibitory complex with a second transmembrane protein, the PhoR kinase via the cytoplasmic protein PhoU. If P_i is low, the complex dissociates and the autophosphorylating kinase PhoR phosphorylates PhoB which, in its phosphorylated form, binds DNA to induce gene expression. When $_i$ subsequently increases, the compex with PhoU is formed again and PhoBP is dephosphorylated. The figure was redrawn from Hsieh and Wanner published in Current Opinion in Microbiology [21].

When enough P_i has been formed, this system is switched off again and PhoBP is dephosphorylated.

In [24], we gave a monolithic Petri net of a simplified version of the phosphate regulatory circuitry. To cover the entire functionality we now construct a modular Petri net model which is composed of three types of modules: (1) one protein module representing the reactions of the PhoB protein, (2) gene modules representing the regulated genes, and (3) mRNA modules representing the transcription of the gene, the translation of mRNA into the protein, and the degradation of mRNA. The degradation of the encoded proteins is represented by degradation modules which have been introduced previously [4].

4.1 The PhoB Module

The PhoB module (Figure 4) represents the reactions of the PhoB protein in its phosphorylated (PhoBP) and dephosphorylated (PhoB) states. It also represents the complex formation of PhoB with its regulatory proteins as well as the binding of PhoB to regulatory sequences in the DNA.

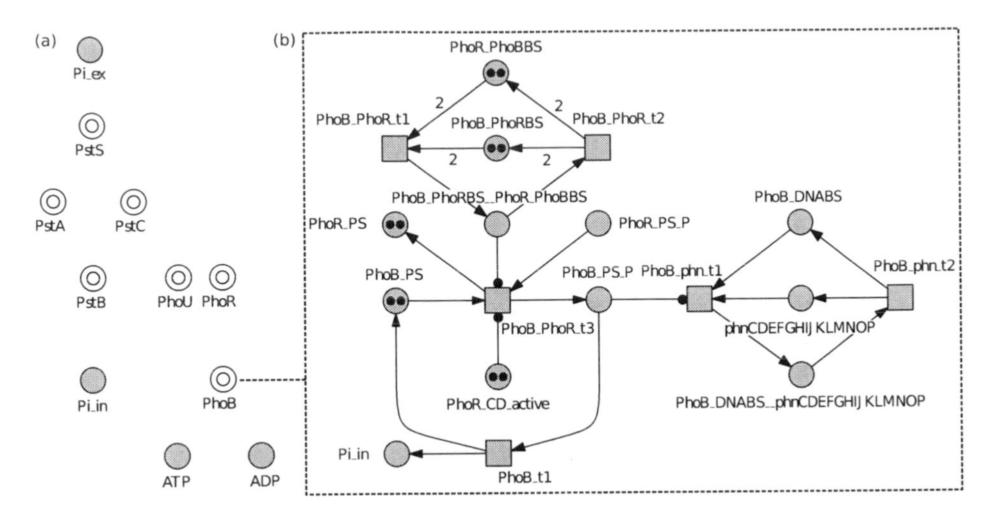

Fig. 4. Petri net representation of the phosphate regulatory network. (a) Top-leve presentation of the modules of the phosphate regulatory system in the form of coarse places as they appear in Snoopy. (b) Protein module of the PhoB protein displaying the direct interactions with binding partners.(BS, binding site).

The PhoB module models the regulatory mechanism schematically shown in Figure 3. Binding and dissociation of PhoBP in its complex with PhoRP is represented separately for each transcriptional unit. Displaying the binding of PhoBP to each regulatory site on the DNA separately, keeps the Petri net graph clearly arranged and allows to reuse the structural motif of binding and unbinding reactions via copy/paste for the various regulated transcriptional units. Accordingly, PhoRP_PhoBP is declared as logical place.

To save space in Figure 4, we only show binding of PhoBP to one of the nine transcriptional units.

4.2 The Gene Modules

A gene module represents the regulation of the gene by other factors (e.g. transcription factors) through the transition between its on and off state (Figure 5). In the quantitative interpretation of the Petri net (as stochastic, continuous, or hybrid Petri net) the regulatory factors influence the equilibrium between the on and the off state of the gene. The on state means that the gene is transcriptionally active and that mRNA molecules can be accordingly formed as modeled in the mRNA module.

In prokaryotes, several functionally related genes can be organized into a single regulatory and transcriptional unit, a so-called operon. The genes of an operon are transcribed in the form of a single mRNA molecule, called a polycistronic message, which may encode several proteins at once. In the case of the *phnCDEFGHIJKLMNOP* transcriptional unit (Figure 6), one polycistronic message encoding 14 different proteins (PhoC to PhoP) is formed upon the

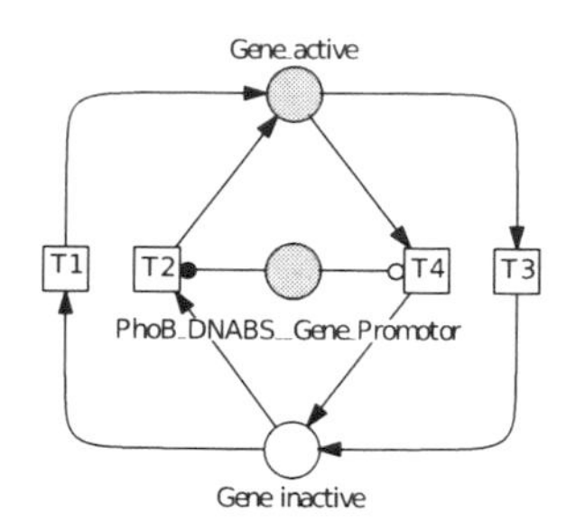

Fig. 5. Prototype of a gene module. The module may represent the activation of any gene responsive to the phosphorylated PhoB protein. Binding of PhoRP_PhoBP to the promotor renders the gene active. These regulatory interactions are modeled through a test arc activating transition T2 and an inhibitory arc blocking T4. The basal activity of the gene in the absence of PhoRP_PhoBP is maintained by T1 and T3.

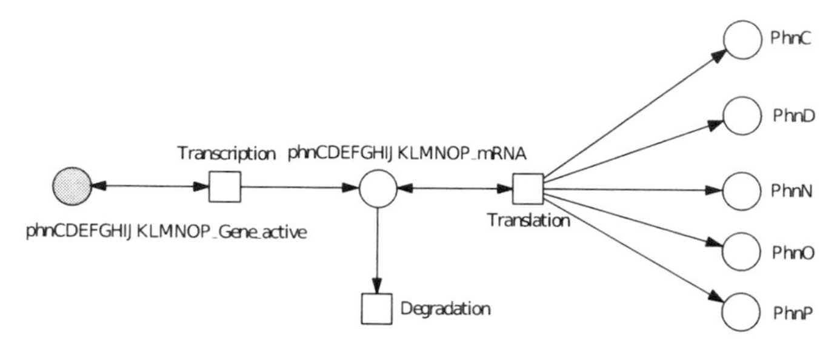

Fig. 6. mRNA module modelling the formation of the polycistronic message synthesized by transcription of the active *phnC-P* gene. The mRNA is translated into the proteins PhnC ... PhnP. For simplicity, the places for only five of the 14 proteins that are formed are shown.

initiation of transcription. The probability per unit of time for the initiation of transcription to occur depends on binding of PhoRP_PhoBP to the regulatory region of the *phnCDEFGHIJKLMNOP* operon on the DNA. From the biological point of view, polycistronic messages provide a simple mechanism for co-regulation of genes encoding proteins that work together in a cellular process. In order to obtain a model including the transcriptional regulation of all 31 genes organized into the 9 transcriptional units that are under control of the PhoB protein, 9 gene modules (*phoA*, *phoBR*, *phoE* etc.) of analogous structure (as shown in Figure 5) are required. For large scale modelling approaches, these network structures could be generated automatically or semi-automatically.

4.3 The mRNA Modules

The mRNA modules models the reactions of the respective mRNA species, namely its biosynthesis by transcription, its degradation, and its translation into the encoded proteins (Figure 6). An mRNA module may in addition represent the binding of regulatory proteins to the RNA, the binding of antisense

RNA influencing the stability of the message, or the processing (e.g. splicing) of the transcript, as it occurs in eukaryotes (shown in Figure 8 and discussed in Section 7.2). The transcription of the bacterial *phnCDEFGHIJKLMNOP* transcriptional unit leads to the formation of a polycistronic message, which encodes for 14 proteins (PhoC ... PhoP). The reactions (e.g. the catalytic activity) of the encoded proteins might then be considered in separate protein modules.

5 Modelling Eukaryotic Gene Regulation with Gene and mRNA Modules

5.1 Eukaryotic Gene Modules

The gene modules designed to model the regulation of eukaryotic genes are very similar to the models of prokaryotic genes as presented in the previous section for the phosphate regulatory network. However, the regulation of eukaryotic genes is typically more complex than in prokaryotes as more protein factors and more binding sites for regulatory proteins on the DNA may be involved. All these factors together may control the on state of a gene.

Gene regulation may involve protein binding sites on the DNA functioning as enhancers or silencers that are located several thousand base pairs distant from the genes they regulate. Proteins bound to these sites may influence the probability for transcription to be initiated through physical interactions with the protein complexes bound to the promotor of the regulated gene. These regulatory sites and the binding of regulatory proteins to these sites are represented as part of the gene module.

A prototype of a module representing the regulatory control of a eukaryotic gene is shown in Figure 7. Making regulatory sites part of the gene module comes with the advantage that potentially cooperative effects in protein binding and gene activation can be considered as part of the module.

5.2 Eukaryotic mRNA Modules

In addition to the biosynthesis of proteins there are several RNA-dependent processes that may be of regulatory importance especially in eukaryotic cells. Typically, the occurrence of these mechanisms depends on the considered RNA species and may also depend on physiological conditions as well as on developmental states. Each of the mechanisms described in the following has been implemented in a basic form using the mRNA module prototype shown in Figure 8.

Alternative Splicing. Primary transcripts in eukaryotes are processed during the maturation of the final protein-encoding mRNA. Processing includes the splicing of the RNA where non-coding introns are excised from the primary transcripts. Due to the occurrence of alternative splicing sites, it may be that differently spliced mRNAs are formed from one and the same primary transcript

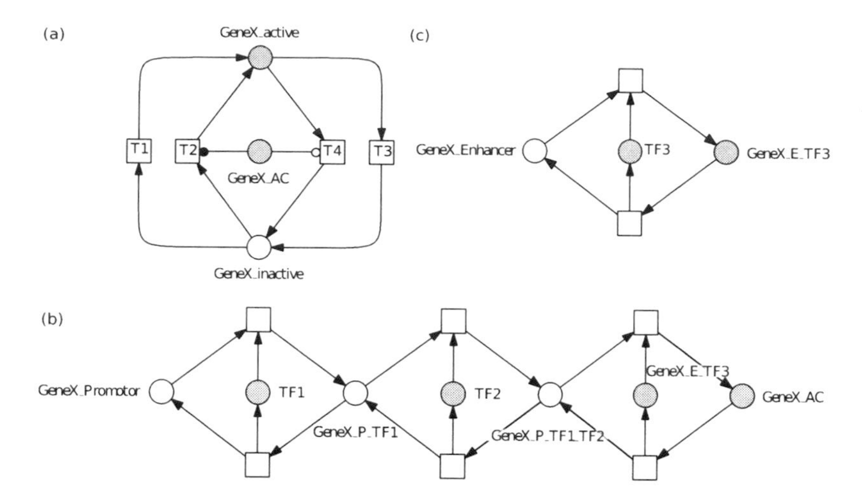

Fig. 7. Prototype of a eurkaryotic gene module. The regulation of the eukaryotic gene (a) depends on more protein factors than the regulation of the prokaryotic gene shown in Figure 5 does. By binding to the promotor region of the gene, these factors form a multimeric protein complex, as shown for transcription factors TF1 and TF2, both of which directly bind to the promotor (b). The third transcription factor, TF3, binds first to an enhancer sequence distant from the promotor (c) and subsequently can bind to the promotor-TF1-TF2 complex to form the gene activating complex GeneX_AC (b) which switches the gene into its transcriptionally active state (a). As in Figure 5, gene activation by protein binding to the promotor is modeled by control arcs and the basal level of gene activity occurs through firing of transitions T1 and T3 (a). According to our module notion the gene module displays all direct molecular interactions of GeneX with the proteins that bind to its regulatory sequences. Note that binding of regulatory proteins may involve cooperative mechanisms which would be represented accordingly in the context of the gene module.

giving rise to proteins of partially different amino acid sequence. This splicing depends on protein factors that may be regulated depending on the physiological or developmental state of the cell. When necessary, the biochemical reactions of these slicing factors (like posttranslational modification or protein-protein interaction) may be represented in the context of protein modules with the help of logical places.

RNA-Binding Proteins. The half-life of mRNA species may vary between minutes and months. This variation may have different reasons in addition to the specific secondary structure of the RNA. One mechanism influencing the half-life of a given mRNA species is the binding to specific RNA-binding proteins that may store or degrade the RNA. Being bound to an RNA-binding protein, like Pumilio for example, the mRNAs can be stored in the cell while being translationally inactive. Upon release from the RNA-binding proteins the mRNA may suddenly become translationally active and hence become available at relatively high concentration [14].

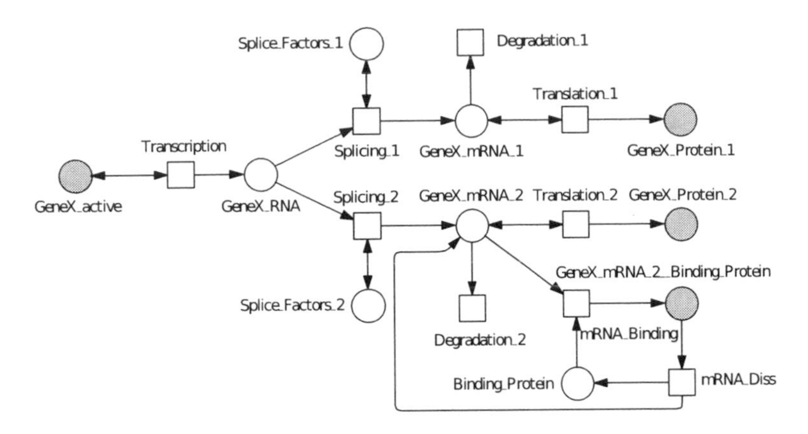

Fig. 8. Prototype of an eukaryotic mRNA module. Transcription of the active Gene (GeneX) leads to the formation of a primary transcript which is processed. In the example shown, the primary transcript is spliced into two alternative mRNAs, GeneX_mRNA_1 and GeneX_mRNA_2, respectively that are subsequently translated into the corresponding proteins. The mature mRNAs may bind to RNA binding proteins that regulate the stability of the mRNA and its availability for the translational maschinery. For simplicity, this reaction is shown for one of the two mRNAs only.

RNA Interference. RNA interference is a natural mechanism for the specific inactivation of the expression of eukaryotic genes, e.g. by small interfering RNAs that bind to the target RNA. The degradation of the RNAs depends accordingly on specific protein factors and on the availability of the interfering RNA [8,33].

These mechanisms may have to be considered in the context of mRNA modules and receive regulatory input from specific cellular proteins.

5.3 Why Is the Integration of Bottom-Up and Top-Down Models Essential, Especially in Eukaryotes?

In some respects systems biology appears to be more difficult for eukaryotic as compared to prokaryotic cells. This may in part be due to the occurrence of fundamentally different regulatory processes in the two domains of life with non-obvious consequences of certain eukaryotic regulatory phenomena.

There is indeed highly detailed knowledge on the canonic pathways of eukaryotic signal transduction which allows the formulation of well-structured bottom-up models representing the biochemical interactions of regulatory components like e.g. the MAP kinase cascade. Such models can be tremendously useful in understanding mechanisms in health and disease [23,27,31] and in finding new and powerful drugs.

However, it is also true that many experimental findings on canonic pathway components seem to be contradictory. This may in part be due to the fact that gene expression patterns in different experimental systems and under different physiological conditions are different leading to a different composition of biochemical reactants within the cell at a given time point or between replicas of a particular experiment.

This certainly restricts the current practical value of bottom-up models without disclaiming their general usefulness. When a certain cellular process, for example the differentiation phenomenon in a eukaryotic cell or the progression through the cell cycle, is to be rigorously analysed at the transcriptomic or the proteomic level, the changes in many perhaps in most of the observed components and their consequences can currently not be explained on the basis of the established bottom-up models of the canonic pathways. This suggests that there are tremendous gaps in our current understanding.

On the other hand, it seems for the time being impossible that the thousands of components accessible through *omics* approaches can be analysed with such experimental effort as invested for the exploration of canonic pathway components. Therefore it seems self-evident that *omics* data are used to rigorously reverse-engineer models. Then, a next and essential step is to integrate these top-down models with relevant bottom-up models to obtain integrated models with predictive power.

6 Case Study: The Sporulation Control Network in *Physarum polycephalum*

6.1 Sporulation Is Controlled by a Gene Regulatory Network

Physarum polycephalum is a unicellular eukaryote belonging to the amoebozoa group of organisms [1,2,15]. During its relatively complex and branched life cycle, *Physarum* develops into various cell types. These cell types occur in temporal order and differ in morphology (shape), molecular composition, and physiological function ([7] and references therein).

One of these cell types is the plasmodium, a multinucleate macroscopic single cell. Differentiation of the plasmodium can be easily studied under lab conditions as the response can be experimentally triggered by applying a brief pulse of far-red light. The light pulse sets a defined starting point on the time axis on which subsequent events are observed. During about 18 hours after the trigger, the entire plasmodial cell is extensively remodeled and fruiting bodies are formed that give rise to mononucleate haploid spores that are precursor cells of amoebal gametes which will develop at later stages of the cycle [7]. This process is called sporulation. Please note that sporulation in bacteria, although the name is the same, is biochemically a completely different process than in eukaryotic cells.

Five to six hours after an inductive far-red pulse, the cell is irreversibly committed to sporulation. The associated molecular events are of scientific interest as the plasmodium loses some stem cell-like capabilities during commitment. The expression pattern of hundreds of genes changes [15,7]. These changes in gene expression that normally occur can be compared to the changes that are seen in mutant cells that have lost their ability to be committed to sporulation [3,20].

A widely-used method in the biosciences is the genetic dissection of gene regulatory networks by generating mutants which are altered in the regulatory control and analysing the phenotype which a mutation produces. Mutants may be produced through forward and/or reverse genetic approaches. In forward genetics, randomly mutated cells or organisms are screened for phenotypes of interest and the gene which causes the phenotype is identified subsequently. In the reverse genetic approach, a gene of interest is mutated and the phenotypic consequence of the generated mutation is analysed. Mutation of a gene, both in forward and reverse genetics, may either cause the loss of a protein or a change in its function. Mutation may change the activity of a protein (up or down) or it may change the specificity of its catalytic activity. In many cases, the molecular mechanisms of how a given mutation translates into the observed phenotype remain unknown for a number of years. Despite this ambiguity, the genetic approach is powerful as it rigorously establishes causal dependencies within the living organism. In most cases biochemistry alone could not fulfill this task.

A powerful way to employ genotype/phenotype relationships for modelling and simulation is the reverse engineering of genetic data. Reverse engineering of gene expression data provides a direct link to bottom-up models of protein-protein interactions. By reverse engineering one can establish effects, which a mutated gene (the allele of a gene) exerts on a cellular process. We define allelic influence modules to represent these influences. We will now show how allelic influence modules are built and how they are useful for the integration of top-down and bottom-up model parts into one coherent model.

6.2 Linking Genotype to Phenotype: Allelic Influence Modules

As gene modules, allelic influence modules are centred around the allele of a given gene. However, allelic influence modules differ from gene modules in representing the regulatory influences exerted by the allele on cellular processes by controlling the firing activities of respective transitions through read or inhibitory arcs. In reality, these influences can be rather indirect by involving numerous other, potentially unknown components. Accordingly, the allelic influence module may represent the control of molecular events like the biosynthesis of RNA by transcription or even more complex processes of potentially unknown molecular mechanism as inferred from functional studies. To make this more clear, let us consider the case study.

In response to far-red light, *Physarum* plasmodia differentially express a large number of genes several hours before the cell is irreversible committed to sporulation [20]. Genes with both up- and down-regulated RNAs have been identified at a genomic scale [15,3], and the precise expression kinetics of some of them have been investigated in detail [20]. Currently, we do neither know the molecular mechanisms through which these genes are controlled nor do we know which causal consequences the change in expression level in detail have. However, the majority of the differentially expressed genes encodes proteins with high sequence similarity to proteins of important regulatory function.

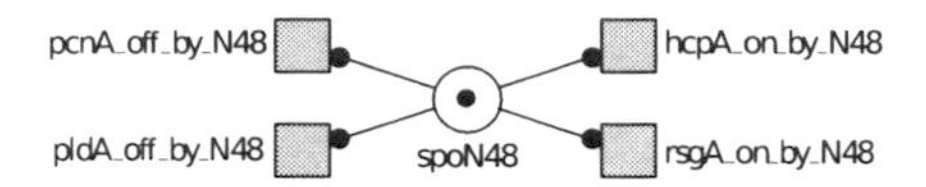

Fig. 9. Allelic influence module. The module represents the differential regulation of four genes, *pcnA*, *pldA* (down regulated), *hcpA*, and *rsgA* (up regulated) by the *spoN48* allele of the *spoN* gene in *Physarum polycephalum*. The logic transitions shown here are also part of the gene modules of the four differentially regulated genes (not shown). Genes and names of their orthologs in the *UniProt* database: *pcnA*, proliferating cell nuclear antigen; *pldA*, Phosphatidylinositol-glycan-specific phospholipase D; *hcpA*, Histone chaperone ASF1A; *rsgA*, Regulator of G-protein signalling 2.

We have genetically dissected the underlying regulatory network with the help of mutants that are altered in the photocontrol of sporulation, as isolated in phenotypic screens ([32]; Rätzel et al., unpublished results). Representatives of one class of these mutants have lost their ability to be committed to sporulation and remain forever in a proliferative state. Although these mutants do not respond to far-red light by sporulation, they clearly respond at the transcriptional level. However the pattern of genes that are differentially expressed in response to the stimulus significantly differs in the mutants as compared to the wild-type and also differs between mutants. The altered gene expression patterns clearly reveal the regulatory influence of the mutated genes and can be used to infer the network of regulatory control in which the different regulators inactivated in each of the mutants are interwoven. The changed pattern of differentially expressed genes can be used to reverse engineer the regulatory influence of the gene mutations. In Figure 9 the allelic influence module obtained for the *phoN48* allele of the *phoN* mutant gene is shown. The proteins encoded by some of the differentially expressed genes have well-known biochemical functions in the developmental control of eukaryotic cells. In a bottom-up approach, these genes can be linked to corresponding protein modules in terms of changed protein concentrations as predicted by the model.

7 Versatility of the Approach and Future Perspectives

We have described a strictly modular approach to Petri net modelling based on clearly defined types of modules corresponding to the different types of molecular entities around which each module is centred: genes, RNAs, and proteins. The small number of module types and the few easy-to-follow rules for creating a module are expected to encourage community efforts in creating a collection of modules in analogy to how Wikipedia collects pages. In Petri net modules, cross references to other modules allow the automatic composition of large models that are directly executable [4]. A web-accessible database was constructed to manage different versions of each module which is available as a prototype [4]. It allows that different explanations of molecular mechanisms directly translate into alternative computational models that predict experimental findings. Moreover, working with modules provides several options for the engineering of biomodels and their scalable application to systems and synthetic biology.

7.1 Regulatory Interactions Appear in Clear Graphical Structure

Because each module summarizes all functional interactions of a given molecular component including its influences on other components, even complex regulatory interactions can be always displayed in the form of an easy to perceive graphical layout. Certainly, not all of the functional interactions that appear in a module necessarily have to be part of a composed model. With the support of a database, modules can be selected according to user-defined criteria and then automatically linked to give a functional model. Those interactions that do not find a counterpart in one of the selected modules remain inactive because the respective places remain unmarked in the composed Petri net.

7.2 Modules May Be Added, Removed, or Exchanged: *in silico* Mutation of Networks

The modular structure allows to remove or exchange modules when automatically composing a model without touching or even considering modules that remain unchanged. This is a crucial advantage as compared to the manual re-engineering of monolithic models which requires careful consideration of how the components are wired up with each other since overlooking connections might accidentally introduce modelling errors. With the modular approach and the built-in version control of the database, different versions of a given module can be easily exchanged. This can be very helpful to analyse how alternative kinetic mechanisms of molecular interactions would influence the overall behaviour of the system.

For example for the phosphate regulatory network, one might wish to analyse whether or not different activation mechanisms of the PhoB protein would change the gene expression response and the performance of the feed-back loop. When working with really complex models, the modeller investigating a local mechanism has not to care about the inner life of all the numerous modules, in analogy to programming where the procedures of an approved library of subroutines do not have to be reconsidered each time they are used for building a new program.

At the moment where the database will contain a relatively high number of modules, automatically generated models might reveal nonobvious regulatory interactions of molecular components, bring them into a quantitative context and predict nonobvious and eventually counterintuitive consequences of network activation or perturbation. This will definitely be the case when regulatory interactions at a genome-wide scale change gene expression levels that translate into an updated marking of the places of protein modules due to the change in cellular concentration of perhaps many proteins. Even without modelling, this is already obvious by just looking at the gene expression data mentioned in the *Physarum* case study.

One might systematically probe components for their global role in the biomolecular system by simply removing modules from a model. This is the *in silico* complement to systematic mutant screens that are performed in genetic model organisms. *In silico* mutational studies may turn out to be of great benefit for synthetic biology in all cases where systematic mutant screens cannot be applied for what reasons soever.

7.3 Modules Can Be Automatically Generated at Large Scale

For genome scale models where the regulatory control of hundreds or thousands of genes or proteins is to be considered, automatic generation of models becomes an issue. By creating multiple copies of the module prototypes described here, modules can be generated fully automatically simply by assigning names to places and transitions. This is especially straightforward for the gene and mRNA modules but also for protein degradation modules. We expect that automatically generated large scale Petri nets will transmute into helpful tools for the reverse engineering of models from transcriptomic and proteomic data sets.

7.4 Allelic Influence Modules Integrate Forward and Reverse Approaches to Biomodel Engineering

Allelic influence modules were designed to represent regulatory influences of mutated genes (the alleles of a gene) onto the system. Unlike in the other module types, these influences may be rather indirect and may involve a number of potentially unknown components. As we have shown, defining these modules allows to reverse engineer networks from data collected on mutants. These reverse engineered networks are indeed fully compatible with the molecule-centred modules through transitions that control the active states of a gene as shown in the case study.

Acknowledgement. We thank Mostafa Herajy, Fei Liu, Christian Rohr and Martin Schwarick for their continuous support in developing Snoopy.

References

1. Baldauf, S.L., Doolittle, W.F.: Origin and evolution of the slime molds (Mycetozoa). Proc. Natl. Acad. Sci. USA 94, 12007–12012 (1997)
2. Baldauf, S.L., Roger, A.J., Wenk-Siefert, I., Doolittle, W.F.: A kingdom-level phylogeny of eukaryotes based on combined protein data. Science 290, 972–976 (2000)
3. Barrantes, I., Glöckner, G., Meyer, S., Marwan, W.: Transcriptomic changes arising during light-induced sporulation in Physarum polycephalum. BMC Genomics 11, 115 (2010)
4. Blätke, M.A., Dittrich, A., Rohr, C., Heiner, M., Schaper, F., Marwan, W.: JAK/STAT signalling - an executable model assembled from molecule-centred modules demonstrating a module-oriented database concept for systems- and synthetic biology (submitted 2012)

5. Blätke, M.A., Meyer, S., Marwan, W.: Pain signaling - A case study of the modular Petri net modeling moncept with prospect to a protein-oriented modeling platform. In: Proceedings of the 2nd International Workshop on Biological Processes & Petri Nets (BioPPN 2011), Newcastle upon Tyne, United Kingdom, pp. 1–19 (2011)
6. Blätke, M.A.: BioModelKit (2012), `http://www.biomodelkit.org`
7. Burland, T.G., Solnica-Krezel, L., Bailey, J., Cunningham, D.B., Dove, W.F.: Patterns of inheritance, development and the mitotic cycle in the protist *Physarum polycephalum*. Adv. Microb. Physiol. 35, 1–69 (1993)
8. Cerutti, H., Casas-Mollano, J.A.: On the origin and functions of RNA-mediated silencing: from protists to man. Current Genetics 50(2), 81–99 (2006)
9. Chia, N.Y., Chan, Y.S., Feng, B., Lu, X., Orlov, Y.L., Moreau, D., Kumar, P., Yang, L., Jiang, J., Lau, M.S., Huss, M., Soh, B.S., Kraus, P., Li, P., Lufkin, T., Lim, B., Clarke, N.D., Bard, F., Ng, H.H.: A genome-wide RNAi screen reveals determinants of human embryonic stem cell identity. Nature 468(7321), 316–320 (2010)
10. Cuomo, A., Bonaldi, T.: Systems biology "on-the-fly": SILAC-based quantitative proteomics and RNAi approach in *Drosophila melanogaster*. Methods in Molecular Biology 662, 59–78 (2010)
11. Durzinsky, M., Marwan, W., Ostrowski, M., Schaub, T., Wagler, A.: Automatic network reconstruction using ASP. Theory and Practice of Logic Programming 11, 749–766 (2011)
12. Durzinsky, M., Wagler, A., Marwan, W.: Reconstruction of extended Petri nets from time series data and its application to signal transduction and to gene regulatory networks. BMC Systems Biology 5(1), 113 (2011)
13. Fisher, J., Henzinger, T.A.: Executable cell biology. Nature Biotechnology 25(11), 1239–1249 (2007)
14. Gerber, A., Luschnig, S., Krasnow, M., Brown, P.: Genome-wide identification of mRNAs associated with the translational regulator PUMILIO in *Drosophila melanogaster*. Proceedings of the National Academy of Sciences 103(12), 4487–4492 (2006)
15. Glöckner, G., Golderer, G., Werner-Felmayer, G., Meyer, S., Marwan, W.: A first glimpse at the transcriptome of *Physarum polycephalum*. BMC Genomics 9, 6 (2008)
16. Hecker, M., Lambeck, S., Toepfer, S., Van Someren, E., Guthke, R.: Gene regulatory network inference: data integration in dynamic models-a review. Bio Systems 96(1), 86–103 (2009)
17. Heiner, M., Gilbert, D., Donaldson, R.: Petri Nets for Systems and Synthetic Biology. In: Bernardo, M., Degano, P., Zavattaro, G. (eds.) SFM 2008. LNCS, vol. 5016, pp. 215–264. Springer, Heidelberg (2008)
18. Heiner, M., Lehrack, S., Gilbert, D., Marwan, W.: Extended stochastic Petri nets for model-based design of wetlab experiments. Transactions on Computational Systems Biology XI, 138–163 (2009)
19. Heiner, M., Herajy, M., Liu, F., Rohr, C., Schwarick, M.: Snoopy – A Unifying Petri Net Tool. In: Haddad, S., Pomello, L. (eds.) PETRI NETS 2012. LNCS, vol. 7347, pp. 398–407. Springer, Heidelberg (2012)
20. Hoffmann, X.K., Tesmer, J., Souquet, M., Marwan, W.: Futile attempts to differentiate provide molecular evidence for individual differences within a population of cells during cellular reprogramming. FEMS Microbiology Letters 329(1), 78–86 (2012)
21. Hsieh, Y.J., Wanner, B.L.: Global regulation by the seven-component Pi signaling system. Current Opinion in Microbiology 13(2), 198–203 (2010)

22. Ideker, T., Krogan, N.J.: Differential network biology. Molecular Systems Biology 8, 565 (2012)
23. Kolch, W., Calder, M., Gilbert, D.: When kinases meet mathematics: the systems biology of MAPK signalling. FEBS Letters 579(8), 1891–1895 (2005)
24. Marwan, W., Rohr, C., Heiner, M.: Petri nets in Snoopy: a unifying framework for the graphical display, computational modelling, and simulation of bacterial regulatory networks. Methods in Molecular Biology 804, 409–437 (2012)
25. Moffat, J., Sabatini, D.M.: Building mammalian signalling pathways with RNAi screens. Nature Reviews Molecular Cell Biology 7(3), 177–187 (2006)
26. Neidhardt, F., Ingraham, J., Schaechter, M.: Physiology of the Bacterial Cell. A Molecular Approach. Sinauer Associates, Sunderland (1990)
27. Orton, R., Adriaens, M., Gormand, A., Sturm, O., Kolch, W., Gilbert, D.: Computational modelling of cancerous mutations in the EGFR/ERK signalling pathway. BMC Systems Biology 3(1), 100 (2009)
28. Pinney, J.W., Westhead, R.D., McConkey, G.A.: Petri net representations in systems biology. Biochem. Soc. Trans. 31, 1513–1515 (2003)
29. Rohr, C., Marwan, W., Heiner, M.: Snoopy–a unifying Petri net framework to investigate biomolecular networks. Bioinformatics 26(7), 974–975 (2010)
30. Schwanhäusser, B., Busse, D., Li, N., Dittmar, G., Schuchhardt, J., Wolf, J., Chen, W., Selbach, M.: Global quantification of mammalian gene expression control. Nature 473(7347), 337–342 (2011)
31. Sturm, O.E., Orton, R., Grindlay, J., Birtwistle, M., Vyshemirsky, V., Gilbert, D., Calder, M., Pitt, A., Kholodenko, B., Kolch, W.: The mammalian MAPK/ERK pathway exhibits properties of a negative feedback amplifier. Science Signaling 3(153), 90 (2010)
32. Sujatha, A., Balaji, S., Devi, R., Marwan, W.: Isolation of *Physarum polycephalum* plasmodial mutants altered in sporulation by chemical mutagenesis of flagellates. Eur. J. Protistol. 41, 19–27 (2005)
33. Uhlmann, S., Mannsperger, H., Zhang, J.D., Horvat, E.A., Schmidt, C., Küblbeck, M., Henjes, F., Ward, A., Tschulena, U., Zweig, K., Korf, U., Wiemann, S., Sahin, O.: Global microRNA level regulation of EGFR-driven cell-cycle protein network in breast cancer. Molecular Systems Biology 8, 570 (2012)

A Simple Model to Control Growth Rate of Synthetic E. coli during the Exponential Phase: Model Analysis and Parameter Estimation

Alfonso Carta, Madalena Chaves, and Jean-Luc Gouzé

INRIA Sophia Antipolis - Méditerranée,
2004 route des Lucioles - BP 93 06902 Sophia Antipolis Cedex, France
{alfonso.carta,madalena.chaves,jean-luc.gouze}@inria.fr
http://www.inria.fr/en/teams/biocore

Abstract. We develop and analyze a model of a minimal synthetic gene circuit, that describes part of the gene expression machinery in *Escherichia coli*, and enables the control of the growth rate of the cells during the exponential phase. This model is a piecewise non-linear system with two variables (the concentrations of two gene products) and an input (an inducer). We study the qualitative dynamics of the model and the bifurcation diagram with respect to the input. Moreover, an analytic expression of the growth rate during the exponential phase as function of the input is derived. A relevant problem is that of identifiability of the parameters of this expression supposing noisy measurements of exponential growth rate. We present such an identifiability study that we validate *in silico* with synthetic measurements.

1 Introduction

Synthetic biology has nearly emerged as a new engineering discipline. The goal of synthetic biology [1,2,3] is to develop and apply engineering tools to control cellular behavior—constructing novel biological circuits in the cell—to perform new and desired functions.

Most recent synthetic designs have focused on the cell transcription machinery, which includes the genes to be expressed, their promoters, RNA polymerase and transcription factors, all serving as potential engineering components. Indeed, synthetic bio-molecular circuits are typically fabricated in *Escherichia coli* (*E. coli*), by cutting and pasting together coding regions and promoters (natural and synthetic) according to designed structures and specific purposes ([4,5,6]).

Along these lines, synthetic biology ultimately aims at developing synthetic bio-molecular circuitry that may help in producing bio-pharmaceuticals, bio-films, bio-fuels, novel cancer treatments and novel bio-materials (see [2] for a review on synthetic biology applications).

In the present work we focus on the gene expression machinery of the bacterium *Escherichia coli*, with the aim of controlling the growth rate of the cells. *E. coli* is a model organism that is easy to manipulate and much knowledge is available about its regulatory networks.

D. Gilbert and M. Heiner (Eds.): CMSB 2012, LNCS 7605, pp. 107–126, 2012.

In the presence of a carbon source—such as glucose—*E. coli* grows in an exponential manner until it exhausts the nutrient sources, and then enters a stationary phase with practically zero growth [7]. The wild-type (namely the genetically unmodified) bacteria grow at different rates in the presence of carbon sources of different types [8]. Notably, glucose is the preferred substrate because it leads to a higher growth rate in wild type. Our control objective is to force the bacterium to significantly modify its response to glucose so as to tune the cells' growth rates. To this end, we take into account the recent applications of synthetic biology which allow us to fabricate engineered promoters which in turn can be externally controlled by inducers [9].

Notably, we will study an open loop configuration of a bi-dimensional model of a mutant *E. coli* inspired by the experiments in [10]. The two basic variables of our model, which describe the gene expression machinery that is responsible for bacterial growth are (see Fig.1):

1. the concentration of a *Component of the Gene Expression Machinery* (CGEM), proteins responsible for global growth (ribosomes and RNA polymerase). Without this CGEM, the bacteria cannot produce any proteins and thus cannot grow.
2. the concentration of CRP, a protein involved in the formation of the complex cAMP-CRP whose level positively correlates with less preferred carbon sources and slower growth [11].

We will assume that an engineered inducible-promoter is used to express the CGEM. Moreover it is assumed that the mutant CGEM activates its own expression. The number and location of equilibria can thus be controlled by means of an input control function of the inducer and, in particular, there can be regions of bi-stability, as observed in [10].

The type of growth rate control we present—which directly acts upon the GEM—could be useful in creating bacterial cells that divert resources used for growth towards the production of a target compound. Thus, the analysis of the simple model presented here is an attempt to help guide the construction of synthetic gene networks, which improves product yield and productivity.

This paper is structured as follows: in Section 2 we describe the open-loop model, providing some biological motivations for the terms forming the differential equations. Next, in Section 3 we qualitatively analyze the open-loop model by means of phase-plane and bifurcation diagram, showing how the steady states of the CGEM can be controlled by the external input (inducer). In Section 4 we derive a mathematical expression of the growth rate during the exponential phase as a function of the amount of the inducer. Finally, in Section 5 we present an *in silico* practical identifiability analysis of such expression.

2 The Open-Loop Model

The principal modeling challenges come from incomplete knowledge of the networks, and the dearth of quantitative data for identifying kinetic parameters

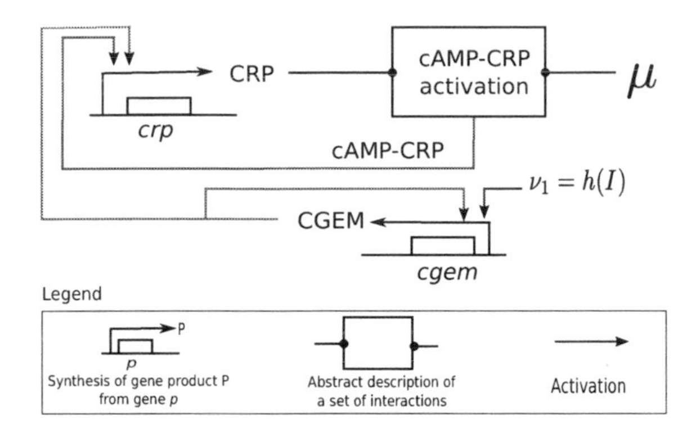

Fig. 1. Regulatory network of the open-loop model in the *mutant E. coli*. The model consists of genes *crp* and synthetic-*cgem* (modified promoter of a component of the gene expression machinery (CGEM) in *E. coli*). The synthetic-*cgem* promoter is positively regulated by the inducer I—according to the input function $\nu_1 = h(I)$—and CGEM. CGEM, being responsible for the bacterial gene expression, positively regulates *crp* gene too. Moreover, *crp* transcription is induced by cAMP-CRP, a metabolite whose formation relies on CRP protein abundance and low level of bacterial growth rate μ.

required for detailed mathematical models. Qualitative methods overcome both of these difficulties and are thus well-suited to the modeling and simulation of genetic networks ([12,13]).

In this work we used a novel *piecewise non-linear* formalism—derived from piece wise affine (PWA) systems (see [14,15,16,17,18] for more details)—to model gene expression affected by dilution due to growth rate.

The open-loop model depicted in Fig. 1—similarly to PWA models of regulatory genetic networks—is built with discontinuous (step) functions. The use of step functions has been motivated by the experimental observation that the activity of certain genes changes in a switch-like manner at a threshold concentration of a regulatory protein [19]. The non linearity is concentrated in the removal term of differential equations, which takes into account the protein degradation and the dilution due to growth.

The open-loop model, expressed by (1), describes the qualitative dynamics of a CGEM responsible for bacterial growth and another protein that reflects growth, such as CRP. The CGEM is assumed to be externally controlled by an inducer I (such as IPTG (Isopropyl β-D-1-thiogalactopyranoside), Tc (tetracycline) etc). This model of ODE exhibits bi-stability in CGEM expression for some parameter sets, as experimentally verified in [10]. We shall take into account this bi-stability to control the model's state to the "low" or to the "high" CGEM stable steady state. Let x_c, $x_p \in \mathbb{R}_{\geq 0}$ be the CRP and CGEM concentrations respectively. Thus, the open-loop model graphically depicted in Fig. 1, can be mathematically translated into:

$$
\begin{cases}
\dot{x}_c\,(t) = k_c^0\ s^+(x_p, \theta_p^1) + k_c^1\ s^+(x_p, \theta_p^2)\ s^+(x_c, \theta_c^1)\ s^-(x_p, \theta_{\bar{\mu}}) \\
\qquad\qquad - (\bar{\mu}\ x_p(t) + \gamma_c)\ x_c(t) \\
\dot{x}_p\,(t) = \nu_1\ k_p^0\ s^+(x_p, \theta_p^1) + \nu_1\ k_p^1\ s^+(x_p, \theta_p^2) \\
\qquad\qquad - (\bar{\mu}\ x_p(t) + \gamma_p)\ x_p(t)
\end{cases}
\tag{1}
$$

where:

- $k_i^0 > 0$ $(i = c, p)$ is the basal synthesis rate constant;
- $k_i^1 > 0$ $(i = c, p)$ is the main synthesis rate constant;
- ν_1 is a positive input accounting for the inducer I; it will be a function $\nu_1(v)$, v being the concentration of I;
- $\gamma_i > 0$ $(i = c, p)$ is the degradation rate constant;
- $\theta_i^j > 0$ $(i = c, p;\ j = 1, 2)$ is the x_i threshold concentration for activation/inhibition;
- $\theta_{\bar{\mu}} > 0$ is a growth threshold depending on which substrate is used;
- $\bar{\mu} > 0$ is a growth constant depending on which substrate is used.

and s^+, s^- denote the step-like functions, defined as

$$
s^+(x_i, \theta_i^j) = \begin{cases} 1 & \text{if } x_i > \theta_i^j \\ 0 & \text{if } x_i < \theta_i^j \end{cases} ; \quad s^-(x_i, \theta_i^j) = 1 - s^+(x_i, \theta_i^j)\ ,
$$

which are used to model the switch-like promoters' regulation carried out by the generic protein x_i. These s^+, s^- are not defined at the threshold values so, to define solutions on the surfaces of discontinuity, i.e. $x_i = \theta_i^j$, we use the approach of Filippov [20], which extends the vector field to a differential inclusion.

In what follows, we will explain the main assumptions adopted in building the system equations (1), which were inspired by the models in [10,12] and the literature on *E. coli.*

2.1 Growth Rate

In bacteria, growth rate is intimately interwined with gene expression ([21,22]) and with the type of substrate [8]. Hence, to keep model complexity to a minimum, we assume growth rate μ to be proportional—with a constant $\bar{\mu}$ depending on the quality of medium—to the concentration of the CGEM which is responsible for bacterial growth:

$$
\mu(t) = \bar{\mu}\ x_p(t)\ . \tag{2}
$$

2.2 cAMP-CRP Activation

The cAMP-CRP complex is formed from cAMP, a small metabolite, which binds the protein CRP. The cAMP concentration is higher at low growth rate and rapidly decreases at high growth rate [11]. Thus, cAMP abundance in cells can be well captured by a negative step function of μ, i.e. $s^-(\mu, \theta_\mu)$. Moreover, being cAMP association with or dissociation from CRP much faster than the synthesis and degradation of proteins [12], we have assumed that as soon as CRP reaches a certain threshold, i.e. θ_c, CRP instantly binds to cAMP in a switch-like fashion. For these reasons, the positive regulation carried out by cAMP-CRP reads as:

$$b^+_{cAMP-CRP} = s^+(x_c, \theta_c)\ s^-(\mu, \theta_\mu).$$

Focusing on the negative step function $s^-(\mu, \theta_\mu)$ and taking into account the expression of μ in (2), we can rewrite $b^+_{cAMP-CRP}$ as:

$$b^+_{cAMP-CRP}(x_c, x_p) = s^+(x_c, \theta_c)\ s^-(x_p, \theta_{\bar{\mu}}) \quad (3)$$

where $\theta_{\bar{\mu}}$ is a threshold concentration of CGEM which depends on the type of carbon source.

2.3 CRP Synthesis

We have assumed that a lower value of x_p, i.e. θ_p^1, induces the basal synthesis $(k_c^0\ s^+(x_p, \theta_p^1))$ of x_c while a higher value of x_p, i.e. θ_p^2, is needed to stimulate its main expression $(k_c^1 s^+(x_p, \theta_p^2))$. Moreover, the *crp* gene is regulated both positively and negatively by cAMP-CRP. However, in order to simplify, we omit the negative control of *crp*, because this mechanism only plays a role when the CRP concentration is low [12][1]. Thus, only one concentration threshold of CRP, i.e. θ_c^1, is required in the model, to allow production of the cAMP-CRP complex. In conclusion, taking into account the regulation function of cAMP-CRP in (3), the CRP synthesis reads:

$$f_c(x) = k_c^0\ s^+(x_p, \theta_p^1) + k_c^1\ s^+(x_p, \theta_p^2)\ b^+_{cAMP-CRP}(x_c, x_p), \quad (4)$$

with

$$0 < \theta_c^1 < max_c, \quad (5)$$

where max_c is the maximum concentration value for CRP.

2.4 CGEM Synthesis

In this bi-dimensional model, since the CGEM is the main factor which determines growth of the cell, it is also responsible for its own synthesis. We have thus assumed that a low concentration (θ_p^1) is sufficient to stimulate its basal production $k_p^0\ s^+(x_p, \theta_p^1)$ while its main production $k_p^1\ s^+(x_p, \theta_p^2)$ is stimulated only above the θ_p^2 threshold. Thus, we can order the thresholds for x_p as:

$$0 < \theta_p^1 < \theta_p^2 < max_p, \quad (6)$$

[1] We found that a model involving the negative control of *crp* by cAMP-CRP does not have any effect on the conclusion of this study.

where max_p is the maximum concentration value.

Moreover, the inducer effect is modeled by input ν_1. For a general formulation of the activation of x_p by an inducer I, we will later on assume that ν_1 is a positive increasing function of I. Consequently, x_p synthesis reads:

$$f_p(x) = \nu_1\ k_p^0\ s^+(x_p, \theta_p^1) + \nu_1\ k_p^1\ s^+(x_p, \theta_p^2). \quad (7)$$

2.5 Proteins Removal

The negative terms in $\dot{x}_c$ and $\dot{x}_p$ of (1) take into account the fact that cells remove proteins by two processes: degradation and dilution due to cell growth [23]. Notably, these terms can generally be expressed as $(\mu(t) + \gamma_i)x_i$ (for $i = c, p$) where $\mu(t) = \bar{\mu}\ x_p(t)$, which is the bacterial growth rate in (2), is responsible for the proteins' dilution while γ_i stands for protein's degradation.

3 Qualitative Analysis of the Open-loop Model

In this section we will qualitatively study, by means of phase-planes and bifurcation diagrams, model (1) in the case that cells are grown in glucose. This will elucidate how qualitative dynamics—in terms of equilibria' location and their stability—is intertwined with biological phenomena. Moreover, we shall show how—through the external input ν_1—the stability of equilibria in (1) can be controlled, pointing out a reciprocal influence between growth rate and gene expression.

3.1 Open-Loop Model in Glucose Growth

If cells are grown in glucose, then parameters depending on the substrate become $\theta_{\bar{\mu}} = \theta_p^G$ and $\bar{\mu} = \mu_G$ in model (1). Moreover, in the presence of glucose or other PTS sugars, adenylate cyclase[2] activity decreases, leading to a drop in the cellular level of cAMP [24,25]. Thus, we have modeled this effect assuming:

$$0 < \theta_p^1 < \theta_p^G < \theta_p^2 < max_p. \quad (8)$$

Therefore, during growth on glucose, the state space of model (1) can be partitioned into eight *regular domains*, where the vector field is uniquely defined:

$$\begin{aligned}
D_1^G &= \{x \in \mathbb{R}^2_{\geq 0} : 0 \leq x_c < \theta_c^1,\ 0 \leq x_p < \theta_p^1\} \\
D_2^G &= \{x \in \mathbb{R}^2_{\geq 0} : \theta_c^1 < x_c \leq max_c,\ 0 \leq x_p < \theta_p^1\} \\
D_3^G &= \{x \in \mathbb{R}^2_{\geq 0} : 0 \leq x_c < \theta_c^1,\ \theta_p^1 < x_p < \theta_p^G\} \\
D_4^G &= \{x \in \mathbb{R}^2_{\geq 0} : \theta_c^1 < x_c \leq max_c,\ \theta_p^1 < x_p < \theta_p^G\} \\
D_5^G &= \{x \in \mathbb{R}^2_{\geq 0} : 0 \leq x_c < \theta_c^1,\ \theta_p^G < x_p < \theta_p^2\} \\
D_6^G &= \{x \in \mathbb{R}^2_{\geq 0} : \theta_c^1 < x_c \leq max_c,\ \theta_p^G < x_p < \theta_p^2\} \\
D_7^G &= \{x \in \mathbb{R}^2_{\geq 0} : 0 \leq x_c < \theta_c^1,\ \theta_p^2 < x_p \leq max_p\} \\
D_8^G &= \{x \in \mathbb{R}^2_{\geq 0} : \theta_c^1 < x_c \leq max_c,\ \theta_p^2 < x_p \leq max_p\}.
\end{aligned}$$

[2] Enzyme that catalyzes the conversion of ATP to cAMP and pyrophosphate.

In addition, there are also *switching domains*, where the model is defined only as a differential inclusion [20], corresponding to the segments where each of the variables is at a threshold ($x_i = \theta_i$ and $x_j \in [0, max_j]$).

In general, for any regular domain D, the synthesis rates (4) and (7) are constant for all $x \in D$, and it follows that model (1) can be written as

$$\begin{cases} \dot{x}_c\,(t) = f_c^D - (\bar{\mu}\; x_p(t) + \gamma_c)\; x_c(t) \\ \dot{x}_p\,(t) = f_p^D - (\bar{\mu}\; x_p(t) + \gamma_p)\; x_p(t) \end{cases} \tag{9}$$

with $f_c^D, f_p^D, \bar{\mu}, \gamma_c, \gamma_p$ positive real constants. For any initial condition $x(t_0) \in D$ the unique solution of (9) can be found explicitly by solving first the x_p-equation of (9), which is an autonomous differential equation, and then solving the x_c-equation, having substituted $x_p(t)$ into it. Thus, it can be shown that $x_c(t)$ is given by:

$$x_c(t) = \frac{1}{b(t)} \left(b(t_0) x_c(t_0) + f_c^D \int_{t_0}^{t} b(s) ds \right)$$

where $b(t) = exp\left(\int_{t_0}^{t} (\bar{\mu}\; x_p(\tau) + \gamma_p) d\tau\right)$. Moreover, defining $\Phi(D) = (\bar{x}_c, \bar{x}_p)^T$ with

$$\begin{aligned} \bar{x}_c &= \frac{f_c^D}{\bar{\mu}\bar{x}_p + \gamma_c}, \\ \bar{x}_p &= \frac{-\gamma_p + \sqrt{\gamma_p^2 + 4\bar{\mu} f_p^D}}{2\bar{\mu}}, \end{aligned} \tag{10}$$

(it is easy to check that $\bar{x}_p$—in (10)—is the only positive solution of $\dot{x}_p = 0$) it turns out that either $x(t) \to \Phi(D)$ as $t \to \infty$ or $x(t)$ reaches the boundary of D.

Definition 1. *Given a regular domain D, the point $\Phi(D) = (\bar{x}_c, \bar{x}_p)^T$ (defined by* (10)*) is called the* focal point *for the flow in D.*

We will group into regions R_j those domains D_i^G where model (1)—in glucose growth— has the same dynamics and thus the same focal points. Considering Definition 1, we have the following focal points:

- $\forall x \in R_1 = \{x \in \mathbb{R}^2_{\geq 0} : x \in D_1^G \cup D_2^G\}$

$$x_c \to 0 \quad \wedge \quad x_p \to 0$$

  Thus, $\Phi_0^G = (0, 0)$ is the focal point of region R_1.
- $\forall x \in R_2 = \{x \in \mathbb{R}^2_{\geq 0} : x \in D_3^G \cup D_4^G \cup D_5^G \cup D_6^G\}$

$$x_c \to \frac{k_c^0}{\mu_G\; \bar{x}_{p,G}^1 + \gamma_c} = \bar{x}_{c,G}^2$$

$$x_p \to \frac{-\gamma_p + \sqrt{\gamma_p^2 + 4\; \nu_1\; k_p^0\; \mu_G}}{2\mu_G} = \bar{x}_{p,G}^1$$

  Thus, $\Phi_1^G = (\bar{x}_{c,G}^2, \bar{x}_{p,G}^1)$ is the focal point of region R_2^G.

– $\forall x \in R_3 = \{x \in \mathbb{R}^2_{\geq 0} : x \in D_7^G \cup D_8^G\}$

$$x_c \to \frac{k_c^0}{\mu_G\ \bar{x}^2_{p,G} + \gamma_c} = \bar{x}^1_{c,G}$$

$$x_p \to \frac{-\gamma_p + \sqrt{\gamma_p^2 + 4\ \nu_1(k_p^0 + \ k_p^1)\mu_G}}{2\mu_G} = \bar{x}^2_{p,G}$$

Thus, $\Phi_2^G = (\bar{x}^1_{c,G}, \bar{x}^2_{p,G})$ is the focal point of region R_3.

The focal points Φ_i^G $(i = 0, 1, 2)$ are equilibrium points of model (1) provided that they belong to their respective regular domain, i.e. $\Phi(D) \in D$. The local stability of equilibrium points is given by the following theorem.

Theorem 1. *Let D be a regular domain and $\Phi(D)$ be the focal point of D. If $\Phi(D) \in D$, then $\Phi(D)$ is a locally stable point of model* (1).

Proof. Model (1) restricted to D is given by (9). In order to assess the stability of $\Phi(D)$, we compute the Jacobian matrix of (9) calculated in $\Phi(D) = (\bar{x}_c, \bar{x}_p)^T$:

$$J(\bar{x}_c, \bar{x}_p) = \begin{pmatrix} -\bar{\mu}\bar{x}_c & -(\bar{\mu}\bar{x}_p + \gamma_p) \\ 0 & -(2\bar{\mu}\bar{x}_p + \gamma_p) \end{pmatrix}.$$

Since all the eigenvalues of J, which are the diagonal entries as J is triangular, are negative, $\Phi(D)$ turns out to be a locally stable point.

Hence, there can be at most three locally stable steady states during growth on glucose.

Fig. 2 depicts the phase-plane of model (1). It can be seen that Φ_0^G, Φ_1^G, Φ_2^G, (for the parameter values used) are locally stable steady states since they are within their respective regular domains (Theorem 1). Notably, it is easy to verify that Φ_0^G is locally stable for any set of parameters. It represents absence of growth and can happen when the initial condition $x_p(t_0)$, is too low—specifically $x_p(t_0) < \theta_p^1$—to initiate gene transcription or when the control input ν_1 does not sufficiently induce CGEM expression, that is when $\bar{x}^1_{p,G} < \theta_p^1$. We refer to Φ_0^G as the *trivial* fixed point. Φ_1^G represents CGEM basal level—leading to a low growth rate (see (2))— while CRP is at a high level, which is in agreement with high *crp* gene expression (by cAMP-CRP) at lower growth rate. Thus, because of the low growth rate achieved, we refer to Φ_1^G as the *low* fixed point. Conversely, at Φ_2^G, CRP is at low level while CGEM , as well as μ, have reached their highest stable values. Thus, Φ_2^G is named the *high* fixed point.

Since $\bar{x}^1_{p,G}(\nu_1)$ and $\bar{x}^2_{p,G}(\nu_1)$ are function of ν_1, it turns out that the location of focal points Φ_1^G and Φ_2^G, and thus the number of equilibria of model (1), depend on the control input ν_1. Hence, choosing appropriate values of ν_1 it is possible to control model (1) towards Φ_1^G or Φ_2^G. To illustrate this, we have depicted in Fig. 3 the x_p-bifurcation diagram when parameter ν_1 varies from 0 to 1 while the other parameter values are the same as those used in Fig. 2.

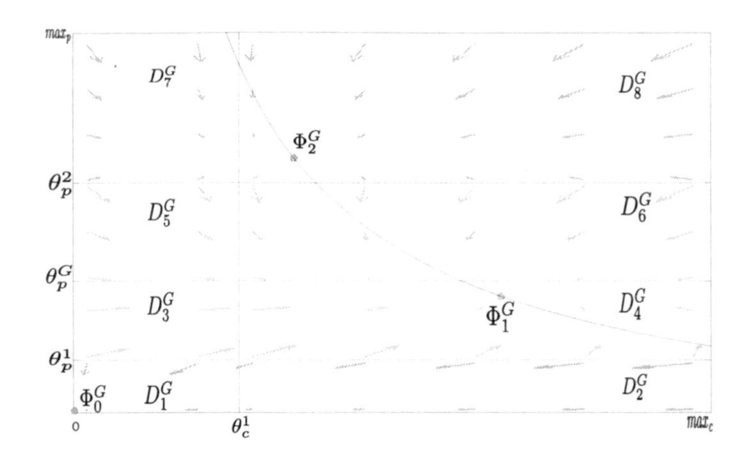

Fig. 2. Phase plane of model (1) during growth in glucose. Parameter values used: $\theta_c^1 = 0.6$, $\theta_p^1 = 0.8$, $\theta_p^G = 2$, $\theta_p^2 = 3.5$, $k_c^0 = 7$, $k_c^1 = 10$, $k_p^0 = 40$, $k_p^1 = 50$, $\gamma_c = 1$, $\gamma_p = 1$, $\mu_G = 2$ e $\nu_1 = .5$. The black curve is the x_c-nullcline: $x_p = \dfrac{k_c^0}{x_c\ \mu_G} - \dfrac{\gamma_c}{\mu_G}$. Stable fixed points: Φ_0^G, Φ_1^G, Φ_2^G.

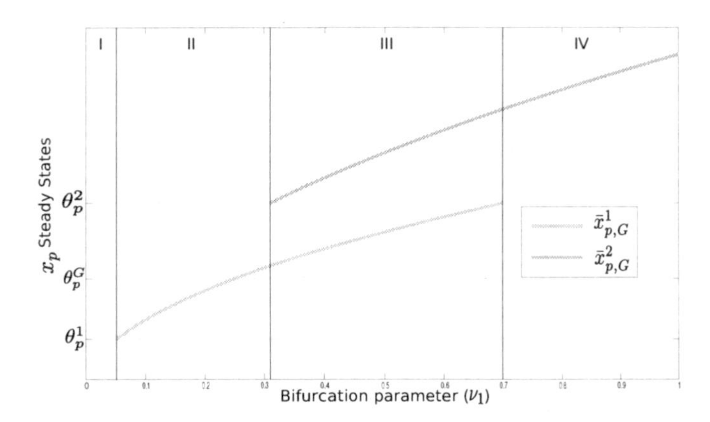

Fig. 3. Bifurcation diagram for model (1) during growth in glucose, showing the non trivial locally stable steady states of x_p as a function of the control input ν_1. Other parameter values used are the same as those in Fig. 2. See Proposition 1 for more details.

We notice that Fig. 3 is divided into four parts in which x_p stability changes significantly. In part I, for those values of ν_1 such that $\bar{x}_{p,G}^1 < \theta_p^1$ and $\bar{x}_{p,G}^2 < \theta_p^2$, neither Φ_1^G nor Φ_2^G are stable steady states. In this case, model (1) during growth on glucose converges towards the only stable point Φ_0^G (not depicted in Fig. 3). So, in I, the control input is too small to allow CGEM to reach a basal level, and prevents bacterial growth.

In part II, when $\bar{x}_{p,G}^1(\nu_1) > \theta_p^1$ and $\bar{x}_{p,G}^2(\nu_1) < \theta_p^2$ hold, only Φ_1^G is a stable steady state (besides the trivial one) according to Theorem 1. Hence, it turns out that choosing an initial condition of CGEM $x_p(t_0) > \theta_p^1$ and ν_1 such that $\bar{x}_{p,G}^1(\nu_1) > \theta_p^1$ and $\bar{x}_{p,G}^2(\nu_1) < \theta_p^2$, we can control model (1) to the stable point Φ_1^G.

In part III, characterized by $\theta_p^1 < \bar{x}_{p,G}^1(\nu_1) < \theta_p^2$ and $\bar{x}_{p,G}^2(\nu_1) > \theta_p^2$, both Φ_1^G and Φ_2^G are stable steady states: this is a region of bi-stability. Moreover, the phase plane corresponding to this configuration is depicted in Fig. 2, where we can also observe the presence of two separatrices $x_p = \theta_p^1$ and $x_p = \theta_p^2$. Is is clear that, depending on $x_p(t_0)$, the model can converge to Φ_1^G (if $\theta_p^1 < x_p(t_0) < \theta_p^2$) or to Φ_2^G (if $x_p(t_0) > \theta_p^2$).

In part IV, when $\bar{x}_{p,G}^1(\nu_1) > \theta_p^2$ holds, only Φ_2^G is a stable steady state and thus, whenever $x_p(t_0) > \theta_p^1$, model (1) converges to Φ_2^G.

The open-loop control in glucose growth can be summarized as follows.

Proposition 1. *Consider model* (1) *with control input* ν_1 *and initial condition* $x_p(t_0)$ *such that:*

- *if* $(\bar{x}_{p,G}^1(\nu_1) < \theta_p^1 \wedge \bar{x}_{p,G}^2(\nu_1) < \theta_p^2) \vee x_p(t_0) < \theta_p^1$, *then model* (1) *converges to the* trivial focal point Φ_0^G *(region I in Fig. 3);*
- *if* $\bar{x}_{p,G}^1(\nu_1) > \theta_p^1 \wedge \bar{x}_{p,G}^2(\nu_1) < \theta_p^2 \wedge x_p(t_0) > \theta_p^1$, *then model* (1) *converges to the* low focal point Φ_1^G *(region II in Fig. 3);*
- *if* $\theta_p^1 < \bar{x}_{p,G}^1(\nu_1) < \theta_p^2 \wedge \bar{x}_{p,G}^2(\nu_1) > \theta_p^2 \wedge x_p(t_0) > \theta_p^1$, *then model* (1) *is* bistable *(region III in Fig. 3) and notably:*
  - *if* $\theta_p^1 < x_p(t_0) < \theta_p^2$, *then model* (1) *converges to the* low focal point Φ_1^G*;*
  - *if* $x_p(t_0) > \theta_p^2$, *then model* (1) *converges to the* high focal point Φ_2^G
- *if* $\bar{x}_{p,G}^1(\nu_1) > \theta_p^2 \wedge x_p(t_0) > \theta_p^1$, *then model* (1) *converges to the* high focal point Φ_2^G *(region IV in Fig. 3).*

4 Growth Rate Expression for Exponential Phase

Here, to account for varying dosage of inducer, we make an assumption to analytically characterize the function $\nu_1 = h(v)$. Notably, to describe the regulation of CGEM gene expression by the inducer, we employ a function typically used in synthetic biology [9]:

$$\nu_1(v) = \alpha + (1 - \alpha)\frac{v^n}{K_v^n + v^n} \tag{11}$$

where v denotes inducer concentration and α accounts for the basal transcriptional activity. Controlled gene expression follows Hill-type dosage-response curve with promoter-activator affinity K_v and cooperative (Hill) coefficient n. During *exponential phase*—the period characterized by cell doubling— the bacterial culture shows a constant growth rate [7]. This means that, according to (2), a stable fixed point of the CGEM has to be reached. Hence, our expression of growth rate during exponential phase reads:

$$\mu = \mu_G \bar{x}_p \tag{12}$$

where $\bar{x}_p$ is the CGEM concentration at steady state, which can be either $\bar{x}^1_{p,G}$ or $\bar{x}^2_{p,G}$—depending on the amount of inducer which determines the level of CGEM expression. Thus, our expression of growth rate during exponential phase can assume the two values below:

$$\mu(v) = \begin{cases} \mu_G \bar{x}^1_{p,G} = \dfrac{-\gamma_p + \sqrt{\gamma_p^2 + 4\,\nu_1\,k_p^0\,\mu_G}}{2} \\ \mu_G \bar{x}^2_{p,G} = \dfrac{-\gamma_p + \sqrt{\gamma_p^2 + 4\,\nu_1\,(k_p^0 + k_p^1)\,\mu_G}}{2} \end{cases} . \tag{13}$$

Specifically, we assumed there is a particular value of inducer, i.e. v^*, such that for an appropriate choice of initial condition and for all $v \leq v^*$ the CGEM steady state is $\bar{x}^1_{p,G}$ while for all $v > v^*$ the steady state is $\bar{x}^2_{p,G}$. Thus, considering that, and substituting (11) into (13) we obtain the theoretical expression for growth rate during exponential phase:

$$\mu(v) = \begin{cases} -\dfrac{\gamma_p}{2}\left[1 - \sqrt{1 + \dfrac{4k_p^0\mu_G\alpha}{\gamma_p^2} + \dfrac{4k_p^0\mu_G(1-\alpha)}{\gamma_p^2}\dfrac{v^n}{K_v^n + v^n}}\right] & \text{if, } v \leq v^* \\ -\dfrac{\gamma_p}{2}\left[1 - \sqrt{1 + \dfrac{4(k_p^0 + k_p^1)\mu_G\alpha}{\gamma_p^2} + \dfrac{4(k_p^0 + k_p^1)\mu_G(1-\alpha)}{\gamma_p^2}\dfrac{v^n}{K_v^n + v^n}}\right] & \text{if, } v > v^* \end{cases} \tag{14}$$

It is worthy to notice that expression (14) directly relates the growth rate μ during exponential phase to the amount of the inducer v. Hence, using (14) we can fine tune—by means of appropriate level of the inducer—the growth rate of the cells during the exponential phase.

5 *In silico* Identifiability Analysis of Growth Rate

Our collaborators (Jérôme Izard and Hans Geiselmann[3]) are currently performing an ongoing experiment on a synthetic *E. coli* – implementing the open-loop model depicted Fig. 1 – which relates the level of growth rate during the exponential phase to the amount of the inducer. In the future, these dose-response curves will be useful to calibrate and validate the growth rate expression (during exponential phase) (14).

Here, we used simulated data to fit the growth rate model (14) and to study the identifiability of the parameters.

[3] *Laboratoire Adaptation et Pathogénie des Microorganismes*, (CNRS UMR 5163), Université Joseph Fourier, Bâtiment Jean Roget, Faculté de Médecine-Pharmacie, La Tronche, France.

5.1 Problem Statement

Given a parametric non-linear model, such as (14), the relationship between a response variable (output) and one or more predictor variables (input) can be represented by the expression:

$$y = \eta(v, p) + \epsilon \ ,$$

where

- y is an $n \times 1$ vector of observations of the response variable,
- v is an $n \times m$ matrix of predictors,
- p is a $q \times 1$ vector of unknown parameters to be estimated,
- η is any function of v and p,
- ϵ is an $n \times 1$ vector of independent, identically distributed random disturbances.

The nonlinear regression problem consists of finding a vector $\hat{p}$ minimizing a scalar cost function $J(p)$, which is generally a measurement of the agreement of experimental data with the outputs predicted by the model. The cost function that we have considered in this work is a weighted least squares criterion:

$$J(p) = \sum_{i=1}^{n} \frac{(y_i - \eta(v_i, p))^2}{y_i^2} \tag{15}$$

where y_i denotes the i-th data-point of the observable y, measured at input-points v_i, and $\eta(v_i, p)$ the i-th observable as predicted by the parameters p. The parameters can be estimated numerically by:

$$\hat{p} = \arg\min \left[J(p) \right] \ . \tag{16}$$

Determining the parameter vector $\hat{p}$ which minimizes $J(p)$ is only a part of the parameter estimation problem. In fact, when preparing to fit a mathematical model or expression to a set of experimental data, the prior assessment of parameter identifiability is a crucial aspect [26]. However, the structural identifiability analysis for non-linear models in systems biology is still a challenging question [27]. Whether or not parameters can be estimated uniquely depends on the model structure, the parameterization of the model and the experiment used to get the data [28].

Regarding this problem, we briefly recall two important definitions on identifiability [29]:

- the parameter p_i, $i = 1, ..., q$ is **structurally globally identifiable** if assuming ideal conditions (error-free model structure and unlimited noise-free observations (v, y)) and if for almost any $p^* \in \mathcal{P}$ (admissible parametric space $\mathcal{P}$),
$$y(p, v) = y(p^*, v), \forall v \Rightarrow p_i = p_i^* .$$

– the parameter p_i, $i = 1, ..., q$ is **structurally locally identifiable** if assuming ideal conditions (error-free model structure and unlimited noise-free observations (v, y)) and if for almost any $p^* \in \mathcal{P}$ (admissible parametric space $\mathcal{P}$), there exists a neighborhood $V(p^*)$ such that

$$p \in V(p^*) \wedge y(p, v) = y(p^*, v), \forall v \Rightarrow p_i = p_i^*.$$

An important complement to the structural identifiability definitions is the notion of **practical identifiability**. Practical identifiability is in fact related to the quality of experimental data and their information content [30]. The question raised by this notion is the following: in the presence of observation errors and/or few data are reliable estimations of the parameters possible? Thus, once having determined the value of $\hat{p}$ minimizing the cost function $J(p)$, it is very important to find a realistic measure of how $\hat{p}$ is precise. To this end, the confidence intervals[4] of the estimated parameters have to be calculated.

It must be noted that, unlike for the linear case for which an exact theory exists, there is no exact theory for the evaluation of confidence intervals for systems which are nonlinear in the parameters. An approximate method based on a local linearization of the output function $\eta(v, p)$ is generally used [31,32], thus the confidence region is evaluated as a function of the parameter covariance matrix. The applicability of such approximate method requires that the response function $\eta(v, p)$ must be continuous in its arguments (v, p), the first partial derivatives $\frac{\partial}{\partial p_i}\eta(v, p)$ must be continuous in its arguments (v, p), and the second partial derivatives $\frac{\partial^2}{\partial p_i \partial p_j}\eta(v, p)$ must be continuous in its arguments (v, p), but our model (14) does not satisfy these conditions because of the discontinuity in $v = v^*$. Hence, in the remainder of the paper a computational method, based on *in silico* generated data, is suggested to argue the practical identifiability of non-linear discontinuous model such as (14).

5.2 Generation of Simulated Data Sets

In order to assess the quality of parameter estimation and thus the practical identifiability of parameters in (14), artificial data were generated by simulation of (14) from a set of pre-defined parameters (to be considered as true values). The true parameter values (Tab. 5.2) were chosen from physiological parameters of *E.coli* cells [21,33] and were based on similar studies of this type [10].

Thus, the artificial growth rate values have been simulated considering a measurement error proportional to the nominal value of growth rate:

$$y = \mu(v) + \sigma\mu(v)\mathcal{N}(0, 1) \tag{17}$$

where $\mathcal{N}(0, 1)$ is a normally distributed random variable with zero mean and unit variance and $\sigma\mu(v)$ is the standard deviation of the observation errors. Four different types of data sets were considered to account for practical identifiability:

[4] A confidence interval $[\sigma_i^-, \sigma_i^+]$ of a parameter estimate $\hat{p}_i$ to a confidence level α signifies that the true value p_i^* is located within this interval with probability α.

Table 1. Nominal parameter values

k_p^0 $[\mu M \cdot min^{-1}]$	k_p^1 $[\mu M \cdot min^{-1}]$	γ_p $[min^{-1}]$	μ_G $[(\mu M \cdot min)^{-1}]$	α	K_v $[\mu M]$	n	v^* $[\mu M]$
0.02	0.11	0.006	0.0014	0.1	30	2	50

- data set I, with $v = [0, 5, 10, 15, ..., 295, 300, 1000]$ and $\sigma = 10^{-2}$;
- data set II, with $v = [0, 10, 20, 30, ..., 290, 300, 1000]$ and $\sigma = 10^{-2}$;
- data set III, with $v = [0, 5, 10, 15, ..., 295, 300, 1000]$ and $\sigma = 5 \cdot 10^{-2}$;
- data set IV, with $v = [0, 10, 20, 30, ..., 290, 300, 1000]$ and $\sigma = 5 \cdot 10^{-2}$;

Notably, data sets I, II, III and IV, have been generated with different number of points (N_{exp}) and different intensities of noise (σ) to study the practical identifiability of the parameters in four realistic experimental conditions. In particular, data sets I and III have the same number of data points, i.e. $N_{exp} = 62$, but different noise, $\sigma = 10^{-2}$ for data set I and $\sigma = 5 \cdot 10^{-2}$ for data set III. Data set II and IV have less number of points, i.e. $N_{exp} = 32$, while the level of noise considered is $\sigma = 10^{-2}$ for data set II and $\sigma = 5 \cdot 10^{-2}$ for data set IV.

5.3 Model Parameterization and Global Optimization

First, to avoid evident structural identifiability problems we will group together those parameters in (14) which appear as combinations of products and/or quotients between parameters. Thus, after some algebraic manipulations expression (14) reads as:

$$\mu(v) = \begin{cases} -\frac{\gamma_p}{2}\left[1 - \sqrt{1 + \frac{4k_p^0\mu_G\alpha}{\gamma_p^2}\left(1 + \frac{(1-\alpha)}{\alpha}\frac{v^n}{K_v^n + v^n}\right)}\right] & \text{if, } v \leq v^* \\ -\frac{\gamma_p}{2}\left[1 - \sqrt{1 + \frac{4(k_p^0 + k_p^1)\mu_G\alpha}{\gamma_p^2}\left(1 + \frac{(1-\alpha)}{\alpha}\frac{v^n}{K_v^n + v^n}\right)}\right] & \text{if, } v > v^* \end{cases} \tag{18}$$

Moreover, to avoid dependence on physical unit as well as to overcome possible scaling problem and to reduce the number of parameters, we decided to calculate a non-dimensional version of expression (18). Notably, the non-dimensional slope $\mu_N(v)$ is obtained by dividing $\mu(v)$ in (18) for the minimal growth rate, which is achieved at the minimum value of the inducer, i.e. at $v = v_0$, which for our data sets I, II, III, IV consists in $v_0 = 0$. Thus, considering the necessary condition $v_0 < v^*$, the non-dimensional growth rate during the exponential phase reads:

$$\mu_N(v) = \begin{cases} \dfrac{1 - \sqrt{1 + \dfrac{4k_p^0 \mu_G \alpha}{\gamma_p^2}\left(1 + \dfrac{(1-\alpha)}{\alpha}\dfrac{v^n}{K_v^n + v^n}\right)}}{1 - \sqrt{1 + \dfrac{4k_p^0 \mu_G \alpha}{\gamma_p^2}}} & \text{if, } v \leq v^* \\ \\ \dfrac{1 - \sqrt{1 + \dfrac{4(k_p^0 + k_p^1) \mu_G \alpha}{\gamma_p^2}\left(1 + \dfrac{(1-\alpha)}{\alpha}\dfrac{v^n}{K_v^n + v^n}\right)}}{1 - \sqrt{1 + \dfrac{4k_p^0 \mu_G \alpha}{\gamma_p^2}}} & \text{if, } v > v^* \end{cases} \tag{19}$$

Now, considering the following parameterization

$$p_1 = \frac{4k_p^0 \mu_G \alpha}{\gamma_p^2};\ p_2 = \frac{(1-\alpha)}{\alpha};\ p_3 = K_v;\ p_4 = n;\ p_5 = \frac{4k_p^1 \mu_G \alpha}{\gamma_p^2};\ p_6 = v^*$$

the expression (19) can be rewritten as

$$\mu_N(v, p) = \begin{cases} \dfrac{1 - \sqrt{1 + p_1\left(1 + p_2 \dfrac{v^{p_4}}{p_3^{p_4} + v^{p_4}}\right)}}{1 - \sqrt{1 + p_1}} & \text{if, } v \leq p_6 \\ \dfrac{1 - \sqrt{1 + (p_1 + p_5)\left(1 + p_2 \dfrac{v^{p_4}}{p_3^{p_4} + v^{p_4}}\right)}}{1 - \sqrt{1 + p_1}} & \text{if, } v > p_6 \end{cases} \tag{20}$$

where $p = [p_1, p_2, p_3, p_4, p_5, p_6]$ and, considering the true parameters values in Tab 5.2 we obtain the true vector of parameters p^*:

$$p^* = [0.3033, 9, 30, 2, 1.6683, 50]\ . \tag{21}$$

Similarly, the data sets I to IV will also be normalized to their minimal value, i.e., each output-point is divided by the minimal observation value, that is $y_{min} = \mu(v_0)$, where $v_0 = 0$.

Our approach in identifying the unknown parameters of model (19) consists in solving a non-linear least squares minimization problem, using a hybrid optimization approach which makes use of the functions *ga* (Genetic Algorithm [34]) and *GlobalSearch* of the *MATLAB*® *Global Optimization Toolbox*™. To start, we used the Genetic Algorithm (GA) for 10^4 generations to get near an optimum point. The genetic algorithm does not use derivatives to detect descent in its minimization steps. Hence, it is a good choice for non-differentiable and/or discontinuous problems. Moreover, GA does not necessarily need an user supplied initial guess, which in most case leads to local sub-optimal convergence if the initial guess is far from the global optimum. The result obtained with the genetic algorithm is then used as initial point of a hybrid function, to further improve the value of the cost function $J(p)$. We decided to use the *GlobalSearch*[5] command as hybrid function since it searches many basins of attraction near the starting point given by GA, arriving faster at an even better solution.

5.4 In Silico Practical Identifiability Analysis

The practical identifiability of model (20) has been tested using data sets I, II, III and IV, which have different values of errors' measurement and different data points. Hence, these artificial data are suitable to mimic realistic experimental set-ups.

For each data set mentioned above, parameters' confidence intervals have been computed following a *Monte Carlo*-like approach.

Notably, $N_{simul} = 200$ runs of the previously described hybrid optimization were performed. Where, at each of the N_{simul} runs, a new realization of the artificial measurements—according to the inputs and noise statistic of each data set—is considered. This optimization yields N_{simul} estimated values for each parameter p_i, $i = 1, \ldots, 6$. Then, for each i, an average value, $\hat{m}_i$, and a standard deviation, $\hat{s}_i$, were computed by fitting a Gaussian distribution $\mathcal{N}(\hat{m}_i, \hat{s}_i^2)$ to the histogram of the N_{simul} values of p_i. Thus, the 95% confidence interval (CI_i) for the p_i parameter is calculated as:

$$CI_i = \hat{m}_i \pm 1.96\hat{s}_i \tag{22}$$

This leads to the confidence intervals listed in Table 2.

As we can see in Table 2, parameters p_i for $i \in \{2, 3, 4, 6\}$ do not show any practical identifiability issues, as the true value is contained in the respective CI with sufficiently precision. On the contrary, the CIs of parameters $\hat{p}_1$ and $\hat{p}_5$ tend to become very large at increasing values of the measurement's errors (σ) and at decreasing numbers of data points, indicating that in real experimental conditions (that is, limited and noisy data), the precise identification of these parameters might be impracticable. Moreover, we found that the correlation coefficient (R) between the two vectors of estimated parameters parameters $\hat{p}_1$ and $\hat{p}_5$ is $R = 0.99$, for all data sets. Recall that the correlation coefficient measures the interrelationship between $\hat{p}_1$ and $\hat{p}_5$ quantifying the compensation effects of changes in the parameter values on the model output. In fact, when two parameters are highly correlated, a change in the model output caused by a change in a model parameter can be balanced by a appropriate change

[5] *GlobalSearch* first runs *fmincon* from the start point you give. If this run converges, GlobalSearch records the start point and end point for an initial estimate on the radius of a basin of attraction. Then, *GlobalSearch* solver starts a local solver (*fmincon*) from multiple starting points and store local and global solutions found during the search process. Notably, the *GlobalSearch* solver first uses a scatter-search algorithm to randomly generate multiple starting points, then filters non-promising start points based upon objective and constraint function values and local minima already found, and finally runs a constrained nonlinear optimization solver to search for a local minimum from the remaining start points.

Table 2. Confidence intervals of estimated parameters $\hat{p}_i$ when (20) is fitted to (non-dimensionalized) data sets I, II,III,IV. The confidence intervals for parameters become larger at increasing values of the measurement error and at decreasing numbers of data points, indicating possible practical identifiability problems especially for $\hat{p}_1$ and $\hat{p}_5$.

	DATA SET I $\sigma = 10^2$ $N_{exp} = 62$	DATA SET II $\sigma = 10^{-2}$ $N_{exp} = 32$	DATA SET III $\sigma = 5 \cdot 10^{-2}$ $N_{exp} = 62$	DATA SET IV $\sigma = 5 \cdot 10^{-2}$ $N_{exp} = 32$
CI_1	0.3328 ± 0.4939	0.3738 ± 0.5441	0.2631 ± 0.4220	0.32 ± 0.49
CI_2	9.23 ± 3.45	9.36 ± 3.88	8.63 ± 3.06	9.21 ± 4.67
CI_3	30.16 ± 3.55	30.00 ± 3.55	29.39 ± 5.15	30.33 ± 7.52
CI_4	2.002 ± 0.079	2.011 ± 0.089	2.006 ± 0.232	2.01 ± 0.33
CI_5	2.053 ± 4.192	2.39 ± 4.51	1.53 ± 3.59	1.93 ± 3.99
CI_6	53.32 ± 4.48	55.98 ± 6.99	53.06 ± 3.58	56.70 ± 6.79

Table 3. Confidence intervals of the ratio $\hat{p}_5/\hat{p}_1$ when (20) is fitted to (non-dimensionalized) data sets I, II,III,IV

	DATA SET I $\sigma = 10^{-2}$ $N_{exp} = 62$	DATA SET II $\sigma = 10^{-2}$ $N_{exp} = 32$	DATA SET III $\sigma = 5 \cdot 10^{-2}$ $N_{exp} = 62$	DATA SET IV $\sigma = 5 \cdot 10^{-2}$ $N_{exp} = 32$
$CI_{\hat{p}_5/\hat{p}_1}$	5.29 ± 2.39	5.54 ± 2.43	4.99 ± 1.15	5.2 ± 1.3

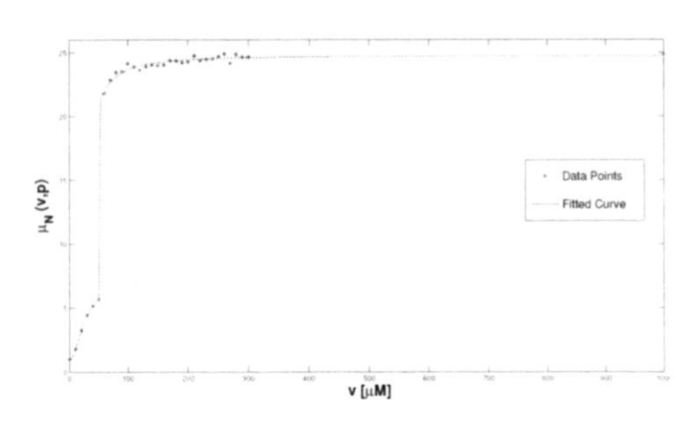

Fig. 4. Fitting the growth rate function (20) using one realization of the non-dimensional data set II. The blue points are the normalized artificial data generated according to specification of data set II. The red curve is the function (20) when $\hat{p}$ is used.

in the other parameter value. Thus, instead of considering the CIs of $\hat{p}_1$ and $\hat{p}_5$ separately—which are not significant—we have computed the confidence interval of their ratio, i.e. $\hat{p}_5/\hat{p}_1$. These results are presented in Table 3. As we can notice in Table 3, the CIs of $\hat{p}_5/\hat{p}_1$ are accurate, since they contain the true value of the ratio $p_5^*/p_1^* = 5.5$, and more precise since their relative width is smaller than the relative width of CI_1 and CI_5.

It must be noted that a further reduced model which takes into account the correlation between p_5 and p_1 can not be achieved. This because expression (20) can be rewritten in terms of the ratio and either p_5 or p_1. Fig 4 shows the fitting of model (20) to one realization of data set II.

6 Conclusions

In this paper, a minimal model consisting of two variables (the concentrations of two gene products) and an input (an inducer) was analyzed and used to describe one possible mechanism to control the growth rate of *E. coli* cells during exponential phase. This model is based on the piecewise affine formalism but a new, non-linear, term was added to account for the dilution effect during growth. The qualitative dynamics of the model can thus be studied, and the bifurcation diagram with respect to the input is obtained. Moreover, this mathematical formalism allows derivation of an analytic expression for the growth rate as function of the input. This expression has two applications:

- it can be directly fitted to experimental data to estimate a set of parameters (this is an advantage relative to the typical "indirect" parameter estimation by fitting to the numerical solutions of the differential equations);
- it provides an indication of how to control the growth rate to a desired value by adding a given quantity of inducer.

Finally, practical identifiability analysis based on numerical simulations is presented, which shows that some issues may arise with noisy measurements. In this case, our analysis suggests that the original growth rates' measurements should be adimensionalized and unknown parameters grouped into a new set of "lumped" parameters in order to obtain local identifiability. Notably, we found that only the ratio between the estimated parameters $\hat{p}_1$ and $\hat{p}_5$ can be estimated with sufficient precision in the case when only limited and noisy data are available. This study and the conclusions on identifiability will be most useful to help dealing with and solving parameter estimation problems with real data sets.

Acknowledgments. This work was supported by ANR GeMCo (French national research agency).

References

1. Andrianantoandro, E., Basu, S., Karig, D., Weiss, R.: Synthetic biology: new engineering rules for an emerging discipline. Molecular Systems Biology 2(1) (2006)
2. Khalil, A., Collins, J.: Synthetic biology: applications come of age. Nature Reviews Genetics 11(5), 367–379 (2010)
3. Mukherji, S., Van Oudenaarden, A.: Synthetic biology: understanding biological design from synthetic circuits. Nature Reviews Genetics 10(12), 859–871 (2009)

4. Elowitz, M., Leibler, S., et al.: A synthetic oscillatory network of transcriptional regulators. Nature 403(6767), 335–338 (2000)
5. Gardner, T., Cantor, C., Collins, J.: Construction of a genetic toggle switch in Escherichia coli. Nature 403, 339–342 (2000)
6. Tigges, M., Marquez-Lago, T., Stelling, J., Fussenegger, M.: A tunable synthetic mammalian oscillator. Nature 457(7227), 309–312 (2009)
7. Monod, J.: The growth of bacterial cultures. Annual Review of Microbiology 3(1), 371–394 (1949)
8. Marr, A.G.: Growth rate of Escherichia coli. Microbiological Reviews 55(2), 316–333 (1991)
9. Kaern, M., Blake, W., Collins, J.: The engineering of gene regulatory networks. Annual Review of Biomedical Engineering 5(1), 179–206 (2003)
10. Tan, C., Marguet, P., You, L.: Emergent bistability by a growth-modulating positive feedback circuit. Nature Chemical Biology 5(11), 842–848 (2009)
11. Bettenbrock, K., Sauter, T., Jahreis, K., Kremling, A., Lengeler, J.W., Gilles, E.D.: Correlation between growth rates, EIIACrr phosphorylation, and Intracellular Cyclic AMP levels in Escherichia coli K-12. J. Bacteriol. 189(19), 6891–6900 (2007)
12. Ropers, D., de Jong, H., Page, M., Schneider, D., Geiselmann, J.: Qualitative simulation of the carbon starvation response in Escherichia coli. Biosystems 84(2), 124–152 (2006)
13. de Jong, H., Geiselmann, J., Hernandez, C., Page, M.: Genetic network analyzer: qualitative simulation of genetic regulatory networks. Bioinformatics 19(3), 336–344 (2003)
14. Casey, R., Jong, H., Gouzé, J.: Piecewise-linear models of genetic regulatory networks: Equilibria and their stability. Journal of Mathematical Biology 52(1), 27–56 (2006)
15. Chaves, M., Gouzé, J.-L.: Piecewise Affine Models of Regulatory Genetic Networks: Review and Probabilistic Interpretation. In: Lévine, J., Müllhaupt, P. (eds.) Advances in the Theory of Control, Signals and Systems with Physical Modeling. LNCIS, vol. 407, pp. 241–253. Springer, Heidelberg (2010)
16. De Jong, H., Gouzé, J., Hernandez, C., Page, M., Sari, T., Geiselmann, J.: Qualitative simulation of genetic regulatory networks using piecewise-linear models. Bulletin of Mathematical Biology 66(2), 301–340 (2004)
17. Gouzé, J., Sari, T.: A class of piecewise linear differential equations arising in biological models. Dynamical Systems 17(4), 299–316 (2002)
18. Grognard, F., De Jong, H., Gouzé, J.: Piecewise-linear models of genetic regulatory networks: theory and example. Biology and Control Theory: Current Challenges, 137–159 (2007)
19. Yagil, G., Yagil, E.: On the relation between effector concentration and the rate of induced enzyme synthesis. Biophysical Journal 11(1), 11–27 (1971)
20. Filippov, A., Arscott, F.: Differential equations with discontinuous righthand sides. In: Mathematics and its Applications Series. Kluwer Academic Publishers (1988)
21. Klumpp, S., Zhang, Z., Hwa, T.: Growth rate-dependent global effects on gene expression in bacteria. Cell 139(7), 1366–1375 (2010)
22. Scott, M., Gunderson, C.W., Mateescu, E.M., Zhang, Z., Hwa, T.: Interdependence of cell growth and gene expression: Origins and consequences. Science 330(6007), 1099–1102 (2010)
23. Eden, E., Geva-Zatorsky, N., Issaeva, I., Cohen, A., Dekel, E., Danon, T., Cohen, L., Mayo, A., Alon, U.: Proteome half-life dynamics in living human cells. Science 331(6018), 764–768 (2011)

24. Krin, E., Sismeiro, O., Danchin, A., Bertin, P.N.: The regulation of Enzyme IIAGlc expression controls adenylate cyclase activity in Escherichia coli. Microbiology 148(5), 1553–1559 (2002)
25. Notley-McRobb, L., Death, A., Ferenci, T.: The relationship between external glucose concentration and cAMP levels inside Escherichia coli: implications for models of phosphotransferase-mediated regulation of adenylate cyclase. Microbiology 143(6), 1909–1918 (1997)
26. Vajda, S., Rabitz, H., Walter, E., Lecourtier, Y.: Qualitative and quantitative identifiability analysis of nonlinear chemical kinetic models. Chemical Engineering Communications 83(1), 191–219 (1989)
27. Chis, O., Banga, J., Balsa-Canto, E.: Structural identifiability of systems biology models: A critical comparison of methods. PloS one 6(11), e27755 (2011)
28. Raue, A., Kreutz, C., Maiwald, T., Bachmann, J., Schilling, M., Klingmüller, U., Timmer, J.: Structural and practical identifiability analysis of partially observed dynamical models by exploiting the profile likelihood. Bioinformatics 25(15), 1923–1929 (2009)
29. Walter, É., Pronzato, L.: Identification of parametric models from experimental data. Communications and Control Engineering, Springer (1997)
30. Dochain, D., Vanrolleghem, P.: Dynamical Modelling and Estimation in Wastewater Treatment Processes. IWA Publishing (2001)
31. Seber, G., Wild, C.: Nonlinear regression, vol. 503. Libre Digital (2003)
32. Gallant, A.: Nonlinear regression. The American Statistician 29(2), 73–81 (1975)
33. Bremer, H., Dennis, P., et al.: Modulation of chemical composition and other parameters of the cell by growth rate. Escherichia Coli and Salmonella: Cellular and Molecular Biology 2, 1553–1569 (1996)
34. Goldberg, D.: Genetic algorithms in search, optimization, and machine learning. Addison-Wesley (1989)

Multi-objective Optimisation, Sensitivity and Robustness Analysis in FBA Modelling

Jole Costanza[1], Giovanni Carapezza[1], Claudio Angione[2], Pietro Liò[2], and Giuseppe Nicosia[1]

[1]Department of Mathematics and Computer Science, University of Catania
[2]Computer Laboratory, University of Cambridge
{costanza,carapezza,nicosia}@dmi.unict.it
{claudio.angione,pietro.lio}@cl.cam.ac.uk

Abstract. In this work, we propose a computational framework to design in silico robust bacteria able to overproduce multiple metabolites. To this end, we search the optimal genetic manipulations, in terms of knockout, which also guarantee the growth of the organism. We introduce a multi-objective optimisation algorithm, called Genetic Design through Multi-Objective (GDMO), and test it in several organisms to maximise the production of key intermediate metabolites such as succinate and acetate. We obtain a vast set of Pareto optimal solutions; each of them represents an organism strain. For each solution, we evaluate the fragility by calculating three robustness indexes and by exploring reactions and metabolite interactions. Finally, we perform the Sensitivity Analysis of the metabolic model, which finds the inputs with the highest influence on the outputs of the model. We show that our methodology provides effective vision of the achievable synthetic strain landscape and a powerful design pipeline.

Keywords: Cell Metabolism, Biological CAD, Sensitive and Fragile Pathways, Genetic Design, Multi-Objective optimisation, Flux balance analysis, Sensitivity and Robustness Analysis.

1 Introduction

Metabolic engineering is central in Biotechnology and has impact also in basic cellular biology. The aim of metabolic engineering is to direct specifically a flux through a metabolic pathway, for instance a product made during the fermentation. To this end, one needs a deep understanding not merely of the genetics of a microorganism, but also of its metabolic capacity (i.e. the amount of all the intermediates). Remarkably, through genetic manipulations (in terms of knockouts) carried out on bacteria, one can overproduce one or more metabolites of interest. A gene knockout is a genetic technique in which one gene in an organism is made inoperative through a base mutation or a deletion. Sometime the inactivation of one gene results in the inactivation of all the downstream genes of the operon. These manipulations are very useful for classical genetic studies

D. Gilbert and M. Heiner (Eds.): CMSB 2012, LNCS 7605, pp. 127–147, 2012.

as well as for modern techniques including functional genomics. Recently, many organisms have been used to analyse their metabolite production potential and to identify the metabolic interventions to produce the metabolite of interest. Indeed, strains have been systematically designed in vivo to overproduce target metabolites such as lycopene [1], ethanol [2], isobutanol [3] and many others.

Metabolic engineering requires mathematical models for accurate metabolic reconstruction of strains, as well as for seeking non-native synthesis pathways. A recent research methodology, called Flux Balance Analysis (FBA) [4], studies biochemical networks, in particular the genome-scale metabolic network reconstructions. These network reconstructions contain all of the known metabolic reactions in an organism and the genes that encode each enzyme. FBA calculates the flow of metabolites through this metabolic network, thereby making it possible to predict the growth rate of an organism or the rate of production of a biotechnologically important metabolite at steady state. Being at steady state, FBA manages large networks very quickly, since it does not require kinetic parameters. This makes it well suited to research on perturbations and genetic manipulations (knockouts) that bacteria might undergo. One of the major advantages of performing computational analysis of stoichiometric models is that the pathways are system proprieties emerging under particular genetic background and nutritional conditions. In other words, the FBA provides better treatment of metabolism than classical biochemistry drawings of metabolic pathways.

By using computational metabolic engineering methods, it is possible to explore the reaction network and search for the genetic interventions to optimise the objectives. By making inoperative the genes, the enzymes that are normally synthesised by those genes are not present anymore in the biological system. In this way, also the corresponding biochemical reactions, normally catalysed by these enzymes, do not occur. Then, the chemical species that constitute the reagents and products of these reactions do not undergo the transformations. The aim is to find the genetic manipulations that change the metabolic process in an organism, in order to increase the flow of desired metabolites, chosen according to biotechnological purposes. Additionally, changing the natural genetic function in an organism may cause the death of the growth cell. Therefore, finding genetic manipulations is a hard problem of search and optimisation.

For all the above reasons and since designing gene knockout in laboratory is very expensive and time-consuming, in the past years a variety of methods has been implemented in order to predict in silico the best knockout strategies that optimise a cellular function of interest. These methods are based on evolutionary algorithms [5], simulated annealing [6], bi-level optimisation framework [7], and mixed-integer linear programming (MILP) [8,9]. All have been tested in FBA organism models, but they require high computational efforts, since the execution times grow exponentially [8,6,5] or linearly [7] as the number of manipulations allowed in the final designs increases. Moreover, cellular metabolism is composed of a large number of reactions, thus the dimension of the solution space is very large and finding genetic manipulations is computationally expensive.

In this work, we present a novel Multi-Objective optimisation algorithm denoted by Genetic Design through Multi-Objective (GDMO), in order to search for the genetic manipulations that optimise multiple cellular functions of interest. Our idea is to use the Pareto optimality to obtain not only a wide range of Pareto optimal solutions, but also the *best trade-off design*. In this context, the multiple biological functions are represented by desired productions, e.g., vitamins, proteins, biofuel, biomass formation, antibodies, electron productivity, or the energetic yield of the organism. For this application, a Pareto solution represents a strain with a particular genetic manipulation (genotype), and that is specialised to overproduce selected metabolites (phenotype), with respect to the wild type (i.e., a strain with genes that are all operative). We test our knockout-based multi-target optimisation on the most recent metabolic data concerning *Escherichia coli*, *Geobacter* [10], *Methanosarcina barkeri* [11], and *Yersinia pestis* [12]. We report that multi-objective optimisation provides more insights than single optimisation on the capability of these organisms to adapt to the simultaneous presence of different conditions and constraints. Furthermore, our method is able to explore effectively the whole space of knockouts. We tested the performance of GDMO by maximising acetate and succinate production rates, and other multiple biological functions in *E. coli*, *i*AF1260, and comparing it against previous methods.

GDMO is accompanied by a robustness analysis that performs the local, global robustness and the Normalised Feasible Parameter Volume of the genetic manipulation proposed by GDMO. For each strain, we compute the robustness indexes, in order to estimate how robust is a strain obtained by GDMO when it undergoes small perturbations, external (changes in the nutrients) or internal (changes in the metabolism). This way, we are able to choose the most robust strain proposed by GDMO. Finally, the Sensitivity Analysis investigates the species solution space and determines the influence of each specie on the output of the FBA model.

2 Methods

2.1 GDMO: Genetic Design through Multi-Objective Optimisation

GDMO is a combinatorial global search method that finds the genetic manipulation strategies to simultaneously optimise multiple cellular functions (i.e., objective functions) in metabolic networks modelled with Flux Balance Analysis (FBA) and Gene-Protein-Reaction (GPR) map. The simultaneous optimisation of multiple objectives differs from the single-objective optimisation because the solution is not unique when the objectives are in conflict with each other. In a maximisation problem objectives are in conflict when the increment of an objective, causes the decrement of at least another one.

The solution of a multi-objective problem is a potentially infinite set of points, called *non-dominated* solutions or *Pareto front*. In a maximisation problem, a solution is Pareto optimal if there exist no feasible solutions that increase some objective without causing a simultaneous decrease in at least one other objective.

In our problem, the genotype of a bacterium is mathematically represented by a string of bits $y \in \{0,1\}^L$. Each bit in y is a gene set that distinguishes between single and multi-functional enzymes, isozymes, enzyme complexes and enzyme subunits; this way, it captures the complexity and diversity of the biological relationships through a Boolean approach. For example, when the genes of the l-th set are all necessary to catalyse the corresponding reactions (a single gene set can linked to more reactions), genes are related by "AND"; otherwise if it is necessary at least a gene, genes are linked by "OR". When the l-th element of y is set to 1, the corresponding gene set is inoperative. Therefore, y represents the vector of decision variables to be found, in order to obtain the higher values of objective functions, satisfying particular constraints (for instance a maximum number of gene knockouts allowable). A point y^* in the solution space is said to be Pareto optimal if there does not exist a point y such that $F(y)$ dominates $F(y^*)$, where F is the vector of r objective functions. The variable space, (i.e., the domain of y) is defined in a discrete space.

The method we present implements a genetic algorithm inspired by NSGA-II [13] and is composed of 4 key steps. We start with the *initialisation* of the population *Pop* and the computation of the fitness score. The population *Pop* is a set of individuals, i.e., a set of feasible solutions. *Pop* is represented by a $I \times (L + r + 2)$ matrix, where I is the number of individuals, L is the number of the decision variables and r is the number of the objective functions, obtained solving the problem (2). The last two columns are used to store two parameters of the algorithm linked to each individual and useful to evaluate the quality of the solution. Each individual is composed of the proposed knockout strategy $\tilde{y}$ and the corresponding objective function values. Each generation select the individuals that are maximal with respect to the product ordering. The individuals of the initial population can be initialised in different ways: either randomly, assigning present status to all genes or selecting a set of knocked out genes.

Successively, three steps are iteratively carried out. In a *binary tournament selection* process, two individuals are selected at random, and their fitness is compared. The individual with the best fitness is selected as a parent. The algorithm selects a number of parents (i.e. the best individuals) equal to the half of the population. Parents are mutated using a *combinatorial mutation operator* convenient to create an offspring population. Mutation represents a switch, from 0 to 1, or from 1 to 0 for the l-th gene set. The process is randomly executed and for each parent individual ten offspring have been formed and only the best is chosen. Mutation can achieve the maximum knockouts number equal to the parameter C (fixed to 50 by default). A novel population of size *Pop* is formed selecting the best individuals from the parents of the previously generation and the current offspring. The new population undergoes a new round of evaluation. For each generation of the algorithm, Pareto optimal solutions are provided. Finally, a *selection* operator is performed in order to reach the last front.

This cycle is repeated until the solution set does not improve, or until an individual with a desired phenotype is achieved or when the number of generation is bounded out. The number of generations D and individuals of the population I are parameters chosen by the user. The time-complexity of the genetic algorithm is $O(2DI^r)$, where D is the number of generations, I is the population size and r is the number of the objectives. GDMO finds a set of Pareto optimal solutions (non-dominated solutions) for a combinatorial multi-objective optimisation problem, which is also a NP-complete problem.

Pareto Optimality is very useful for the analysis of metabolism, as reported in the previous works [14,15,16], where authors used multi-objective approaches to evaluate the fluxes distributions and genetic manipulations in metabolic networks. In our work, we remark the usefulness of Pareto optimality and adopt an effective and state-of-the-art algorithm to investigate the knockout space. Additionally, for the first time, we used the ϵ-dominance optimality, to do an accurate search in a neighbourhood of the edge of the Pareto region.

2.2 Pareto ϵ-Dominance

Another analysis that we perform is inspired by the idea described in [17]. They use a condition of approximated dominance for their evolutionary multi-objective algorithm with the aim of improving the diversity of solutions and convergence. We, however, use this idea to perform a post-processing analysis in order to calculate an approximated Pareto front. This calculation is designed to search for new solutions and, in particular, solutions that may have been discarded, but they are dominated by an amount that, for our purposes, can be considered negligible. Therefore, once the optimisation routine has been carried out, all the sampled points are revisited. Then, a new set of solutions is built, called "ϵ-non-dominated" set, by applying a "relaxed" condition of dominance, called ϵ-dominance. Formally, assuming that all the objective functions must be maximised, given $\epsilon > 0$, we seek all the points (solutions) belonging to the set: $\{w : w_z + \epsilon \geq u_z, \ \forall \ z = 1, ..., r\}$. Remarkably, this set contains both the new "ϵ-non-dominated" solutions and the previous non-dominated ones.

2.3 FBA Modelling and the Combinatorial Optimisation Problem

FBA is a modelling framework used for studying biochemical networks and in particular the m metabolites and n reactions that are involved (e.g., their formation and degradation, transport and cellular utilisation). For each metabolite X_i, $i = 1, \ldots, m$ a material balance is $\frac{dX_i}{dt} = \sum_{j=1}^{n} S_{ij} v_j$, where S_{ij} is the stoichiometric coefficient associated with each flux v_j, $j = 1, \ldots, n$. At steady state, $\sum_{j=1}^{n} S_{ij} v_j = 0$ holds. This balance equation can be written in matrix form $Sv = 0$, where S is the stoichiometric matrix of m rows and n columns, and v is the vector of fluxes (metabolic and transport fluxes). The matrix S is not square and $n > m$, so we have a plurality of solutions. Each solution is a flux distribution representing a particular metabolic state, depending on the genotype and the transport fluxes. The FBA approach finds the metabolic state in order to optimise a particular objective function, given as a linear combination of fluxes (e.g., growth rate, ATP production). Consequently, the problem can be formulated as a linear programming problem:

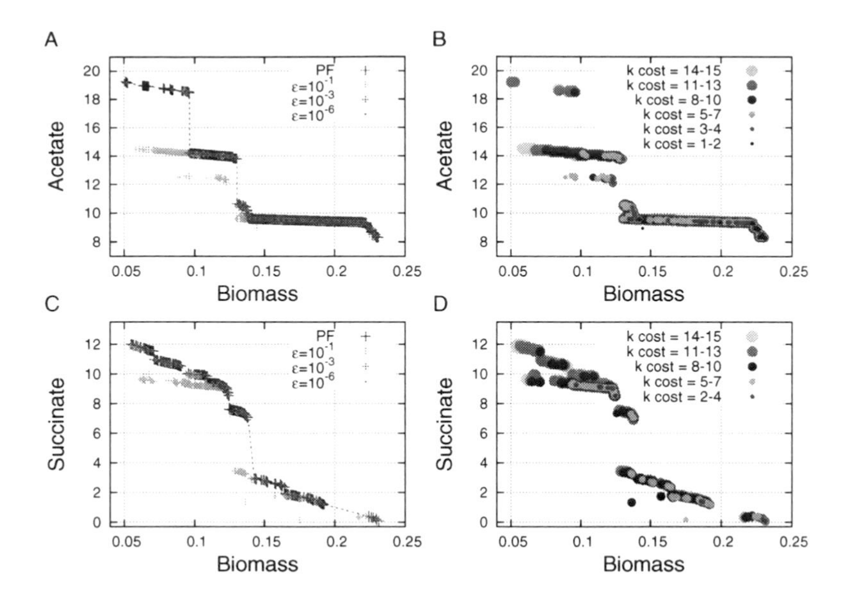

Fig. 1. ϵ-dominance analysis results in *E. coli* network for acetate (A) and succinate (C) multi-objective optimisation. Figures B and D report the knockout cost associated with the solutions reported respectively in Figures A and C, and the dimension of circles reflects the knockout cost associated with the solution point.

$$\begin{aligned} &\text{maximise (or minimise)} && f'v \\ &\text{such that} && Sv = 0 \\ & && v_j^L \leq v_j \leq v_j^U, j = 1, \ldots, n, \end{aligned} \tag{1}$$

where f is a vector of weights (n dimensional). All the elements in f are either 0 or 1. In our work, f_i is equal to 1 if v_i is the biomass core, but it is possible any combination of fluxes as an objective functions. v_j^L and v_j^U are the lower and upper bound values (thermodynamic constraints) of the generic flux v_j (in our analysis, we consider $v_j^U = 100$ and $v_j^L = -100$ for the fluxes that represent reversible reactions). The output of FBA is a particular distribution of fluxes, denoted by v, which optimises the objective function. Remarkably, FBA does not describe how a certain flux distribution is realised (by kinetics or enzyme regulation), but which flux distribution is optimal for the cell.

We propose the gene-protein-reaction (GPR) mappings to allow our algorithm to work at the genetic level. GPR mappings provide links between each gene and all the reactions v_j depending on it, and define how certain genetic manipulations affect reactions in the metabolic network. For a set of L genetic manipulations, the GPR mappings are represented by a $L \times n$ matrix G, where the *(l,j)*-th element is 1 if the *l*-th genetic manipulation maps onto the reaction j, and is 0 otherwise.

Our approach is based on the technique adopted in OptKnock [8], which finds the fluxes distribution in the metabolic network in order to reproduce the desired productions (synthetic objectives) and achieve the maximal growth.

Unlike Optknock, we are able to optimise more than one objective. The bi-level problem [8] is represented by the following formulation:

$$
\begin{array}{lll}
\max & g'v & \\
\text{such that} & \sum_{l=1}^{L} y_l \leq C & \\
 & y_l \in \{0,1\} & \\
 & \max & f'v \\
 & \text{such that} & Sv = 0 \\
 & & (1-y)'G_j v_j^L \leq \; v_j \leq \; (1-y)'G_j v_j^U, \\
 & & j = 1, \ldots, n,
\end{array}
\tag{2}
$$

where g is a vector of weights (n dimensional) associated with the synthetic objectives, and g' is its transpose. For example, when the synthetic objectives v_j and v_h have to be maximised, the weights g_j and g_h are equal to 1. y is the knockout vector (L dimensional). If there are no impaired reactions in the metabolic network, y contains only zeros. Conversely, when $y_l = 1$, the gene set involved in the manipulation l is turned off, and the corresponding reactions are in the absent status (the lower and upper bounds are set to zero, resulting in a modified metabolic network). C is an integer representing the maximum number of knockouts allowed. The bi-level problem can be converted to a MILP problem as described in [8] (for a detailed description, see the original work [8]). We implemented and solved the problem using the GLPK solver.

2.4 Sensitivity Analysis

In modelling, Sensitivity Analysis (SA) is a method used to discover which inputs play a key role on the output of the model. In the last years, scientists used SA indexes in systems biology interrogating the reactions space (RoSA - Reactions oriented Sensitivity Analysis) [18], [19] and species space (SoSA - Species oriented Sensitivity Analysis) to find their influence on the outputs of the system [20]. We perform SA to find the most sensitive inputs in FBA model of *E. coli* using the Matlab SensSB Toolbox [21].

The *E. coli* model analysed in this work contains 2382 fluxes, 299 of which represent exchange fluxes, 2082 represent inner metabolic reactions, and 1 the growth rate or biomass.

The n_{ex} exchange reactions ($n_{ex} < n$) described by the vector $v_{ex} \subset v$, allowing nutrients to enter and leave the system, are unconstrained in the forward direction (v_{ex}^U, upper bound vector), while are constrained in reverse directions (v_{ex}^L, lover bound vector) to zeros when uptake rates is not allowed. Moreover, the *"EX glc"* is an exchange reaction for glucose and has a lower bound of "-10" indicating a potential glucose uptake rate of 10 mmol gDW^{-1} h^{-1}.

We performed the SA method considering as inputs of the model the v_{ex}^L lower bound vector of exchange fluxes. For each of n_{ex} exchange fluxes we varied each element of v_{ex}^L in the interval [-100, 0] of the region of interest Ω, n_{ex}-dimensional unit hypercube. We adopted the Morris [22] method in order to identify the uptake rates whose tuning results in a major system response. SA is based on the calculation of the *elementary effect* due to the variation of each input. For a given value of v_{ex}^L, we define the elementary effect of the h-th input as:

$$d_h(v_{ex}^L) = \frac{F(v_{ex}^L(1), \ldots, v_{ex}^L(h-1), v_{ex}^L(h) + \Delta, v_{ex}^L(h+1), \ldots, v_{ex}^L(n_{ex})) - F(v_{ex}^L)}{\Delta}. \tag{3}$$

We considered the vector of fluxes v as output $F(v_{ex}^L)$ for *E. coli* model, calculated by solving the problem (1). For each of the n_{ex} exchange fluxes, the attention is restricted to a region of experimentation ω, n_{ex}-dimensional k-level grid, where each v_h^{ex} may take a value from $\omega = \left\{-100, -100\frac{k-2}{k-1}, \ldots, -100\frac{2}{k-1}, -100\frac{1}{k-1}, 0\right\}$. Δ is a predetermined multiple of $\frac{1}{(k-1)}$ and represents the *perturbation* of the input v_h^{ex}. The distribution of elementary effects EE_h for the input v_h^{ex} is obtained by randomly sampling Q points from ω. The estimation of the mean μ^* and standard deviation σ^* of those distributions EE_h will be used as indicator of which inputs should be considered

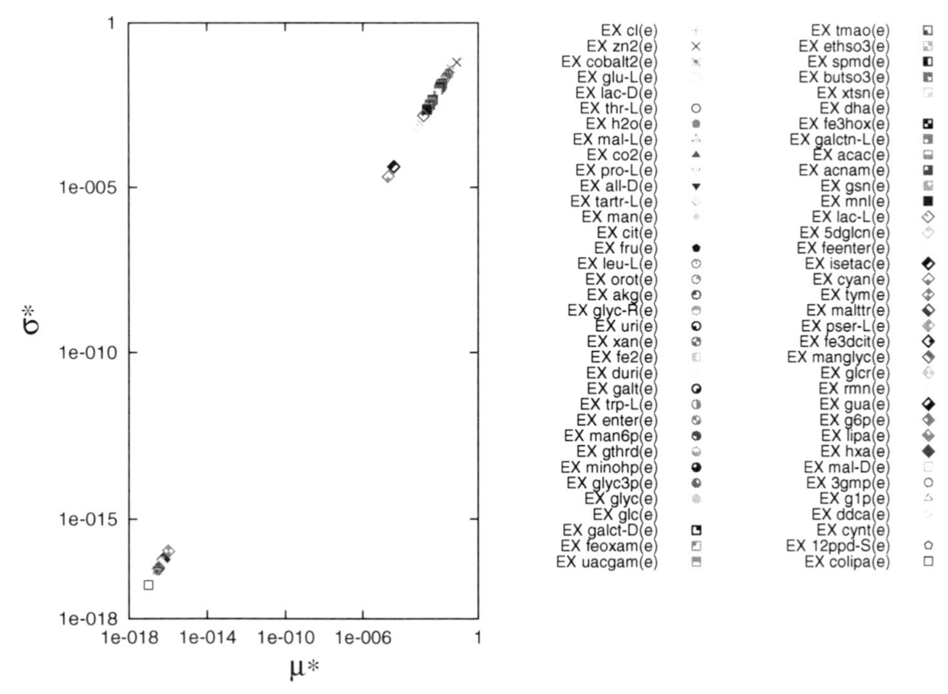

Fig. 2. Uptake Rate-oriented Sensitivity Analysis for the *E. coli* model iAF1260. In this analysis we investigate the input fluxes of the model (299 nutrients) and evaluate their sensitivity with respect to all fluxes of the model. We find that only 70 fluxes (reported in the key) out of 299 are influent, the other ones have sensitivity indexes equal to zero. Results have been obtained averaging over 3000 evaluates of function F.

important. A high μ^* mean indicates an input with an important "overall" influence on the output. A large measure of σ^* variance indicates an input whose influence is highly dependent on the values of the inputs.

2.5 Robustness Analysis

The ability of a system to adapt to perturbations due to internal or external agents, aging, temperature, environmental changes and, in our case, also due to molecular noise and mutation is one of evolutionism guidelines and should also be a fundamental design principle. To optimise the production of a specific metabolite (and simultaneously the formation of biomass, which is necessary to maintain the survival of the bacteria), we used GDMO that obtain a *strain* that maximises the feature required by us. At this point, the validity of the biological strain, designed in-silico, must be tested as regards robustness and sensitivity to endogenous and exogenous perturbations, and this is done by the *robustness analysis*. In this way, we also know the ability of a strain to adapt to small perturbations that can occur at any stage of the biochemical processes within the bacterium, or caused by the environment in which it reproduces. As we shall see, by the term "adaptive capacity" we mean the ability to maintain "acceptable" the performances relative to the metabolite production and biomass formation previously optimised.

There are numerous methods that can be used to fulfil this task. Among these, in [23] the authors consider a big network (in this case, however, considered the Internet network) and use the *theory of percolation on random graphs* to test the robustness of the network in case of random or targeted node deletion, or in case of random link deletion. They associate occupation of nodes or links with their functioning, and for occupation probability they mean the probability of operation of them. They consider that this probability is uniform or depends on the degree of each node (that is the number of connections at that node) distribution. So they analyse the robustness of the network connectivity as the occupation probability is varied. Through this analysis, they highlight that a network with these characteristics is robust to random removal of nodes or links, but not if they are targeted nodes with highest degree. In another work [24], the relationship between the general characteristics of a chemical reaction network and the sensitivity of his equilibrium is investigated according to changes in the overall supply of reagents. The authors define *the sensitivity of a species* as the variation of it with respect to the element concentration one, and they find a lower bound to such sensitivity that depends on the network structure alone. In particular, they argue that a strong robustness of the equilibrium against element variations is likely only if the various species are constructed from building block highly gregarious (i.e. each one binds with many others) or present in some species with high multiplicity. Finally, in [25] the authors use a combined approach of global and local robustness that they call *Glocal Robustness*. The global analysis investigates the parameter space with the aim of finding where a circuit cell shows experimental observed features (global), while the local one determines the robustness of parameter sets sampled during the previous phase.

Similar work making use of the robustness analysis for parameter estimation are also present in [26] and in [27]. In our work, however, we use very simple robustness analysis that shows a high degree of transversality because easily applicable in other fields, as was done in [28] and in [29].

The basic principle of this analysis is as follows. Firstly, we define the perturbation as a function $\tau = \gamma(\Psi, \sigma)$ where γ applies a stochastic noise σ to the system Ψ and generates a trial sample τ. The γ-function is called γ-perturbation. Without loss of generality, we assume that the noise is defined by a random distribution. In order to make statistically meaningful the calculation of robustness, we generate a set T of trial samples τ. Each element τ of the set T is considered robust to the perturbation, due to stochastic noise σ, for a given property (or metric) ϕ if the following condition is verified:

$$\rho(\Psi, \tau, \phi, \epsilon) = \begin{cases} 1, & \text{if } |\phi(\Psi) - \phi(\tau)| \leq \epsilon \\ 0, & \text{otherwise} \end{cases} \tag{4}$$

where Ψ is the *reference system*, ϕ is a *metric* (or property), τ is a *trial sample* of the set T and ϵ is a *robustness threshold*. The definition of this condition makes no assumptions about the function ϕ. It can be anything (not necessarily related to properties or characteristics of the system); however, it is implicitly assumed that it is quantifiable. The robustness of a system Ψ is the number of robust trials of T, with respect to the property ϕ, over the total number of trials ($|T|$). In formal terms:

$$\Gamma(\Psi, T, \phi, \epsilon) = \frac{\sum_{\tau \in T} \rho(\Psi, \tau, \phi, \epsilon)}{|T|} \tag{5}$$

where Γ is a dimensionless quantity that states, in general, the robustness of a system and, in this case, of a strain.

Robustness index is a function of ϵ, so the choice of this parameter is crucial and not a trivial task. Since we are interested in the behaviour of strain when subjected to small perturbations and because the behaviour is acceptable when the deviations from the original value is as small as possible, we choose the values of epsilon equal to 1% of the metric and sigma equal 1% of the perturbed variable.

Based on this principle, we evaluate two values of robustness, the *Global Robustness* value (GR) and the *Local Robustness* value (LR). Also we evaluated the *Normalised Feasible parameter Volume* ([25]) to give a comparison between these results and the GR/LR values. The first two values only differ in the perturbation kind, in particular, chosen σ, it will differ the set of variables that will be perturbed.

Global Robustness. As regards the Global Robustness of a strain, we perturbed the upper and lower bounds of each metabolic flux. Hence, a trial τ is created by perturbing all the upper v_j^U and lower bounds v_j^L, $j = 1, \ldots, n$ of the metabolic flux. We create a set T_τ of trials, and for each of them we perturb all the bounds and evaluate the property $\phi(\tau)$ (by flux balance analysis), which in our case can be the value of acetate, succinate, biomass or a combination of them; and then, we calculate the function ρ. Once a value of ρ is obtained for each of the trials, we compute the value of robustness (Equation 5), which in this case we call *Global Robustness* because all the parameters are perturbed.

Local Robustness. In this case, we perturb again the upper v_j^U and lower bounds v_j^L, $j = 1, \ldots, n$, of a metabolic flux, but we create a sample trial perturbing a single flux, we evaluate the property $\phi(\tau)$ and we calculate the function ρ. After creating a set T_τ of trials, we calculate the robustness (Equation 5), which in this case we call *Local Robustness*. Hence, we calculate a LR value for each metabolic flux.

Normalised Feasible parameter Volume. We also implemented the analysis described in [25] to compare the results obtained by GR and LR. In this analysis, the authors implement a procedure that calculates the volume occupied by those parameters such that the system maintains the desired characteristics. The volume is computed in the 2n-dimensional parameter space. In our case, the volume is such that Equation 4 holds. Since this research requires a huge computational effort, given the high number of dimensions (R^{2n}, where 2n is the number of parameters), it is guided by an iterative procedure that involves the Principal Component Analysis (PCA). In the second part, they calculate local coefficients, and from these they derive which parameters are influential on the robustness (by Spearman's partial correlation coefficient).

In particular, the first part requires two steps. The first is a Monte Carlo sampling obtained with 2n-dimensional Gaussian random variations centred around a parameter vector (known in advance). In our case this vector is represented by the 2n parameters: v_j^U and v_j^L. Then a set $T_\tau^{(1)}$, $2n \times K$ is created, that contains K parameter vectors. Among these, only a fraction will satisfy the Equation 4, the set comprising this fraction is the set of the feasible parameter vectors $V^{(1)}$. The second step begins with a principal component analysis of $V^{(1)}$; this analysis allows to identify statistical linear structures within high-dimension data sets. Here, instead, it is used to guide the sampling of the parameter vectors in subsequent iterations. In particular, $T_\tau^{(2)}$ and the subsequent sets $T_\tau^{(h)}$ are generated from $V^{(1)}$ and, in general, from $V^{(h-1)}$, where $h = 1, \ldots, H$ are the iterations number. In particular the generic element $\tau_{j,k}$ of $T_\tau^{(h)}$ is generated as:

$$\tau_{j,k} = \frac{\sum_{t^*=1}^{T^*} V_{j,t^*}^{(h-1)}}{|T^*|} + \lambda^{(h-1)} \cdot \xi_{j,k}, \tag{6}$$

where $j = 1, \ldots, 2n$, since the columns of $T_\tau^{(h)}$ contain the perturbed values of the parameters v_j^U and v_j^L, $j = 1, \ldots, n$; $k = 1, \ldots, K$ is the cardinality of $T_\tau^{(h)}$; the first term, on the right side, is the average of the elements for each perturbed parameter (that is the average for each row) of the set $V^{(h-1)}$ obtained in the previous iteration; $\xi_{j,k}$ is a Gaussian noise with zero mean and standard deviation equals to the $(j, k)^{th}$–element of the covariance matrix $\Sigma^{(h-1)}$, i.e. the pair wise covariance calculated for all vectors τ_a and τ_b of $V^{(h-1)}$ (the eigenvectors of this matrix are the principal axes of the $V^{(h-1)}$ set by PCA); finally, the real value $\lambda^{(h-1)}$ guides the hth Gaussian process by scaling the standard deviations of the distribution along the PCA directions. The purpose of Equation 6 is to avoid unnecessary sampling in a parameter space region where there are no probably feasible vectors. At the end of this procedure, a hyper-box B is constructed in the parameter space, whose axes are parallel to the PCA axes of the last iteration. The bounds of this box, for each direction, are given by the more extreme elements in the set V^H of the last iteration. Then B is uniformly sampled constructing the final set T_τ; a subset V of T_τ will verify the Equation 4. Finally, the feasible parameter volume will be calculated as $R^{2n} = (|V| \setminus |T_\tau|) * Vol(B)$, where $|.|$ determines the cardinality. The logic of this measure is that as the value of R^{2n} increases as the likelihood that perturbing a parameter vector, another feasible parameter vector is generated increases. Finally, for comparing systems with different number of parameters the *normalised feasible parameter volume* R is defined as $R = \sqrt[2n]{R^{2n}}$. R can be considered as the *permissible average variation per-parameter* that leaves intact the system performance.

The second part of this analysis is connected to the global part. The authors take into account the final set of the feasible parameter vectors V and for each parameter vector produces Q sample trials perturbing the $2n$ parameters by Gaussian noise with zero mean and sigma equal to 0.2; then, they calculate the fraction of robust trials; after that, they repeat the calculations for all vectors. Finally, for the 2n-parameters, they calculate the Spearman partial correlation coefficient with respect to the robust trial fraction values and the different values assumed by the parameters $\delta_j(V(j), X)$, where $j = 1, \ldots, 2n$; $V(j)$ is the jth row of V (containing the observations of the jth-parameter) and X is a vector (containing the values of the robust trial fractions).

3 Results

We tested the performance of GDMO to maximise the production of acetate and succinate in the recent model of *E. coli* K-12 MG1655, iAF1260 [30], and we compared it with GDLS [7], OptFlux [6] and OptGene [5]. In Table 1, we report the productions in wild type, and the results obtained by previous methods and in particular the greater level of acetate and succinate we reach. Pareto and ϵ-optimality present several suitable solutions, which are reported in details in Table 3. Our method reaches interesting results in terms of acetate, succinate, biomass and, mostly, in knockout cost. The knockout cost is defined according to the Boolean relationship between genes. For example, if a gene set is composed by two genes linked by an "AND" relation, the cost to ensure the catalysis of the corresponding reactions is 2, since both genes are necessary to turn on the reactions. Instead, the cost to ensure the turning off of the corresponding reactions (knockout cost) is 1. In our optimisation, indeed, we select as third objective the minimisation of the knockout cost, since in vivo knockout is an expensive and a difficult biotechnological procedure. In all the simulations we initialised the network with an empty set of knockouts, in order to compare our results with GDLS.

Table 1. Comparison between GDMO and previously genetic design methods. We compare OptFlux [6], OptGene [5], GDLS [7], OptKnock [8] and our multi-objective optimisation algorithm (GDMO) for maximising acetate (Ac) and succinate (Suc) production [mmolh^{-1} gdW^{-1}]. The third and fourth rows provide the biomass (Bm) [h^{-1}] and the knockout cost (kc). We report two candidate solutions for acetate optimisation: the first strain, named A_5 (Table 3), provides a low kc equal to 3, and the second one (A_2) reaches an elevated value of acetate, +130.7%, outperforming the previous methods. For succinate production, we obtain +13659% with respect to wild type, deleting only 8 genes (B_3). The last three rows provide the robustness indexes. R values [25] and GR values are global robustness indexes. For LR we report the minimum value found that is associated with the less robust flux (glucose uptake rate). "n.a." stands for *not applicable*.

	Wilde Type	OptFlux	OptGene	GDLS	GDLS	OptKnock	OptKnock	**GDMO**	**GDMO**	**GDMO**
Ac	8.30	15.129	15.138	15.914	n.a.	n.a.	12.565	13.791	19.150	n.a.
		(+82.3%)	(+82.4%)	(+91.7%)	n.a.	n.a.	(+51.4%)	***(+66.13%)***	***(+130.7%)***	n.a.
Suc	0.077	10.007	9.874	n.a.	9.727	9.069	n.a.	n.a.	n.a.	10.610
		(+12877%)	(+12704%)	n.a.	(+12514%)	(+12362%)	n.a.	n.a.	n.a.	***(+13659%)***
Bm	0.23	n.a.	n.a.	0.0500	0.0500	0.1181	0.1165	0.130	0.053	0.087
		n.a.	n.a.	(-78.4%)	(-78.4%)	(-77.9%)	(-49.6%)	(-43.72%)	(-77.10%)	(-62%)
kc	n.a.	n.a.	n.a.	14	26	54	53	***3***	10	***8***
GR	54.76%	n.a.	n.a.	13.76%	16.6%	43.24%	43.08%	45.32%	27.6%	40.40%
LR	54.0%	n.a.	n.a.	16.0%	21.33%	40.0%	40.60%	39.33%	24.0%	46.0%
R	1.30	n.a.	n.a.	1.45	1.45	1.18	1.02	0.78	0.44	1.32

For each solution, we also calculate the Robustness indexes. In particular the Global Robustness index (GR) can be seen as an index to discriminate a strain from each other to choose the best, as regard the robustness with respect to the selected metrics. Moreover, if GR is high, the likelihood of the strain to maintain the performance increases, even if subjected to perturbations. R values indicate the *permissible average variation per-parameter* that leaves intact the system performance. Therefore, also these values can be seen as an index to discriminate a strain from each other. If we consider the GR and R values, we can see a similar behaviour in most cases. *Local Robustness* (LR) index represents the absolute and relative minimum of the results obtained for each strain. Only for the flux related to *D-glucose exchange (Ex glc)*, we obtain LR values less than 100%. Finally, we evaluated Spearman partial correlation coefficients. Since this procedure requires a considerable computational effort, the results of this analysis are calculated for only one strain ($A4$, Table 2). The results indicate that the highest value is $\delta i^* = -0.24$ (the other values are smaller at least by one order of magnitude) that corresponds to the *D-glucose exchange (EX glc)* reaction. It indicates that the Robustness of the strain is more correlated to this reaction. The result is identical to that we obtained with the Local Robustness analysis. Also in this case, the *D-glucose exchange (EX glc)* is the fragile reaction of the strain. Results are shown in Table 2.

The study of genes and reactions of *E. coli* has involved inferring several Pareto trade-offs in anaerobic and aerobic conditions (Figure 3 and Figure 4). The experiments reported in Figure 3 show that GDMO overcomes all the above mentioned methods, and in particular GDLS. The latter performs a single-objective optimisation maximising the synthetic objective function acetate

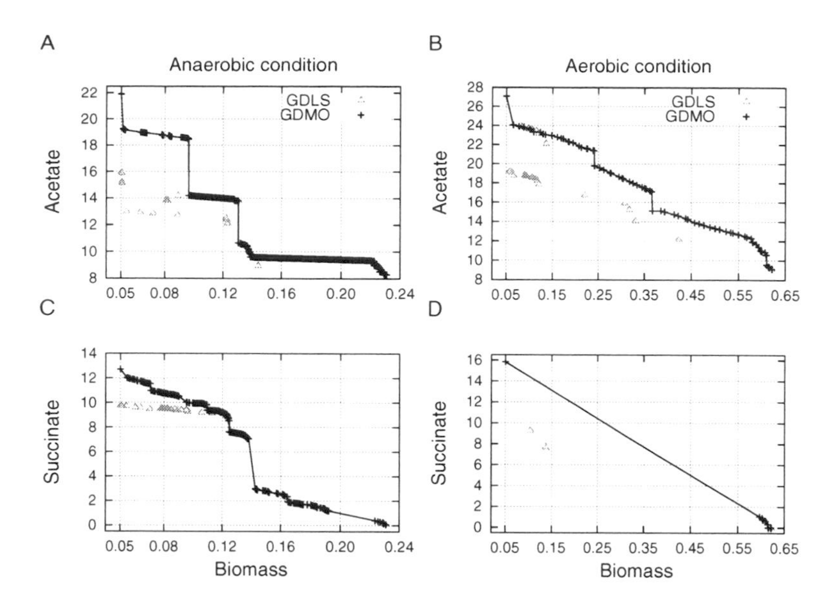

Fig. 3. Maximisation of biomass and acetate production in anaerobic and aerobic conditions (A,B), and maximisation of biomass formation and succinate production in anaerobic and aerobic Condition (C,D), with glucose uptake rate 10 mmolh^{-1} gDW^{-1} in iAF1260. In black the Pareto solutions obtained by GDMO, and in red the optimal results obtained by GDLS [7].

production with knockout cost equal to 14, or optimising succinate production with knockout cost 26, as shown in Table 1 and Figure 3.

The Pareto front strategy is useful to investigate the biological and statistical complexity in several organisms. Figure 5 reports four Pareto curves obtained optimising the acetate/succinate production and the biomass formation in different organisms: the *E. coli* [30], the *Methanosarcina barkeri* [11], *Geobacter sulfurreducens* [10] and *Yersinia pestis* [12]. *M. barkeri* is an archaea able to live in anaerobic condition and produce methane using three known metabolic pathways for methanogenesis. *Geobacter* species are of ecological importance due to bioremediation capabilities. The organism can metabolise uranium, has the ability to generate electricity and can decompose petroleum contaminants in polluted groundwater. For all the organisms, glucose uptake rate is fixed at a maximum of 10 $mmolh^{-1}gDW^{-1}$ as a carbon source (except for *M. barkery*, which does not features the glucose exchange flux in its metabolic network). Some external metabolites (e.g., calcium, ammonia, sulfate, phosphate, oxygen, water, proton, iron (II-III), potassium, sodium, copper, chloride and carbon dioxide) are allowed to both enter and leave the system, while the others are allowed only to leave the system. GDMO highlights the response of different systems and the ability of the organisms to produce the desired metabolite. In the same conditions, *M. barkeri* and *G. sulfurreducens* reach higher levels of acetate than *E. coli*.

Table 2. Robustness analysis results. For each strain we report: the *Global Robustness Value* (GR), the *normalised feasible parameter volume* (R) and the *Local Robustness* (LR) values. For the LR values are shown the minimum associated with the glucose uptake rate. For all other fluxes, we obtained 100% of local robustness.

Strain	GR(%)	R	LR(%)
A_1	28.72	0.39	26.67
A_2	27.60	0.44	24.00
A_3	40.72	1.27	35.33
A_4	41.52	1.74	36.0
A_5	45.32	0.78	39.33
B_1	44.60	0.15	44.67
B_2	43.48	0.92	42.0
B_3	40.40	1.32	46.0
B_4	44.64	1.30	44.0

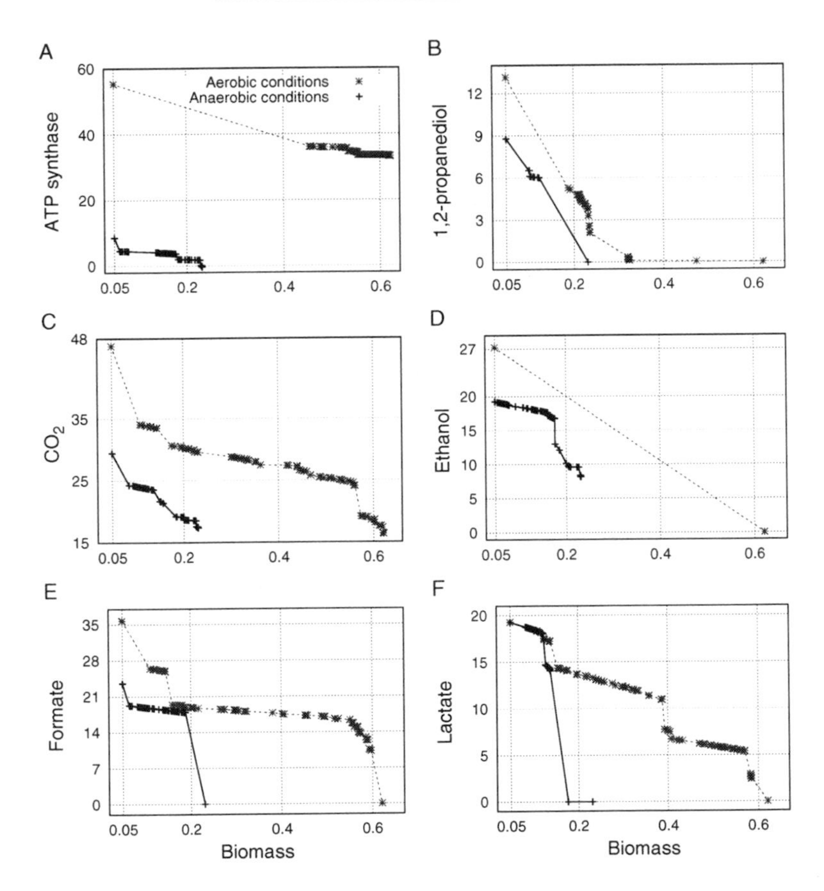

Fig. 4. Pareto fronts for six optimisation problems. We simultaneously maximise biomass formation [h^{-1}] and A) ATP synthase rate, B) 1,2-propanediol, C) CO_2, D) ethanol, E) formate , F) lactate production rates [mmolh^{-1} gDW^{-1}]. In blue we simulate aerobic conditions with O_2=10 mmolh^{-1} gDW^{-1}, in black anaerobic conditions.

Table 3. We report some of the proposed solutions obtained by GDMO to maximise acetate and succinate productions [mmolh^{-1} gDW^{-1}] in *E. coli* network. For each strategy, we report the biomass formation [h^{-1}], the knockout cost (k cost) and the corresponding genes and reactions switched off. The variation of acetate, succinate and biomass in comparison with the wild type is enclosed in brackets.

Strain	Acetate	Biomass	k cost	Knocked out Genes	Deleted Reactions
A_1	19.198	0.052	12	(b0351) OR (b1241)	acetaldehyde dehydrogenase (acetylating)
	(131.26%)	(-77.38%)		(b0910)	cytidylate kinase (CMP)
					cytidylate kinase (dCMP)
				(b2975) OR (b3603)	D-lactate transport via proton symport
					glycolate transport via proton symport, reversible
					L-lactate reversible transport via proton symport
				(b4381)	deoxyribose-phosphate aldolase
				(FdhF and Hyd4) or (FdhF and HycB)*	Formate-hydrogen lyase
				(b0243)	glutamate-5-semialdehyde dehydrogenase
				(b3617)	glycine C-acetyltransferase
				(b0963)	methylglyoxal synthase
				Nuo*	NADH dehydrogenase
A_2	19.150	0.053	10	(b0351) OR (b1241)	acetaldehyde dehydrogenase (acetylating)
	(130.7%)	(-77.10%)		(b3945)	aldose reductase (acetol)
					Glycerol dehydrogenase
					D-Lactaldehyde:NAD+ 1-oxidoreductase
				(b4381)	deoxyribose-phosphate aldolase
				(FdhF and Hyd4) or (FdhF and HycB)*	Formate-hydrogen lyase
				(b3617)	glycine C-acetyltransferase
				(b1380) OR (b2133)	D-lactate dehydrogenase
				(b3236)	malate dehydrogenase
A_3	18.532	0.096	9	(b0351) OR (b1241)	acetaldehyde dehydrogenase (acetylating)
	(123.2%)	(-58.6%)		(b0910)	cytidylate kinase (CMP)
					cytidylate kinase (dCMP)
				(b2975) OR (b3603)	D-lactate transport via proton symport
					glycolate transport via proton symport, reversible
					L-lactate reversible transport via proton symport
				(b4381)	deoxyribose-phosphate aldolase
				(b3617)	glycine C-acetyltransferase
				(b0963)	methylglyoxal synthase
				Nuo	NADH dehydrogenase
A_4	14.046	0.104	5	(b0351) OR (b1241)	acetaldehyde dehydrogenase (acetylating)
	(69.20%)	(-55.14%)		(b3617)	glycine C-acetyltransferase
				(b4025)	glucose-6-phosphate isomerase
				(b3708)	Tryptophanase (L-tryptophan)
A_5	13.791	0.130	3	(b0351) OR (b1241)	acetaldehyde dehydrogenase (acetylating)
	(66.13%)	(-43.72%)		(b1539)	L-allo-threonine dehydrogenase
					D-serine dehydrogenase
					L-serine dehydrogenase
Strain	**Succinate**	**Biomass**	**k cost**	**Knocked out Genes**	**Deleted Reactions**
B_1	12.012	0.055	15	(b0351) OR (b1241)	acetaldehyde dehydrogenase (acetylating)
	(15476%)	(-76.33%)		(b2587)	2-oxoglutarate reversible transport via symport
				(b0870) OR (b2551)	D-alanine transaminase
					alanine transaminase
					L-allo-Threonine Aldolase
					Threonine aldolase
				(b1852)	glucose 6-phosphate dehydrogenase
				(b1849)	GAR transformylase-T
				(b1380) OR (b2133)	D-lactate dehydrogenase
				(b2463)	malic enzyme (NADP)
				(b0963)	methylglyoxal synthase
				(b4388)	phosphoserine phosphatase (L-serine)
				(b2661)	succinate-semialdehyde dehydrogenase (NADP)
				(b1602 AND b1603)	NAD(P) transhydrogenase (periplasm)
				(b3708)	Tryptophanase (L-tryptophan)
B_2	11.530	0.070	10	(b0351) OR (b1241)	acetaldehyde dehydrogenase (acetylating)
	(14875%)	(-69.3%)		(b2587)	2-oxoglutarate transport via symport
				(b3945)	aldose reductase (acetol)
					Glycerol dehydrogenase
					D-Lactaldehyde:NAD+ 1-oxidoreductase
				(b1852)	glucose 6-phosphate dehydrogenase
				(b1380) OR (b2133)	D-lactate dehydrogenase
				(b2463)	malic enzyme (NADP)
				(b2661)	succinate-semialdehyde dehydrogenase (NADP)
				(b1602 AND b1603)	NAD(P) transhydrogenase
B_3	10.610	0.087	8	((b0351)OR(b1241))	acetaldehyde dehydrogenase (acetylating)
	(13659%)	(-62%)		((b3945))	aldose reductase (acetol)
					Glycerol dehydrogenase
					D-Lactaldehyde:NAD+ 1-oxidoreductase
				((b1380)OR(b2133))	D-lactate dehydrogenase
				(b2463)	malic enzyme (NADP)
				(b0767)	6-phosphogluconolactonase
				((b1602ANDb1603))	NAD(P) transhydrogenase
B_4	9.284	0.093	5	((b0356) OR (b1241) OR (b1478))	alcohol dehydrogenase (ethanol)
	(11939%)	(-59.55%)		(b4025)	glucose-6-phosphate isomerase
				(b2501)	polyphosphate kinase
					polyphosphate kinase

*We report the protein 1) "Nuo" associated to the gene set:(b2276 AND b2277 AND b2278 AND b2279 AND b2280 AND b2281 AND b2282 AND b2283 AND b2284 AND b2285 AND b2286 AND b2287 AND b2288),
and 2) "(FdhF and Hyd4) or (FdhF and HycB)" associated to (b4079 AND (b2481 AND b2482 AND b2483 AND b2484 AND b2485 AND b2486 AND b2487 AND b2488 AND b2489 AND b2490) or (b4079 AND (b2719 AND b2720 AND b2721 AND b2722 AND b2723 AND b2724)))

In order to study the favourable environmental conditions, i.e. nutrients for *E. coli*, we performed the simultaneous optimisation of acetate, succinate and biomass on the complete network, i.e. without knockouts. We consider the anaerobic and aerobic condition (O_2 uptake rate $= 10\ mmolh^{-1}gDW^{-1}$) by keeping fixed the glucose uptake rate to 10 $mmolh^{-1}gDW^{-1}$. We use the Non-Dominated Sorting Genetic Algorithm II [13] to perform optimisation by exploring the continuous space of exchange fluxes. The algorithm implements the *Simulated Binary Crossover* operator for crossover and the *polynomial mutation*. In our analysis, the decision variable vector is the lower bound vector of the flux values that constitute the 297 exchange fluxes (glucose and oxygen are kept constant) in the FBA model of *E. coli*. The decision variables are real values from 0 to -100 (0 when the uptake is not allowed, -100 when the potential uptake rate is 100 mmol gDW^{-1} h^{-1}). The algorithm parameters are the population size (set as 100 individuals) and the generation number (set at 500).

Our method reaches the maximum value of acetate (+100 mmol gDW^{-1} h^{-1}), and highlights conflictive behaviour of biomass and succinate (see their maximisation in the Pareto fronts of Figure 6). In anaerobic condition, we found 100 mmolh^{-1} gDW^{-1} h^{-1} of acetate, 42.918 mmolh^{-1} gDW^{-1} of succinate and 3.6204 h^{-1} of biomass (the trade-off). In this condition, we individuated a significant increment in the L-Aspartate, Citrate, Lactose, Fumarate and Malate uptake rates. Instead, in aerobic condition, we found 100 mmolh^{-1} gDW^{-1} h^{-1} of acetate, 21.889 mmolh^{-1} gDW^{-1} of succinate and 4.16 h^{-1} of biomass and a significant increment in the L-Asparagine, 1, 4-alpha-D-glucan, Fe(III)dicitrate, 2-Oxoglutarate uptake rates. In our analysis, we perturbed simultaneously almost

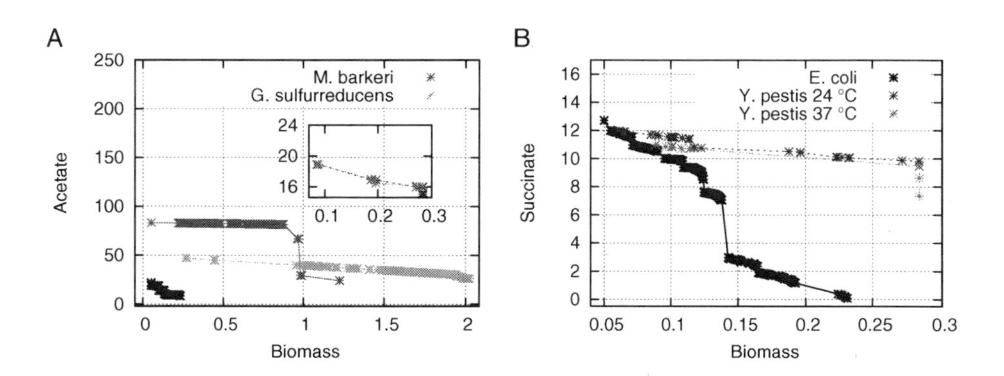

Fig. 5. Pareto fronts obtained by optimising the acetate production [mmolh^{-1} gDW^{-1}] (A), succinate production (B) and the biomass formation [h^{-1}] using GDMO algorithm in four organisms models: *E. coli*, *M. barkeri*, *G. sulfurreducens* and *Y. Pestis*. For *Y. Pestis* we consider two biomass compositions: at 24-28 °C and 37 °C. The significance of these two temperatures stems from the two types of hosts that *Y. Pestis* infects in the natural environment, namely insect vectors at ambient temperature and mammalian hosts with regulated body temperatures of about 37°C. In *M. barkeri*, *G. sulfurreducens* and *Y. Pestis*, the yield of acetate and biomass is larger than *E. coli* due to the lower number of reactions in the metabolic reconstructions.

all the exchange fluxes, but it is possible to select a smaller set of nutrients to study according to experimental requirement.

Sensitivity analysis results are shown in Figure 2, revealing that only 70 out of 299 are influent in the output of the model, i.e., the remaining do not change significantly the metabolic network. In particular, *Chloride, Zinc, Co2+, L-Glutamate exchanges* are the most sensitive (the complete list is reported in Figure 2).

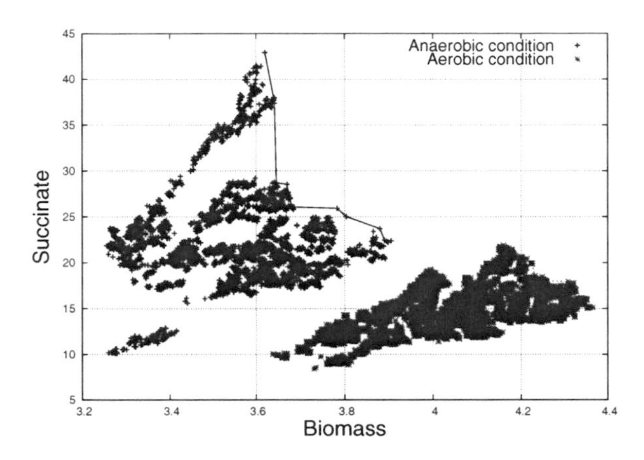

Fig. 6. Feasible regions for acetate production, succinate production (y axis) and biomass formation (x axis). We consider the wild-type bacteria (i.e. knockout zero) and perform the optimisation in aerobic ($O2 = 10$ mmolh^{-1} gDW^{-1}, in blue) and anaerobic (black) conditions on a basis of 10 mmolh^{-1} gDW^{-1} glucose fed to identify favourable nutrients. In both conditions, the algorithm reaches the maximum production of acetate (100 mmolh^{-1} gDW^{-1}).

4 Conclusions

This paper highlights that the Pareto front has a close link with the biotechnology productivity. For the biosynthesis, Pareto optimality is important to obtain not only a wide range of Pareto optimal solutions, but also the *best trade-off design*. Pareto front provides not merely the visualisation of the optimisation process, but also significant information in metabolic design automation. For instance, the size of non-dominated solutions, the first derivative and the area under the Pareto curve could play a key role for the best design within the same organism or between different organisms. Remarkably, the reduced size of the Pareto front could indicate the incompleteness of the model in terms of the number of reactions modelled; in this case, the Pareto optimality could be thought of as a parameter describing the improvement of a model for a bacterium with respect to a previous model for the same bacterium.

Exploratory analysis and comparative metabolic models suggest that the area underlying the Pareto provides an estimate of the number of intermediates which may be exploited for biotechnological purposes (optimisation of an additional objective) or to build synthetic pathways (synthetic biology). Given two bacteria or two conditions for the same bacterium, a larger area under the Pareto front probably represents the best conditions for adding or optimising pathways leading to new biotechnological products. The slope of the Pareto front reflects the progressive lack of pathways able to sustain the production of one component when we are optimising the metabolism to maximise the other. The anaerobic Pareto front has also many more jumps (quick decreases) than the aerobic one. Jumps mark the sudden loss of pathways due to the critical unavailability of an enzymatic step. In other words they correspond to sudden decreases in the availability of entire pathways and subnetworks when a crucial hub is eliminated (e.g., the Krebs cycle). The region of the Pareto front nearby a jump suggests that slight changes of conditions, or a handful of genetic mutations, may result in a large change in the amount of product. Hence, the first derivative, and in particular its discontinuity, indicates the preferable conditions for the metabolites production as highlighted in the Figure 3-C-D regarding succinate and biomass optimisation. In fact, Pareto front in aerobic condition presents a wide jump, confirming that anaerobic condition is favourable for succinate fermentation as given in literature, while in aerobic condition succinate is used as intermediate to produce energy and is totally consumed. GDMO scales effectively as the size of the metabolic system and the number of genetic manipulations increase. Moreover, our results show that the multi-objective approach is very suitable for the genetic design strategies (GDS) discovering. We believe that the algorithm could be further extended and tuned to specific cases.

In the framework we propose in our work, the robustness analysis allows currently to discriminate the strains based on GR or R value: the higher these values, the greater the possibility that bacteria, reproduced in laboratory, maintain the desired performance. In future works, the local robustness analysis and other statistical connected analysis will enable us to reach a better understanding of the metabolic network fragility and this could help the GDMO algorithm to find more robust strains.

References

1. Alper, H., Miyaoku, K., Stephanopoulos, G.: Construction of lycopene-overproducing E. coli strains by combining systematic and combinatorial gene knockout targets. Nature Biotechnology 23(5), 612–616 (2005)
2. Jarboe, L.R., Zhang, X., Wang, X., Moore, J.C., Shanmugam, K.T., Ingram, L.O.: Metabolic engineering for production of biorenewable fuels and chemicals: Contributions of synthetic biology. Journal of Biomedicine and Biotechnology (2010)
3. Atsumi, S., Wu, T.Y., Eckl, E.M., Hawkins, S.D., Buelter, T., Liao, J.C.: Engineering the isobutanol biosynthetic pathway in Escherichia coli by comparison of three aldehyde reductase/alcohol dehydrogenase genes.. Applied Microbiology and Biotechnology 85(3), 651–657 (2010)
4. Orth, J.D., Thiele, I., Palsson, B.Ø.: What is flux balance analysis? Nature Biotechnology 28(3), 245–248 (2010)
5. Patil, K.R., Rocha, I., Förster, J., Nielsen, J.: Evolutionary programming as a platform for in silico metabolic engineering. BMC Bioinformatics 6(1), 308 (2005)

6. Rocha, M., Maia, P., Mendes, R., Pinto, J.P., Ferreira, E.C., Nielsen, J., Patil, K.R., Rocha, I.: Natural computation meta-heuristics for the in silico optimization of microbial strains. BMC Bioinformatics 9(1), 499 (2008)
7. Lun, S.D., Rockwell, G., Guido, N.J., Baym, M., Kelner, J.A., Berger, B., Galagan, J.E., Church, G.M.: Large-scale identification of genetic design strategies using local search. Mol. Syst. Biol. 5(296) (2009)
8. Burgard, A.P., Pharkya, P., Maranas, C.D.: Optknock: a bilevel programming framework for identifying gene knockout strategies for microbial strain optimization. Biotechnology and Bioengineering 84(6), 647–657 (2003)
9. Pharkya, P., Maranas, C.: An optimization framework for identifying reaction activation/inhibition or elimination candidates for overproduction in microbial systems. Metabolic Engineering 8(1), 1–13 (2006)
10. Sun, J., Sayyar, B., Butler, J.E., Pharkya, P., Fahland, T.R., Famili, I., Schilling, C.H., Lovley, D.R., Mahadevan, R.: Genome-scale constraint-based modeling of *Geobacter metallireducens*. BMC Systems Biology 3(1), 15+ (2009)
11. Feist, A.M., Scholten, J.C.M., Palsson, B.Ø., Brockman, F.J., Ideker, T.: Modeling methanogenesis with a genome-scale metabolic reconstruction of Methanosarcina barkeri. Mol. Syst. Biol. 2 (January 2006)
12. Charusanti, P., Chauhan, S., McAteer, K., Lerman, J.A., Hyduke, D.R., Motin, V.L., Ansong, C., Adkins, J.N., Palsoon, B.Ø.: An experimentally-supported genome-scale metabolic network reconstruction for *Yersinia pestis* co92
13. Deb, K., Pratap, A., Agarwal, S., Meyarivan, T.: A fast and elitist multiobjective genetic algorithm: NSGA-II. IEEE Transactions on Evolutionary Computation 6(2), 182–197 (2002)
14. Sendin, J.O., Alonso, A., Banga, J.: Multi-objective optimization of biological networks for prediction of intracellular fluxes. In: Corchado, J., De Paz, J., Rocha, M., Rocha, M., Fernández Riverola, F. (eds.) 2nd International Workshop on Practical Applications of Computational Biology and Bioinformatics (IWPACBB 2008). AISC, vol. 49, pp. 197–205. Springer, Heidelberg (2009)
15. Schuetz, R., Zamboni, N., Zampieri, M., Heinemann, M., Sauer, U.: Multidimensional optimality of microbial metabolism. Science 336(6081), 601–604 (2012)
16. Xu, M., Bhat, S., Smith, R., Stephens, G., Sadhukhan, J.: Multi-objective optimisation of metabolic productivity and thermodynamic performance. Computers & Chemical Engineering 33(9), 1438–1450 (2009)
17. Laumanns, M., Thiele, L., Deb, K., Zitzler, E.: Combining convergence and diversity in evolutionary multiobjective optimization. Evol. Comput. 10(3), 263–282 (2002)
18. Stracquadanio, G., Umeton, R., Papini, A., Liò, P., Nicosia, G.: Analysis and optimization of c3 photosynthetic carbon metabolism. In: Rigoutsos, I., Floudas, C.A. (eds.) Proceedings of 10th IEEE International Conference on Bioinformatics and Bioengineering (IEEE BIBE), Philadelphia, PA, USA, May 31-June 3, pp. 44–51. IEEE Computer Society (2010)
19. Umeton, R., Stracquadanio, G., Papini, A., Costanza, J., Lio, P., Nicosia, G.: Identification of sensitive enzymes in the photosynthetic carbon metabolism. Advances in Experimental Medicine and Biology 736, 441–459 (2012)
20. Zhang, H.X., Goutsias, J.: A comparison of approximation techniques for variance-based sensitivity analysis of biochemical reaction systems. BMC Bioinformatics 11(246) (2010)
21. Rodriguez-Fernandez, M., Banga, J.R.: Senssb: a software toolbox for the development and sensitivity analysis of systems biology models. Bioinformatics 26(13), 1675–1676 (2010)

22. Morris, M.D.: Factorial sampling plans for preliminary computational experiments. Technometrics 33(2), 161–175 (1991)
23. Callaway, D.S., Newman, M.E.J., Strogatz, S.H., Watts, D.J.: Network robustness and fragility: Percolation on random graphs. Physical Review Letters 85, 5468–5471 (2000)
24. Shinar, G., Alon, U., Feinberg, M.: Sensitivity and robustness in chemical reaction networks. SIAM Journal of Applied Mathematics 69(4), 977–998 (2009)
25. Hafner, M., Koeppl, H., Hasler, M., Wagner, A.: Glocal robustness analysis and model discrimination for circadian oscillators. PLoS Comput. Biol. 5(10) (2009)
26. Donaldson, R., Gilbert, D.: A Model Checking Approach to the Parameter Estimation of Biochemical Pathways. In: Heiner, M., Uhrmacher, A.M. (eds.) CMSB 2008. LNCS (LNBI), vol. 5307, pp. 269–287. Springer, Heidelberg (2008)
27. Lodhi, H., Gilbert, D.: Bootstrapping Parameter Estimation in Dynamic Systems. In: Elomaa, T., Hollmén, J., Mannila, H. (eds.) DS 2011. LNCS, vol. 6926, pp. 194–208. Springer, Heidelberg (2011)
28. Umeton, R., Stracquadanio, G., Sorathiya, A., Papini, A., Lio, P., Nicosia, G.: Design of robust metabolic pathways. In: Design Automation Conference (DAC), 2011 48th ACM/EDAC/IEEE, pp. 747–752 (June 2011)
29. Nicosia, G., Rinaudo, S., Sciacca, E.: An evolutionary algorithm-based approach to robust analog circuit design using constrained multi-objective optimization. Knowledge-Based Systems 21(3), 175 (2008), The 27th SGAI International Conference on Artificial Intelligence
30. Feist, A.M., Henry, C.S., Reed, J.L., Krummenacker, M., Joyce, A.R., Karp, P.D., Broadbelt, L.J., Hatzimanikatis, V., Palsson, B.Ø.: A genome-scale metabolic reconstruction for escherichia coli k-12 mg1655 that accounts for 1260 orfs and thermodynamic information. Mol. Syst. Biol. 3(121), 291–301 (2007)

Analysis of Modular Organisation of Interaction Networks Based on Asymptotic Dynamics*

Franck Delaplace[1,**], Hanna Klaudel[1], Tarek Melliti[1], and Sylvain Sené[1,2]

[1] Université d'Evry – Val d'Essonne, IBISC, EA 4526, 91000 Evry, France
[2] Institut rhône-alpin des systèmes complexes, IXXI, 69007 Lyon, France
franck.delaplace@ibisc.univ-evry.fr

Abstract. This paper investigates questions related to modularity in biological interaction networks. We develop a discrete theoretical framework based on the analysis of the asymptotic dynamics of biological interaction networks. More precisely, we exhibit formal conditions under which agents of interaction networks can be grouped into modules, forming a *modular organisation*. Our main result is that the conventional decomposition into strongly connected components fulfills the formal conditions of being a modular organisation. We also propose a modular and incremental algorithm for an efficient equilibria computation. Furthermore, we point out that our framework enables a finer analysis providing a decomposition in elementary modules, possibly smaller than strongly connected components.

Keywords: modularity, interaction networks, discrete dynamics, equilibria.

1 Introduction

The analysis of the relations between the *structure* of a biological system and the *related biological functions* that identify specific states describing particular behaviours is among the most challenging problems [1] at the frontier of theoretical computer science and biology. Let us introduce an illustration of these structure/function relations. On the one hand, gene regulation may be structured into a directed graph, called the *interaction graph*, from which a dynamics is computed. On the other hand, the *attractors* (*i.e.*, stable configurations and/or sustained oscillations) of such a dynamics identify the functions of the system. For instance, for the bacteriophage λ, the reciprocal regulations between genes Cro and cI induce two biological functions, namely the lysis and the lysogeny [2,3,4], each corresponding to a distinct attractor.

* This work is supported by the project SYNBIOTIC of French National Agency for Research, ANR BLAN-0307-01.

** Corresponding author.

D. Gilbert and M. Heiner (Eds.): CMSB 2012, LNCS 7605, pp. 148–165, 2012.

Generally, studying complex biological functions relies on their decomposition into sub-functions identifying some basic behaviours. Each sub-function is supported by a part of the structure. In the context of gene regulation, this part corresponds to a sub-graph of the interaction graph. Thus, the whole system can be viewed in a modular way, where modularity establishes the link between the parts and their related sub-functions. The module composition refers to a structural composition as well as a dynamical one.

In the literature, methods related to module discovery in interaction networks are generally based on both the analysis of the network structures (a field close to graph theory) and the study of their associated dynamics [5]. *Structural analysis* identifies sub-networks with specific topological properties motivated either by a correspondence between topology and functionality [6,7] or by the existence of statistical biases with respect to random networks [8]. Specific topologies like cliques [9], or more generally strongly connected components (SCCs) are commonly used to reveal modules by structural analysis. Particular motifs [10,11] may also be interpreted as modules viewed as basic components. They represent over-represented biological sub-networks with respect to random ones. Moreover, *dynamical analysis* lays on the hypothesis that expression profiles provide insights on the relations between regulators, modules being possibly revealed from correlations between the expressions of biological agents. For instance, using yeast gene expression data, the authors of [12,13] inferred modules from co-regulated genes and the condition under which the regulation occurs. As a consequence, the discovery of a modular organisation in biological interaction networks is closely related to the influence of agents on one another and needs to investigate their expression dynamics [14,15].

In [16,17,18], the authors point out the need to relate structure and function to deal with modular organisations. The objective of this article is to define formally the notion of *modular organisation* as a list of modules together with a composition operation so that the dynamics of the module composition meets the global dynamics of the system. Indeed, modularity is somehow related to an invariance property of module asymptotic dynamics against regulatory perturbations of other modules [19], which supports the idea of viewing the global dynamics as the composition of the module's dynamics. Thus, using a discrete model of biological interaction networks [20,21], we propose an approach that analyses the conditions of module formation and characterises the relations between the global behaviour of a network and the local behaviours of its components. We show under which conditions interaction networks can be divided into modules. As main results, we propose a modular and incremental algorithm to compute equilibria and we show that the conventional structural decomposition into strongly connected components fulfils the formal conditions of being a modular organisation. Furthermore, we show that our framework enables a finer analysis providing a decomposition in elementary modules, possibly smaller than strongly connected components.

The paper is structured as follows: First, Section 2 introduces the main definitions and notations used throughout the paper. Section 3 presents the central notion of a modular organisation of a network along with its structural and dynamical properties. Section 4 defines elementary modular organisation and the conditions leading to obtain it. Some concluding remarks and perspectives are provided in Section 5.

2 The Interaction Network and Its Associated Dynamics

This section introduces the discrete based asynchronous dynamics, modelling the dynamics of biological networks.

Relation. First, we introduce basic notations. Let $\rightharpoonup \subseteq S \times S$ be a binary relation on a set S, given $s, s' \in S$ and $S' \subseteq S$, we denote by $s \rightharpoonup s'$ the fact that $(s, s') \in \rightharpoonup$, by $(s \rightharpoonup) \triangleq \{s' \mid s \rightharpoonup s'\}$ the image of s by $\rightharpoonup$, and by $(S' \rightharpoonup)$ its generalisation to the state set S'. Similarly, we denote by $(\rightharpoonup s)$ and $(\rightharpoonup S')$ the corresponding preimages. The composition of two binary relations will be denoted by $\rightharpoonup \circ \rightharpoonup'$ and the reflexive and transitive closure by $\rightharpoonup^* = \bigcup_{i \in \mathbb{N}} \rightharpoonup^i$, with $\rightharpoonup^0$ as the identity relation.

States and Operations on States. Given a set $A = \{a_1, \ldots, a_n\}$ of agents of interest, each $a_i \in A$ has a *local state*, denoted by s_{a_i}, taking values in some nonempty finite set S_{a_i}. In the examples, all S_{a_i} are Boolean sets $\{0, 1\}$, but the proposed framework is not restricted to it. A *state* of A (or a configuration) is defined as a vector $s \in S$ associating to each $a_i \in A$ a value in S_{a_i}, where $S \triangleq S_{a_1} \times \ldots \times S_{a_n}$ is the set of all possible states. For any $X \subseteq A$ and $s \in S$, we denote by $s|_X$ the restriction of s to the agents in X, and by $s|^X$ the completion of s by all the values of agents in X; these notations extend to sets of states naturally. For example, the completion of the state $s_{a_2} = 0$ by the set of agents $\{a_1, a_3\}$ is $s_{a_2}|^{\{a_1,a_3\}} = \{000, 001, 100, 101\}$, and the restriction of $s = 101$, $s \in S_{a_1} \times S_{a_2} \times S_{a_3}$, on $\{a_2\}$ is 0. The X-equivalence defines an equivalence relation on states with regard to the state restriction on the agent set X.

Definition 1. *Two states $s_1, s_2 \in S$ are said to be* X-equivalent *and denoted by $s_1 \sim_X s_2$, for some $X \subseteq A$, if and only if $s_1|_X = s_2|_X$, i.e., if they cannot be distinguished in $S|_X$.*

Evolution and Asynchronous Dynamics. An *evolution* is a relation on states $\rightharpoonup$. Each $s \rightharpoonup s'$ is a transition meaning that s evolves to s' by $\rightharpoonup$. Thus, the *global evolution* of η can be represented by a directed graph $\mathcal{G} = (S, \rightharpoonup)$ called the *state graph*. In this work, we pay particular attention to *local evolutions*, since each agent $a \in A$ has its own evolution $\rightharpoonup_a$. The collection of all these local evolutions results in the *asynchronous* view of the global evolution of η, *i.e.*, $\rightharpoonup = \bigcup_{a \in A} \rightharpoonup_a$.

Definition 2. *The* asynchronous dynamics *(or* dynamics *for short) of a network η is the triple $\langle A, S, (\rightharpoonup_a)_{a \in A}\rangle$, where A is a set of agents, S is a set of states, and for each $a \in A$, $\rightharpoonup_a \subseteq S \times S$ is a total or empty relation characterising the evolution of agent a such that for any $s \rightharpoonup_a s'$, either $s = s'$ or s differs from s' only on the a-th component.*

Interaction Network and Interaction Graph. We are now in a position to introduce formally the *interaction network* as a family of functions $\eta = \{\eta_a\}_{a \in A}$, such that each $\eta_a : S \to S_a$ defines the next state $\eta_a(s)$ with respect to the asynchronous evolution of a from s. Network η allows to deduce a directed *interaction graph* $G \triangleq (A, \longrightarrow)$ such that $a_i \longrightarrow a_j$ if a_i occurs in the definition of η_{a_j}. When $\rightharpoonup_a$ is empty for some a (*i.e.*, the local state of a remains invariant), then a plays the role of an *input*, which means that no other agents of A influence it (*i.e.*, there are no arcs towards a in G), see Figure 1.

Orbit and Equilibrium. Given a set $S' \subseteq S$, we introduce the following notions:

- an *orbit* of S', $\Omega(S')$, is the set of states comprising S' and all the states reachable from S' by $\rightharpoonup$;
- an *equilibrium* $e \in S$ is a state reachable infinitely often by $\rightharpoonup$; $\Psi(S')$ denotes the set of equilibria reachable from S';
- an *attractor* is a set of equilibria $E \subseteq S$ such that $\forall e \in E\colon \Psi(\{e\}) = E$. In a state graph, an attractor is the set of states comprised in one terminal strongly connected component[1] that can be of two kinds:
  - a *stable state* is a singleton $E \subseteq S$;
  - a *limit set* is an attractor E such that $|E| > 1$.

Moreover, the restriction of $\rightharpoonup$ to $X \subseteq A$ is defined as: $\rightharpoonup_X = \bigcup_{a \in X} \rightharpoonup_a$. Orbits and equilibria are determined by two operators having two arguments, an agent set and a state set.

Definition 3. *The* orbit operator Ω *and the* equilibrium operator Ψ, *are defined as follows for $X \subseteq A$ and $S' \subseteq S$:*

- $\Omega_X(S') = (S' \rightharpoonup_X^*)$;
- $\Psi_X(S') = \{s \in \Omega_X(S') \mid \forall s' \in S : s \rightharpoonup_X^* s' \implies s' \rightharpoonup_X^* s\}$.

The equilibrium operator Ψ_X is idempotent, upper-continuous and monotone (see Proposition 2 in Appendix).

The example in Figure 1 illustrates the dynamics of an interaction network. It is defined by η, each η_a being the local transition function of agent a. Given a state s, the evolution $s \rightharpoonup_a s'$ means that s' is obtained by applying η_a to s, *i.e.*, $s \rightharpoonup_a s' \triangleq s'_a = \eta_a(s) \wedge (\forall a' \in A \setminus \{a\} : s'_{a'} = s_{a'})$. Let us remark that:

- the *orbit* of $\{1111\}$ is $\Omega(\{1111\}) = \{1111, 1101, 1100\}$;
- 1100 is an *equilibrium*, as well as 0101 and 0010 are;
- the set of equilibria reachable from 0000 is $\{0xyz \mid x, y, z \in \{0, 1\}\}$;
- two attractors exist: a stable state $\{1100\}$ and a limit set $\Psi(\{0000\})$.

[1] Recall that, in a terminal strongly connected component each path starting from a vertex of the component remains in this component.

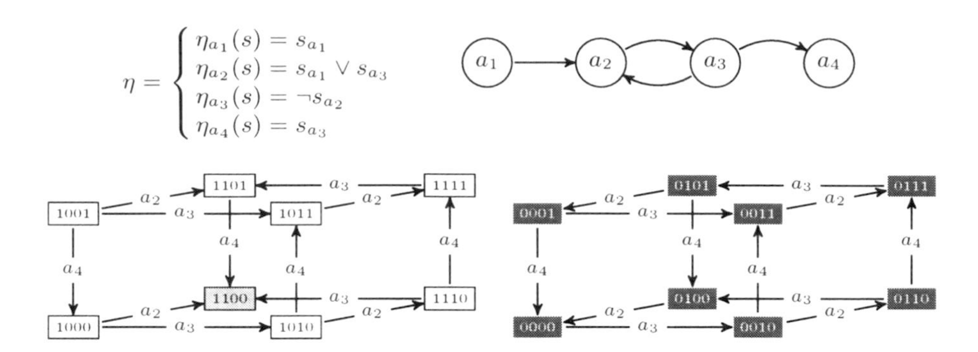

Fig. 1. An interaction network η (top left), its graphical representation (top right), and the state graph $\mathcal{G}$ of η composed of two disconnected components (bottom). $\mathcal{G}$ represents the dynamics of η for each state $s \in \{0,1\}^4$, in which by convention, self loops are omitted, stable states are depicted in gray while limit sets are in black.

Regulation. The regulation is a sub-relation of the interaction specifying a dynamics-based dependence between two agents. Agent a_k *regulates* agent a_ℓ, if at least one modification of a state of a_ℓ requires a modification of a state of a_k.

Definition 4. *an interaction,* $a_k \longrightarrow a_\ell$ *is a* regulation *if and only if there exist two states* $s, s' \in S$ *such that* $(s \sim_{A\setminus\{a_k\}} s') \wedge ((s \rightharpoonup_{a_\ell}) \nsim_{a_\ell} (s' \rightharpoonup_{a_\ell}))$. *By extension, given* $X_i, X_j \subseteq A$, $X_i \longrightarrow X_j$ *if and only if* $\exists a_k \in X_i, \exists a_\ell \in X_j :$ $a_k \longrightarrow a_\ell$.

It may arise that the graph of interaction differs from the graph of regulation because the interaction depends on the syntactic definition of a network whereas the regulation relies on a property of the dynamics. In Figure 1, the sets of regulators of agents a_1, a_2, a_3 and a_4 are respectively $\{a_1\}$, $\{a_1, a_3\}$, $\{a_2\}$ and $\{a_3\}$. Notice also that there are the following relations on sets of agents: $\{a_1, a_2\} \longrightarrow \{a_3, a_4\}, \{a_3\} \longrightarrow \{a_2, a_4\}$ and $\{a_1\} \longrightarrow \{a_2, a_3\} \longrightarrow \{a_4\}$. Another example of a regulation graph is given in Figure 2 (right). It shows that interaction (a_1, a_2) in the interaction network is actually not a regulation because no modification of a_1 influences the state of a_2. All other interactions are effective, meaning that the underlying regulation graph contains all interactions from the network but (a_1, a_2).

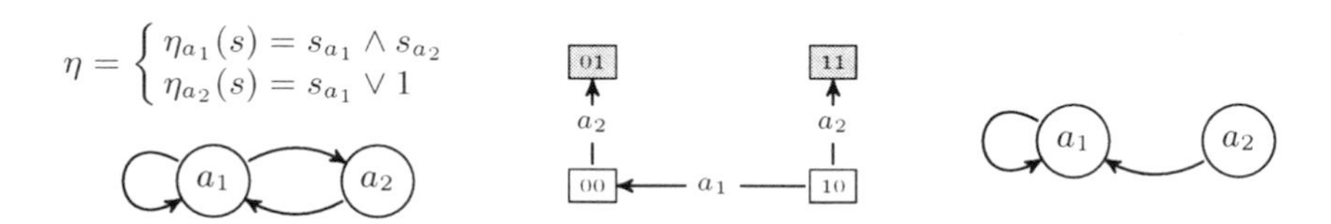

Fig. 2. An interaction network, its state graph and the corresponding regulation graph

3 Composition of Equilibria

In this section, a relation between the equilibria of an interaction network and that of its parts is presented. It allows to consider a modular view of the system in which each part is seen as a *module*, *i.e.*, a subset of agents. It means that modules, which influence each other, reveal the underlying *biological functions* materialised by their equilibria.

3.1 Modular Organisation

Our objective is to find a decomposition of the set of agents A into modules, *i.e.*, a partition[2] of A, together with a composition operator $\oslash$ of the equilibria of these modules, allowing to retrieve the global equilibria of the network. Finding an adequate operator $\oslash$ is a challenging question. Basically, a modular organisation $(X_1, \ldots, X_m)$ should satisfy the following equation characterizing the composition of the module equilibria:

$$\Psi_{X_1} \oslash \ldots \oslash \Psi_{X_m} = \Psi_{\bigcup_{i=1}^m X_i}. \tag{1}$$

One can easily see that, in general, taking $\oslash = \cup$ for example is not a solution. If we consider the interaction network in Figure 3 and a partition into two parts $\{a_1, a_2\}$ and $\{a_3\}$, then the corresponding sets of equilibria are respectively $\{110, 111\}$ and $\{000, 001, 100, 101, 011, 111\}$, while the set of global equilibria is $\{111\}$, which is not the union of the previous ones. However, one may see that, for the same parts, the computation of the equilibria of $\{a_3\}$ from the equilibria of $\{a_1, a_2\}$, gives the expected property $\Psi_{\{a_3\}} \circ \Psi_{\{a_1,a_2\}} = \Psi_{\{a_1,a_2,a_3\}}$, whereas $\Psi_{\{a_1,a_2\}} \circ \Psi_{\{a_3\}} \neq \Psi_{\{a_1,a_2,a_3\}}$. This suggests that the order in which parts are taken into account plays an important role in the definition of the composition operator. Unfortunately, in general none of the usual operators such as $\cup$, $\cap$ and $\circ$ can be used as the modular composition operator as one can check in the following network: $\{\eta_{a_1}(s) = s_{a_1} \wedge \neg s_{a_3}, \eta_{a_2}(s) = s_{a_2} \wedge \neg s_{a_3}, \eta_{a_3}(s) = s_{a_2}\}$, while a modular decomposition exists: $\{a_1\}$ followed by $\{a_2, a_3\}$.

Thus, we will focus on an *ordered partition* $\pi = (X_1, \ldots, X_m)$ of A, *i.e.*, a partition of A provided with a strict total order and represented by a sequence, called a *modular organisation*, preserving (1). Furthermore, we would like to be able to "fold" contiguous modules in π in order to deal with them as with a single module[3], while preserving the result of the composition of equilibria. As a consequence, we require a modular organisation to support *folding* and to be such that the composition operator $\oslash$ is associative according to the order in π.

In order to form a modular organisation, the modules and their order in π should satisfy some conditions related to their dynamics. Intuitively, two disjoint sets of agents X_i and X_j, $i < j$, can be modules in π, either if they do not regulate each other, or if X_i regulates X_j. In both cases, we can remark that the equilibria of X_i should embed the asymptotic evolution of X_j, which leads to encompass

[2] A *partition* of a set A is a set of nonempty disjoint subsets of A which covers A.

[3] The folding of modules corresponds to the union of these modules.

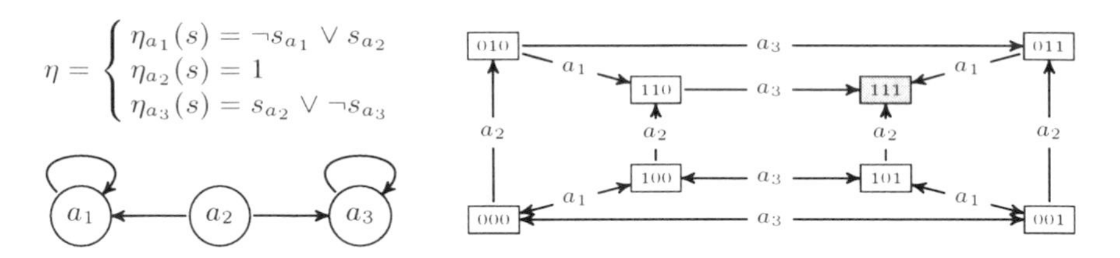

Fig. 3. An interaction network and its state graph

the equilibria of X_j in the equilibria of X_i. These conditions are expressed by the *modularity relation* (M-relation).

Definition 5. *The* M-relation $\rightsquigarrow \subseteq \mathcal{P}(A) \times \mathcal{P}(A)$ *is defined as:*

$$X_i \rightsquigarrow X_j \triangleq \forall S' \subseteq S : (\Psi_{X_i} \circ \Psi_{X_i \cup X_j}(S'))_{\rightharpoonup X_j} \subseteq (\Psi_{X_i} \circ \Psi_{X_i \cup X_j}(S')).$$

Some fundamental properties of the M-relation can be found in Proposition 3 of the Appendix. In this context, a modular organisation can be defined as follows.

Definition 6. *A* modular organisation $(X_1, \ldots, X_m)$ *is an ordered partition of* A *such that for all* $1 < i \leq m$: $(\bigcup_{j=1}^{i-1} X_j) \rightsquigarrow X_i$.

From Definition 6, Proposition 1 states that being a modular organisation is preserved by any folding of its contiguous parts[4].

Proposition 1. *Let* $\pi = (X_1, \ldots, X_m)$ *be a modular organisation. For all* $1 \leq i \leq j \leq m$, $(X_1, \ldots, X_{i-1}, \bigcup_{k=i}^{j} X_k, X_{j+1}, \ldots, X_m)$ *is a modular organisation.*

In the literature [6,10,11], modules are frequently assimilated to SCCs of interaction networks. Although these works focus on structural arguments only, it turns out that they are compatible with Definition 6. Indeed, any topological order[5] of SCCs is actually a modular organisation. Notice that, a topological order on the quotient graph of SCCs always exist since the graph is acyclic. For instance, $(\{a_1\}, \{a_2, a_3\}, \{a_4\})$ is a modular organisation of the interaction network presented in Figure 1. In what follows, we present an approach addressing formally this aspect. As a result, we show that, in particular, the structural decomposition in SCCs makes sense and may be improved by a deeper analysis leading to the decomposition of SCCs in elementary modules (see Section 4), potentially smaller than those coming from SCCs.

3.2 Regulation and Modularity Relation

The regulation and the M-relation are related, as shown below.

Lemma 1. *For any* X_i, X_j *subsets of* A: $\neg(X_j \longrightarrow X_i) \implies X_i \rightsquigarrow X_j$.

[4] Proofs are in Appendix.

[5] A topological order is a total order obtained by topological sorting.

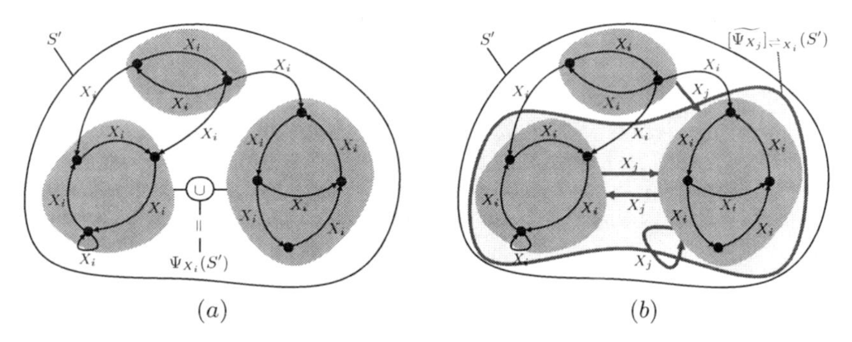

Fig. 4. Successive steps leading to the definition of the composition operator $\oslash$. The SCCs defined for $\rightharpoonup_{X_i}$ are in gray (a). The two terminal SCCs at the bottom correspond to attractors of $\rightharpoonup_{X_i}$ (a). The equilibria $\Psi_{X_i \cup X_j}(S')$ is computed from $\rightharpoonup_{X_j}$ (bold X_j arrows) on these attractors (b).

According to Definition 6, Theorem 1 provides a connection between structural properties of a regulation graph and the corresponding modular organisations (possibly reduced to a single module).

Theorem 1. *Any topological order of the SCC quotient graph of a regulation graph is a modular organisation.*

3.3 Composition Operator

In this section, we present the successive steps leading to the definition of the composition operator $\oslash$. From (1), $\oslash$ is a binary operator that applies on the equilibria of parts X_i and X_j of π, with $i < j$. Thus, its definition is based on the attractors of $\rightharpoonup_{X_i}$ which correspond to terminal nodes (terminal SCCs) of the SCC quotient graph of $\rightharpoonup_{X_i}$, namely $\mathcal{G}/_{\rightleftharpoons X_i}$, where $\rightleftharpoons_{X_i}$ is the equivalence relation identifying states belonging to the same SCC and defined as $s \rightleftharpoons_{X_i} s' \triangleq (s \rightharpoonup^*_{X_i} s') \wedge (s' \rightharpoonup^*_{X_i} s)$. For any $S' \subseteq S$, an attractor of $\rightharpoonup_{X_i}$ coincides with $[s]_{\rightleftharpoons X_i} \subseteq \Psi_{X_i}(S')$ (see Figure 4.a). Moreover, for all $S' \subseteq S$, we denote by:

- $[S']_{\rightleftharpoons X_i} = \{[s]_{\rightleftharpoons X_i} | s \in S'\}$ the set of equivalence classes of $\rightleftharpoons_{X_i}$ in S';
- $[s]_{\rightleftharpoons X_i} \; [\rightharpoonup_{X_j}]_{\rightleftharpoons X_i} \; [s']_{\rightleftharpoons X_i} \triangleq \exists s \in [s]_{\rightleftharpoons X_i}, \exists s' \in [s']_{\rightleftharpoons X_i} : s \rightharpoonup_{X_j} s'$ the evolution by agents of X_j on these equivalence classes.

We define an operator $[\widetilde{\Psi_{X_j}}]_{\rightleftharpoons X_i}$, similar to the equilibria operator, computing the set of equilibria of $[\rightharpoonup_{X_j}]_{\rightleftharpoons X_i}$ in $\mathcal{G}/_{\rightleftharpoons X_i}$ (see Figure 4.b and Section 3.4 for the algorithm) as follows:

$$[\widetilde{\Psi_{X_j}}]_{\rightleftharpoons X_i}(S') \triangleq \{[s]_{\rightleftharpoons X_i} \in [S']_{\rightleftharpoons X_i} \mid (([s]_{\rightleftharpoons X_i}[\rightharpoonup_{X_j}]^*_{\rightleftharpoons X_i}) \subseteq [S']_{\rightleftharpoons X_i}) \wedge$$
$$\forall [s']_{\rightleftharpoons X_i} \in [S]_{\rightleftharpoons X_i} : [s]_{\rightleftharpoons X_i}[\rightharpoonup_{X_j}]^*_{\rightleftharpoons X_i}[s']_{\rightleftharpoons X_i} \implies [s']_{\rightleftharpoons X_i}[\rightharpoonup_{X_j}]^*_{\rightleftharpoons X_i}[s]_{\rightleftharpoons X_i}\}.$$

Operator $\oslash$ is thus defined as:

$$\Psi_{X_i} \oslash \Psi_{X_j} \triangleq \mathsf{Flat} \circ [\widetilde{\Psi_{X_j}}]_{\rightleftharpoons X_i} \circ \Psi_{X_i}, \tag{2}$$

where, for any set $E \subseteq \mathcal{P}(S)$, $\mathsf{Flat}(E) = \bigcup_{e \in E} e$ flattens the set. Thus, if applied to a set of attractors, Flat gives the underlying set of equilibria. For example, $\mathsf{Flat}(\{\{00, 01\}, \{10\}\}) = \{00, 01, 10\}$. As a result, one can see that $\oslash$ does compute the set of states belonging to the attractors of X_i which are also the equilibria of $\rightharpoonup_{X_j}$.

Lemma 2 below shows that the global equilibria of $X_i \cup X_j$ are obtained using the composition $\Psi_{X_i} \oslash \Psi_{X_j}$.

Lemma 2. *For all X_i, X_j disjoint subsets of A:*

$$X_i \rightsquigarrow X_j \implies \forall S' \subseteq S : \Psi_{X_i \cup X_j}(S') = (\Psi_{X_i} \oslash \Psi_{X_j}) \circ \Omega_{X_i \cup X_j}(S').$$

As a main result, the computation of global equilibria of an interaction network can be obtained modularly (Theorem 2).

Theorem 2. *Let $A' = \bigcup_{i=1}^{m} X_i \subseteq A$ be a set of agents, if $(X_1, \ldots, X_m)$ is a modular organisation then we have:*

$$\Psi_{A'} = (\Psi_{X_1} \oslash \ldots \oslash \Psi_{X_m}) \circ \Omega_{A'}.$$

3.4 Modular and Incremental Computation of Equilibria

Modularity allows an incremental and efficient computation of the equilibria avoiding the generation of the complete state space S. The algorithm presented in Figure 5 is an application of Theorem 2 introducing incremental processing based on the fact that the evolution of an agent only depends on the current states of its regulators. Indeed, the equilibria of a set of agents X in a state space $S|_{X \cup Y}$ where X and Y are disjoint sets and $X \cup Y$ contains the regulators $R_X = (\longrightarrow X)$ of X, can be computed from the restriction to $X \cup R_X$ completed by $Y \setminus R_X$. In other words, the equilibria computation operator and the completion operator commute for an appropriate selection of sets of agents, as shown by the following lemma.

Lemma 3. $\Psi_X(S|_{X \cup Y}) = \Psi_X(S|_{X \cup R_X})|^{Y \setminus R_X}$,
with $R_X = (\longrightarrow X)$, $R_X \subseteq X \cup Y$ and $X \cap Y = \emptyset$.

Each step of the algorithm is seen as the computation of equilibria for the following modular organisation (X, X_i) where $X = X_1 \cup \ldots \cup X_{i-1}$ is the folding of the modules preceding X_i in the initial modular organisation and corresponds to the following equation:

$$\Psi_{X \cup X_i}(S) = \mathsf{Flat}\left(\left[\widetilde{\Psi_{X_i}}\right]_{\rightleftharpoons X} \left(\Psi_X \left(S|_{X \cup R_X}\right)|^{X_i \setminus R_X}\right)\right). \tag{3}$$

The algorithm of Figure 5 is divided in two parts: PART I corresponding to the completion of states by X_i, and PART II corresponding to the computation of

Input: $(X_1, \ldots, X_m)$ a modular organisation of A.
Result: the set of equilibria, $\Psi_A(S)$.
Function: TermSCCs $(\mathcal{G})$ computes the set of terminal strongly connected components of a graph $\mathcal{G}$.
Variables:
- A': set of agents already processed;
- R_{X_i} : regulators of X_i;
- N : set of new agents;
- $\tilde{\psi}, \tilde{\psi}_{tmp}$: set of attractors;
- $\mathcal{G}$: quotient graph with attractors as vertices.

States are encoded by words, with ϵ as the empty word.
// initialisation
$A' = \emptyset$;
$\tilde{\psi} = \{\{\epsilon\}\}$;
for $i = 1$ **to** m **do**

PART I
 // Extension of attractor states
 $R_{X_i} = (\longrightarrow X_i)$; *// regulators of* X_i
 $N = (X_i \cup R_{X_i}) \setminus A'$; *// new agents for evolution computation*
 $\tilde{\psi}_{tmp} = \emptyset$;
 foreach $Att \in \tilde{\psi}$ **do**
  // structure preserving completion of attractors ;
  foreach $s_N \in S|_N$ **do**
  $\tilde{\psi}_{tmp} = \tilde{\psi}_{tmp} \cup \{\{s|\ s|_{A'} \in Att \wedge s|_N = s_N\}\}$;
 end
 $\tilde{\psi} = \tilde{\psi}_{tmp}$;

PART II
 // Attractors computation

PART II.1
 // Quotient graph computation with attractors as vertices
 $\tilde{\psi}_{tmp} = \emptyset$;
 foreach $Att \in \tilde{\psi}$ **do**
  // computation of the core of attractors ;
  if $\left(Att[\rightharpoonup_{X_i}]^*_{\rightleftharpoons_{A'}}\right) \subseteq \tilde{\psi}$ **then** $\tilde{\psi}_{tmp} = \tilde{\psi}_{tmp} \cup \{Att\}$;
 end
 $\mathcal{G} = (\tilde{\psi}_{tmp}, [\rightharpoonup_{X_i}]_{\rightleftharpoons_{A'}})$; *// quotient graph of* $\rightharpoonup_{X_i}$ *defined on the core*
 $\tilde{\psi}_{\mathcal{G}} =$ TermSCCs$(\mathcal{G})$; *// equilibria computation on* $\mathcal{G}$

PART II.2
 // Computation of the set of attractors for $X_1, \ldots, X_i$
 $\tilde{\psi} = \emptyset$;
 foreach $Att \in \tilde{\psi}_{\mathcal{G}}$ **do** $\tilde{\psi} = \tilde{\psi} \cup \{\mathsf{Flat}(Att)\}$; *// flatten each attractor* $\rightarrow \tilde{\psi}$
 $A' = A' \cup X_i$;

end
return Flat$(\tilde{\psi})$; *// flatten the attractor set* $\rightarrow$ *equilibria set*

Fig. 5. Algorithm of modular and incremental computation of equilibria

the attractors taking into account $\rightharpoonup_{X_i}$. In PART I the attractors are duplicated by completing their state values while preserving the structure of attractors. PART II is divided into two subparts. PART II.1 computes the quotient graph with attractors as vertices and the quotiented evolution by X_i as arcs. Notice that some attractors are removed during this step. Indeed, from some states belonging to attractors, $\rightharpoonup_{X_i}$ may reach states located outside the attractors set. By definition of the M-relation (Definition 5), they cannot be considered as equilibria and are not included. The remaining set of attractors, called the *core*, is used in PART II.2 to compute the equilibria as the terminal SCCs of the core graph. Since TermSCCs returns a set of "attractors of attractors", they are finally flattened to retrieve the structure of a set of equilibria.

The complexity of the algorithm in the product of the maximal number of equilibria by the number of agents is exponential in general. It is however linear for acyclic regulatory graphs. Let α_i be the number of equilibria computed at step i, and N the set of newly introduced agents. The computation time is bounded by $k \cdot \sum_{i=1}^{m} \alpha_i \cdot 2^{|N|}$, for $k \in \mathbb{N}$. The exponential time corresponds to the computation of the completion and of the quotient graph. $|N|$ is bounded by β, the number of agents in the greatest SCC, under the assumption that all modular organisations are subdivisions of topological orders of the SCC quotient graphs. Indeed, in the modular organisation the regulators of X_i always precede X_i unless they are also regulated by X_i. Hence, the computation time is bounded by $k \cdot 2^{\beta} \cdot m \cdot \alpha$ where α stands for the maximal number of equilibria for all steps, leading to a complexity in $\mathcal{O}(2^{\beta} \cdot m \cdot \alpha)$.

In the worst case, notably corresponding to a regulatory graph reduced to a single SCC, the complexity is the same as the brute-force algorithm computing equilibria from the whole state graph (*i.e.*, $\pi = \{A\}, \beta = |A|$). The algorithm is more efficient in practice. In particular, for networks whose interaction graph is acyclic, each module corresponds to a single agent ($\beta = 1$) leading to a complexity in $\mathcal{O}(|A| \cdot \alpha)$ with $\alpha \leq 2^{\Delta_0}$ where Δ_0 is the number of all input agents having no regulators but possibly themselves. Hence, for regulatory path-graphs, the algorithm is linear in the number of agents because $\Delta_0 = 1$.

A modular organisation based on SCCs is computed by first identifying the quotient graph of SCCs and then obtaining a topological order, whose complexity is in $\mathcal{O}(|A|^2)$.

4 Elementary Modular Organisation

Informally, a module is *elementary* if it is not separable, *i.e.*, if the equilibria of each of its agents depend entirely on the equilibria of all the others. For instance, consider negative circuits that lead to asymptotic sustained oscillations [21]. In such regulation patterns, the equilibria of an agent cannot be encompassed into that of the others because, in order to reach its own equilibria, each agent evolves from the equilibria of all the others.

In this context, a modular organisation provided by some topological orders of the SCC quotient graph (see Theorem 1) does not always provide an elementary decomposition.

Figure 6 depicts an interaction network η composed of three agents a_1, a_2 and a_3 with the associated strongly connected regulation graph. Its underlying state graph shows that the global dynamics of η leads to two attractors, stable state $\{111\}$ and limit set $\{000, 100, 101, 001\}$. It is easy to see that $\{a_2\} \rightsquigarrow \{a_1, a_3\}$ and that this M-relation is (obviously) preserved by folding, because there are only two modules. Hence, ordered partition $(\{a_2\}, \{a_1, a_3\})$ is a modular organisation of η. However, separability is not possible in general as illustrated in Figure 7.

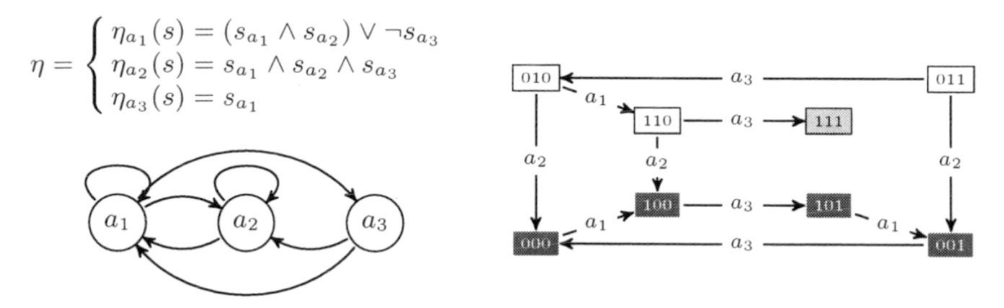

Fig. 6. A separable network with modular organisation $(\{a_2\}, \{a_1, a_3\})$

Indeed, starting from modular organisation $\pi = (\{a_1\}, \{a_2, a_3\})$ obtained from the SCCs, the separation of $\{a_2, a_3\}$ should lead to one of the ordered partitions $\pi' = (\{a_1\}, \{a_2\}, \{a_3\})$ and $\pi'' = (\{a_1\}, \{a_3\}, \{a_2\})$. The condition for π' to be a modular organisation is that $\{a_1, a_2\} \rightsquigarrow \{a_3\}$, *i.e.*, the evolution by a_3 from the equilibria of $\{a_1, a_2\}$ has to be included in the equilibria of $\{a_1, a_2\}$. We can observe that the attractors for $\{a_1, a_2\}$ are $\{001, 101\}$ and $\{010, 110\}$, while the evolution by a_3 from either 101 or 110 leaves the attractors of $\{a_1, a_2\}$, which means that π' is not a modular organisation. As a consequence, $\{a_2, a_3\}$ cannot be separated. Indeed, here, although agents a_2 and a_3 are together M-related and thus can be separated *a priori*, they cannot be in the context of agent a_1. The same reasoning applies for π''.

Hence, the separation condition of a module X_i in π is not local to this module but depends on the module "context", that is the global equilibria (*i.e.*, $\Psi_{\bigcup_{k=1}^{i-1} X_k}$) of the modules that precede. Deciding the separability of X_i into X_i^1 and X_i^2 implies checking two conditions: $\bigcup_{k=1}^{i-1} X_k \rightsquigarrow X_i^1$ and $\bigcup_{k=1}^{i-1} X_k \cup X_i^1 \rightsquigarrow$

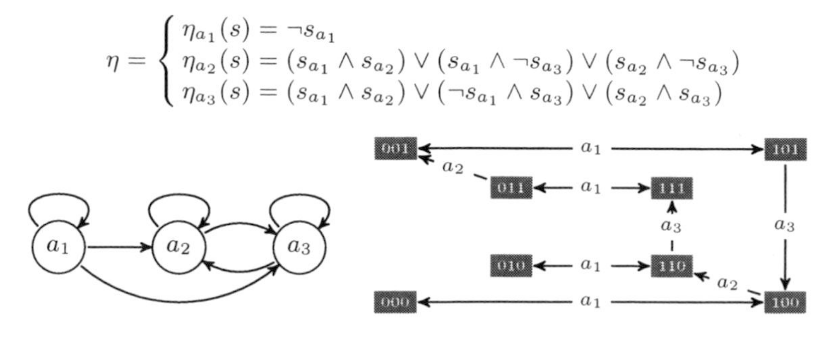

Fig. 7. Example of non-separability of $\{a_2, a_3\}$ in $\pi = (\{a_1\}, \{a_2, a_3\})$

X_i^2. Of course, the complexity of the underlying computation is exponential in the size of π and also depends on the position of X_i in π. Nevertheless, brute-force computation may be used in practice for small interaction networks (of about 15 agents). A more efficient method allowing to go beyond this limitation is, for the moment, an open question.

5 Conclusion

We developed a formal framework for the analysis of the modularity in interaction networks assuming asymptotic dynamics of modules and enabling their composition. We exhibited modularity conditions governing the composition of modules and an efficient computation method such that the global equilibria of interaction networks are obtained from the local ones, leading to an efficient algorithm. Moreover, we confirmed that usual assumptions identifying modules with SCCs have a strong motivation coming from theory.

The next step should be identifying a characteristic property for finding elementary modular organisations. Then, since this work provides a rigorous setting for studying other questions around modularity. For example, a success factor of synthetic biology is to ensure the safety of modular design [22] which in our theoretical framework is guaranteed by construction through the concept of modular organisation. Also, questions related to robustness and evolution could be tackled thanks to the modular knowledge of interaction networks.

References

1. Monod, J.: Chance and necessity: an essay on the natural philosophy of modern biology. Knopf (1970)
2. Delbrück, M. (ed.): Viruses. California Institute of Technology (1950)
3. Lederberg, E.M., Lederberg, J.: Genetic studies of lysogenicity in *Escherichia coli*. Genetics 38, 51–64 (1953)
4. Jacob, F., Monod, J.: Genetic regulatory mechanisms in the synthesis of proteins. Journal of Molecular Biology 3, 318–356 (1961)
5. Qi, Y., Ge, H.: Modularity and dynamics of cellular networks. PLoS Computational Biology 2, e174 (2006)
6. Gagneur, J., Krause, R., Bouwmeester, T., et al.: Modular decomposition of protein-protein interaction networks. Genome Biology 5, R57 (2004)
7. Chaouiya, C., Klaudel, H., Pommereau, F.: A modular, qualitative modeling of regulatory networks using Petri nets. In: Modeling in Systems Biology: The Petri Nets Approach, pp. 253–279. Springer (2011)
8. Rives, A.W., Galitski, T.: Modular organization of cellular networks. Proceedings of the National Academy of Sciences of the USA 100, 1128–1133 (2003)
9. Spirin, V., Mirny, L.A.: Protein complexes and functional modules in molecular networks. Proceedings of the National Academy of Sciences of the USA 100, 12123–12128 (2003)
10. Milo, R., Shen-Orr, S., Itzkovitz, S., et al.: Network motifs: simple building blocks of complex networks. Science 298, 824–827 (2002)

11. Alon, U.: Biological networks: The tinkerer as an engineer. Science 301, 1866–1867 (2003)
12. Bar-Joseph, Z., Gerber, G.K., Lee, T.I., et al.: Computational discovery of gene modules and regulatory networks. Nature Biotechnology 21, 1337–1342 (2003)
13. Segal, E., Shapira, M., Regev, A., et al.: Module networks: discovering regulatory modules and their condition specific regulators from gene expression data. Nature Genetics 34, 166–176 (2003)
14. Thieffry, D., Romero, D.: The modularity of biological regulatory networks. Biosystems 50, 49–59 (1999)
15. Han, J.D.J.: Understanding biological functions through molecular networks. Cell Research 18, 224–237 (2008)
16. Siebert, H.: Dynamical and structural modularity of discrete regulatory networks. In: Proceedings of Comp. Mod. Electronic Proceedings in Theoretical Computer Science, vol. 6, pp. 109–124. Open Publishing Association (2009)
17. Delaplace, F., Klaudel, H., Cartier-Michaud, A.: Discrete causal model view of biological networks. In: Proceedings of CMSB, pp. 4–13. ACM (2010)
18. Demongeot, J., Goles, E., Morvan, M., et al.: Attraction basins as gauges of robustness against boundary conditions in biological complex systems. PLoS One 5, e11793 (2010)
19. Bernot, G., Tahi, F.: Behaviour preservation of a biological regulatory network when embedded into a larger network. Fundamenta Informaticae 91, 463–485 (2009)
20. Thomas, R.: Boolean formalisation of genetic control circuits. Journal of Theoretical Biology 42, 563–585 (1973)
21. Thomas, R.: On the relation between the logical structure of systems and their ability to generate multiple steady states or sustained oscillations. In: Numerical methods in the study of critical phenomena. Springer Series in Synergetics, vol. 9, pp. 180–193. Springer (1981)
22. Purnick, P.E.M., Weiss, R.: The second wave of synthetic biology: from modules to systems. Nature Reviews Molecular Cell Biology 10, 410–422 (2009)

Appendix

Proposition 2 (Properties of Ψ). *Let $\langle A, S, (\rightharpoonup_a)_{a\in A}\rangle$ be an asynchronous dynamics and $X \subseteq A$ be a subset of agents. Ψ_X has the following properties, for all sets S', S'' subsets of states:*

a. *Idempotency: $\Psi_X \circ \Psi_X(S') = \Psi_X(\Psi_X(S')) = \Psi_X(S')$;*
b. *Upper-continuity: $\Psi_X(S' \cup S'') = \Psi_X(S') \cup \Psi_X(S'')$;*
c. *Monotony (order-preserving) : $S' \subseteq S'' \implies \Psi_X(S') \subseteq \Psi_X(S'')$.*

Proof (Proposition 2). Let $Eq_X(s)$ be the predicate meaning that s is an equilibrium for $\rightharpoonup_X$.

a. By expanding $\Psi_X(\Psi_X(S'))$, we have:

$$\begin{aligned}\Psi_X(\Psi_X(S')) &= \{s \in \Omega_X(\Psi_X(S')) \mid Eq_X(s)\}\\ &= \{s \in \Psi_X(S') \mid Eq_X(s)\}\\ &= \Psi_X(S').\end{aligned}$$

b. By definition, $\Psi_X(S' \cup S'') = \{s \in \Omega_X(S' \cup S'') \mid Eq_X(s)\}$, where $\Omega_X(S' \cup S'') = (S' \cup S'')\rightharpoonup^*_X$. Since $\rightharpoonup^*_X$ is upper-continuous on the lattice of state sets, we have:

$$\begin{aligned}\Psi_X(S' \cup S'') &= \{s \in \Omega_X(S') \cup \Omega_X(S'') \mid Eq_X(s)\} \\ &= \{s \in \Omega_X(S') \mid Eq_X(s)\} \cup \{s \in \Omega_X(S'') \mid Eq_X(s)\} \\ &= \Psi_X(S') \cup \Psi_X(S'').\end{aligned}$$

c. An upper-continuous function is monotone. □

Proposition 3 (Properties of the modularity relation). *Let S' be a subset of S. For all X_i, X_j subsets of A, we have the following properties:*

1. $X_i \rightsquigarrow X_j \iff \Psi_{X_i \cup X_j}(S') \subseteq (\Psi_{X_i} \circ \Omega_{X_i \cup X_j}(S'))$;
2. $X_i \rightsquigarrow X_j \iff \Psi_{X_i \cup X_j}(S') = \Psi_{X_i} \circ \Psi_{X_i \cup X_j}(S')$.

Proof. 1. ($\Rightarrow$) Let $S' \subseteq S$, $s \in \Psi_{X_i \cup X_j}(S')$ and $s \notin \Psi_{X_i}(\Omega_{X_i \cup X_j}(S'))$, and $s' \in \Psi_{X_i}(\Omega_{X_i \cup X_j}(\{s\}))$. By definition of equilibrium, $s' \rightharpoonup^*_{X_i \cup X_j} s$. Now, we have:
- $\forall s'' \in (s' \rightharpoonup_{X_i}) : s'' \in \Psi_{X_i} \circ \Psi_{X_i \cup X_j}(\{s'\})$, by definition of equilibria;
- $\forall s'' \in (s' \rightharpoonup_{X_j}) : s'' \in \Psi_{X_i} \circ \Psi_{X_i \cup X_j}(\{s'\})$, by definition of $\rightsquigarrow$.

As a consequence, $s' \not\rightharpoonup^*_{X_i \cup X_j} s$, which leads to a contradiction.

($\Leftarrow$) Let $S' \subseteq S$, $s \in \Psi_{X_i \cup X_j}(S')$, and $\Psi_{X_i \cup X_j}(S') \subseteq \Psi_{X_i}(\Omega_{X_i \cup X_j}(S'))$. Then, by hypothesis, we have:

$$\forall s'' \in (s \rightharpoonup_{X_j}) : s'' \in \Psi_{X_i \cup X_j}(S') \wedge s'' \in \Psi_{X_i}(\Omega_{X_i \cup X_j}(S')),$$

which means that $s, s'' \in \Psi_{X_i} \circ \Psi_{X_i \cup X_j}(S')$. Thus, $X_i \rightsquigarrow X_j$.

2. From Proposition 3.1 and since $\Psi_{X_i \cup X_j}(S') = \Omega_{X_i \cup X_j}(\Psi_{X_i \cup X_j}(S'))$. □

Proof (Proposition 1). Let $\pi = (X_1, \ldots, X_{i-1}, X_i, X_{i+1}, \ldots, X_m)$ be a modular organisation and let $X = \bigcup_{k=1}^{i-1} X_k$. We want to show that $(X_1, \ldots, X_{i-1}, X_i \cup X_{i+1}, \ldots, X_m)$ is a modular organisation. By definition 6, we have:

$$(X \cup X_i) \rightsquigarrow X_{i+1} \tag{4}$$

and:

$$X \rightsquigarrow X_i. \tag{5}$$

We want to show that: (4)$\wedge$(5) $\implies X \rightsquigarrow (X_i \cup X_{i+1})$. First, by Proposition 3.2, we can write:

$$\Psi_{X \cup X_i} \circ \Psi_{X \cup X_i \cup X_{i+1}} = \Psi_{X \cup X_i \cup X_{i+1}} \text{ by (4)}, \tag{6}$$

$$\Psi_X \circ \Psi_{X \cup X_i} = \Psi_{X \cup X_i} \text{ by (5)}. \tag{7}$$

Thus:

$$\begin{aligned}\Psi_X \circ \Psi_{X \cup X_i \cup X_{i+1}} &= \Psi_X \circ (\Psi_{X \cup X_i} \circ \Psi_{X \cup X_i \cup X_{i+1}}) \text{ by (6)} \\ &= (\Psi_X \circ \Psi_{X \cup X_i}) \circ \Psi_{X \cup X_i \cup X_{i+1}} \\ &= \Psi_{X \cup X_i} \circ \Psi_{X \cup X_i \cup X_{i+1}} \text{ by (7)} \\ &= \Psi_{X \cup X_i \cup X_{i+1}} \text{ by (6)},\end{aligned}$$

which is the expected result. From Proposition 3.2, we can deduce that: $X \rightsquigarrow (X_i \cup X_{i+1})$. Iteratively, we show that $X \rightsquigarrow \bigcup_{k=i}^{j} X_k$. As a result, $(X_1, \ldots, X_{i-1}, \bigcup_{k=i}^{j} X_k, X_{j+1}, \ldots, X_m)$ is a modular organisation. □

Proof (Lemma 1). By Definition 4, for any $X_i, X_j \subseteq A$ and for any $s, s' \in S$, we have $\neg(X_j \longrightarrow X_i) \wedge (s \sim_{A \setminus X_j} s') \implies (s \rightharpoonup_{X_i}) \sim_{X_i} (s' \rightharpoonup_{X_i})$. This property is obviously preserved at equilibria. Indeed, for any $s, s' \in S$, we have $\neg(X_j \longrightarrow X_i) \wedge (s \sim_{A \setminus X_j} s') \implies \Psi_{X_i}(s) \sim_{X_i} \Psi_{X_i}(s')$. Thus, the restrictions $\Psi_{X_i}(s)$ and $\Psi_{X_i}(s')$ to X_i are identical. Then, the evolution by X_j from the equilibria of X_i remains in the equilibria of X_i. Hence, we get that $\neg(X_j \longrightarrow X_i) \implies X_i \rightsquigarrow X_j$. □

Proof (Theorem 1). Observe that, in the SCC quotient graph G of a regulation graph, $X_i \longrightarrow X_j$ always implies that $\neg(X_j \longrightarrow X_i)$, because of the acyclicity of G. Thus, folding contiguous modules with respect to any topological order preserves the absence of regulation. As a consequence, if $(X_1, \ldots, X_m)$ is a topological order of G, for all $i, j \in \mathbb{N}$ such that $1 \leq i \leq j \leq m$, we have $\neg(X_j \longrightarrow X_i)$, and by Lemma 1, $X_i \rightsquigarrow X_j$. □

Proof (Lemma 2). Let $X_i, X_j \subseteq A$ such that $X_i \rightsquigarrow X_j$ and S' a subset of S. ($\subseteq$) First, let us show $\Psi_{X_i \cup X_j}(S') \subseteq (\Psi_{X_i} \oslash \Psi_{X_j}) \circ \Omega_{X_i \cup X_j}(S')$. From Proposition 3.2, we know that $\Psi_{X_i} \circ \Psi_{X_i \cup X_j}(S') = \Psi_{X_i \cup X_j}(S')$. Thus, $\forall s \in \Psi_{X_i \cup X_j}(S')$, we have $[s]_{\rightleftharpoons X_i} \subseteq \Psi_{X_i \cup X_j}(S')$. Similarly, an evolution by $X_i \cup X_j$ from an attractor of X_i remain in the same attractor except potentially with evolutions by X_j. We have then, for all $s, s' \in \Psi_{X i \cup X_j}(S')$:

$$[s]_{\rightleftharpoons X_i} [\rightharpoonup_{X_i \cup X_j}]^*_{\rightleftharpoons X_i} [s']_{\rightleftharpoons X_i} \iff \\ ([s]_{\rightleftharpoons X_i} [\rightharpoonup_{X_j}]^*_{\rightleftharpoons X_i} [s']_{\rightleftharpoons X_i}) \vee ([s]_{\rightleftharpoons X_i} = [s']_{\rightleftharpoons X_i}).$$

Now, since both s and s' belong to attractors of $X_i \cup X_j$ (by hypothesis), if there exists an evolution by X_j from $[s]_{\rightleftharpoons X_i}$ to $[s']_{\rightleftharpoons X_i}$, there exists obviously another path labelled by X_j from $[s']_{\rightleftharpoons X_i}$ to $[s]_{\rightleftharpoons X_i}$. Hence, for all $s, s' \in \Psi_{X_i \cup X_j}(S')$, we have:

$$([s]_{\rightleftharpoons X_i} [\rightharpoonup_{X_j}]^*_{\rightleftharpoons X_i} [s']_{\rightleftharpoons X_i}) = ([s]_{\rightleftharpoons X_i} [\rightharpoonup_{X_i \cup X_j}]^*_{\rightleftharpoons X_i} [s']_{\rightleftharpoons X_i}) \implies \\ ([s']_{\rightleftharpoons X_i} [\rightharpoonup_{X_i \cup X_j}]^*_{\rightleftharpoons X_i} [s]_{\rightleftharpoons X_i}) = ([s']_{\rightleftharpoons X_i} [\rightharpoonup_{X_j}]^*_{\rightleftharpoons X_i} [s]_{\rightleftharpoons X_i}).$$

As a result, we have $[s]_{\rightleftharpoons X_i} \in [\widetilde{\Psi_{X_j}}]_{\rightleftharpoons X_i}(\Psi_{X_i \cup X_j}(S'))$, for all $s \in \Psi_{X_i \cup X_j}(S')$. Moreover, since from Proposition 2, operator $[\widetilde{\Psi_{X_j}}]_{\rightleftharpoons X_i}(S')$ is monotone and since $\Psi_{X_i \cup X_j}(S') \subseteq \Omega_{X_i \cup X_j}(S')$, for all $s \in \Psi_{X \cup Y}(S')$, we can write that $[s]_{\rightleftharpoons X_i} \in [\widetilde{\Psi_{X_j}}]_{\rightleftharpoons X_i}(\Omega_{X_i \cup X_j}(S'))$. Now, since $s \in [s]_{\rightleftharpoons X_i}$, $\forall s \in \Psi_{X_i \cup X_j}(S')$, we have $s \in \mathsf{Flat} \circ [\widetilde{\Psi_{X_j}}]_{\rightleftharpoons X_i}(\Omega_{X_i \cup X_j}(S'))$. From (2) and Proposition 3.2, we can write:

$$\begin{aligned} \Psi_{X_i} \oslash \Psi_{X_j} \circ \Omega_{X_i \cup X_j}(S') &= \mathsf{Flat} \circ [\widetilde{\Psi_{X_j}}]_{\rightleftharpoons X_i} \circ \Psi_{X_i} \circ \Omega_{X_i \cup X_j}(S') \\ &= \mathsf{Flat} \circ [\widetilde{\Psi_{X_j}}]_{\rightleftharpoons X_i} \circ \Omega_{X_i \cup X_j}(S'). \end{aligned}$$

Hence, for all $s \in \Psi_{X \cup Y}(S')$, we have: $s \in \Psi_X \oslash \Psi_Y \circ \Omega_{X \cup Y}(S')$, which corresponds to the following inclusion:

$$\Psi_{X \cup Y}(S') \subseteq \Psi_X \oslash \Psi_Y \circ \Omega_{X \cup Y}(S').$$

($\supseteq$) Now, let us show $(\Psi_{X_i} \oslash \Psi_{X_j}) \circ \Omega_{X_i \cup X_j}(S') \subseteq \Psi_{X_i \cup X_j}(S')$. To do so, let us consider a state $s \in (\Psi_{X_i} \oslash \Psi_{X_j}) \circ \Omega_{X_i \cup X_j}(S')$. From (2) and Proposition 3.2, we have $[s]_{\rightleftharpoons X_i} \in [\Psi_{X_j}]_{\rightleftharpoons X_i} \circ [\Psi_{X_i} \circ \Omega_{X_i \cup X_j}(S')]_{\rightleftharpoons X_i}$. Now, consider $[[s]_{\rightleftharpoons X_i}]_{\rightleftharpoons X_j}$. By definition of attractors, for all $s_1, s_2 \in \mathsf{Flat} \circ \mathsf{Flat}([[s]_{\rightleftharpoons X_i}]_{\rightleftharpoons X_j})$, we have $s_1 \rightharpoonup^*_{X_i \cup X_j} s_2$. This means that $[[s]_{\rightleftharpoons X_i}]_{\rightleftharpoons X_j} = [s]_{\rightleftharpoons X_i \cup X_j}$ and, as a consequence, that $s \in \Psi_{X \cup Y}(S')$. As a result, the inclusion $(\Psi_X \oslash \Psi_Y) \circ \Omega_{X \cup Y}(S') \subseteq \Psi_{X \cup Y}(S')$ holds. □

Proof (Theorem 2). This proof is made directly by induction on the modular organisation, using Definition 6 and Lemma 2. Since $\pi = (X_1, \ldots, X_m)$ is a modular organisation, it is folding preserving and $\bigcup_{i=1}^{m-1} X_i \rightsquigarrow X_m$. Then, using (2) and Lemma 2, we have:

$$\begin{aligned}
\Psi_{A'} &= (\Psi_{\bigcup_{i=1}^{m-1} X_i} \oslash \Psi_{X_m}) \circ \Omega_{A'} \\
&= \mathsf{Flat} \circ [\widetilde{\Psi_{X_m}}]_{\rightleftharpoons \bigcup_{i=1}^{m-1} X_i} \circ \Psi_{\bigcup_{i=1}^{m-1} X_i} \circ \Omega_{A'} \\
&= \mathsf{Flat} \circ [\widetilde{\Psi_{X_m}}]_{\rightleftharpoons \bigcup_{i=1}^{m-1} X_i} \circ (\Psi_{\bigcup_{i=1}^{m-2} X_i} \oslash \Psi_{X_{m-1}}) \circ \Omega_{\bigcup_{i=1}^{m-1} X_i} \circ \Omega_{A'} \\
&= \mathsf{Flat} \circ [\widetilde{\Psi_{X_m}}]_{\rightleftharpoons \bigcup_{i=1}^{m-1} X_i} \circ (\Psi_{\bigcup_{i=1}^{m-2} X_i} \oslash \Psi_{X_{m-1}}) \circ \Omega_{A'} \\
&= \mathsf{Flat} \circ [\widetilde{\Psi_{X_m}}]_{\rightleftharpoons \bigcup_{i=1}^{m-1} X_i} \circ [\widetilde{\Psi_{X_{m-1}}}]_{\rightleftharpoons \bigcup_{i=1}^{m-2} X_i} \circ \Psi_{\bigcup_{i=1}^{m-2} X_i} \circ \Omega_{A'} \\
&= \ldots \\
&= \mathsf{Flat} \circ [\widetilde{\Psi_{X_m}}]_{\rightleftharpoons \bigcup_{i=1}^{m-1} X_i} \circ \ldots \circ [\widetilde{\Psi_{X_2}}]_{\rightleftharpoons X_1} \circ \Psi_{X_1} \circ \Omega_{A'}.
\end{aligned}$$

As a result, we obtain: $\Psi_{A'}(S') = (\Psi_{X_1} \oslash \ldots \oslash \Psi_{X_m}) \circ \Omega_{A'}(S')$, which is the expected result. □

Proof (Lemma 3). The evolution is governed by the values of the regulators and the evolution concerns the states of agents of X only. Under the assumptions that X and Y are two disjoint sets and that $R_X \subseteq X \cup Y$, the following property holds by definition of the evolution: $\forall s_1, s_2 \in S|_{X \cup Y} : (s_1 \rightharpoonup_X s_2) \iff (s_1|_{X \cup R_X} \rightharpoonup_X s_2|_{X \cup R_X} \wedge s_1|_{Y \setminus R_X} = s_2|_{Y \setminus R_X})$. This property extends to the transitive closure by induction:

$$\forall s_1, s_2 \in S|_{X \cup Y} : \\ (s_1 \rightharpoonup^*_X s_2) \iff (s_1|_{X \cup R_X} \rightharpoonup^*_X s_2|_{X \cup R_X} \wedge s_1|_{Y \setminus R_X} = s_2|_{Y \setminus R_X}). \quad (8)$$

First we prove that:

$$\forall s \in S|_{X \cup Y} : s \in \Psi_X(S|_{X \cup Y}) \iff s|_{X \cup R_X} \in \Psi_X(S|_{X \cup R_X}).$$

Let $s \in \Psi_X(S|_{X \cup Y})$. By definition of the equilibrium operator (Definition 3), s complies with the following equivalent property:

$\iff \forall s' \in S|_{X \cup Y} : s \rightharpoonup_X^* s' \implies s' \rightharpoonup_X^* s.$

By application of Equation 8, we derive:

$\iff \forall s' \in S|_{X \cup Y} : (s|_{Y \setminus R_X} = s'|_{Y \setminus R_X}) \implies$
$(s|_{X \cup R_X} \rightharpoonup_X^* s'|_{X \cup R_X} \implies s'|_{X \cup R_X} \rightharpoonup_X^* s|_{X \cup R_X}).$

Since $s \rightharpoonup_X^* s'$ insures that $s|_{X \cup R_X} = s'|_{X \cup R_X}$, by Equation 8, we simplify and obtain:

$\iff \forall s' \in S|_{X \cup Y} : s|_{X \cup R_X} \rightharpoonup_X^* s'|_{X \cup R_X} \implies s'|_{X \cup R_X} \rightharpoonup_X^* s|_{X \cup R_X}.$

By definition of the equilibrium operator, this equivalently leads to:

$\iff s|_{X \cup R_X} \in \Psi_X(S|_{X \cup R_X}).$

Now, by definition of the completion, any state s complies with the following property: $s \in (s|_{X \cup R_X})|^{Y \setminus R_X}$. Since $s|_{X \cup R_X} \in \Psi_X(S|_{X \cup R_X})$, we deduce that $s \in \Psi_X(S|_{X \cup R_X})|^{Y \setminus R_X}$. Hence, we conclude that:

$$\forall s \in S|_{X \cup Y} : s \in \Psi_X(S|_{X \cup Y}) \iff s \in \Psi_X(S|_{X \cup R_X})|^{Y \setminus R_X}.$$

□

Concretizing the Process Hitting into Biological Regulatory Networks

Maxime Folschette[1,2], Loïc Paulevé[3], Katsumi Inoue[2], Morgan Magnin[1], and Olivier Roux[1]

[1] LUNAM Université, École Centrale de Nantes, IRCCyN UMR CNRS 6597
(Institut de Recherche en Communications et Cybernétique de Nantes)
1 rue de la Noë - B.P. 92101 - 44321 Nantes Cedex 3, France
Maxime.Folschette@irccyn.ec-nantes.fr
[2] National Institute of Informatics,
2-1-2, Hitotsubashi, Chiyoda-ku, Tokyo 101-8430, Japan
[3] LIX, École Polytechnique, 91128 Palaiseau Cedex, France

Abstract. The Process Hitting (PH) is a recently introduced framework to model concurrent processes. Its major originality lies in a specific restriction on the causality of actions, which makes the formal analysis of very large systems tractable. PH is suitable to model Biological Regulatory Networks (BRNs) with complete or partial knowledge of cooperations between regulators by defining the most permissive dynamics with respect to these constraints.

On the other hand, the qualitative modeling of BRNs has been widely addressed using René Thomas' formalism, leading to numerous theoretical work and practical tools to understand emerging behaviors.

Given a PH model of a BRN, we first tackle the inference of the underlying Interaction Graph between components. Then the inference of corresponding Thomas' models is provided using Answer Set Programming, which allows notably an efficient enumeration of (possibly numerous) compatible parametrizations.

In addition to giving a formal link between different approaches for qualitative BRNs modeling, this work emphasizes the ability of PH to deal with large BRNs with incomplete knowledge on cooperations, where Thomas' approach fails because of the combinatorics of parameters.

1 Introduction

As regulatory phenomena play a crucial role in biological systems, they need to be studied accurately. Biological Regulatory Networks (BRNs) consist in sets of either positive or negative mutual effects between the components. With the purpose of analyzing these systems, they are often modeled as graphs which make it possible to determine the possible evolutions of all the interacting components of the system. Indeed, besides continuous models of physicists, often designed through systems of ordinary differential equations, a discrete modeling approach was initiated by René Thomas in 1973 [1].

D. Gilbert and M. Heiner (Eds.): CMSB 2012, LNCS 7605, pp. 166–186, 2012.

In this approach, the different levels of a component, such as concentration or expression levels, are abstractly represented by (positive) integer values and transitions between these levels may be considered as instantaneous. Hence, qualitative state graphs may be derived from which we are able to formally find out all the possible behaviors expressed as sequences of transitions between these states. Nevertheless, these dynamics can be precisely established only with regard to some discrete parameters, hereafter called "Thomas' parameters", which stand for kinds of "focal points", i.e. the evolutionary tendency from each state and depending on the set of the other currently interacting components.

Thomas' modeling has motivated numerous works around the link between the Interaction Graph (IG) (summarizing the global influences between components) and the possible dynamics (e.g., [2,3]), model reduction (e.g., [4]), formal checking of dynamics (e.g., [5,6]), and the incorporation of time (e.g., [7,8]) and probability (e.g., [9]) dimensions, to name but a few. While the formal checking of dynamical properties is often limited to small networks because of the state graph explosion, the main drawback of this framework is the difficulty to specify Thomas' parameters, especially for large networks.

In order to address the formal checking of dynamical properties within very large BRNs, we recently introduced in [10] a new formalism, named the *"Process Hitting"* (PH), to model concurrent systems having components with a few qualitative levels. A PH describes, in an atomic manner, the possible evolutions of a process (representing one component at one level) triggered by the hit of at most one other process in the system. This framework can be seen as a special class of formalisms like Petri Nets or Communicating Finite State Machines, where the causality between actions is restricted. Thanks to the particular structure of interactions within a PH, very efficient static analysis methods have been developed to over- and under-approximate reachability properties making tractable the formal analysis of BRNs with hundreds of components [11].

PH is suitable to model BRNs with different levels of abstraction in the specification of cooperations (associated influences) between components. This allows to model BRNs with a partial knowledge on precise evolution functions for components by capturing the largest (the most general) dynamics.

The objectives of the work presented in this paper are the following. Firstly, we show that starting from one PH model, it is possible to find back the underlying IG. We perform an exhaustive search for the possible interactions on one component from all the others, consistently with the knowledge of the dynamics that these interactions lead to and that are expressed in PH. The second phase of our work concerns the Thomas' parameters inference. It consists in determining the nesting set (possibly too large) of the parameters which necessarily lead to the satisfaction of the known cooperating constraints. The resulting BRN dynamics is ensured to respect the PH dynamics, i.e. no spurious transitions are made possible by the inference. Answer Set Programming (ASP) [12] turns out to be effective for these enumerative searches.

The outcome of this work is twofold. The first benefit is that such an approach makes it possible to refine the construction of BRNs with a partial and progressively brought knowledge in PH, while being able to export such models in the Thomas' framework. This work thus strengthens the formal link between both modelings. The second feature of our method is that it can be applied on very large BRNs.

Finally, it must be noticed that we are not interested in this paper in the derivation of one PH from a BRN (which was previously described in [10]) but, on the contrary, to finding out a set of BRNs from one PH.

Our work is related to the approach of [13] which relies on temporal logic, and [14,15] which also uses constraint programming. Both aim at determining a class of models which are consistent with available partial data on the regulatory structure and dynamical properties. Our method is based on a model rather than on constraints, which allows to define some properties on the system structure (such as cooperations). Furthermore, we claim that we are able to deal with larger biological networks.

Outline. Sect. 2 recalls the PH and Thomas frameworks; Sect. 3 defines the IG inference from PH; Sect. 4 details the enumeration of Thomas parametrizations compatible with a PH and discuss its implementation in ASP. Sect. 5 illustrates the applicability of our method on simple examples and large biological models.

Notations. $[i;j]$ is the set of integers $\{i, i+1, \ldots, j\}$; we note $[i_1;j_1] \leq_{[]} [i_2;j_2] \overset{\Delta}{\Leftrightarrow} (i_1 \leq i_2 \wedge j_1 \leq j_2)$ and $[i_1;j_1] <_{[]} [i_2;j_2] \overset{\Delta}{\Leftrightarrow} (i_1 < i_2 \wedge j_1 \leq j_2) \vee (i_1 \leq i_2 \wedge j_1 < j_2)$. Given an integer k, $k < [i;j] \overset{\Delta}{\Leftrightarrow} k < i$ and $k > [i;j] \overset{\Delta}{\Leftrightarrow} k > j$.

2 Frameworks

2.1 The Process Hitting Framework

We recall here the definition and semantics of the Process Hitting (PH), and its usage to model cooperation between concurrent components. Two examples of PH modeling a BRN at different abstraction levels are given. They serve as running examples in the rest of this article.

A PH (Def. 1) gathers a finite number of concurrent *processes* grouped into a finite set of *sorts*. A process belongs to a unique sort and is noted a_i where a is the sort and i the identifier of the process within the sort a. At any time, one and only one process of each sort is present; a state of the PH thus corresponds to the set of such processes.

The concurrent interactions between processes are defined by a set of *actions*. Actions describe the replacement of a process by another of the same sort conditioned by the presence of at most one other process in the current state of the PH. An action is denoted by $a_i \rightarrow b_j \Rsh b_k$ where a_i, b_j, b_k are processes of sorts a and b. It is required that $b_j \neq b_k$ and that $a = b \Rightarrow a_i = b_j$. An action $h = a_i \rightarrow b_j \Rsh b_k$ is read as "a_i *hits* b_j to make it bounce to b_k", and a_i, b_j, b_k are called respectively *hitter*, *target* and *bounce* of the action, and can be referred to as $\mathsf{hitter}(h), \mathsf{target}(h), \mathsf{bounce}(h)$, respectively.

Definition 1 (Process Hitting). *A* Process Hitting *is a triple* $(\Sigma, L, \mathcal{H})$*:*

- $\Sigma \triangleq \{a, b, \dots\}$ *is the finite set of* sorts*;*
- $L \triangleq \prod_{a \in \Sigma} L_a$ *is the set of states with* $L_a = \{a_0, \dots, a_{l_a}\}$ *the finite set of* processes *of sort* $a \in \Sigma$ *and* l_a *a positive integer with* $a \neq b \Rightarrow \forall (a_i, b_j) \in L_a \times L_b, a_i \neq b_j$*;*
- $\mathcal{H} \triangleq \{a_i \to b_j \upharpoonright b_k, \dots \mid (a, b) \in \Sigma^2 \wedge (a_i, b_j, b_k) \in L_a \times L_b \times L_b \wedge b_j \neq b_k \wedge a = b \Rightarrow a_i = b_j\}$ *is the finite set of* actions.

$\mathcal{P}$ *denotes the set of all processes* ($\mathcal{P} \triangleq \{a_i \mid a \in \Sigma \wedge a_i \in L_a\}$).

The sort of a process a_i is referred to as $\Sigma(a_i) = a$ and the set of sorts present in an action $h \in \mathcal{H}$ as $\Sigma(h) = \{\Sigma(\mathsf{hitter}(h)), \Sigma(\mathsf{target}(h))\}$. Given a state $s \in L$, the process of sort $a \in \Sigma$ present in s is denoted by $s[a]$, that is the a-coordinate of the state s. If $a_i \in L_a$, we define the notation $a_i \in s \overset{\Delta}{\Leftrightarrow} s[a] = a_i$.

An action $h = a_i \to b_j \upharpoonright b_k \in \mathcal{H}$ is *playable* in $s \in L$ if and only if $s[a] = a_i$ and $s[b] = b_j$. In such a case, $(s \cdot h)$ stands for the state resulting from the play of the action h in s, that is $(s \cdot h)[b] = b_k$ and $\forall c \in \Sigma, c \neq b, (s \cdot h)[c] = s[c]$. For the sake of clarity, $((s \cdot h) \cdot h')$, $h' \in \mathcal{H}$ is abbreviated as $(s \cdot h \cdot h')$.

Example. Fig. 1 represents a PH $(\Sigma, L, \mathcal{H})$ with $\Sigma = \{a, b, c\}$, $L_a = \{a_0, a_1, a_2\}$, $L_b = \{b_0, b_1\}$, $L_c = \{c_0, c_1\}$, and

$$\begin{aligned}\mathcal{H} = \{a_2 \to b_1 \upharpoonright b_0, \quad & b_0 \to a_2 \upharpoonright a_1, & c_0 \to a_2 \upharpoonright a_1,\\ & b_0 \to a_1 \upharpoonright a_0, & c_0 \to a_1 \upharpoonright a_0,\\ & b_1 \to a_0 \upharpoonright a_1, & c_1 \to a_0 \upharpoonright a_1,\\ & b_1 \to a_1 \upharpoonright a_2, & c_1 \to a_1 \upharpoonright a_2\} \ .\end{aligned}$$

The action $h = b_1 \to a_1 \upharpoonright a_2$ is playable in the state $s = \langle b_1, a_1, c_0 \rangle$; and $s \cdot h = \langle b_1, a_2, c_0 \rangle$.

This PH example actually models a BRN where the component a has three qualitative levels and components b and c are boolean. In this BRN, b and c activate a, while a inhibits b. The inhibition of b by a is only effective when a is at level 2; in the other cases, b cannot evolve in any direction. The activation of a by b (c) is encoded by the actions making the level of a increase (resp. decrease) when b (c) is present (resp. absent). It is worth noticing that the activation of a by b (c) is independent from c (b). This may express a lack of knowledge on the cooperation between these two regulators: we thus model an over-approximation of the possible actions.

Modeling cooperation. As described in [10], the cooperation between processes to make another bounce can be expressed in PH by building a *cooperative sort*. Fig. 2 shows an example of cooperation between processes b_1 and c_1 to make a_1 bounce to a_2: a cooperative sort bc is defined with 4 processes (one for each sub-state of the presence of processes b_1 and c_1). For the sake of clarity, the bc processes are indexed using the sub-state they represent. Hence, bc_{01} represents

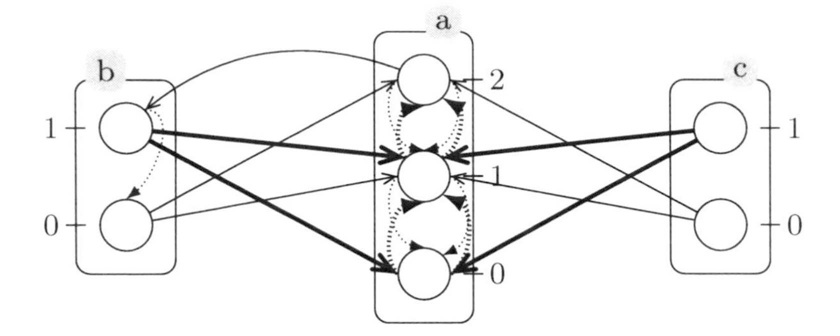

Fig. 1. A Process Hitting (PH) example. Sorts are represented by labeled boxes, and processes by circles (ticks are the identifiers of the processes within the sort, for instance, a_0 is the process ticked 0 in the box a). An action (for instance $b_1 \rightarrow a_1 \upharpoonright a_2$) is represented by a pair of directed arcs, having the hit part (b_1 to a_1) in plain line and the bounce part (a_1 to a_2) in dotted line. Actions involving b_1 or c_1 are in thick lines.

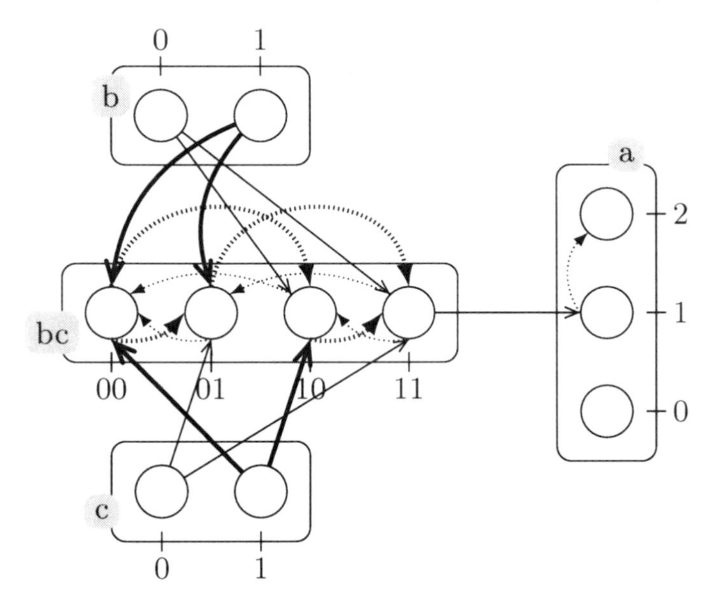

Fig. 2. A PH modeling a cooperativity between b_1 and c_1 to make a_1 bounce to a_2. Actions involving b_1 or c_1 are in thick lines.

the sub-state $\langle b_0, c_1 \rangle$, and so on. Each process of sort b and c hit bc to make it bounce to the process reflecting the status of the sorts b and c (e.g., $b_1 \rightarrow bc_{00} \upharpoonright bc_{10}$ and $b_1 \rightarrow bc_{01} \upharpoonright bc_{11}$). Then, it is the process bc_{11} which hits a_1 to make it bounce to a_2 instead of the independent hits from b_1 and c_1.

We note that cooperative sorts are standard PH sorts and do not involve any special treatment regarding the semantics of related actions.

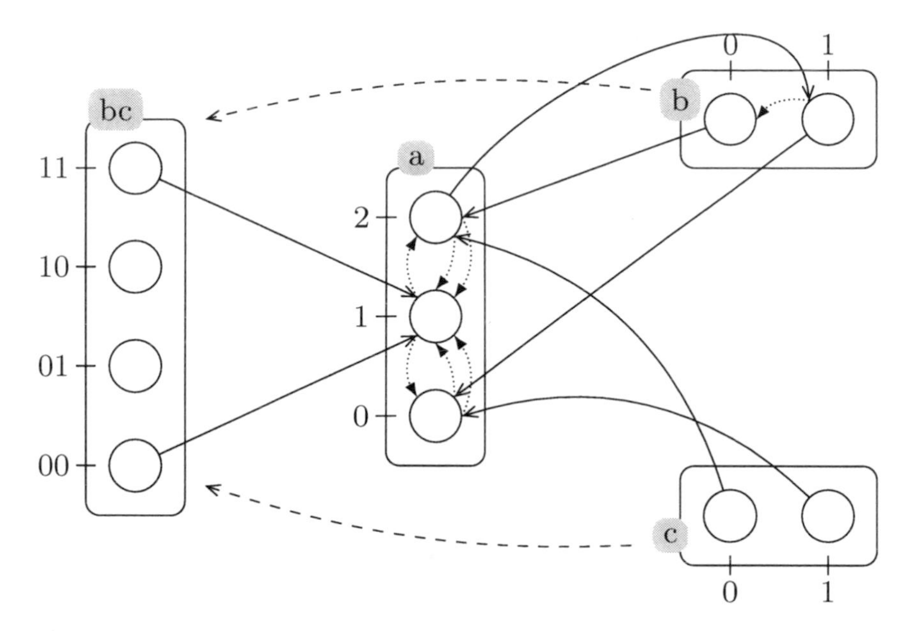

Fig. 3. PH resulting from the refinement of the one in Fig. 1 by the specification of several cooperations. The actions from b and c to the cooperative sort bc are identical to those defined in Fig. 2 and are represented here by a single dashed arc.

When the number of cooperating processes is large, it is possible to chain several cooperative sorts to prevent the combinatoric explosion of the number of processes created within cooperative sorts. For instance, if b_1, c_1, and d_1 cooperate, one can create a cooperative sort bc with 4 processes reflecting the presence of b_1 and c_1, and a cooperative sort bcd with 4 processes reflecting the presence of bc_{11} and d_1. Such constructions are helpful in PH as the static analysis of dynamics developed in [11] does not suffer from the number of sorts, but on the number of processes within a single sort.

While the construction of cooperation in PH allows to encode any boolean functions between cooperating processes [10], it is worth noticing they introduce a temporal shift in their application. This allows the existence of interleaving of actions leading to a cooperative sort representing a past sub-state of the presence of the cooperative processes. The resulting behavior is then an over-approximation of the realization of an instantaneous cooperation.

Example. The PH in Fig. 3 results from the refinement of the PH in Fig. 1 where several cooperations have been specified. In particular, the bounce to a_2 is the result of a cooperation between b_1 and c_1; and the bounce to a_0 of a cooperation between b_0 and c_0. Hence, this PH expresses a BRN where a requires both b and c active to reach its highest level, and a does not become inactive unless both b and c are inactive.

2.2 Thomas' Modeling

We concisely present the Thomas' modeling of BRNs dynamics, merely inspired by [5,16]. In order to enlarge the class of Thomas' models compatible with PH dynamics (w.r.t. the presented inference), we extend the classical formalism by setting parameters to intervals of values instead of single values, and briefly discuss this addition.

Thomas' formalism lies on two complementary descriptions of the system. First, the *Interaction Graph* (IG) models the structure of the system by defining the components' mutual influences. The *parametrization* then specifies the levels to which tends a component when a given configuration of its regulators applies.

The IG is composed of nodes that represent components, and edges labeled with a threshold that stand for either positive or negative regulations (Def. 2). For such a regulation to take place, the expression level of its head component has to be higher than its threshold; otherwise, the opposite influence is expressed. The uniqueness of these regulations makes the following sections simpler. We call $\mathsf{levels}_+(a \to b)$ (resp. $\mathsf{levels}_-(a \to b)$) the levels of a where it is an activator (resp. inhibitor) of b (Def. 3); l_a denotes the maximum level of a.

Definition 2 (Interaction Graph). *An* Interaction Graph *(IG) is a triple* (Γ, E_+, E_-) *where* Γ *is a finite number of* components, *and* E_+ *(resp.* E_-*)* $\subset \{a \xrightarrow{t} b \mid a, b \in \Gamma \wedge t \in [1; l_a]\}$ *is the set of positive (resp. negative)* regulations *between two nodes, labeled with a* threshold.

A regulation from a *to* b *is uniquely referenced: if* $a \xrightarrow{t} b \in E_+$ *(resp.* E_-*),* $\nexists a \xrightarrow{t'} b \in E_+$ *(resp.* E_-*),* $t' \neq t$ *and* $\nexists a \xrightarrow{t'} b \in E_-$ *(resp.* E_+*),* $t' \in \mathbb{N}$.

Definition 3 (Effective levels (levels)). *Let* (Γ, E_+, E_-) *be an IG and* $a, b \in \Gamma$ *two of its components:*

- *if* $a \xrightarrow{t} b \in E_+$, $\mathsf{levels}_+(a \to b) \stackrel{\Delta}{=} [t; l_a]$ *and* $\mathsf{levels}_-(a \to b) \stackrel{\Delta}{=} [0; t-1]$*;*
- *if* $a \xrightarrow{t} b \in E_-$, $\mathsf{levels}_+(a \to b) \stackrel{\Delta}{=} [0; t-1]$ *and* $\mathsf{levels}_-(a \to b) \stackrel{\Delta}{=} [t; l_a]$*;*
- *otherwise,* $\mathsf{levels}_+(a \to b) \stackrel{\Delta}{=} \mathsf{levels}_-(a \to b) \stackrel{\Delta}{=} \emptyset$.

For all component $a \in \Gamma$, $\Gamma^{-1}(a) \stackrel{\Delta}{=} \{b \in \Gamma \mid \exists b \xrightarrow{t} a \in E_+ \cup E_-\}$ is the set of its regulators. We allow any number of levels for the components, without considering the number of outgoing edges, as the number of processes in a PH sort is not constrained in any way.

Example. Fig. 4(left) represents an Interaction Graph (Γ, E_+, E_-) with $\Gamma = \{a, b, c\}$, $E_+ = \{b \xrightarrow{1} a, c \xrightarrow{1} a\}$ and $E_- = \{a \xrightarrow{2} b\}$; hence $\Gamma^{-1}(a) = \{b, c\}$.

A *state* s of an IG (Γ, E_+, E_-) is an element in $\prod_{a \in \Gamma}[0; l_a]$. $s[a]$ refers to the level of component a in s. The specificity of Thomas' approach lies in the use of discrete *parameters* to represent the focal level interval towards which the component will evolve in each configuration of its regulators (Def. 4). Indeed, for each possible state of a BRN, all regulators of a component a can be divided into *activators* and *inhibitors*, given their type of interaction and expression level, referred to as the *resources* of a in this state (Def. 5).

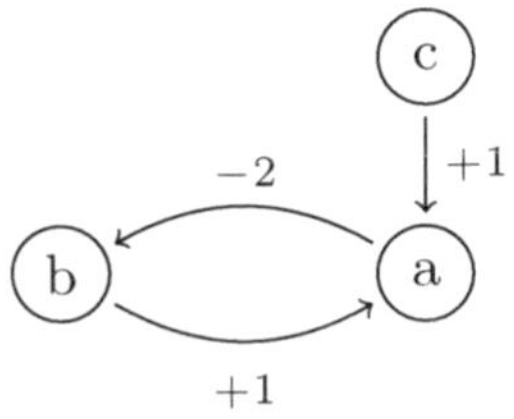

$$K_{a,\{b,c\},\emptyset} = [2;2] \qquad K_{b,\{a\},\emptyset} = [0;1]$$
$$K_{a,\{b\},\{c\}} = [1;1] \qquad K_{b,\emptyset,\{a\}} = [0;0]$$
$$K_{a,\{c\},\{b\}} = [1;1]$$
$$K_{a,\emptyset,\{b,c\}} = [0;0] \qquad K_{c,\emptyset,\emptyset} = [0;1]$$

Fig. 4. (left) IG example. Regulations are represented by the edges labeled with their sign and threshold. For instance, the edge from b to a is labeled $+1$, which stands for: $b \xrightarrow{1} a \in E_+$. (right) Example parametrization of the left IG.

Definition 4 (Discrete parameter $K_{a,A,B}$ and Parametrization K). *For a given component $a \in \Gamma$ and A (resp. B) $\subset \Gamma^{-1}(a)$ a set of its activators (resp. inhibitors) such that $A \cup B = \Gamma^{-1}(a)$ and $A \cap B = \emptyset$, the discrete* parameter *$K_{a,A,B} = [i;j]$ is a non-empty interval towards which a will tend in the states where its activators (resp. inhibitors) are the regulators in set A (resp. B). The complete map K of discrete parameters for $\mathcal{G}$ is called a* parametrization *of $\mathcal{G}$.*

Definition 5 (Resources $\mathsf{Res}_a(s)$). *For a given state s of a BRN, we define the* activators *A and* inhibitors *B of a in s as $\mathsf{Res}_a(s) = A, B$, where:*

$$A = \{b \in \Gamma \mid s[b] \in \mathsf{levels}_+(b \to a)\}$$
$$B = \{b \in \Gamma \mid s[b] \in \mathsf{levels}_-(b \to a)\}$$

We also denote: $\mathsf{Res}_a = \{(A;B) \mid \exists s \in \prod_{a \in \Gamma}[0;l_a], \mathsf{Res}_a(s) = A, B\}$

At last, Def. 6 gives the asynchronous dynamics of a BRN using Thomas' parameters. From a given state s, a transition to another state s' is possible provided that only one component a will evolve of one level towards $K_{a,\mathsf{Res}_a(s)}$.

Definition 6 (Asynchronous dynamics). *Let s be a state of a BRN using Thomas' parameters $(\mathcal{G}, K)$ where $\mathcal{G} = (\Gamma, E_+, E_-)$. The state that succeeds to s is given by the indeterministic function $f(s)$:*

$$f(s) = s' \Leftrightarrow \exists a \in \Gamma, s'[a] = f^a(s) \wedge \forall b \in \Gamma, b \neq a, s[b] = s'[b] \quad \text{, with}$$

$$f^a(s) = \begin{cases} s[a] + 1 & \text{if } s[a] < K_{a,A,B} \\ s[a] & \text{if } s[a] \in K_{a,A,B} \\ s[a] - 1 & \text{if } s[a] > K_{a,A,B} \end{cases} \quad \text{, where } A, B = \mathsf{Res}_a(s).$$

While the use of intervals as parameter values does not add expressivity in boolean networks, it allows to specify a larger range of dynamics in the general case (w.r.t. the above definitions). Indeed, assume that $K_{a,A,B} = [i; i+2]$; we aim at obtaining three different parameters $K_{a,A_1,B_1} = i$, $K_{a,A_2,B_2} = i+1$, $K_{a,A_3,B_3} = i+2$. The only possible modification in resources is to add a as a self-regulator. However, because resources have a boolean definition (a component is either an activator or an inhibitor of a), it is not possible to differentiate the 3 values. We also remark that the use of intervals makes optional some explicit auto-activations in the IG (as for b in Fig. 4, for instance).

Example. In the BRN that consists of the IG and parametrization of Fig. 4, the following transitions are possible given the semantics defined in Def. 6: $\langle a_0, b_1, c_1\rangle \rightarrow \langle a_1, b_1, c_1\rangle \rightarrow \langle a_2, b_1, c_1\rangle \rightarrow \langle a_2, b_0, c_1\rangle \rightarrow \langle a_1, b_0, c_1\rangle$, ending in a steady state, where a_i is the component a at level i. As $K_{b,\{a\},\emptyset} = [0;1]$, no auto-regulation on b is needed to prevent its evolution when a is not at level 2.

3 Interaction Graph Inference

In order to infer a complete BRN, one has to find the Interaction Graph (IG) first, as some constraints on the parametrization rely on it. Inferring the IG is an abstraction step which consists in determining the global influence of components on each of its successors.

This section first introduces the notion of focal processes within a PH (Subsect. 3.1) which is used to characterize well-formed PH for IG inference in Subsect. 3.2, and as well used by the parametrization inference presented in Sect. 4. Finally, the rules for inferring the interactions between components from a PH are described in Subsect. 3.3. We consider hereafter a global PH $(\Sigma, L, \mathcal{H})$ on which the IG inference is to be performed.

3.1 Focal Processes

Many of the inferences defined in the rest of this paper rely on the knowledge of *focal processes* w.r.t. a given context (a set of processes that are potentially present). When such a context applies, we expect to (always) reach one focal process in a bounded number of actions.

For $S_a \subseteq L_a$ and a context (set of processes) ς, let us define as $\mathcal{H}(S_a, \varsigma)$ the set of actions on the sort a having their hitter in ς and target in S_a (Eq. (1)); and the digraph (V, E) where arcs are the bounces within the sort a triggered by actions in $\mathcal{H}(S_a, \varsigma)$ (Eq. (2)). $\mathsf{focals}(a, S_a, \varsigma)$ denotes the set of focal processes of sort a in the scope of $\mathcal{H}(S_a, \varsigma)$ (Def. 7).

$$\mathcal{H}(S_a, \varsigma) \triangleq \{b_i \rightarrow a_j \upharpoonright a_k \in \mathcal{H} \mid b_i \in \varsigma \wedge a_j \in S_a\} \tag{1}$$

$$\begin{aligned} E &\triangleq \{(a_j, a_k) \in (S_a \times L_a) \mid \exists b_i \rightarrow a_j \upharpoonright a_k \in \mathcal{H}(S_a, \varsigma)\} \\ V &\triangleq S_a \cup \{a_k \in L_a \mid \exists (a_j, a_k) \in E\} \end{aligned} \tag{2}$$

Definition 7 ($\mathsf{focals}(a, S_a, \varsigma)$). *The set of processes that are focal for processes in S_a in the scope of $\mathcal{H}(S_a, \varsigma)$ are given by:*

$$\mathsf{focals}(a, S_a, \varsigma) \triangleq \begin{cases} \{a_i \in V \mid \nexists (a_i, a_j) \in E\} & \textit{if the digraph } (V, E) \textit{ is acyclic,} \\ \emptyset & \textit{otherwise.} \end{cases}$$

We note $L(\varsigma)$ the set of states $s \in L$ such that $\forall a \in \Sigma(\varsigma), s[a] \in \varsigma$, where $\Sigma(\varsigma)$ is the set of sorts with processes in ς. We say a sequence of actions $h^1, \ldots, h^n$ is *bounce-wise* if and only if $\forall m \in [1; n-1], \mathsf{bounce}(h^m) = \mathsf{target}(h^{m+1})$. From Def. 7, it derives that:

1. if $\mathsf{focals}(a, S_a, \varsigma) = \emptyset$, there exists a state $s \in L(\varsigma \cup S_a)$ such that $\forall n \in \mathbb{N}$ there exists a bounce-wise sequence of actions $h^1, \ldots, h^{n+1}$ in $\mathcal{H}(S_a, \varsigma)$ with $\mathsf{target}(h^1) \in s$.
2. if $\mathsf{focals}(a, S_a, \varsigma) \neq \emptyset$, for all state $s \in L(\varsigma \cup S_a)$, for any bounce-wise sequence of actions $h^1, \ldots, h^n$ in $\mathcal{H}(S_a, \varsigma)$ where $\mathsf{target}(h^1) \in s$, either $\mathsf{bounce}(h^n) \in \mathsf{focals}(a, S_a, \varsigma)$, or $\exists h^{n+1} \in \mathcal{H}(a, \varsigma)$ such that $\mathsf{bounce}(h^n) = \mathsf{target}(h^{n+1})$. Moreover $n \leq |\mathcal{H}(S_a, \varsigma)|$ (i.e. no cycle of actions possible).

It is worth noticing that those bounce-wise sequences of actions may not be successively playable in a state $s \in L(\varsigma \cup S_a)$. Indeed, nothing impose that the hitters of actions are present in s. In the general case, the playability of those bounce-wise sequences, referred to as *focals reachability* may be hard to prove. However, in the scope of this paper, the particular contexts used with focals ensure this property. Notably, the rest of this section uses only *strict* contexts (Def. 8) which allow at most one hitter per sort in the bounce-wise sequences (and thus are present in s).

Definition 8 (Strict context for S_a). *A context (set of processes) ς is strict for $S_a \subseteq L_a$ if and only if $\{b_i, b_j\} \subset \varsigma \wedge b \neq a \Rightarrow i = j$.*

In other words, assuming focals reachability, if $\mathsf{focals}(a, S_a, \varsigma)$ is empty, there exists a sequence of actions that may be played an unbound number of times (cycle); if it is non-empty, it is ensured that any state in $L(\varsigma \cup S_a)$ converges, in a bounded number of steps, either to a process in S_a that is not hit by processes in ς, or to a process in $L_a \setminus S_a$.

Example. In the PH of Fig. 1, we obtain:

$$\mathsf{focals}(a, L_a, \{b_0, c_0\}) = \{a_0\} \qquad \mathsf{focals}(a, L_a, \{b_1, c_1\}) = \{a_2\}$$
$$\mathsf{focals}(a, L_a, \{b_1, c_0\}) = \emptyset \qquad \mathsf{focals}(a, \{a_1\}, \{b_1, c_0\}) = \{a_0, a_2\}$$

3.2 Well-Formed Process Hitting for Interaction Graph Inference

The inference of an IG from a PH assumes that the PH defines two types of sorts: the sorts corresponding to BRN components, and the cooperative sorts. This leads to the characterization of a *well-formed* PH for IG inference.

The identification of sorts modeling components relies on the observation that their processes represent (ordered) qualitative levels. Hence an action on such a sort cannot make it bounce to a process at a distance more than one. The set of sorts satisfying such a condition is referred to as Γ (Eq. (3)). Therefore, in the rest of this paper, Γ denotes the set of components of the BRN to infer.

$$\Gamma \triangleq \{a \in \Sigma \mid \nexists b_i \to a_j \upharpoonright a_k \in \mathcal{H}, |j - k| > 1\} \tag{3}$$

Any sort that does not act as a component should then be treated as a cooperative sort. As explained in Subsect. 2.1, the role of a cooperative sort υ is to compute the current state of set of cooperating processes. Hence, for each sub-state σ formed by the sorts hitting υ, υ should converge to a focal process. This is expressed by Property 1, where the set of sorts having an action on a given sort a is given by $\Sigma^{-1}(a)$ (Eq. (4)) and $\mathcal{P}(\sigma)$ is the set of processes that compose the sub-state σ.

$$\forall a \in \Sigma, \Sigma^{-1}(a) \triangleq \{b \in \Sigma \mid \exists b_i \to a_j \Rsh a_k \in \mathcal{H}\} \tag{4}$$

Property 1 (Well-formed cooperative sort). A sort $\upsilon \in \Sigma$ is a well-formed cooperative sort if and only if each configuration σ of its predecessors leads υ to a unique focal process, denoted by $\upsilon(\sigma)$:

$$\forall \sigma \in \textstyle\prod_{a \in \Sigma^{-1}(\upsilon) \wedge a \neq \upsilon} L_a, \mathsf{focals}(\upsilon, L_\upsilon, \mathcal{P}(\sigma) \cup L_\upsilon) = \{\upsilon(\sigma)\}$$

Such a property allows a large variety of definitions of a cooperative sort, but for the sake of simplicity, does not allow the existence of multiple focal processes. While this may be easily extended to (the condition becomes $\mathsf{focals}(\upsilon, L_\upsilon, \mathcal{P}(\sigma) \cup L_\upsilon) \neq \emptyset$), it makes some hereafter equations a bit more complex to read as they should handle a set of focal processes instead of a unique focal process.

Finally, Property 2 sums up the conditions for a Process Hitting to be suitable for IG inference. In addition of having either component sorts or well-formed cooperative sorts, we also require that there is no cycle between cooperative sorts, and that sorts being never hit (i.e. serving as an invariant environment) are components.

Property 2 (Well-formed Process Hitting for IG inference). A PH is well-formed for IG inference if and only if the following conditions are verified:

- each sort $a \in \Sigma$ either belongs to Γ, or is a well-formed cooperative sort;
- there is no cycle between cooperative sorts (the digraph $(\Sigma, \{(a, b) \in (\Sigma \times \Sigma) \mid \exists a_i \to b_j \Rsh b_k \in \mathcal{H} \wedge a \neq b \wedge \{a, b\} \cap \Gamma = \emptyset\})$ is acyclic);
- sorts having no action hitting them belong to Γ ($\{a \in \Sigma \mid \nexists b_i \to a_j \Rsh a_k \in \mathcal{H}\} \subset \Gamma$).

Example. In the PH of Fig. 3, bc is a well-formed cooperative sort as defined in Property 1, because:

$$\mathsf{focals}(bc, L_{bc}, \{b_0, c_0\} \cup L_{bc}) = \{bc_{00}\} \quad \mathsf{focals}(bc, L_{bc}, \{b_0, c_1\} \cup L_{bc}) = \{bc_{01}\}$$
$$\mathsf{focals}(bc, L_{bc}, \{b_1, c_0\} \cup L_{bc}) = \{bc_{10}\} \quad \mathsf{focals}(bc, L_{bc}, \{b_1, c_1\} \cup L_{bc}) = \{bc_{11}\}$$

Hence, both Fig. 1 and Fig. 3 are well-formed PH for IG inference with $\Gamma = \{a, b, c\}$.

3.3 Interaction Inference

At this point we can divide the set of sorts Σ into components (Γ, see Eq. (3)) and cooperative sorts ($\Sigma \setminus \Gamma$) that will not appear in the IG. We define in Eq. (5) the set of predecessors of a sort a, that is, the sorts influencing a by considering direct actions and possible intermediate cooperative sorts. The predecessors of a that are components are the regulators of a, denoted $\mathsf{reg}(a)$ (Eq. (6)).

$$\begin{aligned}\forall a \in \Sigma, \mathsf{pred}(a) \triangleq \{b \in \Sigma \mid \exists n \in \mathbb{N}^*, \exists (c^k)_{k\in[0;n]} \in \Sigma^{n+1},\\ c^0 = b \wedge c^n = a \\ \wedge \forall k \in [\![0; n-1]\!], c^k \in \Sigma^{-1}(c^{k+1}) \cap (\Sigma \setminus \Gamma)\}\end{aligned} \tag{5}$$

$$\forall a \in \Sigma, \mathsf{reg}(a) \triangleq \mathsf{pred}(a) \cap \Gamma \tag{6}$$

Given a set g of components and a configuration (i.e. a sub-state) σ, $\varsigma_g(\sigma)$ refers to the set of processes hitting a regulated by any sort in g (Eq. (7)). If $g = \{b\}$, we simple note $\varsigma_b(\sigma)$. This set is composed of the active processes of sorts in g, and the focal process (assumed unique) of the cooperative sorts υ hitting a that have a predecessor in g. The evaluation of the focal process of υ in context σ, denoted $\upsilon(\sigma)$, relies on Property 1, which gives its value when all the direct predecessors of υ are defined in σ. When a predecessor υ' is not in σ, we extend the evaluation by recursively computing the focal value of υ' is σ, as stated in Eq. (8). Because there is no cycle between cooperative sorts, this recursive evaluation of $\upsilon(\sigma)$ always terminates.

$$\forall g \subset \Gamma, \varsigma_g(\sigma) \triangleq \{\sigma[b] \mid b \in g\} \cup \{\upsilon(\sigma) \mid \upsilon \in \Sigma^{-1}(a) \setminus \Gamma \wedge g \cap \mathsf{reg}(\upsilon) \neq \emptyset\} \tag{7}$$

$$\upsilon(\sigma) \triangleq \upsilon(\sigma \uplus \langle \upsilon'(\sigma) \mid \upsilon' \in \Sigma^{-1}(\upsilon) \wedge \upsilon' \in \Sigma \setminus \Gamma \rangle) \tag{8}$$

We aim at inferring that b activates (inhibits) a if there exists a configuration where increasing the level of b makes possible the increase (decrease) of the level of a. This is analogous to standard IG inferences from discrete maps [2].

This reasoning can be straightforwardly applied to a PH when inferring the influence of b on a when $b \neq a$ (Eq. (11)). Let us define $\gamma(b \to a)$ as the set of components cooperating with b to hit a, including b and a (Eq. (9)). Given a configuration $\sigma \in \prod_{c \in \gamma(b \to a)} L_c$, $\mathsf{focals}(a, \{a_i\}, \varsigma_b(\sigma))$ gives the bounces that a given process a_i can make in the context $\varsigma_b(\sigma)$. We note $\sigma\{b_i\}$ the configuration σ where the process of sort b has been replaced by b_i. If there exists $b_i, b_{i+1} \in L_b$ such that one bounce in $\mathsf{focals}(a, \{\sigma[a]\}, \varsigma_b(\sigma\{b_i\}))$ has a lower (resp. higher) level that one bounce in $\mathsf{focals}(a, \{\sigma[a]\}, \varsigma_b(\sigma\{b_{i+1}\}))$, then b as positive (resp. negative) influence on a with a maximum threshold $l = i + 1$.

$$\gamma(b \to a) \triangleq \{a, b\} \cup \{c \in \Gamma \mid \exists \upsilon \in \Sigma \setminus \Gamma, \upsilon \in \mathsf{pred}(a) \wedge \{b, c\} \subset \mathsf{pred}(\upsilon)\} \tag{9}$$

Then, we infer that a has a self-influence if its current level can have an impact on its own evolution at a given configuration σ. We consider here a configuration σ of a group g of sorts having a cooperation on a. This set of sort groups is given by $X(a)$ (Eq. (10)) which returns the set of connected components (noted $\mathcal{C}$) of the graph linking two regulators b, c of a if there is a cooperative sort hitting a regulated by both of them. Given $a_i, a_{i+1} \in L_a$, we pick $a_j \in \mathsf{focals}(a, \{a_i\}, \varsigma_g(\sigma\{a_i\}))$ and $a_k \in \mathsf{focals}(a, \{a_{i+1}\}, \varsigma_g(\sigma\{a_{i+1}\}))$. If $k = j + 1$, we can not conclude as there is no difference in the evolution of both levels. If $k \neq j+1$ and $k - j \neq 0$, then a_i and a_{i+1} have divergent evolutions: we infer an influence of sign of $k - j$ at threshold $i + 1$. We note that some aspects of this inference are arbitrary and may impact the number of parameters to infer in the next section. In particular, in some cases, the use of intervals for Thomas' parameters drops the requirement of inferring a self-activation.

$$X(a) = \mathcal{C}\left((\mathsf{reg}(a), \{\{b, c\} \mid \exists v \in \Sigma^{-1}(a) \setminus \Gamma, \{b, c\} \subset \mathsf{reg}(v)\})\right) \tag{10}$$

Proposition 1 details the inference of all existing influences between components occurring with a threshold l. These influences are split into positive and negative ones, and represent possible edges in the final IG. We do not consider the cases where a component has no visible influence on another.

Proposition 1 (Edges inference). *We define the set of positive (resp. negative) influences $\hat{E}_+$ (resp. $\hat{E}_-$) for any $a \in \Gamma$ by:*

$$\begin{aligned}
&\forall b \in \mathsf{reg}(a), b \neq a, \\
&\qquad b \xrightarrow{l} a \in \hat{E}_s \iff \exists \sigma \in \textstyle\prod_{c \in \gamma(b \to a)} L_c, \exists b_i, b_{i+1} \in L_b, \\
&\qquad\qquad \exists a_j \in \mathsf{focals}(a, \{\sigma[a]\}, \varsigma_b(\sigma\{b_i\})), \\
&\qquad\qquad \exists a_k \in \mathsf{focals}(a, \{\sigma[a]\}, \varsigma_b(\sigma\{b_{i+1}\})), \\
&\qquad\qquad s = \mathsf{sign}(k - j) \wedge l = i + 1
\end{aligned} \tag{11}$$

$$\begin{aligned}
a \xrightarrow{l} a \in \hat{E}_s \iff &\exists g \in X(a), \sigma \in L_a \times \textstyle\prod_{b \in g} L_b, \exists a_i, a_{i+1} \in L_a, \\
&\exists a_j \in \mathsf{focals}(a, \{a_i\}, \varsigma_g(\sigma\{a_i\})), \\
&\exists a_k \in \mathsf{focals}(a, \{a_{i+1}\}, \varsigma_g(\sigma\{a_{i+1}\})), \\
&k \neq j + 1 \wedge s = \mathsf{sign}(k - j) \wedge l = i + 1
\end{aligned} \tag{12}$$

where $s \in \{+, -\}$, $\bar{s} = + \overset{\Delta}{\Leftrightarrow} s = -$, $\bar{s} = - \overset{\Delta}{\Leftrightarrow} s = +$, $\mathsf{sign}(n) = + \overset{\Delta}{\Leftrightarrow} n > 0$, $\mathsf{sign}(n) = - \overset{\Delta}{\Leftrightarrow} n < 0$, *and* $\mathsf{sign}(0) \overset{\Delta}{=} 0$.

We are now able to infer the edges of the final IG by considering positive and negative influences (Proposition 2). We infer a positive (resp. negative) edge if there exists a corresponding influence with the same sign. If an influence is both positive and negative, we infer an unsigned edge. In the end, the threshold of each edge is the minimum threshold for which the influence has been found. As unsigned edges represent ambiguous interactions, no threshold is inferred.

Proposition 2 (Interaction Graph inference). *We infer $\mathcal{G} = (\Gamma, E_+, E_-, E_\pm)$ from Proposition 1 as follows:*

$$\begin{aligned}
E_+ &= \{a \xrightarrow{t} b \mid \nexists a \xrightarrow{t'} b \in \hat{E}_- \wedge t = \min\{l \mid a \xrightarrow{l} b \in \hat{E}_+\}\} \\
E_- &= \{a \xrightarrow{t} b \mid \nexists a \xrightarrow{t'} b \in \hat{E}_+ \wedge t = \min\{l \mid a \xrightarrow{l} b \in \hat{E}_-\}\} \\
E_\pm &= \{a \to b \mid \exists a \xrightarrow{t} b \in \hat{E}_+ \wedge \exists a \xrightarrow{t'} b \in \hat{E}_-\}
\end{aligned}$$

Example. The IG inference from the PH of Fig. 3 gives $\hat{E}_+ = \{b \xrightarrow{1} a, c \xrightarrow{1} a\}$ and $\hat{E}_- = \{a \xrightarrow{2} b\}$, corresponding to the IG of Fig. 4. No self-influence are inferred ($X(a) = \{\{b, c\}\}$, $X(b) = \{\{a\}\}$, and $X(c) = \emptyset$).

4 Parametrization Inference

Given the IG inferred from a PH as presented in the previous section, one can find the discrete parameters that model the behavior of the studied PH using the method presented in the following. It relies on an exhaustive enumeration of all predecessors of each component in order to find attractor processes and returns a possibly incomplete parametrization, given the exhaustiveness of the cooperations. The last step consists of the enumeration of all compatible complete parametrizations given this set of inferred parameters, the PH dynamics and some biological constraints on parameters.

4.1 Parameters Inference

This subsection presents some results related to the inference of independent discrete parameters from a given PH. These results are equivalent to those presented in [10], with notation adapted to be shared with the previous section. In addition, we introduce the well-formed PH for parameter inference property (Property 3), which implies that the inferred IG does not contain any unsigned interactions, and thus can be seen as the regular IG (Γ, E_+, E_-), and that any processes in $\mathsf{levels}_+(b \to a)$ (resp. $\mathsf{levels}_-(b \to a)$) share the same behavior regarding a.

Property 3 (Well-formed PH for parameter inference). A PH is well-formed for parameter inference if and only if it is well-formed for IG inference, and the IG $(\Gamma, E_+, E_-, E_\pm)$ inferred by Proposition 2 verifies $E_\pm = \emptyset$ and if the following property holds:

$$\begin{aligned} \forall b \in \Gamma^{-1}(a), \forall (i, j \in \mathsf{levels}_+(b \to a) \vee i, j \in \mathsf{levels}_-(b \to a)), \\ \forall c, ((b \neq a \wedge c = a) \vee (c \in \mathsf{pred}(a) \wedge b \in \Sigma^{-1}(c))), \\ b_i \to c_k \upharpoonright c_l \in \mathcal{H} \Leftrightarrow b_j \to c_k \upharpoonright c_l \in \mathcal{H} \end{aligned} \tag{13}$$

Let $K_{a,A,B}$ be the parameter we want to infer for a given component $a \in \Gamma$ and $A \subset \Gamma^{-1}(a)$ (resp. $B \subset \Gamma^{-1}(a)$) a set of its activators (resp. inhibitors). This inference, as for the IG inference, relies on the search of focal processes of the component for the given configuration of its regulators.

For each sort $b \in \Gamma^{-1}(a)$, we define a context $C^b_{a,A,B}$ in Eq. (14) that contains all processes representing the influence of the regulators in the configuration A, B. The context of a cooperative sort v that regulates a is given in Eq. (15) as the set of focal processes matching the current configuration. $C_{a,A,B}$ refers to the union of all these contexts (Eq. (16)).

$$\forall b \in \Gamma,\ C^{b}_{a,A,B} \triangleq \begin{cases} \mathsf{levels}_{+}(b \to a) & \text{if } b \in A, \\ \mathsf{levels}_{-}(b \to a) & \text{if } b \in B, \\ L_b & \text{otherwise;} \end{cases} \quad (14)$$

$$\forall v \in \mathsf{pred}(a) \setminus \Gamma,\ C^{v}_{a,A,B} \triangleq \{v(\sigma) \mid \sigma \in \textstyle\prod_{c \in \Sigma^{-1}(v)} C^{c}_{a,A,B}\} \quad (15)$$

$$C_{a,A,B} \triangleq \textstyle\bigcup_{b \in \mathsf{pred}(a)} C^{b}_{a,A,B} \quad (16)$$

The parameter $K_{a,A,B}$ specifies to which values a eventually evolves as long as the context $C_{a,A,B}$ holds, which is precisely the definition of the focals function (Def. 7 in Subsect. 3.1), where the focals reachability property can be derived from Property 3 and Eq. (15). Hence $K_{a,A,B} = \mathsf{focals}(a, C^{a}_{a,A,B}, C_{a,A,B})$ if this latter is a non-empty interval (Proposition 3).

Proposition 3 (Parameter inference). *Let $(\Sigma, L, \mathcal{H})$ be a Process Hitting well-formed for parameter inference, and $\mathcal{G} = (\Gamma, E_{+}, E_{-})$ the inferred IG. Let A (resp. B) $\subseteq \Gamma$ be the set of regulators that activate (resp. inhibit) a sort a. If* $\mathsf{focals}(a, C^{a}_{a,A,B}, C_{a,A,B}) = [a_i; a_j]$ *is a non-empty interval, then $K_{a,A,B} = [i; j]$.*

Example. Applied to the PH in Fig. 1, we obtain, for instance, $K_{b,\{a\},\emptyset} = [0; 1]$ and $K_{a,\{b,c\},\emptyset} = [2; 2]$, while $K_{a,\{b\},\{c\}}$ can not be inferred. For the PH in Fig. 3, this latter is evaluated to $[1; 1]$.

Given the Proposition 3, we see that in some cases, the inference of the targeted parameter is impossible. This can be due to a lack of cooperation between regulators: when two regulators independently hit a component, their actions can have opposite effects, leading to either an indeterministic evolution or to oscillations. Such an indeterminism is not possible in a BRN as in a given configuration of regulators, a component can only have an interval attractor, and eventually reaches a steady-state. In order to avoid such inconclusive cases, one has to ensure that no such behavior is allowed by either removing undesired actions or using cooperative sorts to prevent opposite influences between regulators.

4.2 Admissible Parametrizations

When building a BRN, one has to find the parametrization that best describes the desired behavior of the studied system. Complexity is inherent to this process as the number of possible parametrizations for a given IG is exponential w.r.t. the number of components. However, the method of parameters inference presented in this section gives some information about necessary parameters given a certain dynamics described by a PH. This information thus drops the number of possible parametrizations, allowing to find the desired behavior more easily.

We first delimit the validity of a parameter (Property 4) in order to ensure that any transition in the resulting BRN is allowed by the studied PH. This is verified by the existence of a hit making the concerned component bounce into the direction of the value of the parameter in the matching context. Thus, assuming Property 3 holds, any transition in the inferred BRN corresponds to at least one transition in the PH, proving the correctness of our inference. We remark that any parameter inferred by Proposition 3 satisfies this property.

Property 4 (Parameter validity). A parameter $K_{a,A,B}$ is valid w.r.t. the PH iff the following equation is verified:

$$\forall a_i \in C^a_{a,A,B}, a_i \notin K_{a,A,B} \Longrightarrow (\exists c_k \rightarrow a_i \upharpoonright a_j \in \mathcal{H}, c_k \in C^c_{a,A,B} \\ \wedge a_i < K_{a,A,B} \Rightarrow j > i \wedge a_i > K_{a,A,B} \Rightarrow j < i)$$

Then, we use some additional biological constraints on Thomas' parameters given in [16], that we sum up in the following three properties:

Property 5 (Extreme values assumption). Let $\mathcal{G} = (\Gamma, E_+, E_-)$ be an IG. A parametrization K on $\mathcal{G}$ satisfies the *extreme values assumption* iff:

$$\forall b \in \Gamma, \Gamma^{-1}(b) \neq \emptyset \Rightarrow 0 \in K_{b,\emptyset,\Gamma^{-1}(b)} \wedge l_b \in K_{b,\Gamma^{-1}(b),\emptyset}$$

Property 6 (Activity assumption). Let $\mathcal{G} = (\Gamma, E_+, E_-)$ be an IG. A parametrization K on $\mathcal{G}$ satisfies the *activity assumption* iff:

$$\forall b \in \Gamma, \forall a \in \Gamma^{-1}(b), \exists (A; B) \in \mathsf{Res}_a, K_{b,A,B} <_{[]} K_{b,A\cup\{b\},B\setminus\{b\}}$$
$$\forall b \in \Gamma, \forall a \in \Gamma^{-1}(b), \exists (A; B) \in \mathsf{Res}_a, K_{b,A\setminus\{b\},B\cup\{b\}} <_{[]} K_{b,A,B}$$

Property 7 (Monotonicity assumption). Let $\mathcal{G} = (\Gamma, E_+, E_-)$ be an IG. A parametrization K on $\mathcal{G}$ satisfies the *monotonicity assumption* iff:

$$\forall b \in \Gamma, \forall (A; B), (A'; B') \in \mathsf{Res}_b, A \subset A' \wedge B' \subset B \Rightarrow K_{b,A,B} \leq_{[]} K_{b,A',B'}$$

4.3 Answer Set Programming Implementation Concepts

Answer Set Programming (ASP) [12] has been chosen to address the enumeration of all admissible parametrizations. The motivations are following:

- ASP efficiently tackles the inherent complexity of the models we use, thus allowing a fast execution of the formal tools defined in this paper,
- it is convenient to enumerate a large set of possible answers,
- it allows us to easily constrain the answers according to some properties.

We now synthesize some key points to better make the reader understand our ASP implementation with the enumeration example.

All information describing the studied model (PH and inferred IG & parameters) are expressed in ASP using facts. For functional purposes, we assign a unique label to each couple A, B of activators and inhibitors of a given component, and in the following we note $K^p_{a,A,B}$ the parameter of component a whose regulators A, B are assigned to the label p. Then, to state the existence of a parameter $K^{\mathtt{p}}_{\mathtt{a},A,B}$, we use an atom named `param_label` in the following fact:

```
param_label(a, p).
```

Defining a set in ASP is equivalent to defining the rule for belonging to this set. For example, we define an atom `param_act` that describes the set of active regulators of a given a parameter. Describing the activators of $K^{p}_{a,\{b,c\},\{d\}}$ gives:

```
param_act(a, p, b).
param_act(a, p, c).
```

The absence of such a fact involving d with label p indicates that d is an inhibitor in the configuration of regulators related to this parameter.

Rules allow more detailed declarations than facts as they have a body (right-hand part below) containing constraints and allowing to use variables, while facts only have a head (left-hand part). For instance, in order to define the set of expression levels of a component, we declare:

```
component_levels(X, 0..M) :- component(X, M).
```

where the `component(X, M)` atom stands for the existence of a component `X` with a maximum level `M`. Considering this declaration, any possible answer for the atom `component_levels` will be found by binding all possible values of its two terms with all existing `component` facts: the existence of an answer `component_levels(a, k)` will depend on the existence of a term a, which is bound with `X`, and an integer k, constrained by: $0 \leq k \leq \mathtt{M}$.

Cardinalities are convenient to enumerate all possible parametrizations by creating multiple answer sets. A cardinality (denoted hereafter with curly brackets) gives any number of possible answers for some atoms between a lower and upper bounds. For example,

```
1 { param(X, P, I) : component_levels(X, I) } :-
      param_label(X, P), not infered_param(X, P).
```

where `param(X, P, I)` stands for: $\mathtt{I} \in K^{\mathtt{P}}_{\mathtt{X},A,B}$, means that any parameter of component `X` and label `P` must contain at least one level value (`I`) in the possible expression levels of `X`. Indeed, the lower bound is 1, forcing at least one element in the parameter, but no upper bound is specified, allowing up to any number of answers. The body (right-hand side) of the rule also checks for the existence of a parameter of `X` with label `P`, and constrains that the parametrization inference was not conclusive for the considered parameter (`not` stands for negation by failure: `not L` becomes true if `L` is not true). Such a constraint gives multiple results as any set of `param` atoms satisfying the cardinality will lead to a new global set of answers. In this way, we enumerate all possible parametrizations which respects the results of parameters inference, but completely disregarding the notion of admissible parametrizations given in Subsect. 4.2.

We rely on integrity constraints to filter only admissible parametrizations. An integrity constraint is a rule with no head, that makes an answer set unsatisfiable if its body turns out to be true. Hence, if we suppose that:

- the `less_active(a, p, q)` atom means that $K^{p}_{a,A,B}$ stands for a configuration with less activating regulators than $K^{q}_{a,A',B'}$ (i.e. $A \subset A'$),

– the `param_inf(a, p, q)` atom means: $K^{p}_{a,A,B} \leq_{[]} K^{q}_{a,A',B'}$,

the monotonicity assumption is formulated as the following integrity constraint:

```
:- less_active(X, P, Q), not param_inf(X, P, Q).
```

which removes all parametrization results where parameters $K^{\mathtt{P}}_{\mathtt{X},A,B}$ and $K^{\mathtt{Q}}_{\mathtt{X},A',B'}$ exist such that $A \subset A'$ and $K^{\mathtt{Q}}_{\mathtt{X},A',B'} <_{[]} K^{\mathtt{P}}_{\mathtt{X},A,B}$, thus violating the monotonicity assumption. Of course, other assumptions can be formulated in the same way.

This subsection succinctly described how we write ASP programs to represent a model and solve all steps of Thomas' modeling inference. It finds a particularly interesting application in the enumeration of parameters: all possible parametrizations are generated in separate answer sets, and integrity constraints are formulated to remove those that do not fit the assumptions of admissible parametrizations, thus reducing the number of interesting parametrizations to be considered in the end.

5 Examples

The inference method described in this paper has been implemented as part of PINT[1], which gathers PH related tools. Our implementation mainly consists in ASP programs that are solved using Clingo[2]. The IG and parameters inference can be performed using the command `ph2thomas -i model.ph --dot ig.dot` where `model.ph` is the PH model in PINT format and `ig.dot` is an output of the inferred IG in DOT format. The (possibly partial) inferred parametrization will be returned on the standard output. The admissible parametrizations enumeration is performed when adding the `--enumerate` parameter to the command.

Applied to the example in Fig. 3 where cooperations have been defined, our method infers the IG and parametrization given in Fig. 4. Regarding the example in Fig. 1, the same IG is inferred, as well as for the parametrization except for the parameters $K_{a,\{b\},\{c\}}$ and $K_{a,\{c\},\{b\}}$ which are undefined (because of the lack of cooperativity between b and c). In such a case, this partial parametrization allows 36 admissible complete parametrizations, as two parameters with 3 potential values could not be inferred. If we constrain these latter parameters so that they contain exactly one element, we obtain only 9 admissible parametrizations.

The current implementation can successfully handle large PH models of BRNs found in the literature such as an ERBB receptor-regulated G1/S transition model from [17] which contains 20 components and 15 cooperative sorts, and a T-cells receptor model from [18] which contains 40 components and 14 cooperative sorts[3]. For each model, IG and parameters inferences are performed together in less than a second on a standard desktop computer. After removing the cooperations from these models (leaving only raw actions), the inferences allow to determine 40 parameters out of 195 for the 20 components model, and

[1] Available at `http://process.hitting.free.fr`

[2] Available at `http://potassco.sourceforge.net`

[3] Both models are available as examples distributed with PINT.

77 out of 143 for the 40 components model. As we thus have an order of magnitude of respectively 10^{31} and 10^{73} admissible parametrizations, these models would therefore be more efficiently studied as PH than as BRNs. We note that the complexity of the method is exponential in the number of regulators of one component and linear in the number of components.

A PH model can be built based on information found in the literature about the local influences between components. The precision of this knowledge will determine the precision of the modeled activations and inhibitions, and some information is likely to help in the representation of cooperations.

6 Conclusion and Discussion

This work establishes the abstraction relationship between PH and Thomas' approaches for qualitative BRN modeling. The PH allows an abstract representation of BRNs dynamics (allowing incomplete knowledge on the cooperation between components) that can not be exactly represented in René Thomas' formalism by a single instance of BRN parametrization. This motivates the concretization of PH models into a set of compatible Thomas' models in order to benefit of the complementary advantages of these two formal frameworks.

We first propose an original inference of the Interaction Graph (IG) from a BRN having its dynamics specified in the PH framework. An IG gives a compact abstract representation of the influence of the components between each others. Then, based on a prior inference of René Thomas' parametrization for BRNs from a PH model, we delimit the set of compatible Thomas' parametrizations that are compatible with the PH dynamics, and give arguments for their correctness. A parametrization is compatible with the PH if its dynamics (in terms of possible transitions) is included in the PH dynamics. The enumeration of such parametrizations is efficiently tackled using Answer Set Programming. We illustrate the overall method on simple examples and large biological models.

Several extensions of the presented work are now to be considered. First, we now plan to explore the inference of Thomas' parameters when the inferred IG involves unsigned interactions, as some particular cases having such IGs are known to have correct Thomas' parametrization (not respecting some assumptions imposed in Subsect. 4.2, however). Second, the inference of BRN multiplexes [19] may be of practical interest as they allow to implicitly reduce the possible parametrizations by making cooperations appear in the IG. Because of its atomicity, the PH allows to specify a range of cooperations that can not be completely captured by a single instance of BRN multiplexes, then encouraging the inference of a set of compatible ones. Finally, in order to improve the performances in the IG inference, we will consider projection operations on the PH structure to undo cooperations between components and reduce the cardinality of configurations to explore by making the interactions independent.

Acknowledgement. This work was partially supported by the Fondation Centrale Initiatives.

References

1. Thomas, R.: Boolean formalization of genetic control circuits. Journal of Theoretical Biology 42(3), 563–585 (1973)
2. Richard, A., Comet, J.P.: Necessary conditions for multistationarity in discrete dynamical systems. Discrete Applied Mathematics 155(18), 2403–2413 (2007)
3. Remy, É., Ruet, P., Thieffry, D.: Graphic requirements for multistability and attractive cycles in a boolean dynamical framework. Advances in Applied Mathematics 4(3), 335–350 (2008)
4. Naldi, A., Remy, E., Thieffry, D., Chaouiya, C.: A Reduction of Logical Regulatory Graphs Preserving Essential Dynamical Properties. In: Degano, P., Gorrieri, R. (eds.) CMSB 2009. LNCS, vol. 5688, pp. 266–280. Springer, Heidelberg (2009)
5. Richard, A., Comet, J.P., Bernot, G.: Formal Methods for Modeling Biological Regulatory Networks. In: Modern Formal Methods and App., pp. 83–122 (2006)
6. Naldi, A., Thieffry, D., Chaouiya, C.: Decision diagrams for the representation and analysis of logical models of genetic networks. In: Calder, M., Gilmore, S. (eds.) CMSB 2007. LNCS (LNBI), vol. 4695, pp. 233–247. Springer, Heidelberg (2007)
7. Siebert, H., Bockmayr, A.: Incorporating Time Delays into the Logical Analysis of Gene Regulatory Networks. In: Priami, C. (ed.) CMSB 2006. LNCS (LNBI), vol. 4210, pp. 169–183. Springer, Heidelberg (2006)
8. Ahmad, J., Roux, O., Bernot, G., Comet, J.P., Richard, A.: Analysing formal models of genetic regulatory networks with delays. International Journal of Bioinformatics Research and Applications (IJBRA) 4(2) (2008)
9. Twardziok, S., Siebert, H., Heyl, A.: Stochasticity in reactions: a probabilistic boolean modeling approach. In: Computational Methods in Systems Biology, pp. 76–85. ACM (2010)
10. Paulevé, L., Magnin, M., Roux, O.: Refining Dynamics of Gene Regulatory Networks in a Stochastic π-Calculus Framework. In: Priami, C., Back, R.-J., Petre, I., de Vink, E. (eds.) Transactions on Computational Systems Biology XIII. LNCS, vol. 6575, pp. 171–191. Springer, Heidelberg (2011)
11. Paulevé, L., Magnin, M., Roux, O.: Static analysis of biological regulatory networks dynamics using abstract interpretation. Mathematical Structures in Computer Science (2012) (in press) (preprint), `http://loicpauleve.name/mscs.pdf`
12. Baral, C.: Knowledge Representation, Reasoning and Declarative Problem Solving. Cambridge University Press (2003)
13. Khalis, Z., Comet, J.P., Richard, A., Bernot, G.: The SMBioNet method for discovering models of gene regulatory networks. Genes, Genomes and Genomics 3(special issue 1), 15–22 (2009)
14. Corblin, F., Fanchon, E., Trilling, L.: Applications of a formal approach to decipher discrete genetic networks. BMC Bioinformatics 1(1), 385 (2010)
15. Corblin, F., Fanchon, E., Trilling, L., Chaouiya, C., Thieffry, D.: Automatic Inference of Regulatory and Dynamical Properties from Incomplete Gene Interaction and Expression Data. In: Lones, M.A., Smith, S.L., Teichmann, S., Naef, F., Walker, J.A., Trefzer, M.A. (eds.) IPCAT 2012. LNCS, vol. 7223, pp. 25–30. Springer, Heidelberg (2012)
16. Bernot, G., Cassez, F., Comet, J.P., Delaplace, F., Müller, C., Roux, O.: Semantics of biological regulatory networks. Electronic Notes in Theoretical Computer Science 180(3), 3–14 (2007)

17. Sahin, O., Frohlich, H., Lobke, C., Korf, U., Burmester, S., Majety, M., Mattern, J., Schupp, I., Chaouiya, C., Thieffry, D., Poustka, A., Wiemann, S., Beissbarth, T., Arlt, D.: Modeling ERBB receptor-regulated G1/S transition to find novel targets for de novo trastuzumab resistance. BMC Systems Biology 3(1) (2009)
18. Klamt, S., Saez-Rodriguez, J., Lindquist, J., Simeoni, L., Gilles, E.: A methodology for the structural and functional analysis of signaling and regulatory networks. BMC Bioinformatics 7(1), 56 (2006)
19. Bernot, G., Comet, J.P., Khalis, Z.: Gene regulatory networks with multiplexes. In: European Simulation and Modelling Conference Proceedings, pp. 423–432 (2008)

Abstraction of Graph-Based Models of Bio-molecular Reaction Systems for Efficient Simulation

Ibuki Kawamata, Nathanael Aubert, Masahiro Hamano, and Masami Hagiya

Graduate School of Information Science and Technology, University of Tokyo
7-3-1 Hongo, Bunkyo-ku, Tokyo, 113-8656, Japan
{ibuki,naubert,hamano,hagiya}@is.s.u-tokyo.ac.jp

Abstract. We propose a technique to simulate molecular reaction systems efficiently by abstracting graph models. Graphs (or networks) and their transitions give rise to simple but powerful models for molecules and their chemical reactions. Depending on the purpose of a graph-based model, nodes and edges of a graph may correspond to molecular units and chemical bonds, respectively. This kind of model provides naive simulations of molecular reaction systems by applying chemical kinetics to graph transition. Such naive models, however, can immediately cause a combinatorial explosion of the number of molecular species because combination of chemical bonds is usually unbounded, which makes simulation intractable. To overcome this problem, we introduce an abstraction technique to divide a graph into local structures. New abstracted models for simulating DNA hybridization systems and RNA interference are explained as case studies to show the effectiveness of our abstraction technique. We then discuss the trade-off between the efficiency and exactness of our abstracted models from the aspect of the number of structures and simulation error. We classify molecular reaction systems into three groups according to the assumptions on reactions. The first one allows efficient and exact abstraction, the second one allows efficient but approximate abstraction, and the third one does not reduce the number of structures by abstraction. We conclude that abstraction is a useful tool to analyze complex molecular reaction systems and measure their complexity.

1 Introduction

Developing models to describe bio-molecular reaction systems is important to analyze and predict biological phenomena. Models are also useful for designing complex artificial machines that are composed of bio-molecules such as DNA [32,27,15] and RNA [30,29]. Graphs (or networks) are often used to model such bio-molecules and their reactions. Systems biology methods using graphs have been commonly applied to analyze dynamic biological phenomena at a system level [20]. In such graph-based models, nodes and edges correspond to biological components (mRNA and proteins for example) and reactions (enzymatic reactions for example), respectively.

D. Gilbert and M. Heiner (Eds.): CMSB 2012, LNCS 7605, pp. 187–206, 2012.

To focus on a rather smaller level of chemical reactions such as hybridization of complementary DNA and the phosphorylation cascade of proteins, graph-based models are also useful to represent bio-molecular components. In this kind of naive graph-based models, nodes correspond to molecular groups such as nucleotides and polypeptides, while edges correspond to chemical connections such as hydrogen and covalent bonds. Chemical reactions among components are represented as transitions of a graph from one state to another. By applying chemical kinetics and either solving differential equations or performing stochastic analysis, naive graph-based models provide simulations that predict how bio-molecular systems behave. Provided that transition rules are thoroughly defined, whole reaction pathways are automatically searched because transition rules are iteratively applied. In this paper, we are interested in simulating reaction networks by solving numerical ODEs rather than stochastic analysis.

When combinations of chemical connections generate an exponential or unbounded number of molecules, simulating systems by graph-based models becomes intractable. To overcome this problem, models that focus on local connections of molecules have recently been investigated. For example, enumeration technique to calculate the equilibrium state of hybridization reaction systems was introduced by focusing on the locality of the hybridization of nucleotides [22,21]. Calculating an equilibrium state is to find the distribution of states that is energetically stable. The proposed method defines an enumeration graph, whose nodes correspond to local hydrogen bonds, and the structures of hybridization reaction system are encoded in paths that are defined as combinations of nodes. The distribution of states are calculated from the weights on edges, which can be efficiently determined because the number of local hydrogen bonds are limited even if the number of paths is exponentially big.

For another example, a graph rewriting system that represents protein interactions was proposed to overcome biological combinatorial explosion [10,13,9,16]. The graph rewriting system defines a graph-based model that focuses on the features of proteins that can modify or bind to other proteins. Because combinations of the features are kept implicit in the model, description of protein interactions becomes compact.

In this paper, we propose an abstraction technique which focuses on the local structures of DNA and RNA to overcome the combinatorial explosion problem of bio-molecules. Although the motivation of this paper is similar to the related work in the previous paragraph, we show an original abstraction technique for nucleic acid reactions. We also explain models and simulation results of DNA hybridization and RNAi as case studies to discuss the efficiency and exactness of our abstraction. We will conclude that abstraction is a useful tool to analyze complex molecular reaction systems and measure their complexity.

The organization of this paper is as follows. Section 2 reviews our previous work [18,19] about a naive graph-based model of DNA hybridization systems and the abstraction of it. Section 3 explains the efficiency and exactness of the abstracted model for DNA hybridization systems. Section 4 illustrate the naive graph-based model and the abstraction of the model of RNAi. Section 5 also explains the efficiency and exactness of the abstracted model for RNAi. Section 6 contains related work and section 7 is the discussion and the conclusion of this paper.

2 Naive Graph-Based Model and Abstraction of DNA Hybridization Systems

2.1 DNA Hybridization System

Recently, various kinds of reaction systems using DNA as components have been developed [1,33,26]. In those systems, a simple base paring ability of complementary sequences is thoroughly utilized. Along with experimental investigations, theoretical models to represent and design such systems have been investigated [32,31,23]. To simulate and automatically design DNA logic gates by evolutionary computation, we previously defined a model of DNA hybridization systems [18]. Although detail explanations are given in our previous work [19], we briefly explain the naive graph-based model and its abstraction in this section.

2.2 Naive Model

The way we model DNA hybridization systems is shown in Fig. 1 using the DNA logic gate, which comes from [28], as an example. DNA is chemically a double helical structure composed of two strands of nucleotide bases (the left-most in Fig. 1). Because a nucleotide base is one of adenine (A), thymine (T), guanine (G), and cytosine (C), a DNA strand can be represented as a sequence of 'A', 'T, 'G', and 'C'. Hydrogen bonds between complementary bases and phosphate backbones that run in antiparallel directions are preserved as undirected and directed edges, respectively (the second left in Fig. 1).

Fig. 1. Model of DNA

Sequences of nucleotides are divided into segments that are reaction units of hybridization (the second right in Fig. 1). We allocate lowercase and uppercase letters to the segments, where lowercase and uppercase of the same letter are complementary to each other. Finally we model the system by a graph data structure, whose nodes represent segments, whose undirected edges represent hydrogen bonds, and whose directed edges represent phosphate backbones (the right-most in Fig. 1).

2.3 Kinetic Simulation

We define three kinds of graph transitions (hybridization, denaturation, and branch migration) to represent the chemical reactions of DNA hybridization

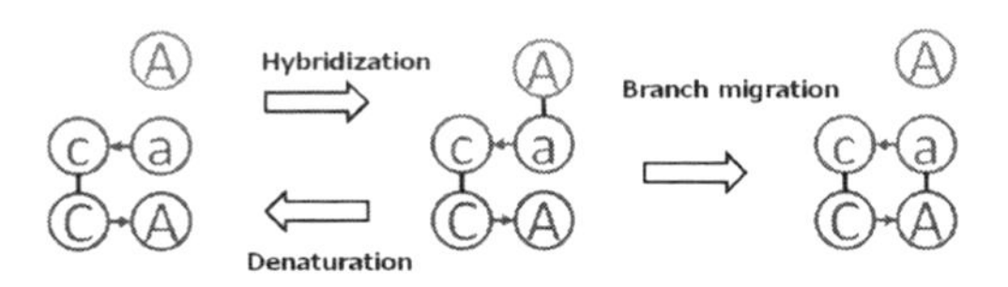

Fig. 2. Reaction rules

systems. An undirected edge between lowercase and uppercase of the same letter is generated by a hybridization reaction (transition from the left to the center in Fig. 2). Inversely, an undirected edge disappears from a graph by a denaturation reaction (transition from the center to the left in Fig. 2). Exchange of undirected edges occurs by a branch migration reaction (transition from the center to the right in Fig. 2).

The above naive graph-based model of DNA hybridization systems allows straight forward simulation by applying chemical kinetics and solving ordinary differential equations. First, we assign variables to all molecular species in the system to represents their concentrations. After we formalize differential equations of all reaction rules by applying chemical kinetics, we solve them to obtain concentration changes as functions of time. Kinetic speeds of reactions are defined by rule of thumb, and unit of concentration and time are arbitrary. Although the kinetic speeds of hybridization and branch migration are fixed, that of denaturation depends on the length of segments that are separating. We actually used a common numerical analysis called Runge-Kutta-Fehlberg-4,5 method to solve ordinary differential equations [12].

2.4 Combinatorial Explosion Problem

Hybridization chain reaction (HCR) is a typical example that generates an unbounded number of structures in the naive model [11]. HCR has two hairpin DNA strands and one initiator strand that starts the cascade of hybridization reactions (Fig. 3). Initiator will first hybridize to Hairpin 1 and branch migration opens the hairpin. After that, hybridization and branch migration occur alternately between Hairpins 1 and 2. Because these reactions continue until the hairpins are thoroughly consumed, the number of structures is essentially unbounded (Fig. 4).

2.5 Abstraction by Local Structure

To avoid the unbounded number of structures, we propose an original abstraction technique [19]. By the abstraction, variables are allocated to *local structures* (Fig. 5), where the number of local structures is finite. A local structure has some information about the strand it belongs to and the information about how each segment is connected. If a segment is not connected to any other segments, its connectivity is defined null. Otherwise its connectivity must identify

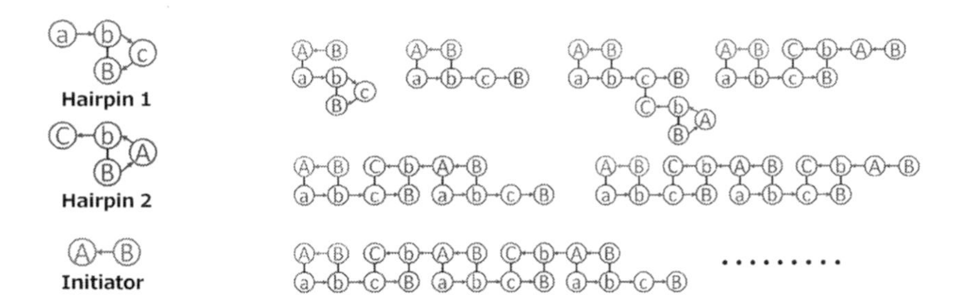

Fig. 3. Components of HCR

Fig. 4. Structures in HCR

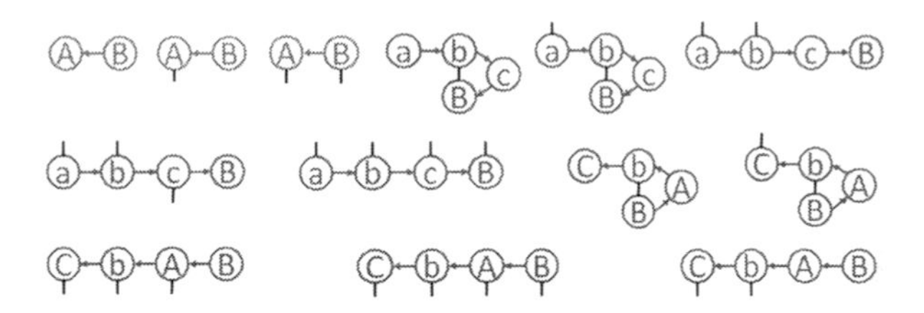

Fig. 5. Local structures

both the strand and the position of complementary segment although this information is omitted in the figure. We can efficiently enumerate the total number of local structures (details are found in [19]).

2.6 Reactions among Local Structures

Three reactions among local structures are defined as transitions from local structures to local structures in a similar way as that of the naive model. For example, the branch migration reaction that exchanges the undirected edge between 'a' of "ac" and 'A' of "BA" to that between 'a' of "ac" and 'A' of "CA"" is shown in Fig. 6. $C_1 \cdots C_8$ in the figure are the variables that represent the concentrations of corresponding local structures. For convenience, we call a local structure with the variable C_i just structure i. Structures 7 and 8 shown in Fig. 7 are the other possible neighbors that can connect to structure 2.

We assume that a local structure proportionally connects to a neighbor local structure. The ratio of each connection is calculated by dividing the concentration of the corresponding neighbor local structure by the sum of all the concentrations of possible neighbor local structures. For example, the concentration of a structure that connects the local structures 1 and 2 is calculated by

$$C_2 \times \frac{C_1}{C_1 + C_7}.$$

$C_1/(C_1 + C_7)$ is the ratio of the concentration of structure 2 connecting to structure 1 against the whole connections of structure 2. We call this assumption *ratio assumption* (the same concept appears in section 4.6).

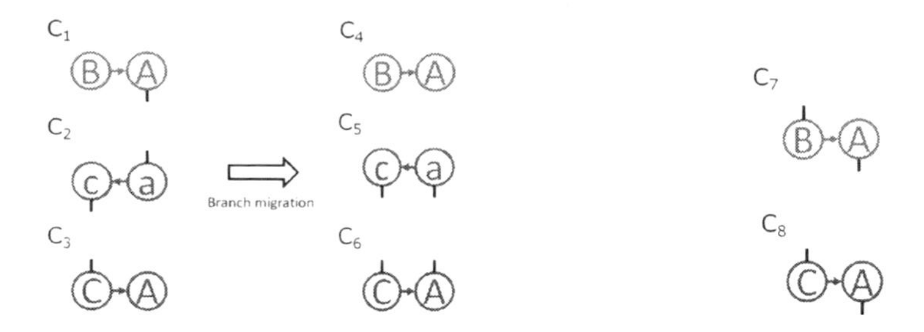

Fig. 6. Branch migration reaction in the local model

Fig. 7. Possible neighbor local structures

Because the branch migration reactions is originally unimolecular in the naive model, differential equations of the figure are formed as follows:

$$\Delta = k_b \times C_2 \times \frac{C_1}{C_1 + C_7} \times \frac{C_3}{C_3 + C_8}$$
$$\frac{d}{dt}C_1 = -\Delta, \frac{d}{dt}C_2 = -\Delta, \frac{d}{dt}C_3 = -\Delta$$
$$\frac{d}{dt}C_4 = \Delta, \frac{d}{dt}C_5 = \Delta, \frac{d}{dt}C_6 = \Delta,$$

where k_b is the kinetic speed of the branch migration reaction. This ratio assumption is also applied to the denaturation reaction because the reaction is originally unimolecular. In contrast, the differential equations of the hybridization are formed as bimolecular, because the local structures are not connected before the reactions. Because these variables can be affected by other reactions, the final form of the differential equations sums up the expressions of all involved reactions.

3 Efficiency and Exactness of Abstracted Model for DNA Hybridization Systems

3.1 Efficient and Approximate Abstraction

Although the number of local structures also increases exponentially as the model becomes bigger because of the combinatorial explosion of connections, it is much smaller than that of the global structures of the naive model [19]. This abstraction is approximate because the ratio assumption is not always satisfied and intra-molecular hybridization is also ignored. Even with these disadvantages, the abstracted simulation is useful for our design purpose because of its efficiency.

3.2 Exact Abstraction

We can develop an exact abstraction for the HCR under some appropriate assumptions. In the naive model, variables are allocated to all the structures depending on the number of opened hairpins (Fig. 8). We allocate C_0, C_{h1}, C_{h2}, and C_i to the initiator strand, Hairpin 1, Hairpin 2, and a structure that has i opened hairpins, respectively. We ignore the denaturation reactions and also assume that hybridization and branch migration occur together. The differential equations for the naive model are formed as follows:

$$\frac{d}{dt}C_0 = -k_b \times C_0 \times C_{\mathrm{h1}}$$

$$\frac{d}{dt}C_{2n+1} = k_b \times C_{2n} \times C_{\mathrm{h1}} - k_b \times C_{2n+1} \times C_{\mathrm{h2}} \quad (n \in \mathbb{N})$$

$$\frac{d}{dt}C_{2n+2} = k_b \times C_{2n+1} \times C_{\mathrm{h2}} - k_b \times C_{2n+2} \times C_{\mathrm{h1}} \quad (n \in \mathbb{N})$$

$$\frac{d}{dt}C_{\mathrm{h1}} = \sum_{i \in \mathbb{N}} -k_b \times C_{2i} \times C_{\mathrm{h1}}$$

$$\frac{d}{dt}C_{\mathrm{h2}} = \sum_{i \in \mathbb{N}} -k_b \times C_{2i+1} \times C_{\mathrm{h2}}$$

where $\mathbb{N}$ denotes the set of nonnegative integers. By ignoring C_i for $i > 5$ to limit the number of structures in the naive model, we obtained the concentration changes from C_1 to C_5 shown in Fig. 10. The initial values for C_0, C_{h1}, and C_{h2} are 0.01, 1.0, and 1.0, respectively (0 for other structures).

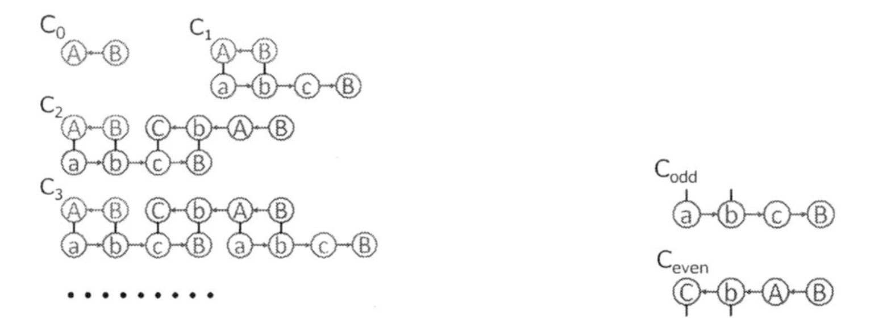

Fig. 8. Variable allocation of the naive model

Fig. 9. Local structures for C_{odd} and C_{even}

We then regard all the opened hairpins that can bind to the next hairpin as one local structure, and allocate C_{odd} and C_{even} for Hairpin 1 and Hairpin 2, respectively (Fig. 9). C_0, C_{h1}, and C_{h2} are allocated as before.

The differential equations for these variables in the local model are formed as follows:

$$\frac{d}{dt}C_0 = -k_b \times C_0 \times C_{\mathrm{h1}}$$

$$\frac{d}{dt}C_{\mathrm{odd}} = k_b \times C_0 \times C_{\mathrm{h1}} + k_b \times C_{even} \times C_{\mathrm{h1}} - k_b \times C_{\mathrm{odd}} \times C_{\mathrm{h2}}$$

$$\frac{d}{dt}C_{\mathrm{even}} = k_b \times C_{\mathrm{odd}} \times C_{\mathrm{h2}} - k_b \times C_{\mathrm{even}} \times C_{\mathrm{h1}}$$

$$\frac{d}{dt}C_{\mathrm{h1}} = -k_b \times C_0 \times C_{\mathrm{h1}} - k_b \times C_{\mathrm{even}} \times C_{\mathrm{h1}}$$

$$\frac{d}{dt}C_{\mathrm{h2}} = -k_b \times C_{\mathrm{odd}} \times C_{\mathrm{h2}}.$$

We can recover each C_i by assuming Possion distribution of the global structures:

$$C_i = (C_{\mathrm{odd}} + C_{\mathrm{even}}) \times \frac{\lambda^i \exp^{-\lambda}}{i!},$$

where λ is a function of time. After simulation of T time units, λ is calculated by

$$k_b \times \int_0^T (C_{\mathrm{h1}} + C_{\mathrm{h2}})dt,$$

which denotes the average number of the occurrences of the branch migration reaction. The simulation result by recovering the concentrations of global structures is shown in Fig. 11. The concentration of local structures are exactly calculated because of the definition of C_{odd} and C_{even}. We can understand the exactness from the property of C_{odd} and C_{even}. Because the local structures with C_{odd} and C_{even} appears only once at the right most position of each global structure of the naive model (Fig. 8), they satisfy the following expressions:

$$C_{\mathrm{odd}} = \sum_{i \in \mathbb{N}} C_{2i+1}$$

$$C_{\mathrm{even}} = \sum_{i \in \mathbb{N}} C_{2i+2}.$$

In contrast, the recovered simulation result is not exact because of the approximation under the assumption of Possion distribution.

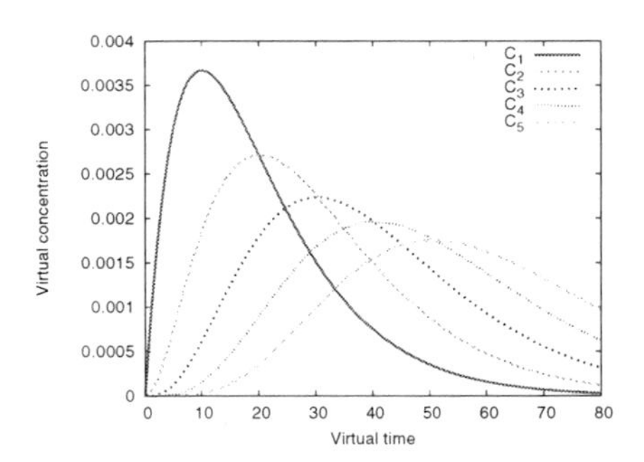

Fig. 10. Original simulation of HCR

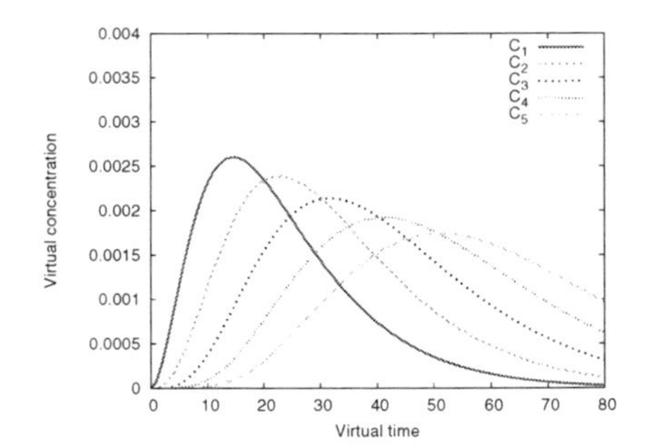

Fig. 11. Global information recovered by local simulation of HCR

4 Models for RNAi

4.1 RNAi

In this section, we apply our graph-based model and abstraction technique to RNA interference (RNAi). RNAi, also known as RNA silencing, is an internal cell mechanism commonly seen in both plants and animals to moderate gene expression [5,2,25]. In this mechanism, small RNAs produced from a double stranded RNA directly control gene expression. Our purpose is to observe the distribution of concentrations of small RNAs as well as the concentration of double stranded RNA. These topics are of importance for experimental biology and some mathematical models can be found in [3,24,14,8]. Because these models are phenomenological and are not based on chemical reactions, it is difficult to understand how each molecular species behaves in RNAi. In contrast, our model is directly based on chemical reactions, hence it clarifies the behavior of each molecular species. This model explains typical phenomena of RNAi and also allows us to apply our abstraction technique.

4.2 Naive Graph-Based Model

Our original naive graph-based model for representing RNAi based on chemical reactions is shown in Fig. 12. In the figure, thin directed arrows represent RNA and short thin undirected lines between them represent hydrogen bonds between complementary bases. Chemical reactions are drawn by box-shaped arrows with their name, which show the changes of molecules from one state to another. Φ is an arbitrary species which may be a waste or constant value for decay and transcription reactions.

In our model, transcription has a continuous activation and generates messenger RNA (mRNA) as in the middle of the figure. Input of double stranded RNA (dsRNA) initiates the cascades of reactions from the top of the figure. DsRNA is first cleaved into smaller segments called small interference RNA (siRNA) by the enzyme called Dicer. As in the case of DNA hybridization systems, sequences of nucleotides are regarded as segments that are reaction units. Because siRNAs have a smaller number of hydrogen bonds, denaturation reactions occur to break the hydrogen bonds and split siRNAs to upper "right-pointing" single stranded RNAs (ssRNAs) and lower "left-pointing" ssRNAs. Upper and lower ssRNAs can hybridize together to compose siRNAs. They can also decay and become wastes. Note that we use the term siRNA only for double stranded small segments, while single stranded sense or anti-sense RNA is simply called upper or lower ssRNA.

Some of the remaining lower ssRNAs are recruited to form RNA-induced silencing complexes (RISCs), which will eventually break down the mRNA by identifying the complementary sequences. Because lower ssRNAs and mRNA have complementary sequences, they hybridize and compose partially double-helical structures. This structure can also denature because of the assumption that denaturation can occur only when the length of hydrogen bonds is one segment long. After hybridization, polymerization reactions will extend the lower part of the double-helical structures and synthesize longer (but still partial) double-helical structures. If the right-most lower ssRNA is used as a primer, polymerization can reproduce the complete dsRNA which can initiate the whole reaction again.

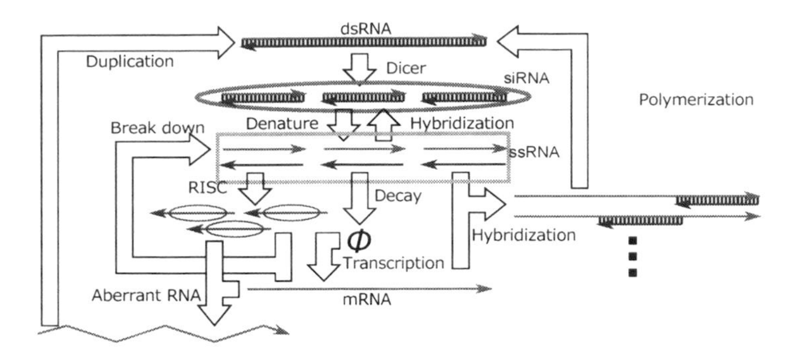

Fig. 12. Schematic explanation of RNAi

There is also another pathway to reproduce complete dsRNA from aberrant RNAs, as seen in plant cells [5]. For the pathway, a target mRNA is aberrated by RISC, and the aberrant RNA is eventually duplicated to reproduce complete dsRNA. Note that because we distinguish all ssRNAs, siRNAs and RISCs based on the original position in dsRNA, we distinguish all double-helical structures based on the positions of double-helices.

4.3 Simulation by Chemical Kinetics

The naive graph-based model of RNAi also provides a straight forward simulation in a similar way as the DNA hybridization systems. The number of segments that is cleaved by the Dicer reaction is a variable of our RNAi model. We assigned initial concentrations of mRNA and dsRNA to be 1.0, where units of concentration and time are arbitrary. To show the advantages of our model, we focus on the concentration change of the complete dsRNA and the concentration distribution of siRNAs among their positions. The concentrations of siRNAs are expected to have a distribution where they gradually decrease from left to right because the polymerization reaction extends lower RNA only toward the left direction.

We carefully decided default kinetic speeds as in Table 1 by checking the parameters of previous investigations [3,24,14,8]. By changing the kinetic speeds of the Dicer and duplication reactions, we successfully obtained three types of behaviors of dsRNA (Fig. 13). x and y-axes of the graph are time and concentration of dsRNA, respectively. For simplicity, we ignore RISC, break down, aberrant, and duplication reactions for the rest of this paper. Formal and detailed definitions of the ODE of RNAi are explained in the supplementary document [17].

4.4 Assumption on Polymerization and Combinatorial Explosion

Assumptions on the polymerization reaction are important because they have a strong influence on the number of molecular species. We assume that the polymerization reaction extends the primer strand for only one segment long though the reaction may happen repeatedly. We also assume that polymerization reaction has neither exonclease nor strand displacement activities. Although the first

Table 1. Kinetics parameters

Reaction	Speed
Hybridization	0.1
Denaturation	0.0078
Polymerization	1.0
Dicer	0.01
Transcription	0.05
Decay	0.01
RISC	0.01
Break down	0.005
Aberrant	0.01
Duplication	0.01

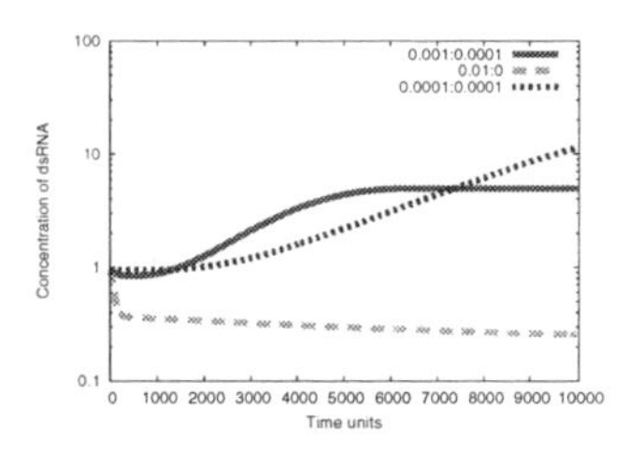

Fig. 13. Concentration change of dsRNA. First and second values in the legend are kinetic speeds of Dicer and duplication, respectively.

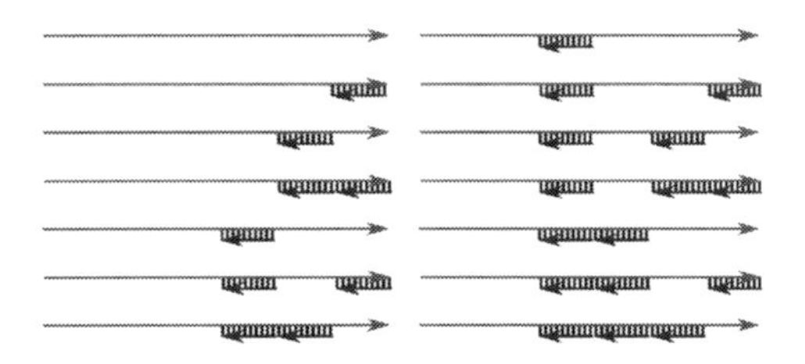

Fig. 14. Combinatorial explosion of RNA structures

one is essential, these assumptions produce an exponential number of partially double-helical structures as illustrated in Fig. 14.

Because for this exponential increase of the number of structures, the number of segments into which the Dicer reaction can cleave dsRNA was limited to 9 segments under our simulations. The number was not enough to verify the expected distribution of siRNAs [25].

4.5 Abstraction by Local Structures

To avoid the exponential increase of the number of structures, we propose an efficient abstraction that is based on the locality of RNA structures. We divide global structures of the naive model into local RNA structures according to their connectivity to theirs neighbor structures. In other word, each global structure in the naive model is represented by a set of local RNA structures. A local RNA structure contains the information about the position of its segment, the presence of a double-helix, and the connections with neighbor local RNA structures. We index the position of a segment from the left-most part of the dsRNA. The presence of a double-helix denotes whether a local structure has complementary RNA hybridized to its position or not. The connections with neighbor structures restrict the local structures in the previous or next position, because information about the presence of a double-helix and the presence of nick of lower RNA are included. If we list all the local structures, there are 26 types for each position (except for the left-most and right-most positions) as in Fig. 15. For example, the upper global structure in Fig. 16 is represented by four local structures at the bottom of the figure. This local RNA structure is formally modeled as a subgraph with annotation that represents the connectivity, the detail of which is explained in the supplementary document [17].

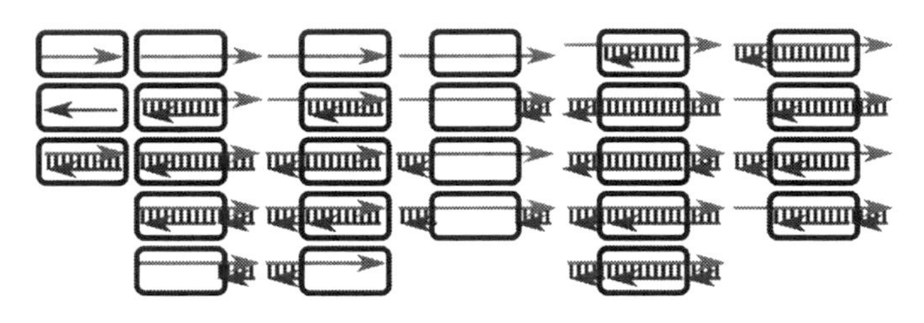

Fig. 15. Local RNA structures for one position in the local model

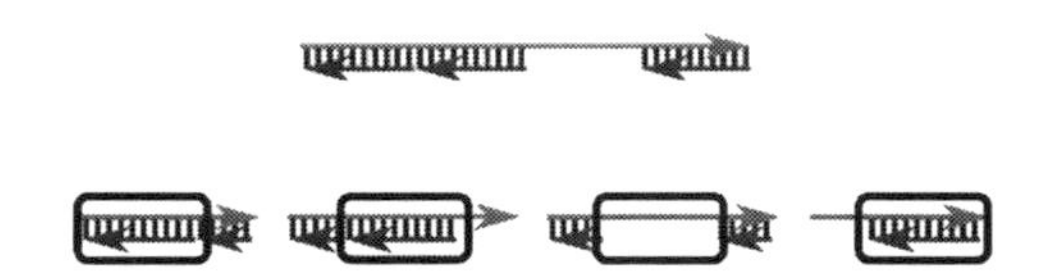

Fig. 16. Example of abstraction by local model

In order to observe the concentration of dsRNA, we refine the local model. Three of the local structures (Fig. 17) are divided into two structures each, depending on the position of the primer of polymerization reaction. If the double-helix of a local structure is extended from the right-most primer, the local structure is distinguished from the others. For example, upper global structure in Fig. 18 is represented by the five local structures at the bottom of the figure. Only the second from the right local structure is extended from the right-most primer. This refined abstraction has 29 types of local structures for each position. By this division, we can distinguish the local structures that are parts of the complete dsRNA. We call this model as the refined local model to distinguish it from the unrefined local model.

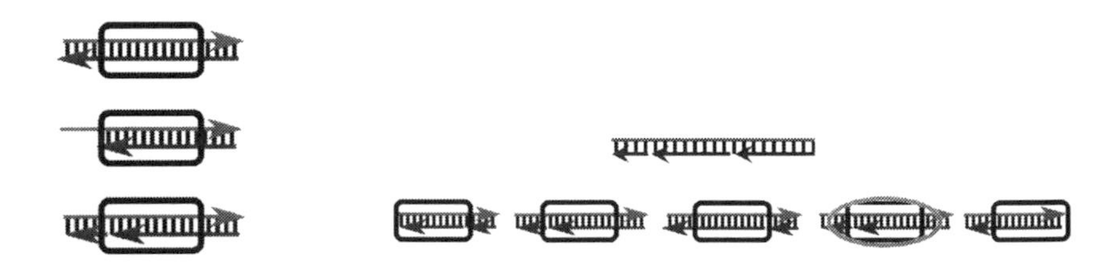

Fig. 17. Three structures that are divided

Fig. 18. Example of abstraction by the refined local model

4.6 Local Reactions

As global structures are abstracted based on local structures, all reaction rules are also abstracted. Fig. 19 shows an example of three local structures becoming four other local structures by a denaturation reaction. Other reactions are similarly modeled as reactions among local structures. Polymerization of the refined local model is carefully defined to preserve the concentration of complete dsRNA.

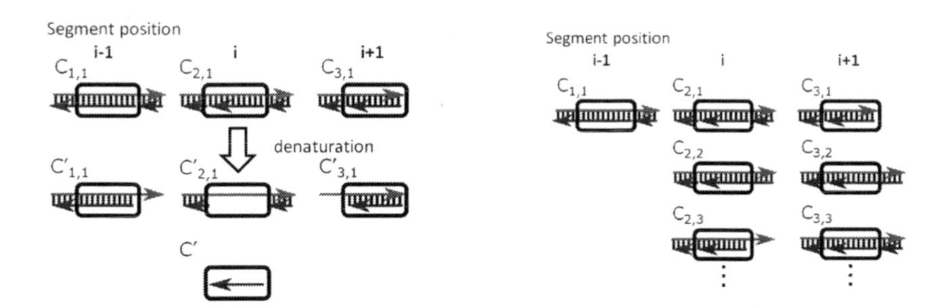

Fig. 19. Example of local reaction **Fig. 20.** Possible neighbor local structures

Similar to the section 2.6, we assume the proportional connections between local structures. Possible local structures are shown in Fig. 20 and variables $C_{1,1}, C_{2,1}, \cdots$ are allocated to the corresponding structures. For convenience, we call a local structure with the variable $C_{i,j}$ just structure (i,j). In the figure, local structure $(1,1)$ whose segment position is $i-1$ can connect to various kinds of local structures such as $(2,1)$, $(2,2)$, $\cdots$ whose segment position is i. Similarly, local structure $(2,1)$ can connect to local structures $(3,1)$, $(3,2)$, $\cdots$. The ratio of each connection is calculated by dividing the concentration of the corresponding neighbor local structure by the sum of all the concentrations of possible neighbor local structures. For example, the concentration of structure that connects between the structures $(1,1)$ and $(2,1)$ is recovered by

$$C_{1,1} \times \frac{C_{2,1}}{C_{2,1} + C_{2,2} + C_{2,3} + \cdots}.$$

$C_{2,1}/(C_{2,1} + C_{2,2} + C_{2,3} + \cdots)$ is the ratio of the connection that connects local structures from 1.1 to 2.2. Though the local models are different from DNA hybridization systems, this assumption is also a ratio assumption.

Differential equation for denaturation reaction is formed as follows.

$$\Delta = k_d \times C_{1,1} \times \frac{C_{2,1}}{C_{2,1} + C_{2,2} + C_{2,3} + \cdots} \times \frac{C_{3,1}}{C_{3,1} + C_{3,2} + C_{3,3} + \cdots}$$

$$\frac{d}{dt}C_{1,1} = -\Delta, \frac{d}{dt}C_{2,1} = -\Delta, \frac{d}{dt}C_{3,1} = -\Delta$$

$$\frac{d}{dt}C'_{1,1} = \Delta, \frac{d}{dt}C'_{2,1} = \Delta, \frac{d}{dt}C'_{3,1} = \Delta, \frac{d}{dt}C' = \Delta,$$

where k_d is the kinetic speed of denaturation reaction. The concentration of a structure that connects three local structure is recovered by the ratio assumption.

4.7 Simulation Result

Because the local model enables to increase the number of segments, we obtained an expected concentration distribution of siRNA (Fig. 21). x, y, and z-axes of the figure correspond to the total number of segments, the position of ssRNA, and the concentration of lower ssRNAs, respectively. For each total number of segments, we simulated 500 time units, and obtained the three-dimensional figure by plotting all distributions. Concentrations of ssRNAs are used because the distribution of ssRNAs are experimentally observed as a clue of siRNAs [25]. We cannot obtain such a distribution with the naive model, because the number of segments that can be simulated was limited by the combinatorial explosion. In the figure, we draw a line to emphasize the distribution of the concentrations for 24-segment simulation. As supported in the supplementary document [17], this simulation is considered to give a good approximation of the naive simulation.

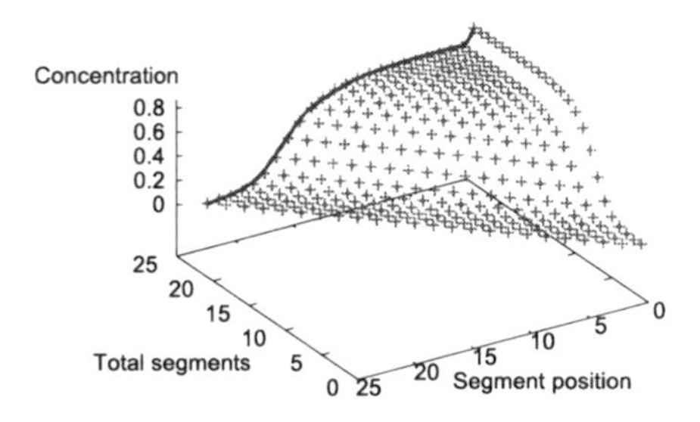

Fig. 21. Distribution of siRNA

5 Efficiency and Exactness of Local Models

5.1 Efficiency

The number of structures and the number of reactions increased exponentially for the naive model (Fig. 22 and Fig. 23). Simulation with more than 9 segments was impossible because of an out-of-memory error (the number of structures and that of reactions exceed 10000 and 800000, respectively). This was the main reason why simulation became intractable by combinatorial explosion. With our abstraction, the number of structures and that of reactions increased linearly, which made the abstract simulation much more efficient. Computation time also decreased dramatically in both local models (Fig. 24).

5.2 Approximation

The difference between exact and approximate abstractions are explained formally by the following diagram [9].

$$\begin{array}{ccc} N & \xrightarrow{\quad R \quad} & N \\ \downarrow\alpha & & \downarrow\alpha \\ A & \xrightarrow{\quad Q \quad} & A \end{array}$$

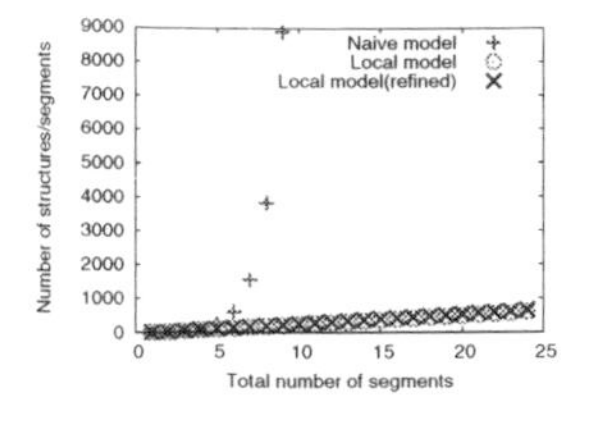

Fig. 22. Comparison of the number of structures

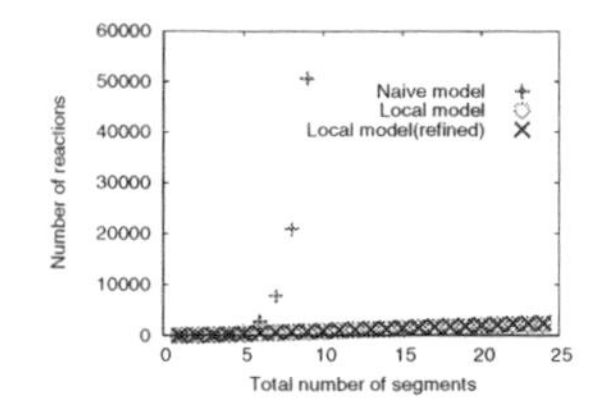

Fig. 23. Comparison of the number of reactions

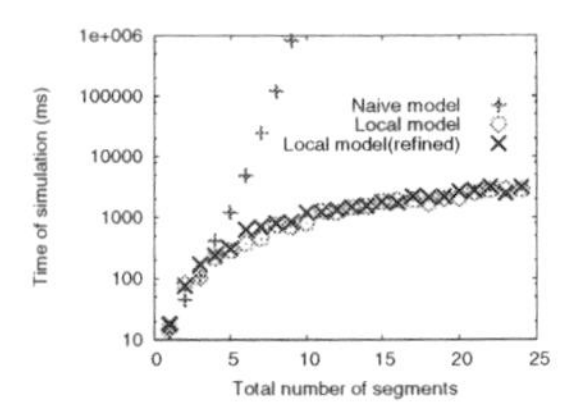

Fig. 24. Comparison of computation time

N and A represent the set of naive models and abstract models, respectively, both of which are sets of maps from molecular species to their concentration. To be more formal, N is the set of maps from G to $\mathbb{R}$, where G denotes the set of global structures and $\mathbb{R}$ the set of non-negative real numbers, while A is the set of maps from L to $\mathbb{R}$, where L denotes the set of local structures. R and Q denote temporal evolution by ODE of naive and abstracted models, respectively. We prepare a function $c(g, l)$ for $g \in G$ and $l \in L$ which denotes the number of occurrences of local structure l in global structure g. The abstraction α, which is a map from N to A, is then defined as $\alpha(n)(l) = \sum_{g \in G} c(g, l)n(g)$, where $n \in N$ and $l \in L$.

The abstraction α is called exact if and only if $\alpha(R(n)) = Q(\alpha(n))$. Otherwise, the abstraction is called approximate. In the case of HCR, where G and L correspond to structures explained in section 2.2 and 2.5, respectively, α is exact although the inverse function to obtain N from A is approximate. In the case of RNAi, where G and L correspond to structures explained in section 4.2 and 4.5, α is approximate, which can be understood from the experimental results in the supplementary document [17]. Note that the abstraction in the next section is exact.

5.3 Exact Abstraction

As far as we have considered, it seems impossible to define an exact abstraction in the current model. An important aspect to define an exact abstraction is to make each reaction satisfy the ratio assumption after the reaction. The ratio assumption is easily broken when a reaction requires particular connection of local structures to occur (polymerization and denaturation are typical examples). In contrast, if a reaction is unimolecular in the local model, it does not break the ratio assumption (decay is a typical example). Moreover, if a bimolecular reaction occurs between any combinations of local structures, the reaction does not break the assumption (hybridization is a typical example).

To define an exact abstraction, some assumptions that restrict the polymerase reaction are required. One assumption is that the reaction speed of polymerization is very fast compared to others. By this assumption, we can define bigger local structures (Fig. 25), which give a new and exact abstraction for RNAi. In this model, all the local structures have one of the two form in the figure, both of which have multiple (or no) short (one segment-long) lower RNA from the

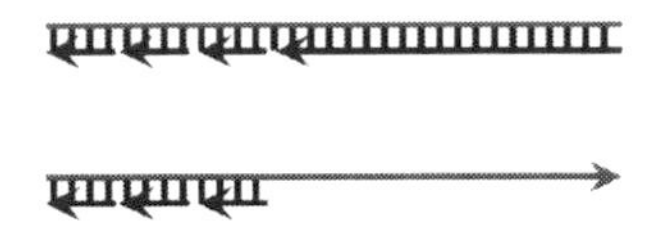

Fig. 25. Model of RNA for exact abstraction

left part of the structure. One form has a long (bigger than one-segment long) lower RNA at the right part. The other does not have any complementary RNA at the right part. A global structure of the naive model is then represented as a set of these local structures.

Reaction rules are also redefined for this model. A denaturation example that separates the third (from the left) short lower RNA is shown in Fig. 26. Even though an intermediate structure is produced (shown in the right side of the figure), it can be ignored because the polymerization reaction occurs immediately. Eventually the reaction is defined as a transition between new local structures. Another denaturation example that separates the second RNA is shown in Fig. 27. After polymerization, the intermediate structure will split into two local structures, which also result in a transition among new local structures. Because these reactions are unimolecular, they do not require the ratio assumption. By similarly defining other reactions, it is possible to define an exact abstraction.

Fig. 26. Reaction rule for exact abstraction

Fig. 27. Another reaction rule for exact abstraction

We can also take into account exonuclease or displacement activities as other assumptions. By these assumptions, only one lower RNA can hybridize to mRNA because polymerization reaction immediately extends the lower RNA toward the left-end of mRNA. The number of structures in the naive model is strongly restricted because combination of lower RNAs is eliminated completely.

5.4 Classification of RNAi Systems

The number of local structures in each exact abstraction can be a measure of computational complexity of a reaction system. According to the abstractions we have explained, we summarize the complexity of RNAi (Table 2) where n denotes the total number of segments. As shown in the table, there are three groups of systems of RNAi. Without any assumption (first group), it seems impossible to define an exact abstraction to allow simulation with polynomial complexity. However, we can efficiently simulate this system by approximate abstraction.

Table 2. Summary of complexity of RNAi. n denotes the number of segments of RNAi explained in 4.3.

Speed of polymerase reaction and its activity	Number of structures in naive model	Number of local structures by exact abstraction
Slow	Exponential	Seems exponential
Fast	Exponential	$O(n^3)$
Fast Exonuclease/displacement activity	$O(n^3)$	$O(n^3)$

With one assumption (second group), it is still impossible to efficiently simulate the system by a naive original model because the number of structures is exponentially big. By the exact abstraction in the previous subsection, however, we can efficiently simulate the system because the number of structures becomes of a polynomial size. If we assume too much on the polymerization reaction (third group), the system becomes simple enough to simulate by a naive model. The number of structures does not decrease by abstraction.

From the aspect of the number of structures, it is possible to compare the computational complexity of bio-molecular reaction systems. As an example of RNAi, we have shown that a system without any assumptions is considered computationally more complex than the systems with assumptions.

6 Related Work

Various approaches for abstracting ODE systems derived from rule-based descriptions of biochemical reaction networks have been studied [7,6,4,13,9]. The domain-oriented approach [7,6,4] introduces a state space transformation that reduces the microscopic model by employing descriptions of macroscopic variables. This approach does not directly apply to our model, due to the presence of feedback in RNAi through which new agents of siRNAs are born by a triggered siRNA.

Since our modelling is based on graphs with annotations, the examples in this paper can be directly encoded in κ-calculus [13,9]. Under the framework of κ, abstraction is defined in terms of fragmentation by local structures, in which the LHS of each rule should not overlap with a local structure or should be entirely contained in a local structure. On the other hand, our fragmentation for DNA and RNA violates this condition for the sake of obtaining smaller local structures. As a trade-off of the violation, our fragmentation imposes the ratio assumption, which is satisfied in HCR and in the abstraction of RNAi in section 5.3. In these two examples, which satisfy the ratio assumption, fragmentation under the framework of κ is also possible by enlarged local structures and results in exact abstraction.

7 Conclusion

To overcome the combinatorial explosion of structures of bio-molecules, we introduced an abstraction technique by presenting the case studies of DNA hybridization systems and RNAi. We succeeded in efficient simulations by reducing the number of structures in both case studies. Simulation becomes approximate when connections among local structures are calculated by the ratio assumption or when intra-molecular reactions are ignored. Although we investigated the errors from approximation by comparing the ratios of connections, it is left as future work to estimate the degree of approximation.

To define an exact abstraction, local structures and reactions must be carefully defined. Unimolecular reactions among local structures provide an exact abstraction because they are not affected by other local structures that are not involved. To satisfy the ratio assumption, multi-molecular reactions must uniformly occur among all the combinations of local structures, otherwise modifications of reactions are required. It is also future work to investigate and develop an exact abstraction technique that can be applied to general graphs because we have shown two case studies in this paper.

Systems of RNAi are classified into three groups according to the number of structures created by abstraction. DNA hybridization systems can also be classified into two groups, both of which allow efficient abstraction. Systems in first group allow only approximate abstraction because of the denaturation and branch migration reactions. In contrast, HCR with appropriate assumptions belongs to the other group, which allows exact abstraction.

Acknowledgment. The author appreciates Richard Potter and anonymous reviewers of CMSB2012 for constructive comments. This work is supported by Grant-in-Aid for JSPS Fellows (11J09247), Exploratory Research (11015189), and Scientific Research on Innovative Areas (24104005).

References

1. Bath, J., Turberfield, A.J.: DNA nanomachines. Nat. Nanotechnol. 2(5), 275–284 (2007)
2. Baulcombe, D.C.: Amplified Silencing. Science 315(5809), 199–200 (2007)
3. Bergstrom, C.T., McKittrick, E., Antia, R.: Mathematical models of RNA silencing: unidirectional amplification limits accidental self-directed reactions. Proc. Natl. Acad. Sci. USA 100(20), 11511–11516 (2003)
4. Borisov, N.M., Chistopolsky, A.S., Faeder, J.R., Kholodenko, B.N.: Domain-oriented reduction of rule-based network models. IET Syst. Biol. 2(5), 342–351 (2008)
5. Brodersen, P., Voinnet, O.: The diversity of RNA silencing pathways in plants. TRENDS in Genettics 22(5), 268–280 (2006)
6. Conzelmann, H., Fey, D., Gilles, E.D.: Exact model reduction of combinatorial reaction networks. BMC Syst. Biol. 2, 78 (2008)

7. Conzelmann, H., Saez-Rodriguez, J., Sauter, T., Kholodenko, B.N., Gilles, E.D.: A domain-oriented approach to the reduction of combinatorial complexity in signal transduction networks. BMC Bioinformatics 7, 34 (2006)
8. Cuccato, G., Polynikis, A., Siciliano, V., Graziano, M., di Bernardo, M., di Bernardo, D.: Modeling RNA interference in mammalian cells. BMC Syst. Biol. 5, 19 (2011)
9. Danos, V., Feret, J., Fontana, W., Harmer, R., Krivine, J.: Abstracting the differential semantics of rule-based models: exact and automated model reduction. In: Proceedings of the Twenty-Fifth Annual IEEE Symposium on Logic in Computer Science, pp. 362–381. IEEE (2010)
10. Danos, V., Feret, J., Fontana, W., Krivine, J.: Abstract Interpretation of Cellular Signalling Networks. In: Logozzo, F., Peled, D.A., Zuck, L.D. (eds.) VMCAI 2008. LNCS, vol. 4905, pp. 83–97. Springer, Heidelberg (2008)
11. Dirks, R.M., Pierce, N.A.: Triggered amplification by hybridization chain reaction. Proc. Natl. Acad. Sci. USA 101(43), 15275–15278 (2004)
12. Fehlberg, E.: Klassische Runge-Kutta-Formeln vierter und niedrigerer ordnung mit schrittweiten-kontrolle und ihre Anwendung auf wärmeleitungsprobleme. Computing 6(1), 61–71 (1970)
13. Feret, J., Danos, V., Krivine, J., Harmer, R., Fontana, W.: Internal coarse-graining of molecular systems. Proc. Natl. Acad. Sci. USA 106(16), 6453–6458 (2009)
14. Groenenboom, M.A.C., Marée, A.F.M., Hogeweg, P.: The RNA Silencing Pathway: The Bits and Pieces That Matter. PLoS Comput. Biol. 1(2), 155–165 (2005)
15. Han, D., Pal, S., Nangreave, J., Deng, Z., Liu, Y., Yan, H.: DNA Origami with Complex Curvatures in Three-Dimensional Space. Science 332(6027), 342–346 (2011)
16. Harmer, R., Danos, V., Feret, J., Krivine, J., Fontana, W.: Intrinsic Information Carriers in Combinatorial Dynamical Systems. Chaos 20(3), 037108 (2010)
17. Kawamata, I.: Formal Definition of the ODE of RNAi and Experimental Results for Approximate Abstraction: supplementary documents (2012), `http://hagi.is.s.u-tokyo.ac.jp/~ibuki/cmsb2012supply.pdf`
18. Kawamata, I., Tanaka, F., Hagiya, M.: Automatic Design of DNA Logic Gates Based on Kinetic Simulation. In: Deaton, R., Suyama, A. (eds.) DNA 15. LNCS, vol. 5877, pp. 88–96. Springer, Heidelberg (2009)
19. Kawamata, I., Tanaka, F., Hagiya, M.: Abstraction of DNA Graph Structures for Efficient Enumeration and Simulation. In: International Conference on Parallel and Distributed Processing Techniques and Applications, pp. 800–806 (2011)
20. Kitano, H.: Systems Biology: A Brief Overview. Science 295(5560), 1662–1664 (2002)
21. Kobayashi, S.: A New Approach to Computing Equilibrium State of Combinatorial Hybridization Reaction Systems. In: Proc. of Workshop on Computing and Communications from Biological Systems: Theory and Applications, pp. 330–335 (2007)
22. Kobayashi, S.: Symmetric Enumeration Method: A New Approach to Computing Equilibria. Technical Report of Dept. of Computer Science, University of Electro-Communications (2008)
23. Lakin, M.R., Parker, D., Cardelli, L., Kwiatkowska, M., Phillips, A.: Design and analysis of DNA strand displacement devices using probabilistic model checking. J. R. Soc. Interface 9(72), 1470–1485 (2012)
24. Marshall, W.F.: Modeling recursive RNA interference. PLoS Comput. Biol. 4(9), e1000183 (2008)
25. Pak, J., Fire, A.: Distinct Populations of Primary and Secondary Effectors During RNAi in C. elegans. Science 315(5809), 241–244 (2007)

26. Pinheiro, A.V., Han, D., Shih, W.M., Yan, H.: Challenges and opportunities for structural DNA nanotechnology. Nat. Nanotechnol. 6(12), 763–772 (2011)
27. Qian, L., Winfree, E.: Scaling Up Digital Circuit Computation with DNA Strand Displacement Cascades. Science 332(6034), 1196–1201 (2011)
28. Seelig, G., Soloveichik, D., Zhang, D.Y., Winfree, E.: Enzyme-Free Nucleic Acid Logic Circuits. Science 314(5805), 1585–1588 (2006)
29. Venkataraman, S., Dirks, R.M., Ueda, C.T., Pierce, N.A.: Selective cell death mediated by small conditional RNAs. Proc. Natl. Acad. Sci. USA 107(39), 16777–16782 (2010)
30. Win, M.N., Smolke, C.D.: Higher-Order Cellular Information Processing with Synthetic RNA Devices. Science 322(5900), 456–460 (2008)
31. Winfree, E.: Algorithmic self-assembly of DNA. Ph.D. thesis, California Institute of Technology (1998)
32. Yin, P., Choi, H.M.T., Calvert, C.R., Pierce, N.A.: Programming biomolecular self-assembly pathways. Nature 451(7176), 318–322 (2008)
33. Zhang, D.Y., Seelig, G.: Dynamic DNA nanotechnology using strand-displacement reactions. Nat. Chem. 3(2), 103–113 (2011)

Parameter Identification and Model Ranking of Thomas Networks*

Hannes Klarner[1], Adam Streck[2], David Šafránek[2], Juraj Kolčák[2], and Heike Siebert[1]

[1] Freie Universität Berlin, Berlin, Germany
Hannes.Klarner@fu-berlin.de
[2] Masaryk University, Brno, Czech Republic
safranek@fi.muni.cz

Abstract. We propose a new methodology for identification and analysis of discrete gene networks as defined by René Thomas, supported by a tool chain: (*i*) given a Thomas network with partially known kinetic parameters, we reduce the number of acceptable parametrizations to those that fit time-series measurements and reflect other known constraints by an improved technique of coloured LTL model checking performing efficiently on Thomas networks in distributed environment; (*ii*) we introduce classification of acceptable parametrizations to identify most optimal ones; (*iii*) we propose two ways of visualising parametrizations dynamics wrt time-series data. Finally, computational efficiency is evaluated and the methodology is validated on bacteriophage λ case study.

Keywords: Thomas network, parameter identification, model checking.

1 Introduction

Discrete modeling frameworks are commonly used in systems biology as a tool that assists in revealing regulatory mechanisms found in biological networks [14,11,20]. A widely used formalism for gene regulatory networks is that of R. Thomas et al. [21] (see [9] for review). The formalism treats changes in gene expression asynchronously, thus bringing a sort of conservatism into the discrete abstraction at the price of large state spaces with many transitions. However, the asynchronous semantics is a natural approach to formalization of concurrent systems in computer science. This enables application of well-established formal methods to Thomas networks [5,16,4,19].

Although discrete regulatory models are very abstract, parameters determining the behavior of regulated components are often unknown. An important problem is therefore inference of these parameters from biological hypotheses and wet-lab measurements e.g. time series data. There is no reliable technique to reveal the regulatory logic, and existing reverse engineering approaches are mostly based on measurement clustering or information theory (see [15] for review).

* This work has been supported by the Czech Grant Agency grant No. GAP202/11/0312.

D. Gilbert and M. Heiner (Eds.): CMSB 2012, LNCS 7605, pp. 207–226, 2012.

Formal methods have been employed to assist in identifying parameters for Thomas networks, utilizing not only time series data but also arbitrary hypotheses formalized in terms of a temporal logic. Naive (bottom-up) approaches [4,12] repeat a procedure deciding for each parametrization whether it satisfies the given temporal constraints or not. That way *acceptable parametrizations* are found. Since the number of possible parametrizations increases exponentially with the number of unknown parameters, such a procedure is intractable in many real cases.

Barnat et al. [2] introduced technique of *colored LTL model checking* (CMC) based on a heuristics reducing the computation effort by means of operating on the parametrization space in a top-down manner. In particular, maximal parametrization sets sharing a required behavior are inferred instead of analyzing each possible parametrization individually. The technique was defined for multi-affine abstractions of continuous models and was based on symbolic representation of parametrization sets thus allowing effective realization of required operations. When employed on Thomas networks, an ideal symbolic representation which would allow effective realization of all required set operations was not found. Therefore the results obtained for Thomas networks were not optimal.

In [12], Klarner et al. developed a workflow for parameter identification of Thomas networks exploiting time series data. Especially notions of *edge constraints* and expression *monotonicity* in between measurements were defined to initially restrict acceptable parametrizations by preliminary known facts about network dynamics.

In this paper, authors of both groups combine their approaches to obtain efficient methods for parameter identification using colored model checking. The result of this collaboration is a comprehensive methodology that further extends the workflow of [12] introducing a *classification of acceptable parametrizations* based on optimal satisfaction of selected criteria. Our methodology guides users towards selection of parametrizations complying with given hypotheses and time series data, and proposes further filtering of obtained parameters based on criteria such as low complexity. Moreover, visualization procedures are proposed that allow a quick and intuitive understanding of the behavior generated by different parametrizations allowing for easy identification of e.g. potential ranges of poor measurement sampling. The workflow is depicted in Fig. 1.

To the best of our knowledge, the only work which attempts to employ some criteria to select most plausible parametrizations in the context of Thomas networks is mentioned in [7]. The approach is a work in progress based entirely on constraint programming. As there are no concrete criteria defined, we currently cannot compare the methodological side.

On the computational side, our approach is supported by a prototype tool chain consisting of three modules: static analyzer, model checker, and parametrization filter. The static analyzer module solves constraints related to the network structure and is implemented on the top of the model checker module. The model checker module implements CMC including computation of compliant behaviors (in model checking terms: generation of all counterexamples for a given time series formula) and parameter ranking. The parametrization filter allows browsing the parameters and filtering them wrt several criteria. Moreover, the filter module gives graphical feedback to the user.

Computational efficiency is obtained by direct distribution and shared enumeration of parametrization sets. To the best of our knowledge, there is only one other efficient approach [3] targeting discrete gene dynamics. It employs a more detailed model – the piece-wise affine framework. The representation of parameter space is specific for the level of abstraction employed. Efficiency is obtained by considering symbolic representation of parametrizations.

The paper, after introducing the basic notions in the next section, is structured according to the workflow mentioned above and depicted in Fig. 1. To illustrate the approach a case study of the bacteriophage λ is considered in Sect. 5. Further information on implementation and performance as well as final remarks conclude the paper.

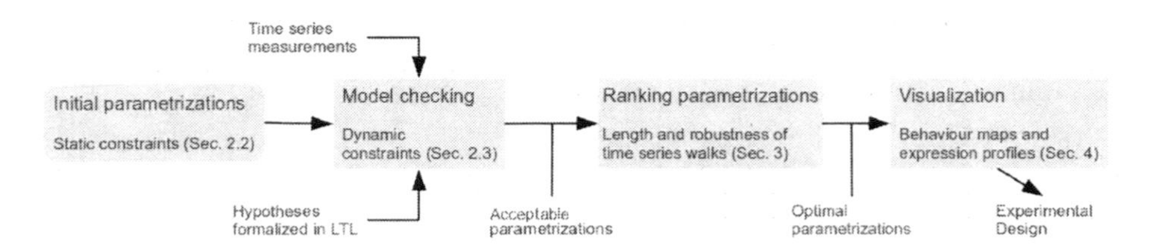

Fig. 1. Parameter identification workflow

2 Background

2.1 Thomas Networks

In the following we recall the logical modeling framework introduced by C. Chaouiya et al. in [5, Section 2], which is a generalization of the formalism of R. Thomas [21].

Regulatory Graphs. The structure of a system, i.e. the components (or species) involved and the dependencies between them, can be captured in a graph. We define an *interaction graph* (V, E) to be a directed graph consisting of $n \in \mathbb{N}_1$ vertices $V = \{v_1, \dots, v_n\}$ called *components* and a set $E \subseteq V \times V$ of ordered pairs of vertices called *interactions*. We use the notation $uv \in E$ for interactions and call u the *regulator of* uv and v the *target of* uv. The in-neighbors $N_E^-(v) := \{u \in V \mid uv \in E\}$ of v are called *regulators of* v and the out-neighbors $N_E^+(v)$ are called *targets of* v.

Since we are not only interested in the structure of the network but also in the dynamics, we interpret the vertices as integer variables whose values signify e.g. the level of concentration of the corresponding substance. Naturally, the impact a regulator has on its target depends on the value of the corresponding variable. This information about the interactions, i.e. the edges in the interaction graph, is also needed to specify the dynamical behavior of the system. This leads to the following definition.

A *regulatory graph* $\mathcal{R} = (V, E, \rho, \theta)$ consists of an interaction graph (V, E) and two functions ρ and θ. The function $\rho : V \to \mathbb{N}_1$ assigns a non-zero natural number $\rho(v)$, called *maximal activity level* of v, to each component. For an *integer interval* $\{k \in \mathbb{N} \mid a \leq k \leq b\}$ with boundaries $a \leq b \in \mathbb{N}$ we use the notation $[a, b]$. The interval $[0, \rho(v)]$ is called *activity interval of component* v and an element of the activity interval is called *activity level* of v.

To a regulatory graph $\mathcal{R}$ we thus associate the *state space* $X := \prod_{i=1}^{n} [0, \rho(v_i)]$. An element $x \in X$ is called a *state* of the regulatory graph and we use the subscript notation x_v to denote the activity of $v \in V$ in state x.

The other function, θ, assigns interaction thresholds $\theta(uv) = (t_1, \ldots, t_k)$ to each interaction $uv \in E$. Each interaction may have a different number $1 \leq k$ of thresholds. The thresholds must be ordered: $t_1 < \cdots < t_k$ and within the non-zero activities of the regulator: $1 \leq t_1$ and $t_k \leq \rho(u)$.

The interaction thresholds $\theta(uv) = (t_1, \ldots, t_k)$ of an interaction uv divide the activities of u into $k + 1$ intervals $[0, t_1 - 1], [t_1, t_2 - 1], \ldots, [t_k, \rho(u)]$ of different *regulation intensity*. Activities of u that belong to the same interval are characterized by being above the same number of thresholds of $\theta(uv)$. We denote the jth interval by I_j^{uv}. The different regulation intervals allow us to distinguish between different effects an interaction between two components can have depending on the activity of the regulator.

Parametrizations. In this subsection we discuss how to parametrize a regulatory graph. Basically, we need to provide all the information necessary to determine effects of any regulators on its target in every state. The effect will not necessarily depend on the exact state, but only on the regulation intervals to which this state belongs. We formalize this idea in the following definitions.

A *regulatory context* ω of a component v assigns an intensity to every interaction $uv \in E$ targeting v. For every regulator $u \in N^-(v)$, there is a regulation intensity I_j^{uv}, such that $\omega(u) = I_j^{uv}$. The set of all combinatorially possible regulatory contexts of v is denoted by C_v.

A *parametrization* P assigns a *target activity value* P_v^ω to every context $\omega \in C_v$ of every component $v \in V$. A priori, the only condition on P is that $P_v^\omega \in [0, \rho(v)]$ is a valid activity of v. The set of all feasible parametrizations is denoted by $\mathcal{P}$.

A parametrized regulatory graph $(\mathcal{R}, P)$ is called *Thomas network* or *model*. Finally, a remark about the scope of the workflow we are going to propose: In Sec. 2.3, we suggest colored model checking to solve the problem of identifying feasible parametrizations. For computational reasons we will consider the values of ρ and θ fixed in a particular problem.

Asynchronous Dynamics. The dynamics of a Thomas model $(\mathcal{R}, P)$ can be captured in a so-called state transition graph, where the finite state space X constitutes the vertex set and edges between states represent state transitions as determined from the logical parameters in the following way.

For every state x and every component v, there is a unique regulatory context $\omega \in C_v$, such that $\forall u \in N^-(v) : x_u \in \omega(u)$. To see this, recall that $\omega(u)$ is a regulatory interval, and that these intervals form a partition of the activities of u.

The parametrization P therefore defines a function F on the state space:

$$F : X \to X, \quad x \mapsto (P_{v_1}^{\omega_1}, \ldots, P_{v_n}^{\omega_n}),$$

where ω_i is the unique regulatory context of component v_i in state x.

The function F can be interpreted as a finite dynamical system, i.e., the dynamics can be derived by iterating an initial state using F. In the resulting state transition graph, each state x has exactly one outgoing edge leading to $F(x)$. Clearly, the synchronicity of the involved processes is a strong idealization, which we want to avoid here.

Instead, the representation should reflect that the time delays associated with the different biological processes corresponding to the updates may vary greatly depending on the corresponding network components. However, the experimental information to determine these time delays is often lacking. This leads to the definition of a non-deterministic transition graph where each outgoing edge from a state corresponds to one of the indicated updates.

The transitions T_P of the *asynchronous and unitary state transition graph* (X, T_P) of a model $(\mathcal{R}, P)$ are derived from F by two rules. A loop $xx \in T_P$ exists, iff $F(x) = x$. An edge $xy \in T_P, x \neq y$ exists, if there is a component v, such that xy is *asynchronous*: $\forall u \neq v : x_u = y_u$ and *unitary*: $y_v - x_v = \text{sign}(F(x)_v - x_v)$. Here *sign* denotes the sign function.

The state transition graph (X, T_P) corresponds naturally to a Kripke structure (KS) $\mathcal{S}(\mathcal{R}, P) := (P, X, X^0, T_P, L)$, which is of interest for formal verification of temporal logical properties. Here, $\mathcal{S}$ consists of states X, initial states X^0, the transition relation T_P and a labeling function L over the atomic propositions AP expressing inequalities $\doteq \in \{=, \leq, \geq, <, >\}$ with

$$\text{AP} := \{v \doteq k \mid v \in V, k \in [0, \rho(v)]\}.$$

If not otherwise noted, all states are considered as initial states, i.e., $X^0 := X$. The labeling function is defined as $L(x) := \{v \doteq k \mid v \in V, k \in [1, \rho(v)], x_v \doteq k\}$.

Finally, the Kripke structure can be generalized to incorporate all possible parametrizations $\mathcal{P}$. For a given regulatory graph $\mathcal{R}$ we consider a *parametrized Kripke structure* (PKS) to be a tuple $\mathcal{S}(\mathcal{R}) := (\mathcal{P}, X, X^0, T_\mathcal{P}, L)$ where $T_\mathcal{P} := \bigcup_{P \in \mathcal{P}} T_P$ and all other elements are defined as above. The PKS $\mathcal{S}(\mathcal{R})$ thus represents all possible behaviors that can be generated by $\mathcal{R}$.

2.2 Constraints

In the following we introduce several notions that allow us to restrict the parameter space to the parametrizations in agreement with all the information we have on the system. We distinguish between static and dynamic constraints as already indicated in Fig. 1. Static constraints refer to information related to the regulatory graph, e.g. existence and character of interactions. In contrast, dynamic constraints capture properties of state transition graphs such as reachability requirements.

Static Constraints. Here we focus on edge labels, which are used to characterize the impact that a regulator has on its target. If there is an effect observable at all, it can be either *activating*, i.e., causing an increase, or *inhibiting*, i.e., causing a decrease in the activity of the target. Formally, several semantics result from combinations of these effects (see [12, Def. 2.9]). Certain edge labels have already been used successfully in case studies of D. Thieffry (see e.g. [18],[10]) and also implemented in analysis tools [17, p. 6].

Since we are dealing with regulatory graphs, whose interactions may have more than one threshold, the concept of edge label must be adjusted accordingly. An edge label is therefore not assigned to a single edge uv, but to a tuple (uv, t_j) where $uv \in E$ and $t_j \in \theta(uv)$. In this paper, we restrict ourselves to unlabeled edges and labels chosen from the set $\{+, -, \text{mon}+, \text{mon}-\}$, where the different notions are defined as follows.

Assume a tuple (uv, t_j) is labeled with mon+. A parametrization P satisfies this label, if for all regulatory contexts $\omega \in C_v$, such that $\omega(u) = I_j^{uv}$ and $\omega' \in C_v$ such that

$$\omega'(w) := \begin{cases} I_{j-1}^{uv} & \text{if } w = u \\ \omega(w) & \text{else} \end{cases}$$

the target value inequality $P_v^{\omega'} \leq P_v^{\omega}$ holds. If instead the label is mon−, then P satisfies this label if for all $\omega, \omega' \in C_v$ as defined above $P_v^{\omega'} \geq P_v^{\omega}$ is true.

The labels + and − correspond to mon+ and mon−, but require observability in addition. *A parametrization P satisfies the observability of* (uv, t_j), if contexts $\omega, \omega' \in C_v$ as defined above, exist, such that the target value inequality $P_v^{\omega'} \neq P_v^{\omega}$ holds.

Dynamic Constraints. In this paper we focus on identifying parametrizations that are in agreement with time series data, which can be interpreted as conditions constraining the dynamical behavior of a system. A *measurement* is a rectangular subset of the state space X. That is, we describe a measurement m by assigning to each component v a *measurement interval* $m_v = [a_v, b_v] \subseteq [0, \rho_v]$. We then identify this description m with the set of all states $x \in X$, such that $\forall v \in V, x_v \in m_v$.

A *time series* is a sequence of measurements $(m^1, \ldots, m^k)$. Notice that measurements may intersect, i.e., there may be states $x \in m^i \cap m^j$ for $i \neq j$.

A state transition graph $S = (X, T)$ *reproduces a time series* $(m^1, \ldots, m^k)$, if it contains a finite walk $(x^i)_{1 \leq i \leq r}, r \in \mathbb{N}_1$, such that there is a mapping $M : [1, k] \to [1, r]$ that is *ordered*: $i < j \implies M(i) \leq M(j)$ and *correct*: $x^{M(i)} \in m^i$.

We call such walk *time series walk*. Notice that we allow $M(i) = M(j)$. The walk can be thought of as a discrete simulation, and the mapping M as describing at which simulation steps the measurements were recorded. We say that a parametrization reproduces a time series, if its transition graph does.

There may of course be multiple walks satisfying these properties. We will discuss this in Section 3, where we introduce a ranking to capture *how well* a model reproduces a time series.

The existence of a time series walk is determined by LTL model checking over the Kripke structure (X, X^0, T, L) associated with the state transition graph (X, T) (see [1] for an introduction). The initial states are chosen in correspondence with a time series $(m^1, \ldots, m^k)$ by $X^0 := m^1$.

A measurement m is translated into the LTL specification

$$\sigma(m) := \bigwedge_{v \in V} \bigvee_{k \in m_v} v \doteq k.$$

A state transition graph reproduces a time series $(m^1, \ldots, m^k)$ if and only if there is a state $x \in X^0$, such that the LTL specification

$$\mathbf{F}(\sigma(m^2) \wedge \mathbf{F}(\sigma(m^3) \wedge \ldots \mathbf{F}(\sigma(m^k)) \ldots) \quad (1)$$

is satisfied in x.

Time series formulae of the form (1) constitute a specific class of properties enabling our analysis method as developed in Section 3. More general LTL formulae are used to specify, e.g., monotonicity of gene expression between two adjacent measurements m^i, m^{i+1} [12] or steady gene activity expected after the last measurement.

2.3 Parameter Identification by LTL Model Checking

In this section we describe the technology of colored model checking used for computing parametrizations satisfying constraints encoded in LTL. This technology is employed in the next sections as a cornerstone for identifying optimal parametrizations. The central notion is the construction of a map (coloring) relating each state x of a regulatory graph to the set of all those parametrizations from $\mathcal{P}$ under which x is reachable.

For a parametrization $P \in \mathcal{P}$ and its corresponding Kripke structure $\mathcal{S}(\mathcal{R}, P) \equiv (P, X_S, X_S^0, T_P, L)$, we define a run, denoted π, as an infinite path in $\mathcal{S}(\mathcal{R}, P)$. The notation π^0 is used to denote a run whose first node is in X_S^0. Since we aim to explore parametrizations which are realizable, i.e. there exists at least one behavior that satisfies given LTL constraints, we consider existential interpretation of LTL. We say that $\mathcal{S}(\mathcal{R}, P)$ satisfies φ, written $\mathcal{S}(\mathcal{R}, P) \models \varphi$, if there exists a run π^0 in $\mathcal{S}(\mathcal{R}, P)$ satisfying φ.

For a given regulatory graph $\mathcal{R}$ and an LTL formula φ, automata-based model checking is employed on $\mathcal{S}(\mathcal{R})$ to identify all parametrizations satisfying φ. As a prerequisite, we assume an alphabet $\Sigma = 2^{AP}$. Then φ is represented by means of a Büchi automaton over Σ, denoted $BA(\varphi)$, and defined $BA(\varphi) := (\Sigma, X_A, X_A^0, \delta_A, F_A)$, where X_A is a set of states, $X_A^0 \subseteq X_A$ is a set of initial states, $\delta_A \subseteq X_A \times \Sigma \times X_A$ is a transition relation, and $F_A \subseteq X_A$ is a set of accepting states. See [1] for techniques of translating φ into $BA(\varphi)$.

We utilize the approach of *colored model checking* (CMC) as introduced in [2]. CMC takes a PKS $\mathcal{S}(\mathcal{R})$, a parametrization space $\mathcal{P}$, and a Büchi automaton $BA(\varphi)$. It returns a *set of all acceptable parametrizations* $\mathcal{P}_\varphi := \{P \in \mathcal{P} \mid \mathcal{S}(\mathcal{R}, P) \models \varphi\}$. The procedure takes the following steps:

– constructing product automaton $BA(\mathcal{R}, \varphi) := \mathcal{S}(\mathcal{R}) \cap BA(\varphi)$
– computing $\mathcal{P}_\varphi$ by executing colored model checking on $BA(\mathcal{R}, \varphi)$

Product Automaton. $BA(\mathcal{R}, \varphi)$ is computed in the standard way [1] as a product of a PKS $\mathcal{S}(\mathcal{R}) \equiv (\mathcal{P}, X_S, X_S^0, T_{\mathcal{P}}, L)$ and $BA(\varphi) \equiv (\Sigma, X_A, X_A^0, \delta_A, F_A)$: $BA(\mathcal{R}, \varphi) := (\mathcal{P} \times \Sigma, X, X^0, \delta, F)$ where

$$X := X_S \times X_B, X^0 := X_S^0 \times X_A^0, F := X_S \times F_A \text{ and}$$

$$((x_s, x_a), (P, \alpha), (x_s', x_a')) \in \delta \text{ iff } x_s x_s' \in T_P \wedge (x_a, \alpha, x_a') \in \delta_A \wedge \alpha \in L(x).$$

If there exists $\alpha \in L(x)$ such that $(x, (P, \alpha), x') \in \delta$, we use the simplifying notation $x \xrightarrow{P} x'$. Transitive and reflexive closure of the relation $\rightarrow$ is denoted $\rightarrow^*$.

$BA(\mathcal{R}, \varphi)$ accepts π^0 - an infinite run through this product automaton - if and only if there is an $x \in F$ that occurs infinitely often on π^0 (projection of π^0 to the second component is an accepting run in $BA(\varphi)$). Hence $BA(\mathcal{R}, \varphi)$ accepts exactly the paths satisfying φ, and the acceptance is always caused by a cycle in $BA(\mathcal{R}, \varphi)$ containing some state in F – therefore we are interested in *accepting cycles* and their *reachability* from initial states.

Our interest is in paths that are realizable in a certain parametrization $P \in \mathcal{P}$. We denote by $BA(\mathcal{R}, \varphi)_P$ the product automaton $BA(\mathcal{R}, \varphi)$ with the alphabet $\{P\} \times \Sigma$ (restricted to the parametrization P). A run in $BA(\mathcal{R}, \varphi)_P$ is denoted π_P. We can conclude that $\mathcal{S}(\mathcal{R}, P)$ satisfies φ iff there exists a run π_P^0 in $BA(\mathcal{R}, \varphi)_P$ that is accepted.

Colored Model Checking. Naive (bottom-up) computation of P_φ by checking each parametrization $P \in \mathcal{P}$ individually suffers from the exponential explosion of $|\mathcal{P}|$ wrt number of unknown parameters. CMC [2] is a heuristic method based on the idea that transitions within PKS are shared by many parametrizations, therefore utilizing a single PKS for a check (top-down) is significantly faster than doing a check on every single KS $\mathcal{S}(\mathcal{R}, P)$.

An important notion is mapping $cl_{\hat{X}}^{\hat{\mathcal{P}}} : X \rightarrow 2^{\mathcal{P}}, \hat{X} \subseteq X, \hat{\mathcal{P}} \subseteq \mathcal{P}$, called *coloring*, in which each state $x \in X$ is assigned a set of parametrizations for which x is reachable from some state in $\hat{X}$, defined and denoted $cl_{\hat{X}}^{\hat{\mathcal{P}}}(x) := \{P \in \hat{\mathcal{P}} \mid \exists \hat{x} \in \hat{X} : \hat{x} \xrightarrow{P}^* x\}$. Using this mapping, the CMC procedure can be described as follows:

For each $x \in F$:
(1) Compute coloring $reach_x \equiv cl_{X^0}^{\mathcal{P}}(x)$ reaching accepting state x.
(2) Compute coloring $cycle_x \equiv cl_{\{x\}}^{reach_x}(x)$ enabling (accepting) cycles on x.

These two steps correspond to traditional LTL model checking [1], where we ask if there exists (1) a path from an initial to a final state and (2) a cycle containing this state, which implies existence of an accepting run. In our case, we do not ask for an existence of a single accepting run for each KS, but directly build a set of parametrizations that have an accepting run in PKS.

To obtain such a set, one has to perform a graph search, which can be done in numerous ways - in Section 6 we explain how to do those steps efficiently. Performance of the algorithm can be also greatly increased by omitting step (2) when using time series formula. This property is within a set of so-called *reachability* properties that can be computed without cycle detection [1].

3 Optimal Parametrizations

In the classical enumerative model checking approach to reverse engineering of Thomas networks, that was introduced by G. Bernot et al. in [4], a given set of parametrizations is divided into acceptable and unacceptable parametrizations depending on whether the transition graph associated to a parametrization satisfies the temporal logic specification or not.

From the perspective of the temporal specification, all acceptable parametrizations are equally suitable and the parameter model checking process ends here.

For the particular class of LTL specifications that we are interested in – the time series constraints as defined in Section 2.2, we introduce a method for ranking acceptable parametrizations.

3.1 The Length Cost

This section starts with a regulatory graph $\mathcal{R}$, a time series $(m^1, \ldots, m^k)$ and a non-empty set of parametrizations $\mathcal{P}' \subseteq \mathcal{P}$ that all reproduce the time series.

Denote by W_P the set of all time series walks of $(m^1, \ldots, m^k)$ in the state transition graph of a *single* parametrization $P \in \mathcal{P}'$. W_P may in general be an infinite set, but most of its walks are not relevant for our purposes. To impose a ranking on the set of time series walks, and through that a ranking on the set of parametrizations, we impose a preference for short walks. Since the walk length can be seen as a measure for the complexity of the behavior in terms of the number of processes that have to be executed to produce the desired result, this approach favors models that provide simple explanations for the observed behavior. In other words, we try to penalize unnecessarily complex realizations of time series data in a model which might also be related to a higher energy cost for the system.

We define the *length cost of a parametrization* $P \in \mathcal{P}'$ *with respect to the time series* as $\mathrm{Cost}(P) := \min\{r \in \mathbb{N} \mid \exists (x^i)_{1 \leq i \leq r} \in W_P\}$, and denote by

$$\mathrm{SW}_P = \{(x^i)_{1 \leq i \leq r} \in W_P \mid r = \mathrm{Cost}(P)\} \subseteq W_P$$

the set of *shortest walks of* P.

The length cost partitions $\mathcal{P}'$ into classes of equal cost, and we are particularly interested in parametrizations with the minimum cost, denoted by $\min_{\mathrm{Cost}}(\mathcal{P}') \subseteq \mathcal{P}'$.

3.2 Robustness

Since the dynamics in the Thomas formalism are non-deterministic, several paths may lead from one state to another and the path corresponding to the actual behavior of the system depends on the time delays associated with the different update processes. If these time delays change, maybe due to environmental influences, the system may follow a different trajectory even when considering the same initial state. However, in some cases, namely when there is only one path between two states in the state transition graph, the behavior of the system is independent of the actual values of the time delays. This can be interpreted as robustness of the system wrt perturbations of the time delays. In the following we will formalize this idea as a property of a given parametrization. Since we are interested in the realization of time series, we will focus our notion of robustness on the time series walks.

We use the standard notion of probability for a finite walk, as defined in [1], where each successor of a node is chosen with equal probability. Then, we say that the probability of a finite time series walk w of length l is

$$\text{Prob}(w) := \prod_{i=1}^{l-1} \frac{1}{deg^{+}(x^i)},$$

where $deg^{+}(x^i)$ is the out-degree of the state x^i of the walk.

We now define *robustness of a parametrization wrt time series* as the sum of probabilities of all distinctive time series walks. For set SW_P and set m^1 of all states that fit the first measurement, we set

$$\text{Robustness}(P) := \frac{\sum_{w \in \text{SW}_P} (\text{Prob}(w))}{|m^1|}.$$

For example, if the time series and the parametrization only allows for a single shortest time series walk, the robustness will be high if the states of the walk have low out-degree. In case the initial measurement has some unknown values, we take the average robustness of walks from all states that fit the measurement.

This notion of robustness is a good starting point for analysis since it distinguishes parametrizations that reproduce the time series with low ambiguity. In addition, it is easy to formalize and compute. A more involved definition should be based not only on the out-degree of a state of a time series walk, but differentiate and weight whether the different successors of the state are themselves states of a time series walk. This would extend the notion of robustness, taking into account not only perturbations of the time delays but also of the states, and will be investigated in future work.

3.3 Computing Optimal Parametrizations

The set of optimal parametrizations is obtained in the following manner:

1. Describe the set $\mathcal{P}$ of all possible parametrizations.
2. Remove parametrizations that do not satisfy edge constraints.

3. Compute the set of acceptable parametrizations using model checking.
4. Take the subset of those that have minimal Cost.
5. Finally, select parametrizations with maximal Robustness.

This way we obtain only parametrizations we have identified to be *optimal*, whose number is usually significantly smaller then the size of $\mathcal{P}$.

Such a procedure can be done automatically. Interpretation and further analysis of the results is left to the user. To support this step, in the following section we suggest two methods for visualization of results.

4 Visualization

In this section we present methods to visualize differences and similarities of parametrizations. To our knowledge, two automated lines of analysis of a set of parametrizations exist. In [8, Sec. 3.2], consensus target value inequalities are derived, while in [12, Sec. 5.1] the focus is on deriving consensus edge labels.

Here we present a novel approach that visualizes the transitions of *a set of acceptable parametrizations* in between measurements, highlighting agreement between parametrizations. We propose, firstly, *behavior maps* that represent state transitions according to the considered parametrizations, and, secondly, *expression profiles* that focus on the activity of a single component.

4.1 Behavior Maps

There is no reason to expect that the time series walks of different parametrizations coincide. However, the information whether certain state transitions are shared by the walks can be immediately exploited for experimental design. For example, new measurements would be most useful if placed between two original measurements that generated many different walks leading from one to the other across the valid parametrizations, since the additional information would then enable us to distinguish between them. The plots proposed in this section aim at making this information about the distribution of state transitions of time series walks easy to assess.

Let W be any finite set of time series walks of $(m^1, \ldots, m^k)$. In each walk we mark the measurements $1, \ldots, k$. We lay out all walks horizontally and align for every $1 \leq i \leq k$, the states marked as the ith measurement vertically. If a measurement is realized by several states in the walk, we choose the state with the smallest index within the walk to represent this measurement. This way we can interpret the horizontal axis as a discrete time axis, progressing from earlier (left) to later (right).

Notice that a state may of course appear in more than one walk, but also multiple times within a single walk. Behavior maps are an attempt to find a compact representation, by removing some of these duplicates, while keeping the acyclic progression from earlier to later states. Therefore we are not interested in the graph defined as the union of walks in W, because it may destroy this progression (by creating cycles).

We treat each pair of successive measurements m^i, m^{i+1} independently and partition the walks into classes of equal length in between m^i and m^{i+1}. Two states are identified as one if they represent the same state of the KS and appear in equal number of steps since last measurement.

We scale the size of a node and width of a transition by the number of parametrizations that produce a walk passing through it. The gray scale shading of a transition is determined by the class it belongs to. Black transitions belong to the overall shortest walks, light gray transitions to the longest walks. E.g., consider following regulatory graph:

The behaviour map of this graph wrt time series $m^1 = \{(0,0)\}, m^2 = \{(2,0)\}$ is:

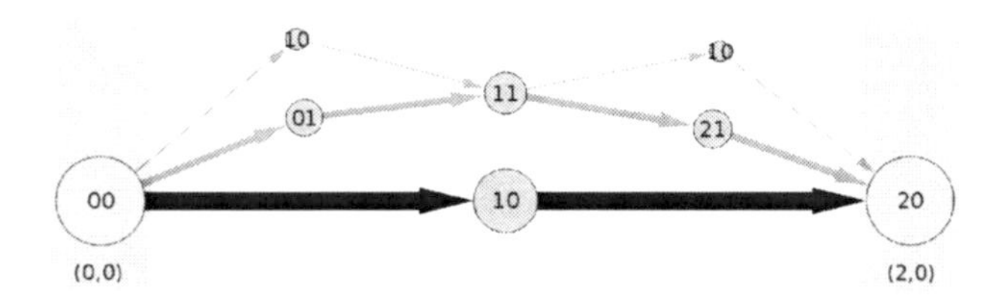

The blue vertical boxes represent measurements. Black transitions belong to length class 3, gray transitions to length class 5. The wider the stroke of a transition or the larger a node, the more walks in W pass through them. Note that the state '10' appears three times. Once in length class 3, and twice in the class with length 5, because it can be visited either early or late in discrete time. State '11' has only one node because it always appears after two transitions.

4.2 Expression Profiles

Behavior maps visualize possible model behaviors in accordance with a time series. For a more refined analysis aiming at experimental design, it is helpful to focus on the behavior of the separate components. The plot of all the ways a component may change its level alongside all time series walks can highlight which components are responsible for differences between walks. In addition, as in the case of behavior maps, we can easily pinpoint for each component between which measurements the component behavior is the most ambiguous.

For single components we suggest to plot sigmoid expression profiles. The transparency of a curve is proportional to the number of shortest walks that share the corresponding transition. Below are the profiles of x and y of the example structure presented in Section 4.1. The most opaque line in the profile of x shows that along most walks it increases evenly from 0 to 2.

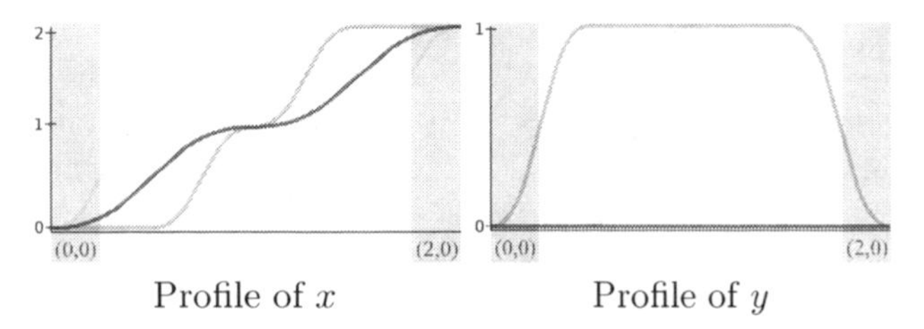

The profile of y indicates that its activity along most walks is constant at 0. In contrast to x, some ambiguity is present here - if this was a real system, experiments should focus on y.

5 Case Study

In this section we apply our workflow to the gene regulatory network of bacteriophage λ. Its discrete version was formulated by Thieffry and Thomas in [20]. The authors also discuss two time series, the lytic (Fig. 5) and lysogenic (Fig. 5) fate of bacteria infected by the bacteriophage (see [20, p.290]). Finally, for comparison, a realistic parametrization denoted R is considered (taken from [20, p.291]). We will judge our results by how close our optimal parametrizations are to R.

Parameter Set Reduction. First, we compute the initial parametrizations with the edge labels of the regulatory graph in Fig. 5. The self-activation of cI at threshold 2 is not observable in R. Since we want R to belong to the initial parametrizations, we also relax this constraint and assign the label mon+ to $(cI\, cI, 2)$. These static constraints reduce the set of feasible parametrizations from $6,879,707,136$ to $82,008$.

Now, we execute the CMC procedure for both time series. The lytic time series is reproduced by $28,043$ parametrizations. Of these, 537 also reproduce the lysogenic time series. During this step we also compute Cost and Robustness functions.

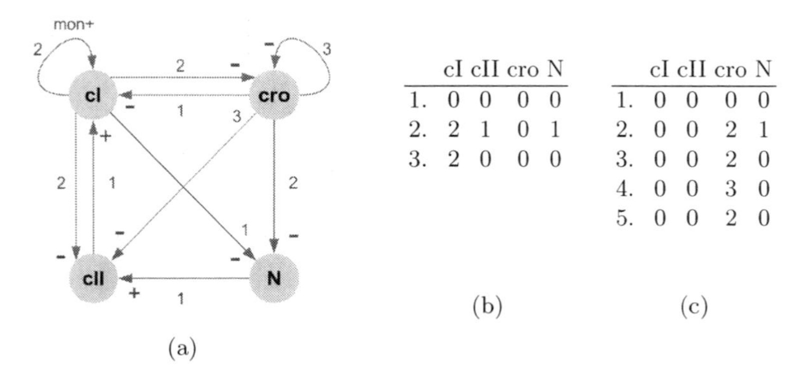

	cI	cII	cro	N
1.	0	0	0	0
2.	2	1	0	1
3.	2	0	0	0

(b)

	cI	cII	cro	N
1.	0	0	0	0
2.	0	0	2	1
3.	0	0	2	0
4.	0	0	3	0
5.	0	0	2	0

(c)

Fig. 2. (a) Regulatory graph of bacteriophage λ with edge constraints. (b) Lysogenic time series. (c) Lytic time series. The last three measurements indicate an oscillation.

Optimal Parametrizations. To illustrate the two step model ranking by Cost and Robustness, we have to focus on one of the time series. We pick the lysogenic series. The theoretical minimum Cost, required to execute the 6 activity changes of the lysogenic series (Fig. 5), is 7. The actual minimum Cost among the 537 feasible parametrizations is 9, and 28 parametrizations contain a walk of this length.

Among those 28, the maximum Robustness is 9.72% and there are 3 parametrizations that attain it. In comparison, R also has a Cost of 9 but its Robustness is only 0.54%.

These 3 parametrizations are equal in all target values except for a single context of *cro*: If the only regulator acting above its threshold is *cro* itself, the target value may be any of 0, 1 or 2 (giving us the 3 remaining parametrizations). Since the threshold of the self-regulation is $\theta(cro\,cro) = 3$, all target values below 3 cause this inhibition to stop itself, giving us identical state transition graphs. Hence the remaining 3 parametrizations agree with R on that the target value must be below 3. Any of the 3 parametrizations is optimal wrt Cost and Robustness, and we denote all of them by O.

We compare the target values of the parametrizations R and O in the table below. Values that differ between the two are bold. The component name in the first row of each column denotes the target component. Each successive row contains a list of regulators that are above their (unique) interaction thresholds. The corresponding target values are in the columns R and O.

cI	R	O	cII	R	O	cro	R	O	N	R	O
$\emptyset$	**2**	**1**	$\emptyset$	**0**	**1**	$\emptyset$	3	3	$\emptyset$	1	1
cI	2	2	cI	0	0	cI	0	0	cI	0	0
cro	0	0	cro	0	0	cro	2	<3	cro	0	0
cII	2	2	N	1	1	cI, cro	0	0	cI, cro	0	0
cI, cII	2	2	cI, cro	0	0						
cI, cro	**0**	**2**	cI, N	0	0						
cII, cro	2	2	cro, N	**0**	**1**						
cI, cII, cro	2	2	cI, cro, N	0	0						

Note that each of the disagreements for cI causes the interaction $cI\,cI$, that is not observable in R, to be observable in O.

Visualization. Using a behavior map including all the shortest time series walks of every acceptable parametrization, we obtain the following graph:

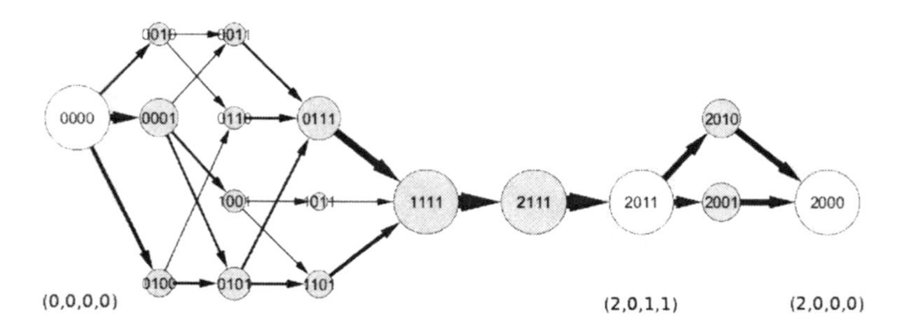

As before, blue boxes mark measurements. The map indicates many possibilities in ordering of activations which always lead from state $(0,0,0,0)$ to state $(1,1,1,1)$, suggesting that it would be reasonable to measure activity levels between these two states.

Number of steps between measurements m^2 and m^3 corresponds to their distance, but for walk from m^1 to m^2 it is not the case.

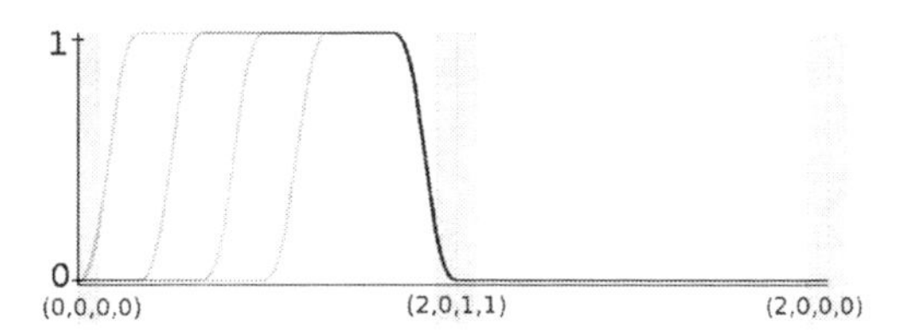

In the cII expression profile above we can see that alongside every path cII is activated and inhibited between measurements m^1 and m^2 – this also explains why in O this component can be activated more often, as this step is necessary for the model to be able to reproduce the time series.

6 Implementation and Evaluation

In this section we briefly describe methodology of synthesis and analysis together with tools deployed for these tasks. Further we focus on description of a time and space-efficient computation of acceptable parametrizations and evaluate it using two different models.

6.1 Usage Description

Our current workflow of analysis is divided into following steps:

1. Creation of a model - regulatory network is described in a single XML file using our own syntax designed for this purpose. In a future work we expect to implement an option to import models from standard formats.
2. Specification of the property - the property (most usually a time series) is currently specified within a model file in the form of Büchi automaton, also using an XML-based syntax.
3. Synthesis - the model is analyzed using the colored model checker *Parsybone* [1], implemented in C++. The tool works in two steps. First, reduction of parametrization space is conducted if there are any initial constrains specified. The reduced parameter space than undergoes the process of parameter synthesis. By default, this step produces only enumeration of acceptable parametrizations. However, for each of the parametrizations we can optionally compute and output its shortest paths or the robustness value.

[1] *Parsybone* – `http://github.com/sybila/Parsybone`

4. Filtering - the amount of data produced by synthesis is vast in most cases, therefore we usually employ a second tool, *ParameterFilter* [2], implemented in C♯. It is a GUI elaborating on the output from the synthesis step allowing to select and compare parametrizations based on their ranking.
5. Plotting - finally, for parametrizations chosen during the previous step we plot their expression profiles (using *ParameterFilter*) or their behavior map which is produced from the synthesis output using our converter, implemented in Java. The converter creates a behaviour map for a given parametrization and allows its visualization using Cytoscape [6].

6.2 CMC Procedure Implementation

Algorithm for colored model checking as presented in [2] does not specify, how distinct parametrizations should be stored and manipulated. For continuous models, we have used bounded intervals of values for each component, creating a parametrization space as a Cartesian product of those. We have later employed this approach for discrete models as well, but it turned out that in this case it suffers from high complexity of often performed operations like set intersection (for more information about the algorithm, see [13]). To tackle this problem, we have moved to explicit representation where all parametrizations are enumerated. We will show that this approach provides numerous advantages and allows for analysis of large parametrization sets.

Encoding. Our approach is based on a computationally efficient encoding of parametrization space. We encode each parametrization set $\mathcal{P}' \subseteq \mathcal{P}$ as a word of length $|\mathcal{P}|$ over alphabet $\Sigma = \{0, 1\}$. Such a word naturally corresponds to a bit vector of the same length and allows fast computation using bitwise operations.

We consider lexicographical ordering of the set $\mathcal{P}$. We denote $P^i \in \mathcal{P}$ an i-th parametrization in $\mathcal{P}$. Now to encode an ordered set $\mathcal{P}' \subseteq \mathcal{P}$, we use the *encoding function* $Code : 2^{\mathcal{P}} \rightarrow \{0,1\}^{|\mathcal{P}|}$ where $Code(\mathcal{P}') = b_1 b_2 ... b_{|\mathcal{P}|}, \forall i (b_i = 1 \Leftrightarrow P^i \in \mathcal{P}')$. This way we encode a coloring of every state as a single word of length $|\mathcal{P}|$.

The encoding function is of a crucial importance, because the idea of the CMC and its main performance improvement lies in the option to create only a single PKS for the whole parametrization space. To create such a structure, we need to be able to label edges of the PKS with transitive parametrizations. This can be done using the encoding function by which we label every transition $x \rightarrow x'$ with a word $Code(\{P | x \xrightarrow{P} x'\})$.

In general, by using such an encoding we reduce the CMC problem to a sequence of bitwise operations.

Splitting. Our coloring algorithm is based on an iterative computation of a fixed point. Complexity of this computation can be improved using multiple heuristics, for complete information we refer to [13]. The most important is the procedure of splitting.

[2] *ParameterFilter* – `http://github.com/sybila/ParameterFilter`

Our idea is based on the assumption that similar parametrizations generate similar KSs [2]. When computing a coloring of a PKS we split its parametrization space to multiple neighbouring regions and work only with a single region at a time. Most of parametrizations within a single region are likely to be either all accepted or all rejected, allowing us to quickly reach the fixed point.

Due to lexicographical ordering of possible parametrizations within a bit vector, we already have similar parametrizations in the neighbouring positions. During the computation we then split the parametrization space by working always with next m bits of the bit vector. Each region is stored within a single *integer* variable, therefore m is equal to size of an integer in bits on a target platform. Note that usage of integers also ensures quick computation of bitwise operations. With this region, we go through the whole process of analysis, output the data, free the memory and continue with another round (ensuring low memory requirements).

Distribution. When using the split parameter space (which we can do only when using explicit data representation), we can easily distribute the computation. This is because every parametrization is completely independent on all others, giving us great potential for a data-parallel distribution. Therefore, we distribute regions of parametrization space between non-communicating processes differing only in their ID.

Each independent worker does its own parsing and pre-computation and then goes through the procedure of parameter identification with a subset of parametrization space that is disjunctive with subsets of other workers.

To achieve as optimal load balance as possible, distribution of regions is interlaced, meaning that in computation of n processes, process with ID $i, 1 \leq i \leq n$ is assigned only regions $i + k \cdot n, k \in \mathbb{N}$. This method is again based on the assumption that similar parametrizations generate similar behaviour, causing acceptable parametrizations to cluster. This way we ensure that such clusters are distributed evenly between processes.

6.3 Evaluation

Mammalian Cell Cycle. To test capabilities of our algorithm, we had it analyze a model of mammalian cell cycle [10] with 9 components. For this model we have defined partial specification, reducing size of parametrization space to final number of $675,584,064$ parametrizations. As a guide for the analysis we have used time series with 8 measurements. More detailed information are presented in Technical report [13].

Parametrization space was evenly distributed between 8 independent process, each one of them having initial set of size $84,448,008$. Computation was run on a Linux server using two processors with four 2.27 GHz cores and took roughly a day with $308,180,639$ acceptable parametrizations computed. During computation each of the processes used less than 15 MB of RAM. Exact results for each process are presented in the Figure 6.3. As can be seen, parametrizations space has been partitioned to sets with almost identical numbers of acceptable parametrizations.

Process ID	Runtime	Result set size	Process ID	Runtime	Result set size
1	29.07 h	38,522,403	5	29.70 h	38,523,691
2	31.08 h	38,521,943	6	28.81 h	38,523,255
3	27.22 h	38,521,656	7	29.55 h	38,522,328
4	32.32 h	38,522,343	8	28.83 h	38,523,020

Fig. 3. Results of distributed analysis of Mammalian cell cycle

Bacteriophage. As a main benchmark we have employed the bacteriophage λ network. Minor utilization was necessary mainly because our old tool (which is employed for comparison) is not able to work with edge labels. We have therefore created unlabelled and partially specified version of the model with $|\mathcal{P}| = 589,824$ out of which $90,112$ parametrizations are acceptable. We ran the analysis five times using each tool. Analysis using the old version took on the average 967 seconds and used at max 50 MB of RAM, analysis using the new version took always less than 6 seconds and did not use more than 3 MB RAM.

To demonstrate scalability we analyze the bacteriophage model using up to 8 independent processes. In Fig. 4 we show average runtime of all processes used. Resulting numbers are taken as an average of three independent experiments on the same platform as in case of mammalian cell cycle. As can be seen from the graph, scaling of our algorithm is roughly linear. This result suggests that in every case where distribution of computation seems sensible, it is possible to achieve almost a linear speedup and therefore we can extend a set of models that can be practically analyzed by using high-performance computational platforms.

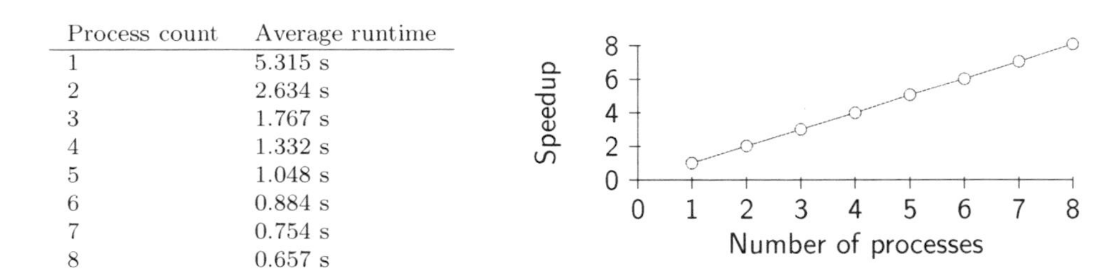

Process count	Average runtime
1	5.315 s
2	2.634 s
3	1.767 s
4	1.332 s
5	1.048 s
6	0.884 s
7	0.754 s
8	0.657 s

Fig. 4. Scalability of algorithm on bacteriophage λ

7 Conclusions

We have contributed to solving the parameter identification problem for Thomas networks in three aspects. First, we have proposed a new methodology based on a colored model checking approach extended with parametrization ranking procedures. Second, we have introduced a new idea of parametrization encoding that allows us to synthesize parametrizations in an efficient manner on distributed platforms. Third, we have implemented a prototype tool chain that supports all steps of our methodology including feasible visualization of obtained results.

By evaluating our algorithms on several biological models, we have demonstrated that the computation achieves good scaling, and moreover, that it copes with larger parameter spaces. Comparing these results with our previous achievements [2,12], possibilities of parameter identification solved by model checking have been significantly improved.

On the methodological side, our achievement brings new insights into applying discrete modeling frameworks to gene networks. The case study has shown that the approach is can help modelers to identify reasonable parametrizations and derive supported suggestions for experimental design.

On the computational side, improving the efficiency of the parameter filtering and visualization part of the tool chain will be a focus of future work.

References

1. Baier, C., Katoen, J.-P.: Principles of Model Checking. The MIT Press (2008)
2. Barnat, J., Brim, L., Krejci, A., Streck, A., Safranek, D., Vejnar, M., Vejpustek, T.: On Parameter Synthesis by Parallel Model Checking. IEEE/ACM Transactions on Computational Biology and Bioinformatics 9(3), 693–705 (2012)
3. Batt, G., Page, M., Cantone, I., Goessler, G., Monteiro, P., de Jong, H.: Efficient parameter search for qualitative models of regulatory networks using symbolic model checking. Bioinformatics 26(18), i603–i610 (2010)
4. Bernot, G., Comet, J.-P., Richard, A., Guespin, J.: Application of formal methods to biological regulatory networks: Extending Thomas' asynchronous logical approach with temporal logic. Journal of Theoretical Biology 229(3), 339–347 (2004)
5. Chaouiya, C., Remy, E., Mossé, B., Thieffry, D.: Qualitative Analysis of Regulatory Graphs: A Computational Tool Based on a Discrete Formal Framework. In: Benvenuti, L., De Santis, A., Farina, L. (eds.) Positive Systems. LNCIS, vol. 294, pp. 119–126. Springer, Heidelberg (2003)
6. Cline, M., et al.: Integration of biological networks and gene expression data using Cytoscape. Nat. Protocols 2(10), 2366–2382 (2007)
7. Corblin, F., Fanchon, E., Trilling, L., Chaouiya, C., Thieffry, D.: Automatic Inference of Regulatory and Dynamical Properties from Incomplete Gene Interaction and Expression Data. In: Lones, M.A., Smith, S.L., Teichmann, S., Naef, F., Walker, J.A., Trefzer, M.A. (eds.) IPCAT 2012. LNCS, vol. 7223, pp. 25–30. Springer, Heidelberg (2012)
8. Corblin, F., et al.: A declarative constraint-based method for analyzing discrete genetic regulatory networks. Biosystems 98(2), 91–104 (2009)
9. de Jong, H.: Modeling and Simulation of Genetic Regulatory Systems: A Literature Review. Journal of Computational Biology 9(1), 67–103 (2002)
10. Fauré, A., Naldi, A., Chaouiya, C., Thieffry, D.: Dynamical analysis of a generic boolean model for the control of the mammalian cell cycle. In: ISMB (Supplement of Bioinformatics) 2006, pp. 124–131 (2006)
11. Helikar, T., Konvalina, J., Heidel, J., Rogers, J.A.: Emergent decision-making in biological signal transduction networks. Proceedings of the National Academy of Sciences 105(6), 1913–1918 (2008)
12. Klarner, H., Siebert, H., Bockmayr, A.: Time series dependent analysis of unparametrized thomas networks. IEEE/ACM Transactions on Computational Biology and Bioinformatics 99(PrePrints) (2012)

13. Klarner, H., Streck, A., Safranek, D., Kolcak, J., Siebert, H.: Parameter identification and model ranking of Thomas networks. Technical Report FIMU-RS-2012-03, Masaryk University (2012)
14. Laubenbacher, R., Mendes, P.: A discrete approach to top-down modeling of biochemical networks. In: Kriete, A., Eils, R. (eds.) Computational Systems Biology, pp. 229–247. Elsevier Academic Press (2005)
15. Lee, W.-P., Tzou, W.-S.: Computational methods for discovering gene networks from expression data. Briefings in Bioinformatics 10(4), 408–423 (2009)
16. Naldi, A., Remy, E., Thieffry, D., Chaouiya, C.: Dynamically consistent reduction of logical regulatory graphs. Theor. Comput. Sci. 412(21), 2207–2218 (2011)
17. Richard, A.: SMBioNet-1.4 User manual (2005)
18. Sánchez, L., Thieffry, D.: A logical analysis of the drosophila gap-gene system. Journal of Theoretical Biology 211(2), 115–141 (2001)
19. Siebert, H., Bockmayr, A.: Incorporating Time Delays into the Logical Analysis of Gene Regulatory Networks. In: Priami, C. (ed.) CMSB 2006. LNCS (LNBI), vol. 4210, pp. 169–183. Springer, Heidelberg (2006)
20. Thieffry, D., Thomas, R.: Dynamical behaviour of biological regulatory networks II. Immunity control in bacteriophage lambda. Bulletin of Mathematical Biology 57, 277–297 (1995)
21. Thomas, R.: Regulatory networks seen as asynchronous automata: A logical description. Journal of Theoretical Biology 153(1), 1–23 (1991)

Investigating Co-infection Dynamics through Evolution of Bio-PEPA Model Parameters: A Combined Process Algebra and Evolutionary Computing Approach

David Marco[1], Erin Scott[1], David Cairns[1], Andrea Graham[2], Judi Allen[3], Simmi Mahajan[3], and Carron Shankland[1,*]

[1] University of Stirling, Stirling, UK
ces@cs.stir.ac.uk
http://www.cs.stir.ac.uk/SystemDynamics/
[2] Princeton University, Princeton NJ 08544, USA
[3] University of Edinburgh, Edinburgh, UK

Abstract. Process algebras are an effective method for defining models of complex interacting biological processes, but defining a model requires expertise from both modeller and domain expert. In addition, even with the right model, tuning parameters to allow model outputs to match experimental data can be difficult. This is the well-known parameter fitting problem. Evolutionary algorithms provide effective methods for finding solutions to optimisation problems with large search spaces and are well suited to investigating parameter fitting problems. We present the Evolving Process Algebra (EPA) framework which combines an evolutionary computation approach with process algebra modelling to produce parameter distribution data that provides insight into the parameter space of the biological system under investigation. The EPA framework is demonstrated through application to a novel example: T helper cell activation in the immune system in the presence of co-infection.

1 Introduction

Process Algebra [2] (PA) is one of a range of formal methods adopted for computational biology [7,27]. The goal is to use models and experimental data together to develop understanding of a given system. This approach relies heavily on the ability to create models which reproduce the behaviour shown in experimental data, despite sensitivity to particular parameter choices. The parameter fitting problem is well known [5]: evolutionary computation approaches provide a potential solution to this problem for models defined using process algebras.

Evolutionary Algorithms (EAs) are an established concept in computing, originally made popular by the work of Fraser [12], Holland [19] and Goldberg [15]. EAs are a class of optimisation algorithm that draw on concepts from evolutionary theory to search for optimal solutions in complex search spaces. They

* Corresponding author.

D. Gilbert and M. Heiner (Eds.): CMSB 2012, LNCS 7605, pp. 227–246, 2012.

generally use a population-based approach where each individual in the population defines a different possible solution for the given problem. The strengths of an EA approach lie in the ability of parallel candidate solutions to a problem to share information and combine it with the results of other solutions. This approach tends to locate useful solutions in less time than running a similar number of trials in sequential order: sharing of information found by one individual will tend to lead to the population as a whole benefitting.

In common with Ross and Imada [28] and Prandi [26], we propose that process algebras and evolutionary algorithms have complementary strengths for modelling bioscience systems [23]. Although evolutionary algorithms are powerful methods for finding solutions to optimisation problems with large search spaces, they require an accurately defined fitness function to provide valid results. By defining the fitness function with respect to a process algebra model, all the advantages of using abstract, modular, individual-based modelling techniques are enhanced with the ability to find suitable parameters for that model. Our focus is on the joint benefits to be gained: the ability to match process algebra models to experimental data, benefitting the systems biology community, and the ability to use the performance of individual-based models to define fitness functions in an EA setting, benefitting the evolutionary computation community.

This paper presents an early step in that work: a framework in which numeric rate parameters for predefined process algebra models are explored using evolutionary computation. The system output is a set of suitable parameters which will cause the behaviour of the model to match target data, together with distribution data on parameter variance, indicating where the model is robust and where it is sensitive to particular parameter values.

The technique can be applied in many domains: our main target is biological problems, due to the wealth of experimental data and the interest in systems biology. To illustrate our approach we choose a topic of current interest in Immunology. Immunological systems are ideally suited to modelling using process algebra. In many ways the immune system is a black box; although many of its inputs and outputs are known, exactly how the system achieves its function is the subject of much investigation. Laboratory experiments have provided large quantities of data, allowing components within the black box to be identified, but there remain many details to be uncovered about the exact nature of how components carry out their functions, or on the behaviour of interactions between components. There are so many potential variables in such systems that exhaustive testing to establish these details is not feasible. The focus of this work is on a particular aspect of how specific components of the immune system, the T helper cell populations, respond to co-infections with parasites making conflicting immunological demands [22].

The paper is organised as follows. In Section 2 we provide background information on process algebra, and more extensive, introductory background on evolutionary computation, together with details of related work. Section 3 presents the Evolving Process Algebra (EPA) framework. This combines process algebra modelling with an evolutionary computation approach. The framework is demonstrated via a novel Bio-PEPA [6] model concerning co-infection in the immune system, presented in Section 4. Details of the experimental method, and results of applying the EPA framework also appear in this section. Finally, Section 5 contains some concluding remarks and thoughts on future work.

2 Background

There are many valuable approaches to formalising models of biology: see [21] for an overview. We adopt process algebra because of its ability to define systems as compositions of individual interacting agents, its concise syntax, the range of analytic techniques available, and for its suitability for other parts of our research programme in evolving the models, not just the parameters [24]. Justification for the use of Bio-PEPA in particular is given, together with an introduction to its syntax. The basic principles of evolutionary computation are presented, particularly those of genetic algorithms. Lastly, details of related work in the area of combining process algebra and evolutionary computation, are presented.

2.1 Process Algebra

For a historical overview of the development of process algebra, see Baeten [2]. The goal behind the development of process algebras was to model distributed computation. To this end, process algebras allow the description of agents, called processes, which can carry out modeller-defined actions. Choices between actions can be made in a variety of ways: deterministically, probabilistically or non-deterministically. Systems are built in a modular and compositional way by combining multiple instances of processes in parallel. Processes may run independently from each other, but, more often, some synchronisation between processes is required. This is achieved through communicating actions. This simple approach to complex systems gives process algebras power and flexibility. These features are also highly suitable for modelling of biological systems [3,7,25,27].

Process algebra is used in this work in two roles. First, process algebra is used to describe the model under investigation, and for which we wish to evolve suitable parameters. There are a wide range of process algebras, any of which would be suitable for this investigation. Earlier work [23] successfully evolved numeric rate parameters for PEPA [17] models. Here we extend that work to accept Bio-PEPA, developed by Ciochetta and Hillston [6], as an input language. In addition to the usual core features (processes, actions, choice and interaction), Bio-PEPA attaches rates to actions which are described by potentially complex algebraic expressions. This facilitates easy description of the collective dynamics of a system, specifically biological models.

Second, process algebras are mathematically based and allow rigorous analysis, meaning it is possible to use process algebra tools for simulation to generate time series data which can then be used in calculating the quality of proposed parameter sets (the fitness of the current solution). The tool used here is the Bio-PEPA Plug-in [9]. This tool automates a range of analyses, including a variety of simulation algorithms, interpretation of models as a set of Ordinary Differential Equations (ODE), inference of invariants, calculation of cumulative distribution function for a given variable, model checking in PRISM, and export to Systems Biology Markup Language (SBML) format.

Bio-PEPA. Bio-PEPA models describe agents, the activities in which they may participate, and the paths they may follow as they execute, focussing on the changing *concentrations*, or population levels, of those agents. Consider the model in Figure 1 which is a simplified part of the model of Section 4.2. The ASCII notation for Bio-PEPA used by the Bio-PEPA Plug-in is adopted here to facilitate reproduction of our experiments.

$$\begin{aligned} divide\_Th1\_rate &= 0.01; \\ divide\_Th2\_rate &= 0.05; \\ death\_rate &= 0.015; \end{aligned}$$

$$\begin{aligned} kineticLawOf\ div1 &: (divide\_Th1\_rate * Th1 * (2 * Th1/(Th1 + Th2))); \\ kineticLawOf\ div2 &: (divide\_Th2\_rate * Th2 * (2 * Th2/(Th1 + Th2))); \\ kineticLawOf\ die1 &: fMA\ (death\_rate); \\ kineticLawOf\ die2 &: fMA\ (death\_rate); \end{aligned}$$

$$\begin{aligned} Th1 &= (div1, 1) \gg Th1 + (die1, 1) \ll Th1; \\ Th2 &= (div2, 1) \gg Th2 + (die2, 1) \ll Th2; \end{aligned}$$

$$Th1[1000] <> Th2[1000]$$

Fig. 1. Simple T helper cell growth Bio-PEPA Model

A system modelled in Bio-PEPA will have numeric rates (at the top here, e.g. *death_rate*), functional rates (introduced next, by *kineticLawOf*) and agent definitions (*Th1* and *Th2*)[1]. There are two cell populations (T helper (Th)1 and T helper (Th)2 cell types). The activities of the *Th1* and *Th2* agents are division (*div1*, *div2*) and death (*die1*, *die2*). Division leads to an increase (operator $\gg$) in cell numbers (by 1 each time), and death leads to a decrease (operator $\ll$) in cell numbers (also by 1 each time). Each agent has a simple choice in each step between the division action and the death action) shown by the choice operator (a plus: '+'). Which path is chosen is dictated by the rates of each action involved: actions with faster rates are more likely to occur, assuming the

[1] Bio-PEPA was developed for biochemical systems, therefore uses kinetic laws to describe the interaction of molecular species. We use agent here for species to avoid confusion with the biological definition of species. Kinetic laws in our model relate to the interaction of cells rather than molecules.

activities in which they take part are both enabled. That is, if the action requires a partner then there is a partner available to participate in the action.

The key abstraction tool in Bio-PEPA lies in the kinetic laws defining rates of activity for a given action based on numeric rates, stoichiometry coefficient, and potentially also agent concentration. Here the rates for division depend on the populations of both cell types, using an arithmetic expression. Trigonometric functions are also available. The rates for death show the use of the built-in mass action law. Other built-in laws include Hill and Michaelis-Menten kinetics.

The last line of the model is the model component, defining communication paths and initial population sizes. In this example, there are one thousand cells of each type, and the cells do not communicate over any action. Interaction is via their influence on the kinetic rate for division.

Bio-PEPA supports compartments: these allow a notion of separate spaces for agents. Agents may move between compartments if desired. Compartments may be used to indicate physical space, but may also represent different kinds of structure (e.g. age). Time and timed events are also supported by Bio-PEPA. These will all be seen in the model of Section 4.2.

2.2 Evolutionary Computation

Evolutionary algorithms have been successfully used in a broad range of areas including determination of protein folding processes, modelling of metabolic pathways [11] and construction of ecological niche models [29]. Different evolutionary approaches are available: the standard Genetic Algorithm (GA) approach first popularised by Holland [19] is used here. This allows us to demonstrate the principle of using an Evolutionary Computation (EC) approach with process algebras.

Genetic algorithms are one of the earlier approaches to using evolutionary computation for optimisation. They start with an initially random set of solutions to a problem. In this case, the problem is expressed as a process algebra model, and each solution is a particular set of numeric values of the parameters to be fitted for this model. Each of these solutions is scored by noting the difference between the desired behaviour and the behaviour produced by the model with those parameters. Note that the concept of desired behaviour can be more complex than simply a target time series trace, and therefore more difficult to capture: the effective use of evolutionary computation can often be determined by the definition of an appropriate fitness function. For example, in biological systems replication of episodic behaviour may be more significant than an exact match to a particular time series trace.

Once we have determined a method for measuring the effectiveness of a given set of parameter values and scored our population of potential solutions using it, we move to the process of selecting individuals as parents for a breeding process. There are a range of different approaches to choose from but they share a common theme of balancing exploration of the search space (via selection of less fit parents) with exploitation of promising leads (e.g. increased tendency to pick relatively good solutions in the current population). For example, we may simply select the best two individuals in the population, or we may just pick two at random. The first choice is effectively a form of "hill climbing" where we follow the current best trend but this has a tendency to get caught in local maxima if the solution surface is deceptive. A random choice of parent will tend to ignore potentially good solutions and effectively leads to a random walk through solution space, hoping to get lucky. An effective and efficient selection method for GAs is *Tournament* selection which enables us to balance these two approaches via the choice of a tournament size.

Tournament selection reflects the concept of competition between individuals in the population. A group or "tournament" of randomly selected individuals is chosen and the fittest individual from this tournament is put forward for breeding with another similarly chosen individual. A large tournament size relative to the population size increases the odds of high fitness solutions being selected. Thus the best solutions in the population are selected for breeding and dominate future generations (an *exploitative* effect). Conversely, a small tournament size leads to weaker solutions getting through that may contain useful parameter values that are being masked by poor choices for other variables (an *explorative* approach).

Having obtained two parents, they are bred to produce offspring, combining the "genetic material" (parameter settings) of the two individuals to produce a new individual. Typical approaches include one point crossover and uniform crossover. One point crossover selects an initial section of parameter values from one parent and the remaining set of parameter values from the second parent; effectively producing new offspring by crossing over parameter information at a single point in the set of parameter values. Uniform crossover considers each parameter value in turn and selects at random a value from either the first parent or the second parent. One point crossover is most effective when there is some inter-dependence in parameter values since it tends to preserve blocks of parameter values. Uniform crossover is a more effective mixing strategy and is most useful when parameters are less tightly coupled.

On completion of the crossover process, a new offspring is produced that is derived entirely from its two parents. In order to inject new "genetic material" into the population pool, an element of mutation is also introduced. If this were not the case, the GA would be limited to the initial parameter values set at the start of the evolutionary process. With low probability (usually less than 0.05), individuals are determined to have mutated or not and if so, randomly selected parameter values are reinitialised. This will have the tendency to weaken the performance of mutated individuals (hence the low probability of performing this operation) but occasionally the random change will place the individual in a new promising location in the parameter search space.

Finally, the new individual is inserted into the population pool. Either the old population pool is completely replaced with new offspring (known as generational replacement) or just one offspring is added at a time (steady state replacement) replacing a weaker individual selected using the inverse of the selection process.

Generational replacement tends to operate more quickly provided we also ensure that the best solution from the previous generation is allowed to go through.

Once a new population pool is produced, the whole process is repeated: the new individuals are scored by running them through the process algebra model with the new crossed-over, and potentially mutated, parameter values. This process is repeated over many generations where an improvement in population fitness is generally observed. The process is stopped when either an acceptable fitness tolerance is reached or a time limit is exceeded. One can also measure the relative change in fitness between generations and decide to stop if minimal improvements have been observed over a given time window.

Further detail on EA approaches may be found in Bäck *et al* [1] and De Jong [8]. A general discussion of its relevance and application to Bioinformatics is available in Fogel and Corne [11].

2.3 Related Work

There is substantial prior work in applying evolutionary approaches to biological systems. See e.g. Fogel and Corne [11]. We do not review that here, focussing instead on the more novel application to process algebra modelling. Relatively few researchers are exploring this promising combination of techniques. The work of Ross and Imada [28] applies a genetic approach to time series data and a subset of the stochastic π-calculus to evolve process algebra models. Their focus is on the best suite of statistical tests to use to measure fitness of candidate solutions. Their method is applied to a chemical reaction, a genetic circuit, and a cyclic process, evolving both parameters and the model itself. Prandi [26] uses Particle Swarm Optimisation (PSO) inside the R system with BlenX to optimise parameter fitting. Prandi demonstrates his system on an idealised ecological food web, matching to a synthetic oscillatory behaviour.

In our previous work we have successfully evolved numeric rate parameters for PEPA models for simple benchmark examples from the literature in computer science and in epidemiology [23]. A genetic algorithm was used and the PEPA Eclipse Plug-in provided the simulated time series data with which to evaluate fitness of solutions. The present paper provides an extension on that work by extending the framework to accept Bio-PEPA, and also through application to a more substantial example. Our work, with that above [26,28], follows the same basic approach of EA applied to process algebra. We all use different EA frameworks and process algebras, and apply our work to quite different examples.

There is a large body of work in applying evolutionary approaches to ODE or CTMC (Continuous Time Markov Chain) models, both of which can be derived from Bio-PEPA models, preserving the semantics of the overall system dynamics. We believe process algebra offers an advantage in model expression. Both ODE and CTMC are more suited to describing general system behaviour with a high level of abstraction. With process algebra the focus is on the many individuals comprising the system, and their interactions. This approach allows for the subtle interaction of a small number of individuals to act as a tipping point for the emergent behaviour of the whole system. This is a behaviour that can be hard to capture in equivalent ODE models. Moreover, our long term goal is to use the framework presented in Section 3 as the core of a system to evolve not just the parameters but the process algebra models themselves [24]. This is a more specialised target than that allowed by more general systems for simulation and optimisation for multiple input languages. See for example the JAMES II system of Himmelspach et al [18], or the use of SBSI tool [4] for Bio-PEPA (via translation to SBML) in which parameter estimation is carried out using a parallelised genetic algorithm.

3 The EPA Framework

We envisage the EPA framework as an experimental system in which process algebras can be combined with a range of evolutionary techniques. This first step combines process algebra with a GA [15,19] for parameter fitting. This has been carried out both with PEPA [23] and with Bio-PEPA (described here). EPA combines the ECJ evolutionary computing framework [20] and the Bio-PEPA Eclipse Plug-in developed at the University of Edinburgh [9]. ECJ is designed for large and complex experimental systems and supports many different evolutionary computing approaches. The Bio-PEPA Plug-in allows direct access to, and control of, Bio-PEPA models and their evaluation.

The main execution loop for the EPA process will now be described. There is much discussion in the literature about the particular choices for the GA parameters mentioned here. It is not our goal to explore the nuances of these choices: the particular settings used for this example are given in Section 4.3.

1. Model. The input to the system is a Bio-PEPA model developed and refined within the Bio-PEPA Plug-in. The user specifies which parameters in the model are fixed and which are to be optimised, with relevant constraints set for each parameter indicating a suitable range of values for these parameters. A population of candidate individuals is generated where each individual is expressed as a set of potential parameter values for the Bio-PEPA model, chosen from within the specified constraints.
2. Evaluate. A time series trace is produced for each candidate solution. The Bio-PEPA Plug-in generates an average time series trace from a number of stochastic simulations, or a time series trace from an ODE interpretation. Currently, we use multiple stochastic simulations, where the duration and granularity of the simulation is fixed by the user. This is because our example models have often not been amenable to ODE analysis: this seems to be linked to the inclusion of events. It would be desirable to offer a choice to the user since when ODE analysis does work it is usually much faster than simulation: this is being considered for future work.

3. Score. A fitness score is assigned to each individual according to a notion of desired behaviour. For example, in cases where the model should produce a set of fixed output values, a simple distance measure (e.g. Euclidean) between the generated time series and the desired time series can be used. We favour a simple fitness measure while the framework is under initial development. More complex fitness measures are part of future planned work. The precise fitness function for this example is presented in Section 4.3.
4. Select. Individuals from the current population are chosen to breed the next generation using a combination of elitism and tournament selection.
5. Breed. A new generation of individuals is produced using one-point crossover and mutation, with a generational approach to population replacement.
6. Finish. Steps 2 to 5 are repeated until either an ideal solution has been found (within some small percentage of fitness to the target data), or a fixed number of generations have been completed, whichever comes first. Note that it is often possible to specify the total range of fitness values, therefore it can be easily decided what "good" fitness means.

The above execution run is repeated a number of times (dependent upon fitness evaluation time and degrees of freedom in the problem) to ensure a proper sample of fitness scores and potential parameter values are obtained for a given model.

4 T Helper Cell Activation

The basic utility of the EPA framework has been previously demonstrated through simple examples from biology and computer science [23]. For this paper we present a more significant challenge through a novel example: modelling components of the immune system, the T helper cell populations, and their response to co-infections with parasites making conflicting immunological demands.

4.1 Biological Background

T helper cells, also known as CD4+ T cells, are central to the modulation of the adaptive immune response. Adaptive immunity is the ability to recognise a pathogen, generate an antigenically tailored response, and remember that pathogen (for a better response next time). There are known to be many different CD4+ cells [30]; here the focus is on T helper (Th)1 and Th2 cell types. It is now well recognised that the control of helminth (parasitic worm) infections depends on the presence of CD4+ T cells with a Th2 cytokine profile [10,14], while most microbial infection is controlled by Th1 CD4+ cells [30]. These contrasting immune responses can interact in helminth-microparasite co-infection and can affect severity and the outcome of disease as Th1 and Th2 responses are mutually inhibitory [16].

Laboratory experiments [22] explore the relationship between Th2 responses induced by immunisation with filarial (worm) antigen and Th1 responses induced by infection with malaria in rodents. One of the questions addressed is the effect of malaria on pre-existing responses to filarial antigens. Three different experimental treatments were designed to expose the behaviour of the Th1 and Th2 populations, as shown in Table 1. Essentially, cells may be "primed" by exposure to worm antigen and transferred (on Day 0 of the experiment) to a congenic mouse. The use of congenic mice allows the behaviour of the transferred CD4+ cells to be tracked in isolation from the rest of the mouse immune system.This mouse might then be exposed to malarial infection (day 3), and/or given an injection of worm antigen (day 16). At the end of the experiment (day 19) cytokine levels are obtained via *ex vivo* intracellular cytokine staining and flow cytometry. The final two columns of Table 1 give the average cytokine measurements on day 19. These values indicate the final Th1 population sizes (defined by the cytokine interferon (IFN)-γ), and Th2 population sizes (defined by the cytokine interleukin (IL)-4). The figures are scaled to treat E1 results as 100% and E2 and E3 in relation to that baseline. Experimental results [22] show there is a statistically significant switch from Th2 to Th1 cytokine profiles.

Table 1. Experimental timeline with treatment variants and final cytokine proportions

Experiment Label	Day -7 Immunisation: Worm Ag	Day 3 Malaria	Day 16 Challenge: Worm Ag	Day 19 IFNγ	Day 19 IL-4
E1: Worm-primed	✔	✘	✔	100.0%	100.0%
E2: Worm-primed + Malaria	✔	✔	✔	161.0%	115.1%
E3: Unprimed + Malaria	✘	✔	✔	122.4%	43.7%

The goal of this work is to establish an initial model of Th1 and Th2 populations and their growth in response to different stimuli, matching the outputs of Table 1. This example throws up considerable problems of modelling which are common to all realistic case studies concerning assumptions which can be made, and values to attach to numeric parameters. Further modelling can then be used to investigate hypotheses about the causes behind the change in Th1 and Th2 behaviour noted above.

4.2 Modelling the Immune System in Bio-PEPA

The three experimental treatments of Table 1 could be represented as three separate models expressing the different behaviours of the Th1 and Th2 populations in response to different stimuli; however, this fails to capitalise on the similarities between the experimental treatments. Instead, we make use of compartments to separate experimental treatments within a single Bio-PEPA model. Each compartment represents one experimental treatment (E1, E2, E3 of Table 1). This allows us to link shared rate parameters affecting behaviour across experimental treatments in the model, instead of having to link rate parameters inside the EPA framework, or worse, to link them manually outside the EPA framework.

We present the whole model here, interspersed with explanatory comments on the basic rates for the model, the specific rates for each compartment, the functional rates and the agent definitions. The complete model can be downloaded from our website http://www.cs.stir.ac.uk/SystemDynamics.

$$\begin{aligned} compsize &= 10000; \\ location\ world\ &:\ size = 30000, type = compartment; \\ location\ E1\ in\ world\ &:\ size = compsize,\ type = compartment; \\ location\ E2\ in\ world\ &:\ size = compsize,\ type = compartment; \\ location\ E3\ in\ world\ &:\ size = compsize,\ type = compartment; \end{aligned}$$

We assume a base recruitment rate for each of *Th1* and *Th2*, and a boosted recruitment rate for *Th1* and *Th2* when challenged by Malaria or Worm Antigen respectively, possibly modified by some additional factor in a particular experimental treatment. The numeric rates for cell division and death can be taken from the literature. The other rate parameters are not known (indicated by ? below). We fix the division and death rates, to preserve homeostasis in the absence of worm or malaria challenges. Timed events (the injection of malaria and helminths) are modelled by the variables *malariaI* and *helminthI*, occurring at day 3 (hour 72) and day 16 (hour 384) respectively. These are used in the kinetic laws to activate either the base recruitment rate or the boosted recruitment rate for that T helper cell type. The Heaviside function (H) is used to give a binary valued function from time. This is used in the definitions below to switch customised behaviours on or off in the kinetic laws.

// Basic constants

$$\begin{aligned} recruit\_Th1\_rate &= ?; \\ boost\_recruit\_Th1\_rate &= ?; \\ inhibition\_Th1\_rate &= 0.1; \\ recruit\_Th2\_rate &= ?; \\ boost\_recruit\_Th2\_rate &= ?; \\ inhibition\_Th2\_rate &= 0.1; \\ divide\_Th1\_rate &= 0.006; \\ divide\_Th2\_rate &= 0.006; \\ death\_rate &= 0.00694; \\ malariaI &= H(time - 72); \\ helminthI &= H(time - 384); \end{aligned}$$

The basic constants of the model are common to all compartments. We fix the inhibition rate, and use the EPA framework to find the others in relation to that arbitrarily chosen fixed rate. The differences between the experimental treatments are largely explained by the customised rates for each compartment. Note that for example, in compartment *E1* the boost rate for *Th1* is the same as the base rate: this is because there is no malaria injection in that experimental treatment. In addition, note that in compartment *E3* the boost rate of *Th1* is suppressed because there is no worm-priming in that experimental treatment. Three parameters are specific to their compartment: *E2_mal_Th2*, *E3_mal_Th2*, and *E3_mal_suppress*.

```
// First compartment is worm-primed + worm boost (E1)
      E1_rec_Th1 = recruit_Th1_rate;
    E1_boost_Th1 = recruit_Th1_rate;
  E1_inhibit_Th1 = inhibition_Th1_rate;
      E1_rec_Th2 = recruit_Th2_rate;
    E1_boost_Th2 = boost_recruit_Th2_rate;
  E1_inhibit_Th2 = inhibition_Th2_rate;
      E1_Th1_div = divide_Th1_rate;
      E1_Th2_div = divide_Th2_rate;
      E1_Th1_die = death_rate;
      E1_Th2_die = death_rate;
// Second compartment is worm-primed + malaria + worm boost (E2)
      E2_rec_Th1 = recruit_Th1_rate;
    E2_boost_Th1 = boost_recruit_Th1_rate;
  E2_inhibit_Th1 = inhibition_Th1_rate;
      E2_rec_Th2 = recruit_Th2_rate;
      E2_mal_Th2 = ?;
    E2_boost_Th2 = boost_recruit_Th2_rate;
  E2_inhibit_Th2 = inhibition_Th2_rate;
      E2_Th1_div = divide_Th1_rate;
      E2_Th2_div = divide_Th2_rate;
      E2_Th1_die = death_rate;
      E2_Th2_die = death_rate;
// Third compartment is unprimed + malaria + worm boost (E3)
             E3_rec_Th1 = 0;
   E3_mal_suppress_Th1 = ?;
           E3_boost_Th1 = boost_recruit_Th1_rate * E3_mal_suppress_Th1;
         E3_inhibit_Th1 = inhibition_Th1_rate;
             E3_rec_Th2 = recruit_Th2_rate;
             E3_mal_Th2 = ?;
           E3_boost_Th2 = boost_recruit_Th2_rate;
         E3_inhibit_Th2 = inhibition_Th2_rate;
             E3_Th1_div = divide_Th1_rate;
             E3_Th2_div = divide_Th2_rate;
             E3_Th1_die = death_rate;
             E3_Th2_die = death_rate;
```

The core of the model is the agent descriptions for *Th1* and *Th2* and kinetic laws. These describe how the populations grow through division, shrink through natural cell death, grow through recruitment from the Naive pool (not modelled directly, and assumed to be infinite for this time-limited set of experiments), and inhibit the recruitment of the other T helper cell type. The kinetic laws describe the rates at which these activities occur. Most of these are simple mass action terms (e.g. division and death), but recruitment is complex, requiring an inhibition term which is scaled by the ratio of the T helper populations to each other. Also, recruitment varies in each compartment, to incorporate the effects of the malaria injection and worm challenge. Observe that this affects only the laws for recruitment of *Th2*; the laws for recruitment of *Th1* are all the same (modulo compartment name).

```
// Functional rates division and death same for all compartments
// Rates shown only for compartment E1
kineticLawOf E1_div1 : (E1_Th1_div * Th1@E1);
kineticLawOf E1_div2 : (E1_Th2_div * Th2@E1);
kineticLawOf E1_die1 : (E1_Th1_die * Th1@E1);
kineticLawOf E1_die2 : (E1_Th2_die * Th2@E1);
kineticLawOf E1_rec1 : (1 - malariaI ) * (E1_rec_Th1 * Th1@E1
                       -E1_inhibit_Th1 * (Th2@E1 + 1)/(Th1@E1 + 1))
                       +(malariaI ) * (E1_boost_Th1 * Th1@E1
                       -E1_inhibit_Th1 * (Th2@E1 + 1)/(Th1@E1 + 1));
kineticLawOf E1_rec2 : (1 - helminthI ) * (E1_rec_Th2 * Th2@E1
                       -E1_inhibit_Th2 * (Th1@E1 + 1)/(Th2@E1 + 1))
                       +(helminthI ) * (E1_boost_Th2 * Th2@E1
                       -E1_inhibit_Th2 * (Th1@E1 + 1)/(Th2@E1 + 1));
kineticLawOf E2_rec1 : (1 - malariaI ) * (E2_rec_Th1 * Th1@E2
                       -E2_inhibit_Th1 * (Th2@E2 + 1)/(Th1@E2 + 1))
                       +(malariaI ) * (E2_boost_Th1 * Th1@E2
                       -E2_inhibit_Th1 * (Th2@E2 + 1)/(Th1@E2 + 1));
kineticLawOf E2_rec2 : (1 - helminthI ) * (((1 - malariaI ) * (E2_rec_Th2 * Th2@E2
                       -E2_inhibit_Th2 * (Th1@E2 + 1)/(Th2@E2 + 1)))
                       +(malariaI * (E2_rec_Th2 * E2_mal_Th2 * Th2@E2
                       -E2_inhibit_Th2 * (Th1@E2 + 1)/(Th2@E2 + 1))))
                       +(helminthI ) * (E2_boost_Th2 * Th2@E2
                       -E2_inhibit_Th2 * (Th1@E2 + 1)/(Th2@E2 + 1));
kineticLawOf E3_rec1 : (1 - malariaI ) * (E3_rec_Th1 * Th1@E3
                       -E3_inhibit_Th1 * (Th2@E3 + 1)/(Th1@E3 + 1))
                       +(malariaI ) * (E3_boost_Th1 * Th1@E3
                       -E3_inhibit_Th1 * (Th2@E3 + 1)/(Th1@E3 + 1));
kineticLawOf E3_rec2 : (1 - helminthI ) * (((1 - malariaI ) * (E3_rec_Th2 * Th2@E3
                       -E3_inhibit_Th2 * (Th1@E3 + 1)/(Th2@E3 + 1)))
                       +(malariaI * (E3_rec_Th2 * E3_mal_Th2 * Th2@E3
                       -E3_inhibit_Th2 * (Th1@E3 + 1)/(Th2@E3 + 1))))
                       +(helminthI ) * (E3_boost_Th2 * Th2@E3
                       -E3_inhibit_Th2 * (Th1@E3 + 1)/(Th2@E3 + 1));
Th1 = (E1_div1, 1) >> Th1@E1 + (E1_die1, 1) << Th1@E1
     +(E1_rec1, 1) >> Th1@E1 + (E1_rec2, 1)(-) Th1@E1
     +(E2_div1, 1) >> Th1@E2 + (E2_die1, 1) << Th1@E2
     +(E2_rec1, 1) >> Th1@E2 + (E2_rec2, 1)(-) Th1@E2
     +(E3_div1, 1) >> Th1@E3 + (E3_die1, 1) << Th1@E3
     +(E3_rec1, 1) >> Th1@E3 + (E3_rec2, 1)(-) Th1@E3;
Th2 = (E1_div2, 1) >> Th2@E1 + (E1_die2, 1) << Th2@E1
     +(E1_rec2, 1) >> Th2@E1 + (E1_rec1, 1)(-) Th2@E1
     +(E2_div2, 1) >> Th2@E2 + (E2_die2, 1) << Th2@E2
     +(E2_rec2, 1) >> Th2@E2 + (E2_rec1, 1)(-) Th2@E2
     +(E3_div2, 1) >> Th2@E3 + (E3_die2, 1) << Th2@E3
     +(E3_rec2, 1) >> Th2@E3 + (E3_rec1, 1)(-) Th2@E3;

              Th1@E1[200] <> Th2@E1[800] <>
Th1@E2[200] <> Th2@E2[800] <> Th1@E3[200] <> Th2@E3[200]
```

The final numeric detail of the model relates to initial populations of T helper cells: these are not available experimentally. Instead, two different proportions of initial T helper cell populations are used. These distinguish the worm-primed cases (20:80 ratio of Th1:Th2) from the unprimed case (50:50 ratio of Th1:Th2).

4.3 Experimental Method: Target Data and Fitness Function

The objective of our experiment was to determine if the EPA framework can be used to find rates for the above model so that simulated data is produced to match the experimental data. The cell numbers in the model are a proxy for the actual cell numbers (which at 10^6 are too large to be simulated in reasonable time in the Bio-PEPA Plug-in). The E1: Worm-primed experiment population results are used as the baseline against which we measure the increase/decrease in population of T helper cells in the other experimental treatments.

The particular GA parameters used here are as follows: we give these to aid reproducibility. The choices are reasonably standard. A population of 100 individuals is used with generational replacement. Selection for breeding is enforced via tournament selection with a tournament of size 10. This is a relatively low selection pressure given that there is only approximately 10% chance of competing with the best individual in the current population in a given tournament. One point crossover is used to produce new offspring from two selected parents and mutation of the offspring parameters is limited to 5% chance of occurrence. Elitism is used to preserve the current best solution across generations.

Evaluation of a given offspring's parameter values is achieved by setting them in the Bio-PEPA model and then simulating the model via the Gibson-Bruck algorithm [13] for the 19 day experiment. Due to the stochastic nature of Bio-PEPA, this is repeated over 20 simulations and the results averaged across the simulations. Ideally, more simulations would be used to give better accuracy, but this is a trade-off with computation time. The final data points of this averaged performance are then compared against the target outcomes and a fitness score calculated based on the difference between the two sets of values according to the fitness function f of Equation (1), where t and s are target and simulated data respectively, and i ranges across $Th1@E2..Th2@E3$. In this study, a low value for the fitness score equates to a high level of fitness since it indicates minimal difference between the target and the simulated result. A score of zero is the best fitness possible. The fitness function also includes preference for solutions in which the boosted recruitment rate is further away from the base recruitment rate (i.e. minimising the terms c_1 and c_2 thus pushing the base and boosted recruitment rates further apart). This constraint is imposed by our biological understanding of the system. The "1 + " appears in Equation (1) to avoid this term dominating fitness.

$$f = \sqrt{\Sigma(t_i - s_i)^2} \,.\, (1 + c_1 + c_2) \tag{1}$$

$$\text{where} \quad c_1 = recruit\_Th1\_rate / boost\_recruit\_Th1\_rate$$
$$c_2 = recruit\_Th2\_rate / boost\_recruit\_Th2\_rate$$

One might expect to execute such a framework, with a given model and target experimental data, to produce a single set of ideal matching parameters. However, it is often the case that different combinations of parameter values will produce identical matches to the target data and the difficulty then becomes the identification of the correct set of values. Biological expertise is crucial, but statistics can assist. In our study, the process was repeated 64 times to establish the robustness of the results and gain an insight into the potential variance of both fitness scores and parameter values. The outcome is a range of parameter values that are equally valid: we know this is the case because the evolved output accurately matches experimental data as expressed by the fitness function.

4.4 Results

The performance of the optimisation process across 64 runs can be seen in Figure 4(a). This graph shows the distribution of best fitness scores within populations across the multiple runs with generations used as a measure of time. The graph has been cropped to 0..200 on the y-axis to better show the interesting behaviour. Fitness scores rapidly improve from the initial random starting point (fitness 611) while maintaining a degree of variability (as indicated by the upper and lower quartile ranges). After around 70 generations, the fitness scores have converged with diminishing returns on further computation. At the end of the 100 generations, the median fitness score across the 64 runs is 6.56 with an upper and lower quartiles of 8.97 and 5.50, indicating a very high level of fit to the target data. This is approximately 1% of the original average fitness for the first generation. This, and the convergence shown in the graph of Figure 4(a), give an indication of the success of the optimisation process.

In any EA there is a tradeoff between various factors. For example, the fitness function could be made more complex to include variance of the 20 simulations in each evaluation as well as the mean. This would result in more computations in each evaluation and potentially require more generations to converge to an acceptably fit level. Here, we have chosen a very simple fitness function, which converges quickly (100 generations). The fact that variance of the fitness across EA runs is low by 100 generations suggests that the generated parameter values are robust to the noise of the genetic process and fitness measure.

Given the above performance, the sets of parameter values that are obtained as a result of these runs can all be considered to be valid solutions for the model to fit the target data. The spread of parameter values obtained at the end of the 64 runs, with median values below each histogram, can be seen in Figures 2 and 3. Each variable is presented with the x-axis showing the full range of permitted values to better illustrate the clustering of fit values. In the majority of cases, they do not conform to a normal distribution, therefore if selecting a particular parameter set, it is preferable to use the medians (as shown), or to choose the set with the best fitness (not shown). This ability to inspect the parameter distribution values provides a form of sensitivity analysis and indicates that the Bio-PEPA model is not driven by a single set of parameters. This information

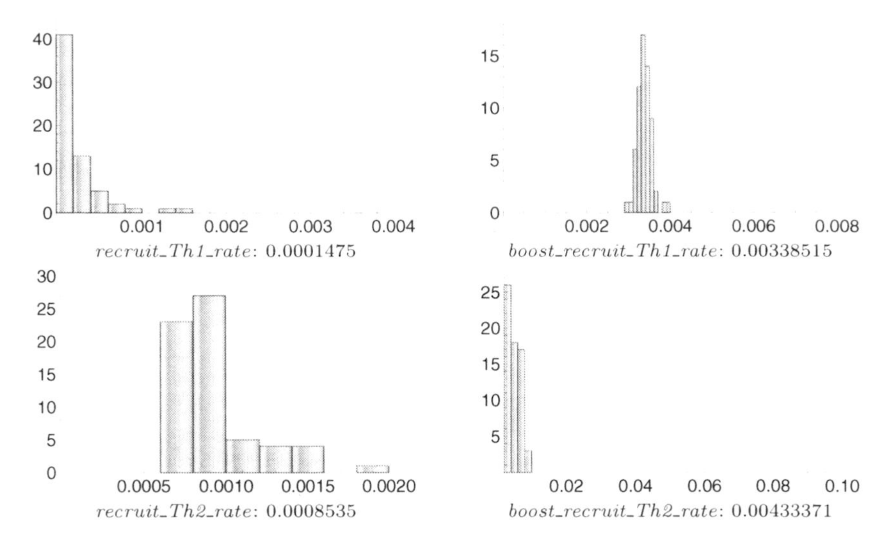

Fig. 2. Histograms for Mouse model recruitment parameters

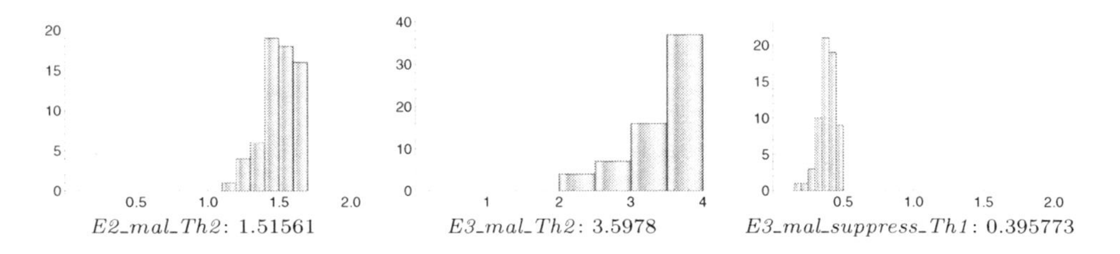

Fig. 3. Histograms for Mouse model scaling parameters

is of value to the modeller since it can offer an insight on the model dynamics, the validity of the model, and potentially the biological system being studied.

For example, from Figures 2 and 3 it can be seen that the fittest values for each parameter cluster at particular points, despite there being a wide range of permitted values for the optimisation algorithm to try. This particularly shows that the scaling variables (*E2_mal_Th2*, *E3_mal_Th2*, and *E3_mal_suppress*) are necessary to explain the experimental data, since their values are not 1. This quantifies the effect of the experimental treatments.

Further analysis (examples shown in Figure 4 (b, c, d)) is to consider variable dependencies: correlations between two or more variables. These are most fruitfully considered for variables showing more variance as indicated in the histograms of Figures 2 and 3. For example, Figure 4(b) shows inverse correlations between the scaling factors *E2_mal_Th2* and *E3_mal_Th2* and the base recruitment rate *recruit_Th2_rate*. This is as expected, since these variables are associated with controlling the growth of *Th2*. In contrast, Figure 4(c) shows these scaling factors against the base recruitment rate for *Th1* cells

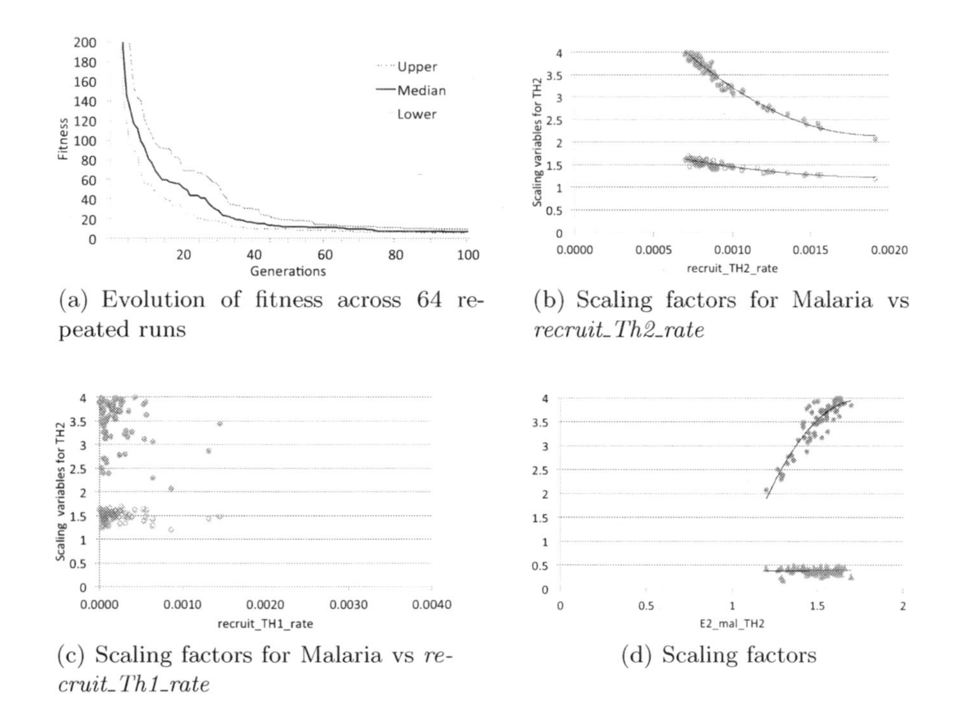

(a) Evolution of fitness across 64 repeated runs

(b) Scaling factors for Malaria vs *recruit_Th2_rate*

(c) Scaling factors for Malaria vs *recruit_Th1_rate*

(d) Scaling factors

Fig. 4. Evolution of fitness, and correlation between selected evolved variables. Scaling factors: *E2_mal_Th2* is shown as empty circles, *E3_mal_Th2* as solid circles, *E3_mal_suppress_Th1* as solid triangles.

recruit_Th1_rate. While there is some clustering, there is no apparent correlation. Figure 4(d) shows the relation between the three scaling factors: *E3_mal_Th2* and *E3_mal_suppress_Th1* plotted against *E2_mal_Th2*. *E3_mal_Th2* grows as *E2_mal_Th2* grows, but *E3_mal_suppress_Th1* is relatively static (as it has less variance, see the histogram of Figure 3).

5 Conclusions and Future Work

We have presented a novel framework in which a standard genetic algorithm is used for parameter estimation of numeric rate parameters in a hand-built Bio-PEPA model. This method was applied to an original immunological example to demonstrate the validity of the modelling choices and to obtain values for parameters not available from wet lab experimentation. While this is a particularly fruitful approach for biological systems (where ample experimental data exists), it can be equally well applied in other application domains. The process of evolutionary parameter fitting allows a modeller to gain an understanding of the overall stability of the solution space. In particular, indications are given as to which parameters the model is sensitive to and which have little or no effect. Such robustness may be expected for aspects of biological systems: if they were sensitive to all parameters the system would fail rather easily. Computing the distribution data in Figures 2 and 3 requires considerable computational effort: the EPA framework ran approximately 13 million simulations (20*100*100*64) to obtain these figures. This takes in the order of 36 hours on a computing cluster of 100 nodes. Compare this with the computational effort required to perform a full parameter sweep (for seven parameters in this case): at least 77 billion simulations (using the same granularity produced by the EPA). A key advantage of the EPA approach is better targeting of our computational resource, focussing on the parameter values of most relevance.

The framework will be useful to systems biology modellers working with experimentalists on systems displaying the following features: the system is amenable to description as a network of interacting components modelled at an individual level, there are several parameters of that model which are difficult to measure in experiments, and there is experimental data for system output which will allow those parameters to be tuned. Having obtained suitable parameter values, and with a greater understanding of the behaviour of the model, these can then be used in further predictive experiments. Different scenarios can be investigated with the confidence that the parameters were well chosen. There is a risk that larger and more realistic examples become intractable. Computation time in the EPA system is related to the time for a single simulation: modellers can employ many abstraction strategies to keep simulation time down, even for larger, complex examples. If this is not possible, the EPA system parameters can be adjusted to bring overall computation time down: there is a trade-off to be made between simulation time and the number of repeated simulations carried out, the numbers of generations evolved, and the required robustness of the results. These can all be customised as required by the example.

There is ample scope for further development of this work, both in the direction of the particular immunological example, and in further development of the EPA framework.

For the immunological model, the first step is to enhance the biological realism of the models, e.g. through direct modelling of the mutually inhibitory effects of Th1 and Th2 via explicit inclusion of cytokine expression. Other lines of inquiry may be to construct process algebra models following completely different experimental design. Results obtained from modelling could then be confirmed by further wet lab experiments, supporting our aim that models are a true representation of the behaviour and function of the immune system.

For the EPA framework, one direction is to explore more fully the additional information gained from the performance of the EA and its implications for parameter sensitivity. There is also potentially much work to be done in extending the flexibility of the framework to utilise other forms of analysis in the Plug-in (most obviously, ODE analysis, but also cumulative distribution function analysis), or in the form of the fitness function in incorporating additional information from simulations (such as variance of multiple simulations). Ross and Imada [28] have already investigated a range of criteria which could be added to the fitness evaluation. Another direction is to explore other EA approaches within the framework, not just the GA approach used here. Lastly, promising new research has shown that rather than just evolve parameters it is possible to evolve the model itself. This development would further integrate the process algebra modelling with the evolutionary approach. The EPA framework has been extended to apply genetic programming techniques to evolving the agent definitions while keeping all other information fixed [24], but much remains to be done.

Acknowledgments. David Marco and Erin Scott are grateful to the Scottish Informatics and Computer Science Alliance (SICSA), a research initiative of the Scottish Funding Council, for financial support of their PhD studies. Initial modelling work was supported by the Carnegie Trust for the Universities of Scotland (ES, Summer 2010), and by the EPSRC through *System Dynamics from Individual Interactions: A process algebra approach to epidemiology* (EP/E006280/1, CS and RN, 2007-2010). The *in vivo* work was funded by the BBSRC (BB/C508734/1 to ALG and JEA). The authors thank the Bio-PEPA Plug-in development team at the University of Edinburgh, particularly Allan Clark, for help with the tool. Lastly, we appreciate the helpful comments of the anonymous reviewers.

References

1. Bäck, T., Fogel, D., Michalewicz, Z.: Evolutionary Computation, vol. 1,2. Taylor & Francis (2000)
2. Baeten, J.: A brief history of process algebra. Theoretical Computer Science 335(2/3), 131–146 (2005)
3. Bernardo, M., Degano, P., Zavattaro, G. (eds.): SFM 2008. LNCS, vol. 5016. Springer, Heidelberg (2008)
4. Centre for Systems Biology at Edinburgh: Systems biology software infrastructure (2011), `http://www.sbsi.ed.ac.uk/`
5. Chou, I.C., Voit, E.O.: Recent developments in parameter estimation and structure identification of biochemical and genomic systems. Mathematical Biosciences 219, 57–83 (2009)
6. Ciocchetta, F., Hillston, J.: Bio-PEPA: a Framework for the Modelling and Analysis of Biochemical Networks. Theoretical Computer Science 410, 3065–3084 (2009)
7. Cohen, J.: The crucial role of CS in systems and synthetic biology. Communications of the Association for Computing Machinery 51, 15–18 (2008)
8. De Jong, K.A.: Evolutionary computation - a unified approach. MIT Press (2006)
9. Duguid, A., Gilmore, S., Guerriero, M., Hillston, J., Loewe, L.: Design and development of software tools for Bio-PEPA. In: Proc. of Winter Simulation Conference 2009, pp. 956–967 (2009)
10. Finkleman, F.D., Shea-Donohue, T., Goldhill, J., Sullivan, C.A., Morris, S.C., Madden, K.B., Gause, W.C., Urban Jr., J.F.: Cytokine regulation of host defense against parasitic gastrointestinal nematodes: lessons from studies with rodent models. Annual Review of Immunology 15, 505–533 (1997)
11. Fogel, G.B., Corne, D.W.: Evolutionary Computation in Bioinformatics. Morgan-Kaufmann (2003)

12. Fraser, A.: Simulation of genetic systems by automatic digital computers. Australian Journal of Biological Sciences 10, 484–491 (1957)
13. Gibson, M.A., Bruck, J.: Efficient Exact Stochastic Simulation of Chemical Systems with Many Species and Many Channels. Journal of Physical Chemistry 104, 1876–1889 (2000)
14. Goff, L.L., Lamb, J., Graham, A., Harcus, Y., Allen, J.E.: IL-4 is required to prevent filarial nematode development in resistant but not susceptible strains of mice. International Journal for Parasitology 32, 1277–1284 (2002)
15. Goldberg, D.: Genetic Algorithms in Search, Optimization & Machine Learning. Addison Wesley (1989)
16. Hartgers, F., Yazdanbakhsh, M.: Co-infection of helminths and malaria: modulation of the immune responses to malaria. Parasite Immunology 28, 497–506 (2006)
17. Hillston, J.: A Compositional Approach to Performance Modelling. Cambridge University Press (1996)
18. Himmelspach, J., Ewald, R., Uhrmacher, A.M.: A Flexible and Scalable Experimentation Layer. In: Mason, S.J., Hill, R.R., Mnch, L., Rose, O., Jefferson, T., Fowler, J.W. (eds.) Proceedings of the 2008 Winter Simulation Conference, pp. 827–835. IEEE (2008)
19. Holland, J.: Adaptation in Natural and Artificial Systems. MIT Press (1992)
20. Luke, S.: Essentials of Metaheuristic, Lulu (2009), `http://cs.gmu.edu/~sean/book/metaheuristics/`
21. Machado, D., Costa, R., Rocha, M., Ferreira, E., Tidor, B., Rocha, I.: Modeling formalisms in systems biology. AMB Express 5, 1–45 (2011)
22. Mahajan, S., Gray, D., Harnett, W., Graham, A., Allen, J.: Helminth-induced unmodified Th2 cells affect malaria-induced immune responses but alter disease little (2010) unpublished draft
23. Marco, D., Cairns, D., Shankland, C.: Optimisation of process algebra models using evolutionary computation. In: Proceedings of 2011 IEEE Congress on Evolutionary Computation, pp. 1296–1301. IEEE (2011)
24. Marco, D., Shankland, C., Cairns, D.: Evolving bio-pepa process algebra models using genetic programming. In: Proceedings of the Genetic and Evolutionary Computation Conference (GECCO 2012), pp. 177–183. ACM (2012)
25. McCaig, C., Begon, M., Norman, R., Shankland, C.: A rigorous approach to investigating common assumptions about disease transmission: Process algebra as an emerging modelling methodology for epidemiology. Theory in Biosciences 130, 19–29 (2011), special issue on emerging modelling methodologies
26. Prandi, D.: Particle swarm optimization for stochastic process calculi. In: Proceedings of the 9th Workshop on Process Algebra and Stochastically Timed Activities, Department of Computing, pp. 77–82. Imperial College, London (2010)
27. Priami, C.: Process calculi and life science. Electronic Notes in Theoretical Computer Science 162, 301–304 (2006)
28. Ross, B.J., Imada, J.: Evolving stochastic processes using feature tests and genetic programming. In: 11th Annual Conference on Genetic and Evolutionary Computation, pp. 1059–1066. ACM (2009)
29. Stockwell, D.: Genetic algorithms II. In: Machine Learning Methods for Ecological Applications, pp. 123–144. Kluwer Academic Publishers (1999)
30. Zhu, J., Paul, W.: Heterogeneity and plasticity of T helper cells. Cell Research 20, 4–12 (2010)

Population Dynamics P Systems on CUDA

Miguel A. Martínez-del-Amor[1], Ignacio Pérez-Hurtado[1], Adolfo Gastalver-Rubio[1], Anne C. Elster[2], and Mario J. Pérez-Jiménez[1]

[1] Research Group on Natural Computing,
Department of Computer Science and Artificial Intelligence,
University of Seville,
Avda. Reina Mercedes s/n, 41012 Sevilla, Spain
{mdelamor,perezh,marper}@us.es, adolaurion@hotmail.co.jp

[2] HPC-Lab, Department of Computer and Information Science,
Norwegian University of Science and Technology,
Sem Sælands vei 9, NO-7491, Trondheim, Norway
elster@ntnu.no

Abstract. Population Dynamics P systems (PDP systems, in short) provide a new formal bio-inspired modeling framework, which has been successfully used by ecologists. These models are validated using software tools against actual measurements. The goal is to use P systems simulations to adopt a priori management strategies for real ecosystems.

Software for PDP systems is still in an early stage. The simulation of PDP systems is both computationally and data intensive for large models. Therefore, the development of efficient simulators is needed for this field. In this paper, we introduce a novel simulator for PDP systems accelerated by the use of the computational power of GPUs. We discuss the implementation of each part of the simulator, and show how to achieve up to a 7x speedup on a NVIDA Tesla C1060 compared to an optimized multicore version on a Intel 4-core i5 Xeon for large systems. Other results and testing methodologies are also included.

Keywords: Ecological Modeling, Population Dynamics, Membrane Computing, Parallel Simulation, GPU Computing, CUDA.

1 Introduction

Membrane Computing [19] is part of the broader field of natural or bio-inspired computing. The related computational models are called *P systems*. They are hierarchically distributed models inspired by how membranes compartmentalize living cells into "protected reactors". The model consists of simultaneous applications of rules (abstraction of chemical reactions) over multisets of objects (abstraction of chemical compounds) [20]. Membrane Computing covers both the study of the theoretical basis for the models as well as the applications of the model to various fields including computational Systems Biology [7,22], and Ecosystem Dynamics [3,4,8]. *Population Dynamics P Systems*, or PDP systems, is a P system based framework for modeling population dynamics [3,17] . It

D. Gilbert and M. Heiner (Eds.): CMSB 2012, LNCS 7605, pp. 247–266, 2012.

enables simultaneous evolution of a high number of species, as well as the management of a large number of auxiliary objects. It also facilitates model development that can be easily interpreted by simulation software.

So far, several algorithms have been developed in order to capture the semantics defined by the modeling framework. A comparison on the performance of these algorithms can be found in [9]. These algorithms select rules according to their associated probabilities, while keeping the maximal parallelism semantics of P systems. In [17], a new simulation algorithm is presented, called *DCBA* (Direct distribution based on Consistent Blocks Algorithm). It overcomes a common problem on the previous algorithms, regarding a distorted selection of rules. Furthermore, the DCBA was initially implemented and validated using the *P-Lingua* software framework [11,24] which resulted in a simulation Java library (pLinguaCore). However, the simulations were slow (taking hours to run a single simulation), since the pLinguaCore library is not focused on efficiency.

A more efficient implementation based on C++ and OpenMP was presented in [16], taking advantage of modern multicore architectures. These preliminary results indicate that the simulation of PDP systems are memory bound. *GPU computing* [15] has been successfully used to implement other P systems simulators [2,5,6]. By using *CUDA* [13,23], the simulators are accelerated taking advantage of the many-core GPU architecture.

This paper, introduces our new CUDA-optimized PDP systems simulator. We describe how the data structures have been optimized, and how also the code is adapted and restructured in different parts. A performance analysis is also provided, comparing the results with further optimized sequential and multicore versions written in C++/OpenMP.

This paper is structured as follows: Section 2 gives an overview of the P systems based framework that our simulator implements. Section 3 explains the simulation algorithm. Section 4 contains some concepts about the CUDA programming model. Section 5 explains the details of the implementation using CUDA. Section 6 shows some performance results by using a random generator of PDP systems. Finally, conclusions and future work are discussed in Section 7.

2 The P Systems Based Framework

Definition 1. *A PDP system of degree (q, m) & time T $(q, m, T \geq 1)$ is a tuple*

$$\Pi = (G, \Gamma, \Sigma, T, R_E, \mu, R, \{f_{r,j} : r \in R, 1 \leq j \leq m\}, \{\mathcal{M}_{ij} : 1 \leq i \leq q, 1 \leq j \leq m\})$$

where:

- $G = (V, S)$ *is a directed graph. Let* $V = \{e_1, \ldots, e_m\}$ *whose elements are called environments;*
- Γ *is the working alphabet and* $\Sigma \subsetneq \Gamma$ *is an alphabet representing the objects that can be present in the environments;*

- *T is a natural number that represents the simulation time of the system;*
- *R_E is a finite set of communication rules between environments of the form $r \equiv (x)_{e_j} \xrightarrow{p_r} (y_1)_{e_{j_1}} \cdots (y_h)_{e_{j_h}}$, where $x, y_1, \ldots, y_h \in \Sigma$, $(e_j, e_{j_l}) \in S$ $(1 \leq l \leq h)$ and p_r is a computable function from $\{1, \ldots, T\}$ to $[0,1]$ depending on $x, j, j_1, \ldots, j_h$. If for any rule, p_r is the constant function 1, then we can omit it. These functions verify the following: for each $e_j \in V$ and $x \in \Sigma$, the sum of functions associated with the rules whose left-hand side is $(x)_{e_j}$ is the constant function 1.*
- *μ is a membrane structure consisting of q membranes injectively labelled by $1, \ldots, q$. The skin membrane is labelled by 1. We also associate electrical charges from the set $\{0, +, -\}$ with membranes.*
- *R is a finite set of evolution rules of the form $r \equiv u[\,v\,]_i^{\alpha} \rightarrow u'[\,v'\,]_i^{\alpha'}$, where $u, v, u', v' \in \Gamma^*$, i $(1 \leq i \leq q)$, $u + v \neq \lambda$ and $\alpha, \alpha' \in \{0, +, -\}$.*
  - *If $(x)_{e_j}$ is the left-hand side of a rule from R_E, then none of the rules of R has a left-hand side of the form $u[v]_1^{\alpha}$, having $x \in u$.*
- *For each $r \in R$ and for each j $(1 \leq j \leq m)$, $f_{r,j} : \{1, \ldots, T\} \longrightarrow [0,1]$ is a computable function verifying the following: for each $u, v \in \Gamma^*$, i $(1 \leq i \leq q)$, $\alpha, \alpha' \in \{0, +, -\}$ and j $(1 \leq j \leq m)$ the sum of functions associated with j and the rules whose left-hand side is $u[v]_i^{\alpha}$ and whose right-hand side has polarization α', is the constant function 1.*
- *For each j $(1 \leq j \leq m)$, $\mathcal{M}_{1j}, \ldots, \mathcal{M}_{qj}$ are strings over Γ, describing the multisets of objects initially placed in the q regions of μ, within the environment e_j. It is usual to manage multisets through strings. A finite multiset $m = \{a_1^{f(a_1)}, \ldots, a_k^{f(a_k)}\}$ can be represented by the string $a_1^{f(a_1)} \ldots a_k^{f(a_k)}$ over the alphabet $\{a_1, \ldots, a_k\}$. Nevertheless, all permutations of this string precisely identify the same multiset m. Throughout this paper, we speak about "the finite multiset m" where m is a string, and meaning "the finite multiset represented by the string m".*

That is, a system defined as above can be viewed as a set of m environments $e_1, \ldots, e_m$ interlinked by the edges from the directed graph G. Each environment e_j contains a P system, $\Pi_j = (\Gamma, \mu, R, \mathcal{M}_{1j}, \ldots, \mathcal{M}_{q,j})$, of degree q, where every rule $r \in R$ has a computable function $f_{r,j}$ associated with it. For each environment e_j, we denote by R_{Π_j} the set of rules with probabilities obtained by coupling each $r \in R$ with the corresponding function $f_{r,j}$.

A *configuration* of the system at any instant t is a tuple of multisets of objects present in the m environments and at each of the regions of each Π_j, together with the polarizations of the membranes in each P system. We assume that all environments are initially empty and that all membranes initially have a neutral polarization. We assume a global clock exists, sychromnizing all membranes and the application of all the rules (from R_E and from R_{Π_j} in all environments).

The P system can pass from one configuration to another by using the rules from $\bigcup_{j=1}^{m} R_{\Pi_j} \cup R_E$ as follows: at each transition step, the rules to be applied are selected according to the probabilities assigned to them, and all applicable rules are simultaneously applied in a maximal way.

An evolution rule $r \in R$, of the form $u[\,v\,]_i^{\alpha} \rightarrow u'[\,v'\,]_i^{\alpha'}$, is applicable to each membrane labelled by i, whose electrical charge is α, and it contains the multiset v, and its father contains the multiset u. When such rule is applied, the objects of the multisets u and v are removed from the father of membrane i and membrane i, respectively. Simultaneously, the objects of the multiset u' are added to the father of membrane i, and objects of multisets v' are introduced in membrane i. The application also replaces the charge of membrane i to α'.

A rule $r \in R_E$, of the form $(x)_{e_j} \xrightarrow{p_r} (y_1)_{e_{j_1}} \cdots (y_h)_{e_{j_h}}$, is applicable to the environment e_j if it contains object x. When such rule is applied, object x passes from e_j to $e_{j_1}, \ldots, e_{j_h}$ possibly modified into objects $y_1, \ldots, y_h$ respectively. At any moment t $(1 \leq t \leq T)$ for each object x in environment e_j, if there exist communication rules whose left-hand side is $(x)_{e_j}$, then one of these rules will be applied. If more than one such a rule can be applied to an object, the system selects one randomly, according to their probability which is given by $p_r(t)$.

For each j $(1 \leq j \leq m)$ there is just one further restriction, concerning the consistency of charges: in order to apply several rules of R_{Π_j} simultaneously to the same membrane, all the rules must have the same electrical charge on their right-hand side.

Following the properties verified by the probabilistic functions, rules in R and R_E can be classified into *blocks of rules*, as showed in definitions 2, 3 and 4.

Definition 2. *The left and right-hand sides of the rules are defined as follows:*

(a) *Given a rule $r \in R$ of the form $u[v]_i^{\alpha} \rightarrow u'[v']_i^{\alpha'}$ where $1 \leq i \leq q$, $\alpha, \alpha' \in \{0, +, -\}$ and $u, v, u', v' \in \Gamma^*$:*
  - *The left-hand side of r is $LHS(r) = (i, \alpha, u, v)$. The charge of $LHS(r)$ is $charge(LHS(r)) = \alpha$. The length of $LHS(r)$ is $|u| + |v|$, what indicates the cooperation degree of the rule.*
  - *The right-hand side of r is $RHS(r) = (i, \alpha', u', v')$. The charge of $RHS(r)$ is $charge(RHS(r)) = \alpha'$. The length of $RHS(r)$ is $|u'| + |v'|$.*

(b) *Given a rule $r \in R_E$ of the form $(x)_{e_j} \xrightarrow{p_r} (y_1)_{e_{j_1}} \cdots (y_h)_{e_{j_h}}$, the left-hand side of r is $LHS(r) = (e_j, x)$, and the right-hand side of r is $RHS(r) = (e_{j_1}, y_1) \cdots (e_{j_h}, y_h)$.*

Definition 3. *Rules from R can be classified into consistent blocks associated with $(i, \alpha, \alpha', u, v)$ as follows:*

$$B_{i,\alpha,\alpha',u,v} = \{r \in R : LHS(r) = (i, \alpha, u, v) \wedge charge(RHS(r)) = \alpha'\}$$

Definition 4. *Rules from R_E can be classified into (consistent) blocks associated with (e_j, x) as follows: $B_{e_j,x} = \{r \in R_E : LHS(r) = (e_j, x)\}$.*

Recall that, according to the semantics of our model, the sum of probabilities of all the rules belonging to the same block is always equal to 1; in particular, rules with probability equal to 1 form individual blocks. Note that rules that have exactly the same left-hand side (LHS) belongs to the same block, but rules with overlapping (but different) left-hand sides are classified into different blocks. The latter leads to object *competition*, what is a critical aspect to manage with the simulation algorithms.

Definition 5. *Two blocks $B_{i_1,\alpha_1,\alpha'_1,u_1,v_1}$ and $B_{i_2,\alpha_2,\alpha'_2,u_2,v_2}$ are mutually consistent with each other, if and only if $(i_1 = i_2 \wedge \alpha_1 = \alpha_2) \Rightarrow (\alpha'_1 = \alpha'_2)$.*

3 The DCBA

The goal of the DCBA (Direct distribution based on Consistent Blocks Algorithm) [17] is to perform a proportional distribution of objects among competing blocks (with overlapping LHS), determining in this way the number of times that each rule in $\bigcup_{j=1}^{m} R_{\Pi_j} \cup R_E$ is applied. *I.e.* the algorithm simulates the computational steps of a PDP systems. Algorithm 3.1 describes the main loop of the DCBA. It follows the same general scheme as its predecessors, *DNDP* and *BBB* [18] where the simulation of a computing step is structured in two stages: The first stage (selection), selects which rules are to be applied (and how many times) on each environment. The second stage (execution), implements the effects of applying the previously selected rules, yielding the next configuration of the PDP system. Note that, although every Π_j has the same set of rules R, the probability functions may be different for each environment. See [17] for a more detailed explanation and examples of how to apply this algorithm.

As shown in Algorithm 3.1, the selection stage consists of three phases: Phase 1 distributes objects to the blocks in a certain proportional way, Phase 2 assures the *maximality* by checking the maximal number of applications of each block, and Phase 3 translates block applications to rule applications by calculating random numbers using the multinomial distribution.

Algorithm 3.1. DCBA MAIN PROCEDURE

Require: A Population Dynamics P system of degree (q, m), $T \geq 1$ (time units), and $A \geq 1$ (*Accuracy*). The initial configuration is called C_0.

1: *INITIALIZATION* ▷ (Algorithm 3.2)
2: **for** $t \leftarrow 1$ to T **do**
3: Calculate probability functions $f_{r,j}(t)$ and $p(t)$.
4: $C'_t \leftarrow C_{t-1}$
5: *SELECTION* of rules:
- *PHASE 1*: distribution ▷ (Algorithm 3.3)
- *PHASE 2*: maximality ▷ (Algorithm 3.4)
- *PHASE 3*: probabilities ▷ (Algorithm 3.5)

6: *EXECUTION* of rules. ▷ (Algorithm 3.6)
7: $C_t \leftarrow C'_t$
8: **end for**

The INITIALIZATION procedure (Alg. 3.2) constructs a static distribution table $\mathcal{T}_j$ for each environment. Two variables, B^j_{sel} and R^j_{sel}, are also initialized, in order to store the selected multisets of blocks and rules, respectively.

Observation 1. *Each column label of the tables $\mathcal{T}_j$ contains the information of the corresponding block left-hand side.*

Observation 2. *Each row of the tables $\mathcal{T}_j$ contains the information related to the object competitions: for a given object, its row indicates which blocks are competing for it (those columns having non-null values).*

Algorithm 3.2. INITIALIZATION

1: Construction of the *static distribution* table $\mathcal{T}$:
   - Column labels: consistent blocks $B_{i,\alpha,\alpha',u,v}$ of rules from R.
   - Row labels: pairs (o, i) and $(x, 0)$, for all object $o \in \Gamma$, $x \in \Sigma$ and membrane i, being 0 the identifier of the environments of the P system.
   - For each row labelled by (o, i) and column labelled by block $B_{i,\alpha,\alpha',u,v}$: place $\frac{1}{k}$ if o appears within i $(0 \le i \le q)$ with multiplicity k in the LHS of $B_{i,\alpha,\alpha',u,v}$.

2: **for** $j = 1$ **to** m **do** ▷ (Construct the *static expanded* tables $\mathcal{T}_j$)
3: $\quad \mathcal{T}_j \leftarrow \mathcal{T}$. ▷ (Initialize the table with the original $\mathcal{T}$)
4: $\quad$ For each rule block $B_{e_j,x}$ from R_E, add a column labelled by $B_{e_j,x}$ to the table $\mathcal{T}_j$; place the value 1 at row $(x, 0)$ for that column.
5: $\quad$ Initialize the multisets $B^j_{sel} \leftarrow \emptyset$ and $R^j_{sel} \leftarrow \emptyset$
6: **end for**

The distribution of objects among the blocks with overlapping LHS (competing blocks) is performed in selection Phase 1 (Algorithm 3.3). The expanded static tables $\mathcal{T}_j$ are used for this purpose in each environment, together with three different filter procedures. FILTER 1 discards the columns of the table corresponding to non-applicable blocks due to mismatch charges in the LHS and in the configuration C'_t. Then, FILTER 2 discards the columns with objects in the LHS not appearing in C'_t. Finally, in order to save space in the table, FILTER 3 discards empty rows. These three filters are applied at the beginning of Phase 1, and the result is a *dynamic table* $\mathcal{T}^t_j$ (for the environment j and time step t).

The semantics of the modeling framework requires a set of mutually consistent blocks before distributing objects to the blocks. For this reason, after applying FILTERS 1 and 2, the mutually consistency is checked. Note that this checking can be easily implemented by a loop over the blocks. If it fails, meaning that an inconsistency was encountered, the simulation process is halted, providing a warning message to the user. Nevertheless, it could be interesting to find a way to continue the execution by non-deterministically constructing a subset of mutually consistent blocks. Since this method can be exponentially expensive in time, it is optional for the user whether to activate it or not.

Once the columns of the *dynamic table* $\mathcal{T}^t_j$ represent a set of mutually consistent blocks, the distribution process starts. This is carried out by creating a temporal copy of $\mathcal{T}^t_j$, called $\mathcal{TV}^t_j$, which stores the following products:

- The normalized value with respect to the row: this is a way to *proportionally* distribute the corresponding object along the blocks. Since it depends on the multiplicities in the LHS of the blocks, the distribution, in fact, penalizes the blocks requiring more copies of the same object. This is inspired in the amount of energy required to gather individuals from the same species.
- The value in the dynamic table (i.e. $\frac{1}{k}$): this indicates the number of possible applications of the block with the corresponding object.

– The multiplicity of the object in the configuration C'_t: this performs the distribution of the number of copies of the object along the blocks.

Algorithm 3.3. SELECTION PHASE 1: DISTRIBUTION

1: **for** $j = 1$ **to** m **do** ▷ (For each environment e_j)
2: Apply filters to table $\mathcal{T}_j$ using C'_t, obtaining $\mathcal{T}_j^t$. The filters are applied as follows:
 a. $\mathcal{T}_j^t \leftarrow \mathcal{T}_j$
 b. FILTER 1 $(\mathcal{T}_j^t, C'_t)$.
 c. FILTER 2 $(\mathcal{T}_j^t, C'_t)$.
 d. Check *mutual consistency* for the blocks remaining in $\mathcal{T}_j^t$. **If** there is at least one inconsistency **then** report the information about the error, and optionally halt the execution (in case of not activating step *3.*)
 e. FILTER 3 $(\mathcal{T}_j^t, C'_t)$.
3: *(OPTIONAL)* Generate a set S_j^t of sub-tables from $\mathcal{T}_j^t$, formed by sets of *mutually consistent* blocks, in a maximal way in $\mathcal{T}_j^t$ (by the inclusion relationship). Replace $\mathcal{T}_j^t$ with a randomly selected table from S_j^t.
4: $a \leftarrow 1$
5: **repeat**
6: **for all** rows X in $\mathcal{T}_j^t$ **do**
7: $RowSum_{X,t,j} \leftarrow$ total sum of the non-null values in the row X.
8: **end for**
9: $\mathcal{TV}_j^t \leftarrow \mathcal{T}_j^t$ ▷ (A temporal copy of the dynamic table)
10: **for all** non-null positions (X, Y) in $\mathcal{T}_j^t$ **do**
11: $mult_{X,t,j} \leftarrow$ multiplicity in C'_t at e_j of the object at row X.
12: $\mathcal{TV}_j^t(X,Y) \leftarrow \lfloor mult_{X,t,j} \cdot \frac{(\mathcal{T}_j^t(X,Y))^2}{RowSum_{X,t,j}} \rfloor$
13: **end for**
14: **for all** not filtered column, labelled by block B, in $\mathcal{T}_j^t$ **do**
15: $N_B^a \leftarrow \min_{X \in rows(\mathcal{T}_j^t)}(\mathcal{TV}_j^t(X, B))$ ▷ (The minimum of the column)
16: $B_{sel}^j \leftarrow B_{sel}^j + \{B^{N_B^a}\}$ ▷ (Accumulate the value to the total)
17: $C'_t \leftarrow C'_t - LHS(B) \cdot N_B^a$ ▷ (Delete the LHS of the block.)
18: **end for**
19: FILTER 2 $(\mathcal{T}_j^t, C'_t)$
20: FILTER 3 $(\mathcal{T}_j^t, C'_t)$
21: $a \leftarrow a + 1$
22: **until** $(a > A) \vee$ *(all the selected minimums at step 15 are* $0)$
23: **end for**

Algorithm 3.4. SELECTION PHASE 2: MAXIMALITY

1: **for** $j = 1$ **to** m **do** ▷ (For each environment e_j)
2: Set a random order to the blocks remaining in the last updated table $\mathcal{T}_j^t$.
3: **for all** block B, following the previous random order **do**
4: $N_B \leftarrow$ number of possible applications of B in C'_t.
5: $B_{sel}^j \leftarrow B_{sel}^j + \{B^{N_B}\}$ ▷ (Accumulate the value to the total)
6: $C'_t \leftarrow C'_t - LHS(B) \cdot N_B$ ▷ (Delete the LHS of block B, N_B times.)
7: **end for**
8: **end for**

After the object distribution process, the number of applications for each block is calculated by selecting the minimum value in each column. This number is then used for consuming the LHS from the configuration. However, this application could be not maximal. The distribution process can eventually deliver objects to blocks that are restricted by other objects. As this situation may occur frequently, the distribution and the configuration update process is performed A times, where A is an input parameter referring to *accuracy*. The more the process is repeated, the more accurate the distribution becomes, but the performance of the simulation decreases. We have experimentally checked that $A = 2$ gives the best accuracy/performance ratio. In order to efficiently repeat the loop for A, and also before going to the next phase (maximality), it is interesting to apply FILTERS 2 and 3 again.

After phase 1, it may be the case that some blocks are still applicable to the remaining objects. This may be caused by a low A value or by rounding artifacts in the distribution process. Due to the requirements of P systems semantics, a maximality phase is now applied (Algorithm 3.4). Following a random order, a maximal number of applications is calculated for each block still applicable.

After the application of phases 1 and 2, a maximal multiset of selected (mutually consistent) blocks has been computed. The output of the selection stage has to be, however, a maximal multiset of selected rules. Hence, Phase 3 (Algorithm 3.5) passes from blocks to rules, by applying the corresponding probabilities (at the local level of blocks). The rules belonging to a block are selected according to a multinomial distribution $M(N, g_1, \ldots, g_l)$, where N is the number of applications of the block, and $g_1, \ldots, g_l$ are the probabilities associated with the rules $r_1, \ldots, r_l$ within the block, respectively.

Algorithm 3.5. SELECTION PHASE 3: PROBABILITY

1: **for** $j = 1$ **to** m **do** ▷ (For each environment e_j)
2: **for all** block $B^{N_B} \in B^j_{sel}$ **do**
3: Calculate $\{n_1, \ldots, n_l\}$, a random multinomial $M(N_B, g_1, \ldots, g_l)$ with respect to the probabilities of the rules $r_1, \ldots, r_l$ within the block.
4: **for** $k = 1$ **to** l **do**
5: $R^j_{sel} \leftarrow R^j_{sel} + \{r_k^{n_k}\}$.
6: **end for**
7: **end for**
8: Delete the multiset of selected blocks $B^j_{sel} \leftarrow \emptyset$. ▷ (Useful for the next step)
9: **end for**

Finally, the execution stage (Algorithm 3.6) is applied. This stage consists on adding the RHS of the previously selected multiset of rules, as the objects present on the LHS of these rules have already been consumed. Moreover, the indicated membrane charge is set.

Algorithm 3.6. EXECUTION

1: **for** $j = 1$ **to** m **do** ▷ (For each environment e_j)
2: **for all** rule $r^n \in R^j_{sel}$ **do** ▷ (Apply the RHS of selected rules)
3: $C'_t \leftarrow C'_t + n \cdot RHS(r)$
4: Update the electrical charges of C'_t from $RHS(r)$.
5: **end for**
6: Delete the multiset of selected rules $R^j_{sel} \leftarrow \emptyset$. ▷ (Useful for the next step)
7: **end for**

4 The CUDA Programming Model

With the commercial sector's demands for video and gaming, it was foreseen by Elster [10] and others that graphics processor development would lead to devices suitable for High Performance Computing (HPC). With the introduction of NVIDIA's CUDA [13] and AMD's Stream SDK environments in 2007, the GPUs became more easily programmable and the era of GPGPU (General-Purpose Computing on Graphics Processing Units) truly began. GPUs can now be considered affordable computing solutions for speeding up computationally demanding applications. GPUs today typically have several hundred computational cores. By parallelizing and optimizing our codes for these cores, we have shown that GPUs can speed up applications ranging from smoothed hydrodynamics (SPH) [14] and real-time gradient vector flows to linear programming [21] and can even be used as an accelerator to compress I/O data for faster I/O speeds [1]. As was mentioned earlier, we have also successfully used GPUs to implement other P systems simulators [2,5,6].

4.1 Compute Unified Device Architecture

NVIDIA's *CUDA* (Compute Unified Device Architecture) provides developers with a high-level programming model that allows developers to take full advantage of the NVIDIA's powerful GPU hardware. To a CUDA programmer [23], the computing system is heterogeneous, consisting of a host (the CPU), and one or more massively parallel many-core devices (GPU systems). In modern software applications, there are often program sections that exhibit rich amount of data parallelism, a property where many arithmetic operations can be safely performed simultaneously on the program data structures. GPUs accelerate the execution of these applications by harvesting a large amount of data parallelism.

CUDA is built around a scalable array of multithreaded Streaming Multiprocessors (SMs). The SM creates, manages, and executes concurrent threads (extra light weight processes) in hardware with with virtually no overhead [23]. This allows very fine-grained decomposition of problems by assigning, for instance, one thread to each data element. Each *threads* has a unique *threadIdx*, and is grouped in 1D, 2D, or 3D *blocks* which are organized in 1D or 2D *grids*. All the threads execute the same code, called *kernel*.

Since all threads of a block are expected to reside on the same processor core and must share the limited memory resources of that core, current GPUs, are limited to a maximum of 1024 threads. The number of thread blocks in a grid is usually dictated by the size of the data being processed or the number of processors in the system, which it can greatly exceed. These features make CUDA an interesting choice for developing simulators for the area of Membrane Computing, as previously demonstrated for active membranes [5], a P systems based solution to SAT [6] and for spiking neural P systems [2].

5 DCBA Implementation on the GPU

The DCBA was first implemented inside the pLinguaCore framework [17,11,24]. This version (hereafter *pdp-plcore-sim*) was validated by a real ecosystem model [17], reproducing the same data as the actual measurements. However, the performance was slow since it as part of the pLinguaCore was written in Java.

Our first approach for making our implementation more efficient, was to develope a *stand alone* simulator written in C++. We then improved performance further by using OpenMP to take advantage of modern multi-core architectures, such as the Intel's i5 Nehalem and i7 Sandy (*pdp-omp-sim*). *Pdp-omp-sim* achieved speedups of up to 2.5x on a 4-core Intel i7. These preliminary results indicated that the simulations of PDP systems are memory bound.

Our previous simulator, *pdp-omp-sim*, is the starting point of the new implementation using CUDA. In this new simulator (let call it *pdp-gpu-sim*), the code and the data structures have been optimized, saving up to 27% of memory. We have also adapted *pdp-omp-sim* to these, achieving better speedups (1.25x for large systems).

Normally, the end user (*i.e.* ecological experts and model designers) runs many simulations on each set of parameters to extract statistical information of the probabilistic model. This can be automated by adding a outermost loop for simulations in Algorithm 3.1. This loop is easily parallelized. Indeed, our tests of *pdp-omp-sim* conclude that parallelizing by simulations or a hybrid technique (simulations plus environments) yields the largest speedups.

At first glance, these two levels of parallelism (simulations and environments) could fit the double parallelism of the CUDA architecture (thread blocks and threads). For example, we could assign each simulation to a block of threads, and each environment to a thread (since they require synchronization at each time step). However, the number of environments depends inherently on the model. Typically, 2 to 20 environments are considered, which is not enough for fulfilling the GPU resources. Number of simulations typically range from 50 to 100, which is sufficient for thread blocks, but still a poor number compared to the several hundred cores available on modern GPUs.

We therefore also parallelize the execution of rule blocks. Our simulator can hence utilize a huge number of thread blocks by distributing simulations (parallel simulations, as memory can store them) and environments in each one, and process each rule block by each thread. Since there are normally more rule blocks (thousand of them) than threads per thread block (up to 512), we create 256 threads which iterate over the rule blocks in tiles. This design is graphically shown on Figure 1. Each phase of the algorithm has been designed following the general CUDA design explained above, and implemented separately as individual kernels. Thus, simulations and environments are synchronized by the successive calls to the kernels.

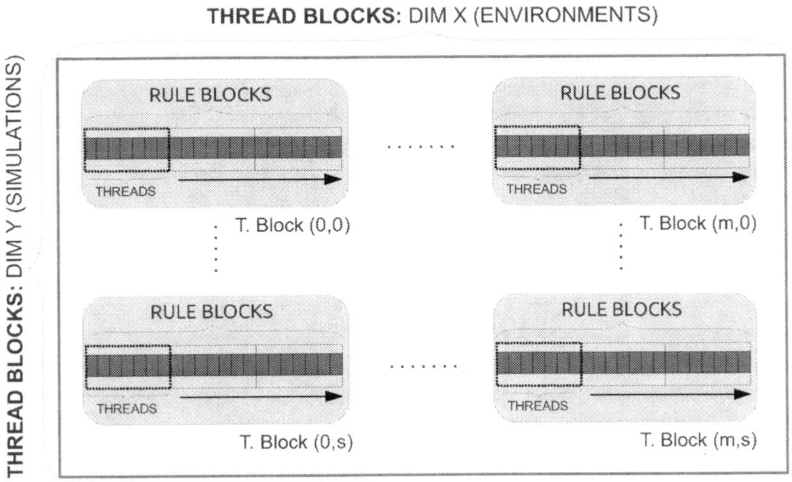

THREAD BLOCKS: DIM X (ENVIRONMENTS)

THREAD BLOCKS: DIM Y (SIMULATIONS)

RULE BLOCKS .... RULE BLOCKS

THREADS THREADS

T. Block (0,0) T. Block (m,0)

RULE BLOCKS .... RULE BLOCKS

THREADS THREADS

T. Block (0,s) T. Block (m,s)

Fig. 1. General design of our CUDA-based simulator: 2D grid, and 1D thread blocks. Threads loop the rule blocks in tiles.

5.1 Implementation of Selection Phase 1

The main challenge at this phase is the construction of the expanded static table $\mathcal{T}_j$. The size of this table is of order $O(|B|\cdot|\Gamma|\cdot(q+1))$, where $|B|$ is the number of rule blocks, $|\Gamma|$ is the size of the alphabet (amount of different objects), and $q+1$ corresponds to the number of membranes plus the space for the environment.

A full implementation of $\mathcal{T}_j$ can be expensive for large PDP systems. Moreover, it is a sparse matrix, having null values in the majority of the positions: competitions for one object appears for a relatively small number of blocks. This problem was overcome in the *pdp-plcore-sim* by using a hash table storing only non-null values. For *pdp-omp-sim*, the idea was to avoid the construction of $\mathcal{T}_j$, by translating the operations over the table to operations directly to the rule blocks information (using the observations made in section 3):

- Operations over columns: they can be transformed to operations for each rule block and the objects appearing in the multisets of the LHS.
- Operations over rows: they can be translated similarly to operations over rows, but the partial results into a global array (one position per row).

Phase 1 can be implemented as described in Algorithm 5.1. Note that FILTER 3 is not needed any more. Although the full table is not created, some auxiliary data structures are used to virtually simulate it (we say it uses a *virtual table*):

- *activationVector*: the information of filtered blocks is stored here as boolean values. The full global size is of order $O(|B| * m * nsim)$, where m is the number of environments and $nsim$ the number of simulations carried out in parallel. This vector is actually implemented passing from boolean to bits.
- *addition*: the total calculated sums for rows are stored here, one number per each pair object and region. Its size is of order $O(|\Gamma| * (q+1) * m * nsim)$.
- *MinN*: the minimum numbers calculated per column are stored here. This is needed in order to substract the corresponding number of applications to C'_t in each loop for the A value. The full global size is of order $O(|B| * m * nsim)$.
- *BlockSel*: the total number of applications for each rule block is stored here. The full global size is of order $O(|B| * m * nsim)$.
- *RuleSel*: the total number of applications for each rule is stored here. The full global size is of order $O(((|R| * m) + |R_E|) * nsim)$, where $|R|$ is the number of rules and $|R_E|$ the number of communication rules.

Algorithm 5.1. Implementation of selection Phase 1 with virtual table

```
1: for j = 1,...,m do                                                      ▷ For each environment
2:     for all block B do
3:         activationVector[B] ← true
4:         if charge(LHS(B)) is different to the one presented C'_t then
5:             activationVector[B] ← false                                 ▷ (Apply FILTER 1)
6:         else if one of the objects in LHS(B) does not exist in C'_t then
7:             activationVector[B] ← false                                 ▷ (Apply FILTER 2)
8:         end if
9:     end for
10:    Check the mutually consistency of blocks.
11:    repeat
12:        for all block B having activationVector[B] = true do          ▷ (Normalization 1)
13:            for each object o^k appearing in LHS(B), associated to region i do
14:                addition[o, i] ← addition[o, i] + k
15:            end for
16:        end for
17:        for all block B having activationVector[B] = true do          ▷ (Normalization 2)
18:            MinN[B] ← Min_{[o^k]_i ∈ LHS(B)}( 1/k^2 * 1/addition[o,i] * C'_t[o, i]).
19:            BlockSel[B] ← BlockSel[B] + MinN[B].
20:        end for
21:        for all block B having activationVector[B] = true do          ▷ (Updating)
22:            C'_t ← C'_t − LHS(B) * MinN[B]
23:        end for
24:        Apply FILTER 2 again (as described in step 6).
25:        a ← a + 1
26:    until a = A or for each active block B, MinN[B] = 0
27: end for
```

The implementation on the device has been constructed directly from Algorithm 5.1. Phase 1 has been implemented using several kernels, avoiding the overload of only one:

- Kernel for Filters (from line 2 to 10 in Algorithm 5.1): FILTERS 1 and 2 are implemented here by using our general CUDA design (Figure 1).
- Kernel for Normalization (from line 11 to 20): the two parts (row additions and minimum calculations) of the normalization step is implemented in a kernel. The two parts are synchronized by *synchtreads* CUDA instruction. The work assigned to threads is divergent; that is, each thread works with one rule block, but writes information for each object appearing in the LHS. Therefore, the writes to *addition* are carried out by atomic operations.
- Kernel for Updating and FILTER 2 (from line 21 to 26). As before, the work of each thread is divergent. Thus, the update of the configuration is also implemented with atomic operations.

5.2 Implementation of Selection Phase 2

Phase 2 is the most challenging part when parallelizing by blocks. The selection of blocks at this phase is performed in an inherently sequential way: we need to know how many objects a block can consume before selecting the next one. In our solution, Phase 2 is implemented by one kernel, using our general CUDA design.

The random order to the blocks is *simulated* by the CUDA thread scheduler: each thread calculates the position in the order of its rule block by using the *atomicInc* operation. Since it does not perform a real random order, random numbers are going to be used soon in next versions. Our first approach (let designate it *ph2-simorder-oneseq*) for phase 2 was to launch 257 threads: 256 threads to calculate the "random" order, and an extra thread to iterate the blocks in that order, selecting and consuming the LHS. Since this approach is still sequentially executed in the GPU, an improved version was constructed.

Our new version (designated *ph2-simorder-dyncomp*) dynamically checks the blocks that are really competing for objects, and calculates which blocks can be selected in parallel, and which depend on the selection of the others. To do this,

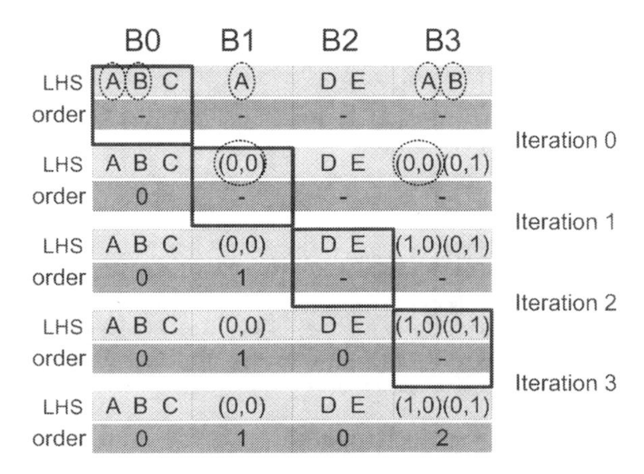

Fig. 2. Sample of our *ph2-simorder-dyncomp* kernel execution

some previous computations are needed. Two arrays are used, one for storing the information of the LHS, and another to store the order of selection (rule blocks having the same order number will be selected in parallel). Both of the arrays are implemented using the GPUs shared memory to speedup this computation. Shared memory on the GPU is one of the on-chip memory spaces which is shared by all the SPs of a SM. Access times to the shared memory are comparable to those of a L1-cache on a traditional CPU. (GPUs also feature high-speed DRAM memory (device memory) with higher latency than on-chip memory (typically hundreds of times slower). The device memory is subdivided in read-write, non-cached (global and local) and read-only, cached (texture) areas.)

Figure 2 shows a sample *ph2-simorder-dyncomp* kernel execution. We iterate for each rule block (using the pre-calculated random order). First, the rule blocks check if they have common objects with block $B0$. In the example, block $B1$ has object A, and block $B3$ has objects A and B. They annotate this competition with the pair $(block, object)$, using the indexes of the array. The current block also calculates the selection order by checking if it has some depending objects. If so, the order is increased by one. For the first iteration, block $B0$ is assigned order 0, but in iteration 1, block $B1$ is assigned order 1 (competing with block $B0$). The rest of the iteration can be seen in Figure 2.

Our experiments shows that *ph2-simorder-dyncomp*, that includes extra computations but allows to execute independent blocks in parallel, achieves up to 20% of performance improvement from *ph2-simorder-oneseq*.

5.3 Implementation of Selection Phase 3

Phase 3 calculates the number of times a rule is applied using a binomial distribution, and the selected block number, both implemented in one kernel.

For random binomial number generation, we have made a *CUDA library* based on *CuRAND*, called *curng_binomial*. This module implements the BINV algorithm proposed by *Voratas Kachitvichyanukul* and *Bruce W. Schmeiser* [12]. Algorithm BINV executes with speed proportional to $n{\cdot}p$ and has been improved by exploiting properties listed in the paper [12]. Also, it has got the best results assuming a normal probability approximation when $n \cdot p > 10$.

In depth, the library implements an *inline device function* which executes binomial randomization (BINV) when $n \cdot p \leq 10$ and normal randomization (CuRAND) otherwise. Our implementation generates binomial random numbers while running the kernel; thus, they are not generated previously.

The implementation of the phase is directly translated from the pseudocode of the DCBA. Also, it has got the best parallelism exploiting until now, comparing to other phases of the algorithm.

5.4 Implementation of Execution (Phase 4)

Phase 4 is implemented as directly shown in the DCBA pseudocode using our general CUDA design. In this case, we go to another level of parallelism for threads, that now works with each rule. As before, threads iterate the rules by tiles, and adding the corresponding RHS (if it has a number of applications $N_r > 0$). Finally, since this operation is divergent (from rules to add objects), we use atomic operations again to update the configuration of the system.

6 Performance Results

In order to test the performance of our simulators, we constructed a random generator of PDP systems (designated *pdps-rand*). These randomly created PDP systems have no biological meaning. The purpose is to stress the simulator in order to analyze the implemented designs with different topologies. *pdps-rand* is parametrized in such a way that it can create PDP systems of a desired size.

We benchmark our *pdp-gpu-sim* and *pdp-omp-sim* (for 1, 2 and 4 cores) by first analyzing the scalability when increasing the size of the system in several ways. We then profile the simulators, showing the percentage of time taken by each phase separately. All experiments are run on a Linux 64-bit server, with a 4-core (2 GHz) dual socket Intel i5 Xeon Nehalem processor, 12 GBytes of DDR3 RAM and two NVIDIA Tesla C1060 graphics cards (240 cores at 1.30 GHZ , 4 GBytes of memory). GPU cores are typically slower than CPU cores.

Figure 3 shows the scalability of the simulator when the number of different objects appearing in the LHS (cooperation degree (see Definition 2)) increases. We can assume that, the greater the cooperation degree, the greater the number of competing blocks generated by *pdps-rand.* The figure shows the simulation time (in milliseconds) for one computation step running 50 simulations of PDP systems with 10 environments, 50000 rule blocks and 5000 different objects. The randomly generated PDP systems are sorted by the mean LHS length, showing that *pdp-gpu-sim* works better for lengths smaller than 3. The speedup achieved by *pdp-gpu-sim* is 6.6x and 2.3x for lengths of 1 and 2 against *pdp-omp-sim* with one core, and 4.5x and 1.9x against *pdp-omp-sim* with 4 cores, respectively.

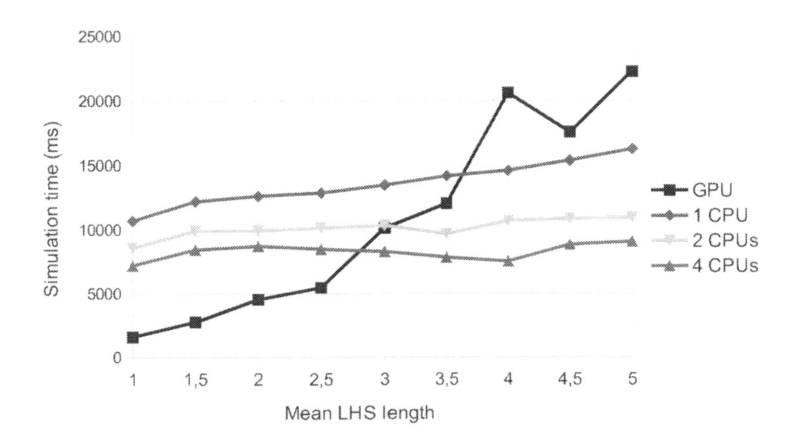

Fig. 3. Scalability when increasing the mean LHS length of rules

The second test analyses the performance when increasing the parallelism level of the CUDA threads within thread blocks, that is, the number of rule blocks. The speedup achieved by *pdp-gpu-sim* versus *pdp-omp-sim* is showed in Figure 4. The number of simulations is fixed to 50, and the environments to 20 (hence, a total of 1000 thread blocks). The number of objects is proportionally increased together with the number of rule blocks, in such a way that the ratio for number of rule blocks and number of objects is always 2. The mean LHS length is 1.5 (this is typical for many real ecosystem models, as seen in the literature). The speedup gets stable to around 7x on the number of rule blocks for the GPU versus CPU. For the multicore versions with 2 and 4 CPUs, the speedups are maintained to 4.3x and 3x, respectively. In our experiments, this number is also achieved when running with 1000000 rule blocks.

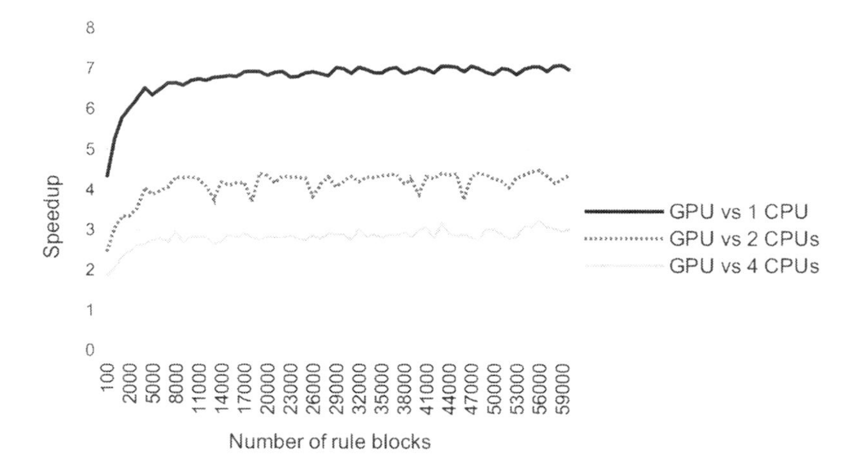

Fig. 4. Scalability of the simulators when increasing the number of rule blocks

The third test is for the second parallelism level in CUDA, concerning thread blocks. It is directly related with the number of environments and simulations. The result is shown in Figure 5. In this experiment, the number of rule blocks is fixed to 10000, the number of objects to 7024 and the mean LHS length is 2. The number of environments is fixed to 1 when increasing the simulations, and vice versa. As it can be seen, for low values, the speedup is demoted below 1. These values come from the fact of insufficient number of thread blocks to fulfill the GPU resources. Another trend shown is that as the number of simulations increases, the advantage of parallelizing by simulations increases. The same effect is observed for environments. This trend is stabilized to 3.5x for high values. However the parallelism over simulations is better carried out by the GPU, giving lower speedups for environments.

As stated in [16], parallelizing by simulations yields the largest speedups on multicore platforms. Therefore, we finalize the first benchmark by comparing these results with the GPU. Rule blocks are fixed to 50000, environments to

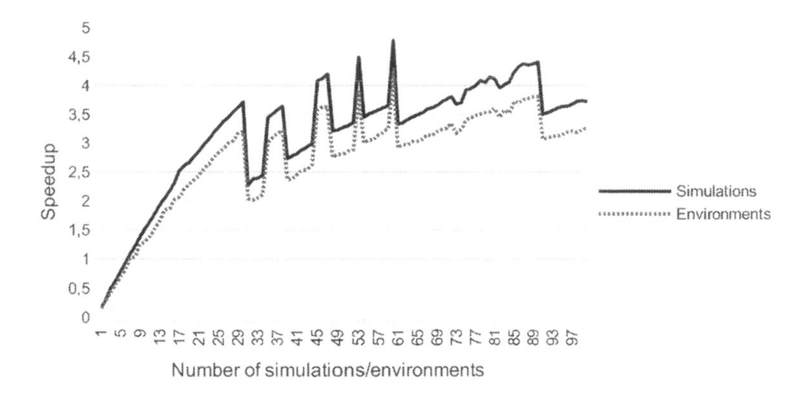

Fig. 5. Scalability – increasing the number of simulations and environments

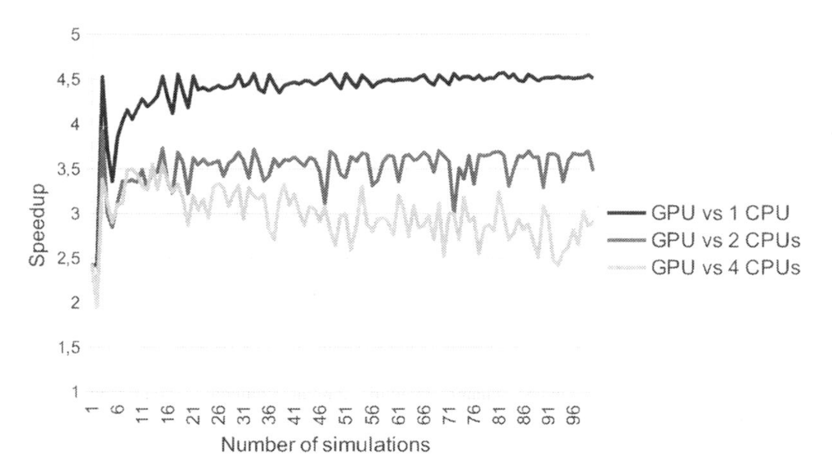

Fig. 6. Scalability of the simulators when increasing the number of simulations

20, objects to 5000 and mean LHS length to 1.5. As shown in Figure 6, the GPU achieves better runtime than the multicore implementations. The speedup is maintained to 4.5x using one core, 3.5x for 2 cores, and 2.7x for 4 cores.

The results of the second benchmark are shown in Table 1. This profile has been calculated running the simulator with 10000 rule blocks, 20 environments, 50 simulations, 5000 objects and two different mean LHS lengths, 1.5 (test A) and 3 (test B). Phase 1 is the most complex part in the simulation (taking more than 50% of the runtime on the CPU). In test A, the GPU implementation offers for phase 1 up to 14x of speedup. Therefore, the percentage of the execution time is decreased to 30%.

Following with Test A, Phase 2 takes only the 12% of the execution time on CPU. However, the GPU can only accelerate this phase by 2x. Therefore, this phase becomes the most expensive when executing the simulator on the GPU (47%). Our novel implementation, *ph2-simorder-dyncomp*, is close (time-wise) to the sequential implementation. Indeed, as mentioned above, this phase

Table 1. Profiling the simulators for GPU and 1 core CPU

	Test A (mean LHS length 1.5)			Test B (mean LHS length 3)		
	% CPU	% GPU	Speedup	% CPU	% GPU	Speedup
Phase 1	53.7%	30.1%	14.23x	55.3%	12%	8.52x
Phase 2	12.6%	47%	2.13x	18.4%	82.8%	0.4x
Phase 3	22.6%	13.7%	13.2x	14%	2.2%	11.72x
Phase 4	11.1%	9.2%	9.7x	12.3%	3%	7.43x

is the most challenging to parallelize. Special efforts have to be considered here. On the other side, Phases 3 and 4 are relatively lightweight, and are successfully accelerated (up to 13x and 9.7x, respectively). Hence, our library random binomial generation based on CuRAND is well suited for Phase 3.

Finally, as shown in Figure 3, the performance of *pdp-gpu-sim* decreases as the mean LHS length is increased. For Test B, the overall speedup decreases from 7.9x (Test A) to 1.8x (Test B). The percentage of time consumed by Phase 2 is dramatically increased for the GPU, taking up to 83%. Thus, the competition degree of rule blocks is a limiting factor in performance, which fully correlates with the achieved results.

7 Conclusions and Future Work

In this paper we have presented the first GPU-based version of a simulator for PDP systems with CUDA. We improved the memory utilization of both our GPU-based version and our previous OpenMP-based version that we benchmarked against. We benchmarked a set of randomly generated PDP systems (without biological meaning), achieving speedups of up to 7x for large sizes on NVIDIA Tesla C1060 GPU over our multi-core version.

We used a general CUDA design for the GPU part of our simulator: environments and simulations are distributed through thread blocks, and rule blocks among threads. Phases 1, 3 and 4 were efficiently executed on the GPU, however Phase 2 was poorly accelerated, since it is inherently sequential.

For future work, the simulator is going to be reconnected with the P-Lingua framework. The pLinguaCore library is able to parse P-Lingua files, simulate P systems computations, and translate P-Lingua files to other file formats. In this respect, a new file format is going to be designed to become the input of our simulator. That way, the same P-Lingua files can contain the input data for the Java simulator inside pLinguaCore and the CUDA simulator. Moreover, we would like to validate our simulator with real ecosystems models. This would also help us to adopt model-oriented heuristics to improve the CUDA design. Finally, it would be useful to design a communication protocol and a Graphics User Interface in order to connect the simulation pipeline with the end-user.

This simulator represents our first attempt on simulating PDP systems using the GPU. Hence, improvements can be done to our new CUDA designs (i.e. iterating from objects to blocks (gather strategy)), optimizing the code, and/or making more efforts on Phase 2. Future work will also take advantage of the new GPU architectures, such as the NVIDIA Kepler, and AMD APU, which deliver better performance than the Tesla C1060 we used for our experiments here.

Acknowledgments. The authors acknowledge the support of "Proyecto de Excelencia con Investigador de Reconocida Valía" of the "Junta de Andalucía" under grant P08-TIC04200, and the support of the project TIN2009-13192 of the "Ministerio de Educación y Ciencia" of Spain, both co-financed by FEDER funds. NVIDIA's support and donations to our HPC-Lab at NTNU is also gratefully acknowledged.

References

1. Aqrawi, A.A., Elster, A.C.: Bandwidth Reduction through Multithreaded Compression of Seismic Images. In: IEEE International Symposium on Parallel & Distributed Processing (IPDPSW 2011), pp. 1730–1739 (2011)
2. Cabarle, F., Adorna, H., Martínez-del-Amor, M.A., Pérez-Jiménez, M.J.: Improving GPU Simulations of Spiking Neural P Systems. Romanian Journal of Information Science and Technology 15(1), 5–20 (2012)
3. Cardona, M., Colomer, M.A., Margalida, A., Palau, A., Pérez-Hurtado, I., Pérez-Jiménez, M.J., Sanuy, D.: A computational modeling for real ecosystems based on P systems. Natural Computing 10(1), 39–53 (2011)
4. Cardona, M., Colomer, M.A., Margalida, A., Pérez-Hurtado, I., Pérez-Jiménez, M.J., Sanuy, D.: A P System Based Model of an Ecosystem of Some Scavenger Birds. In: Păun, G., Pérez-Jiménez, M.J., Riscos-Núñez, A., Rozenberg, G., Salomaa, A. (eds.) WMC 2009. LNCS, vol. 5957, pp. 182–195. Springer, Heidelberg (2010)
5. Cecilia, J.M., García, J.M., Guerrero, G.D., Martínez-del-Amor, M.A., Pérez-Hurtado, I., Pérez-Jiménez, M.J.: Simulation of P systems with Active Membranes on CUDA. Briefings in Bioinformatics 11(3), 313–322 (2010)
6. Cecilia, J.M., García, J.M., Guerrero, G.D., Martínez-del-Amor, M.A., Pérez-Hurtado, I., Pérez-Jiménez, M.J.: Simulating a P system based efficient solution to SAT by using GPUs. Journal of Logic and Algebraic Programming 79(6), 317–325 (2010)
7. Cheruku, S., Păun, A., Romero-Campero, F.J., Pérez-Jiménez, M.J., Ibarra, O.H.: Simulating FAS-induced apoptosis by using P systems. Progress in Natural Science 17(4), 424–431 (2007)
8. Colomer, M.A., Lavín, S., Marco, I., Margalida, A., Pérez-Hurtado, I., Pérez-Jiménez, M.J., Sanuy, D., Serrano, E., Valencia-Cabrera, L.: Modeling Population Growth of Pyrenean Chamois (Rupicapra p. pyrenaica) by Using P-Systems. In: Gheorghe, M., Hinze, T., Păun, G., Rozenberg, G., Salomaa, A. (eds.) CMC 2010. LNCS, vol. 6501, pp. 144–159. Springer, Heidelberg (2010)
9. Colomer, M.A., Pérez-Hurtado, I., Pérez-Jiménez, M.J., Riscos, A.: Comparing simulation algorithms for multienvironment probabilistic P system over a standard virtual ecosystem. Natural Computing, doi:10.1007/s11047-011-9289-2
10. Elster, A.C.: High-Performance Computing: Past, Present, and Future. In: Fagerholm, J., Haataja, J., Järvinen, J., Lyly, M., Råback, P., Savolainen, V. (eds.) PARA 2002. LNCS, vol. 2367, pp. 433–444. Springer, Heidelberg (2002)

11. García-Quismondo, M., Gutiérrez-Escudero, R., Pérez-Hurtado, I., Pérez-Jiménez, M.J., Riscos-Núñez, A.: An Overview of P-Lingua 2.0. In: Păun, G., Pérez-Jiménez, M.J., Riscos-Núñez, A., Rozenberg, G., Salomaa, A. (eds.) WMC 2009. LNCS, vol. 5957, pp. 264–288. Springer, Heidelberg (2010)
12. Kachitvichyanukul, V., Schmeiser, B.W.: Binomial random variate generation. Communications of the ACM 31(2), 216–222 (1988)
13. Kirk, D., Hwu, W.: Programming Massively Parallel Processors: A Hands on Approach, MA, USA (2010)
14. Krog, Ø.E., Elster, A.C.: Fast GPU-Based Fluid Simulations Using SPH. In: Jónasson, K. (ed.) PARA 2010, Part II. LNCS, vol. 7134, pp. 98–109. Springer, Heidelberg (2012)
15. Harris, M.: Mapping computational concepts to GPUs. In: ACM SIGGRAPH 2005 Courses, NY, USA (2005)
16. Martínez-del-Amor, M.A., Karlin, I., Jensen, R.E., Pérez-Jiménez, M.J., Elster, A.C.: Parallel Simulation of Probabilistic P Systems on Multicore Platforms. In: Proceedings of the Tenth Brainstorming Week on Membrane Computing, vol. II, pp. 17–26 (2012)
17. Martínez-del-Amor, M.A., Pérez-Hurtado, I., García-Quismondo, M., Macías-Ramos, L.F., Valencia-Cabrera, L., Romero-Jiménez, A., Graciani, C., Riscos-Núñez, A., Colomer, M.A., Pérez-Jiménez, M.J.: DCBA: Simulating Population Dynamics P Systems with Proportional Object Distribution. In: Proceedings of the Tenth Brainstorming Week on Membrane Computing, vol. II, pp. 27–56 (2012)
18. Martínez-del-Amor, M.A., Pérez-Hurtado, I., Pérez-Jiménez, M.J., Riscos-Núñez, A., Sancho-Caparrini, F.: A simulation algorithm for multienvironment probabilistic P systems: A formal verification. International Journal of Foundations of Computer Science 22(1), 107–118 (2011)
19. Păun, G.: Computing with membranes. Journal of Computer and System Sciences 61(1), 108–143 (2000), TUCS Report No 208
20. Păun, G., Rozenberg, G., Salomaa, A. (eds.): The Oxford Handbook of Membrane Computing. Oxford University Press (2010)
21. Spampinato, D.G., Elster, A.C.: Linear optimization on modern GPUs. In: IEEE International Symposium on Parallel & Distributed Processing (IPDPS 2009), pp. 1–8 (2009)
22. Terrazas, G., Krasnogor, N., Gheorghe, M., Bernardini, F., Diggle, S., Cámara, M.: An Environment Aware P-System Model of Quorum Sensing. In: Cooper, S.B., Löwe, B., Torenvliet, L. (eds.) CiE 2005. LNCS, vol. 3526, pp. 479–485. Springer, Heidelberg (2005)
23. NVIDIA CUDA programming guide 4.0 (2012), `http://www.nvidia.com/cuda`
24. The P-Lingua web page, `http://www.p-lingua.org`

Approximate Bisimulations for Sodium Channel Dynamics

Abhishek Murthy[1], Md. Ariful Islam[1], Ezio Bartocci[2], Elizabeth M. Cherry[5], Flavio H. Fenton[3], James Glimm[4], Scott A. Smolka[1], and Radu Grosu[2]

[1] Department of Computer Science, Stony Brook University
[2] Department of Computer Engineering, Vienna University of Technology
[3] Department of Biomedical Sciences, Cornell University
[4] Department of Applied Mathematics and Statistics, Stony Brook University
[5] School of Mathematical Sciences, Rochester Institute of Technology

Abstract. We show that in the context of the Iyer et al. 67-variable cardiac myocycte model (IMW), it is possible to replace the detailed 13-state probabilistic model of the sodium channel dynamics with a much simpler Hodgkin-Huxley (HH)-like two-state sodium channel model, while only incurring a bounded approximation error. The technical basis for this result is the construction of an *approximate bisimulation* between the HH and IMW sodium channel models, both of which are input-controlled (voltage in this case) CTMCs.

The construction of the appropriate approximate bisimulation, as well as the overall result regarding the behavior of this modified IMW model, involves: (1) Identification of the voltage-dependent parameters of the m and h gates in the HH-type channel via a two-step fitting process, carried out over more than 22,000 representative observational traces of the IMW channel. (2) Proving that the distance between observations of the two channels is bounded. (3) Exploring the sensitivity of the overall IMW model to the HH-type sodium-channel approximation. Our extensive simulation results experimentally validate our findings, for varying IMW-type input stimuli.

1 Introduction

The emergence of high throughput data acquisition equipment has changed cell biology from a purely wet lab-based science to also an engineering and information science. The identification of a mathematical model from cellular experimental data, and the use of this model to predict and control the cell's behavior, are nowadays indispensable tools in cell biology's arsenal [35,5].

Continual progress in data acquisition has also led to the creation of increasingly sophisticated partial Differential Equations Models (DEMs) for cardiac cells (myocytes). These are similar in spirit to the DEMs used in physics: their main purpose is to elucidate the biological laws governing the electric behavior of cardiac myocytes, i.e., their underlying cellular and ionic processes [9].

D. Gilbert and M. Heiner (Eds.): CMSB 2012, LNCS 7605, pp. 267–287, 2012.

Inspired by the squid-neuron DEM [19] developed by Hodgkin and Huxley (HH), Luo and Rudy (LR) devised one of the first myocyte DEMs, for guinea pig ventricular cells [29]. Adapting this model to human myocytes led to the ten Tusscher-Noble^2-Panfilov (TNNP) DEM [40], which has 17 state variables and 44 parameters. Based on updated experimental data, Iyer, Mazhari and Winslow (IMW) subsequently developed a DEM comprising of 67 state variables and 94 parameters [20]. This DEM reflects a highly detailed physiological view the electrochemical behavior of human myocytes.

From 17 to 67 variables, all such DEMs capture myocytic behavior at a particular level of abstraction, and hence all of them play an important role in the modeling hierarchy. It is essential, however, to maintain focus on the purpose of a particular DEM; that is, of the particular cellular and ionic processes whose behavior the DEM is intended to capture. Disregarding this purpose may lead to the use of unnecessarily complex DEMs, which may render not only analysis, but also simulation, intractable.

If the only entity of interest is the myocyte's transmembrane voltage, co-authors Cherry and Fenton have experimentally shown that a minimal DEM (MM) consisting of only 4 variables and 27 parameters can accurately capture voltage propagation properties in 1D, 2D, and 3D networks of myocytes [4]. The MM has allowed us to obtain dramatic simulation speedups [1], and to use its linear hybridization as the basis for formal symbolic analysis [18].

Since new technological advances are expected to lead to further insights into myocytic behavior, it is likely that the IMW model will be further refined by adding new variables. As in model checking and controller synthesis, one would therefore like to compute the smallest approximation of the State Of the Art DEM (SOA) that is observationally equivalent to the SOA with respect to the property of interest, modulo some bounded approximation error. This, however, is not easily accomplished, as it implies the automatic approximation of very large nonlinear DEMs.

A first step toward the desired automation is to identify a set of approximation techniques that allow one to systematically remove unobservable variables from, say, the SOA to end up with the MM, if the only observable variable is the voltage. This is one of the goals of the project Computational Modeling and Analysis of Complex Systems (CMACS) [36]. A byproduct of this work is to establish a long-missing formal relation among the existing myocyte DEMs, facilitating the transfer of properties established at one layer of abstraction to the other layers. Building such *towers of abstraction* is becoming increasingly prevalent in systems biology [22,11].

The main focus of this paper is on *sodium channel approximations*. In the HH DEM and Noble's DEM of [37], the transmembrane sodium channel is assumed to consist of four independent Markovian gates, whose opening and closing rates depend on the transmembrane voltage. The probability of each of the three identical activating (m-type) gates being open, i.e. a state favoring ion flow, is denoted by m, and the probability of the fourth inactivating (h-type) gate being open is denoted by h. The sodium channel conducts when all the four gates are in the open state.

The IMW model uses the formulation of Irvine et al. [28], where experimental data is used to show the existence of five interdependent gates. This leads to a considerably larger Markovian model for the sodium channel, consisting of 13 state variables.

The main question posed in this paper is the following: *Assuming that the conductance of the sodium channel is the only observable, is the behavior of the HH channel equivalent to the behavior of the IMW channel, modulo a well-defined approximation error?* Rather than dealing with behavioral equivalence explicitly, we ask if it is possible to construct an approximate bisimulation [12,14,13,15] between the discrete-time versions of the HH and IMW channel models? This notion of equivalence is stronger than the conventional behavioral equivalence, which compares the observed behaviors (trajectories) of two systems.

Moreover, proving the two models to be approximately bisimilar ensures that when the 13-state sodium channel model is replaced by the 2-state HH-type abstraction in the overall IMW cardiac cell model, the modified IMW model retains the properties of interest (in discrete time). Thus, the reduced 2-state model is a valid reduction in the context of the whole-cell IMW model.

The answer to the above-posed question is of broad interest, as it reduces to showing the existence of an approximate bisimulation between two Continuous-Time Markov Decision Processes (CT-MDPs); that is, two input-controlled (voltage in this case) continuous-time Markov chains (CTMCs). We answer this question in the positive, by explicitly constructing such a bisimulation.

The construction involves: (1) The identification of the voltage-dependent parameters of the m and h-type gates of the HH-type abstraction, based on the observations of the IMW channel. (2) Proving that the distance between the observations of the two channels never exceeds a given error. (3) Exploring the sensitivity of the overall IMW DEM to the HH-type sodium-channel approximation.

The identification of the voltage-dependent parameters is performed via a two-step fitting process. In the first step, which we call *Parameter Estimation from Finite Traces*, more than 22,000 observational traces of the IMW channel are fit to obtain the parameter values at constant voltage. The second step, which we call *Rate Function Identification*, combines the step-1 constant-voltage parameter values to obtain the voltage-dependent parameters defining the HH-type channel. Finally, the resulting two-state HH-type channel is proved to be approximately bisimilar to the IMW channel and the error between the two systems is bounded. See Fig. 1 for an overview of our approach.

The rest of the paper is organized as follows. Section 2 introduces the relevant background for the HH and the IMW DEMs and their sodium-channel MDP formulations. Section 3 presents our parameter identification technique and the resulting HH-type MDP for the sodium channel. Section 4 proves the existence of an approximate bisimulation between the HH and IMW sodium-channel MDPs. Sections 5 and 6 discuss related work, our conclusions, and future directions.

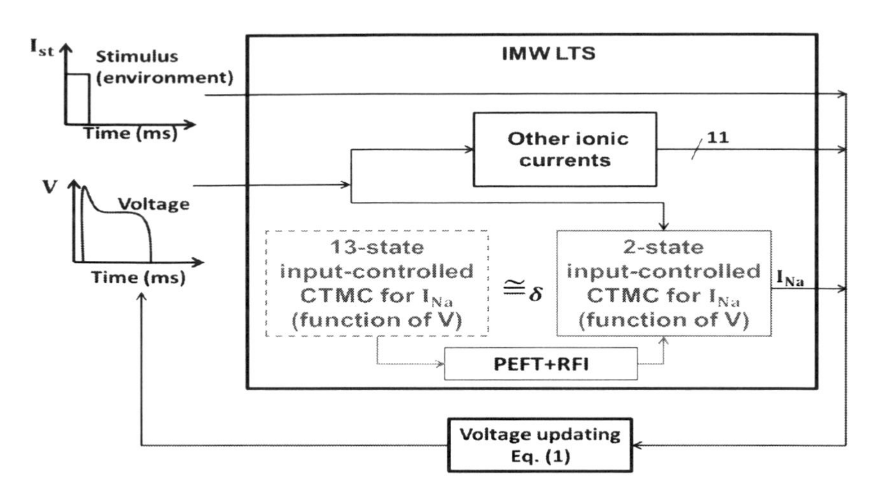

Fig. 1. A Labeled Transition System (LTS)-based view of the IMW DEM, composed of various concurrently evolving subsystems corresponding to the different ionic currents. We replace the 13-state I_{Na} subsystem with a 2-state HH-type abstraction. PEFT and RFI are the two steps of identifying the abstract model. As the 2-state model is proved to be approximately bisimilar (denoted by $\cong_\delta$) to the detailed model, composing it with the other concurrently evolving ionic current models (subsystems) retains the cell-level behaviors of the IMW model. Note that the subsystems ignore the stimulus input during their respective evolution and only depend upon the voltage input. The LTS outputs the 13 currents in Eq. (1).

2 Background

The heart is the central organ of the circulatory system and is responsible for pumping blood in the pulmonary and systemic circulation loops [8]. Pumping is achieved through the synchronized contraction of around four billion myocytes, which constitute the cardiac tissue. This is controlled in a distributed fashion, through the propagation and reinforcement of an electric pulse (clock). The pulse originates in the sino-atrial node of the heart and diffuses from one myocyte to the other through a sophisticated communication infrastructure.

Cardiac myocytes belong to the class of excitable cells, which also includes neurons. Such cells respond to an external electrical stimulus in the form of an Action Potential (AP), which measures the change of the transmembrane potential with time in response to the stimulus. A typical ventricular myocyte AP and its associated phases are shown in Fig. 2(Right). Starting from the resting state, a myocyte can either be excited by an external stimulus or by the diffusing charge of the neighboring myocytes. In this paper, we will restrict our focus on the upstroke phase of the AP.

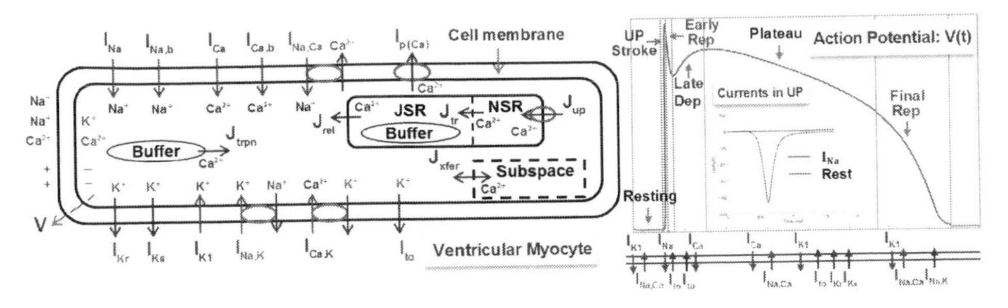

Fig. 2. (Left) Currents in IMW: Blue and brown arrows show ionic currents flowing through channels. Blue circles and arrows correspond to ionic exchanger currents and green circles denote ionic pumps. Intra-cellular currents are shown in Magenta. (Right) The Action Potential (AP), its phases and associated currents. (Right-Inlay) Sodium current in red, and the sum of all other currents in blue, in Upstroke Phase (UP).

2.1 The IMW Cellular Model

The IMW DEM is a physiologically detailed model capturing the ionic processes responsible for the generation of an AP in human ventricular myocytes:

$$\begin{aligned} -C\dot{V} = I_{Na} + I_{Na_b} + I_{Ca} + I_{Ca_b} + I_{K_r} + I_{K_s} + I_{K_1} + I_{to_1} + I_{p(Ca)} + \\ I_{NaCa} + I_{NaK} + I_{CaK} + I_{st} \end{aligned} \quad (1)$$

where V is the membrane's potential, $\dot{V}$ is its first-order time derivative, C is the membrane's capacitance, and I_v are the ionic currents shown in Fig. 2(Left), except for I_{st}. This is the stimulus current, which could be either an external stimulus or the diffused charge from neighboring cells.

The remaining currents are the result of the flow of the sodium (Na^+), potassium (K^+) and calcium (Ca^{2+}) ions, across the myocyte's membrane. Three types of transport mechanisms are responsible for the ion flows: channels, pumps and exchangers. Channels are special proteins that penetrate the membrane's lipid bi-layer, and are selectively permeable to ions species. Depending on the conformation of the constituent protein, the channel either allows or inhibits the unidirectional movement of an ion specie.

The protein conformation is voltage dependent, thus the name voltage-gated channels. All the transmembrane currents in Fig. 2 result from voltage-gated ionic channels, except for I_{NaK}, I_{NaCa} and $I_{p(Ca)}$, which are exchanger or pump currents. The concentration of calcium is regulated by a sophisticated intracellular mechanism, and is out of scope of this paper.

Fig. 2(Right-inlay) plots the sodium current I_{Na} and the sum of all the other ionic currents during the upstroke phase (UP), of a typical AP of the IMW DEM. The sodium current I_{Na} dominates all the other. *The behavior of the sodium channel, which regulates the flow of I_{Na}, chiefly contributes to the upstroke phase, and will be the focus in the remainder of the paper.* In the HH DEM the situation is similar, and in MM the role of I_{Na} is played by the abstract fast inward current J_{fi}.

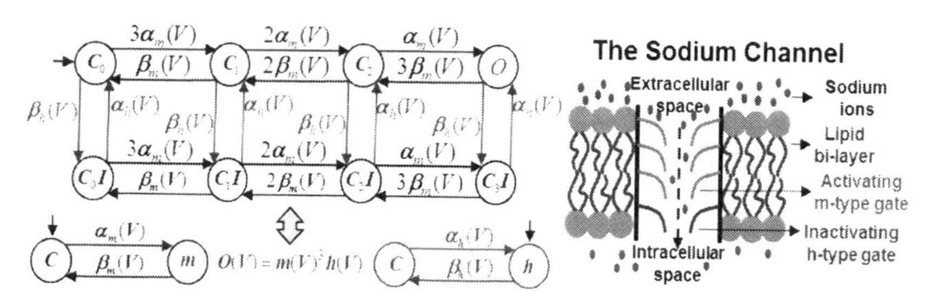

Fig. 3. (Left-top) Sodium channel MDP in 8 states counting the number of independent open/closed gates, and observation function $O(t)$. (Left-bottom) The open-closed MDPs for the m and h-type gates. The equivalent sodium channel behavior is obtained as $O(t) = m(V)^3h(V)$. (Right) The schematic representation of the sodium channel with its associated independent gates.

2.2 The HH Sodium Current

The sodium current I_{Na} in the HH DEM is defined by the following equation:

$$I_{Na} = \overline{g}_{Na}\, m^3(V)\, h(V)\, (V - V_{Na})$$

where $\overline{g}_{Na}$ is the maximum conductance of the sodium channel, V_{Na} is the sodium's channel Nernst potential, $m(V)$ and $h(V)$ are the probabilities of the voltage-dependent activation gate and the inactivation gates being open respectively.

A graphic illustration of the sodium channel is given in Fig. 3 (Right). It consists of four independent voltage-controlled gates, three of which are identical activation gates (m-type), and one of which is an inactivation gate (h-type).

The activation and inactivation gates are shown in Fig. 3 (Left-bottom). They are Continuous Time Markov Decision Processes (CT-MDP). Both CT-MDPs have a closed and an open state, respectively, and the rates of transitioning between these two states are given by the voltage-dependent parameters $\alpha(V)$ and $\beta(V)$. The 8-state CT-MDP for the whole channel is shown in the left-half of Fig. 3. Evolution of the state variables (occupancy probabilities of the 8 states) of this model is governed by Kolmogorov equations[21], which form an 8-state DEM. It turns out that any of the 8-variables can be observed using the two gates m and h as they form a stable invariant manifold of the 8-state DEM [23]. At rest the m-gate is closed and the h-gate is open. Their DEM is as follows:

$$\dot{m} = \alpha_m(V)(1 - m) - \beta_m(V)m, \qquad \dot{h} = \alpha_h(V)(1 - h) - \beta_h(V)h$$

We refer to this DEM as M_H. The linear system obtained by fixing $V = v$ will be denoted as M_H^v. At any point in time the occupancy probability of the open state O in the 8-state DEM is given by $m(V)^3h(V)$. Thus the observation function O of this DEM will be $m(V)^3h(V)$. We now introduce the following notation:

$$x = [m, h]', \quad A = diag(-(\alpha_m + \beta_m), -(\alpha_h + \beta_h)), \quad B = [\alpha_m, \alpha_h]'$$

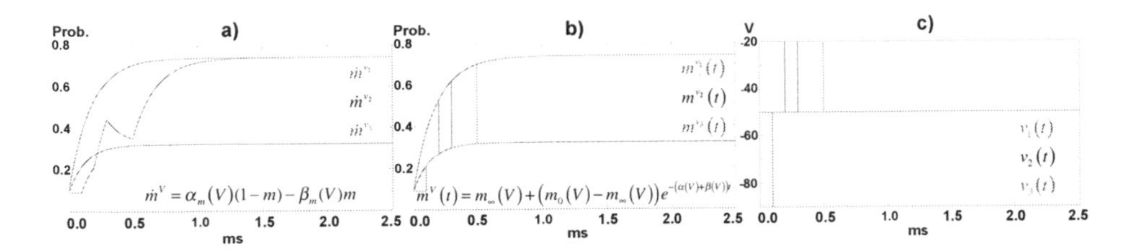

Fig. 4. Probability for the m-gate to be open in HH: a) Numerical integration of m for different voltage changes; b) Analytical solution of m for different voltage changes; c) Voltage changes applied for the analytical and the numerical integration solutions.

The independence of the gates also implies that the DEM is in diagonal form, and it can be therefore written as follows:

$$\dot{x} = Ax + B, \qquad x_0 = [m_0, h_0]'$$

Despite the linear-looking form, this equation is nonlinear, as A and B depend on the voltage. For example, Fig. 4(a) shows its numeric solution for the input in Fig. 4(c). However, HH computed an approximate closed form solution as follows. In the resting state, defined as $V = 0$, and in the equilibrium state, for a fixed $V = v$, the gates m and h, and the rates τ have the following values:

$$\begin{array}{ll} m_0 = \alpha_{m_0}/(\alpha_{m_0} + \beta_{m_0}), & m_\infty = \alpha_m/(\alpha_m + \beta_m) \\ h_0 = \alpha_{h_0}/(\alpha_{h_0} + \beta_{h_0}), & h_\infty = \alpha_h/(\alpha_h + \beta_h) \\ \tau_m = 1/(\alpha_m + \beta_m), & \tau_h = 1/(\alpha_h + \beta_h) \end{array}$$

Then solving the DEM above as if A and B were constant and the differential equation therefore linear, Hodgkin and Huxley derived the following solution:

$$x = [m_\infty - (m_\infty - m_0)e^{-t/\tau_m}, h_\infty - (h_\infty - h_0)e^{-t/\tau_h}]'$$

As shown in Fig. 4(b) this closed-form solution jumps for a changing input shown in Fig. 4(c) between the solutions obtained for constant input. This behavior is however not problematic when replaced in the cellular model, as the voltage only jumps at the beginning, when the stimulus is applied, and then varies in a continuous way.

2.3 The IMW Sodium Current

The sodium current I_{Na} in the IMW DEM is defined by the following equation:

$$I_{Na} = \overline{g}_{Na}\,(O_1(V) + O_2(V))\,(V - V_{Na}) \qquad (2)$$

where $\overline{g}_{Na}$ and V_{Na} have the same meaning as in the HH DEM, $O_1(V)$ and $O_2(V)$ are occupancy probabilities of the two states of the MDP shown in Fig. 5.

The IMW view of the sodium channel is shown Fig. 5 [26,28], with transition rates in Table 1. There are now four identical m-type gates, and the transition

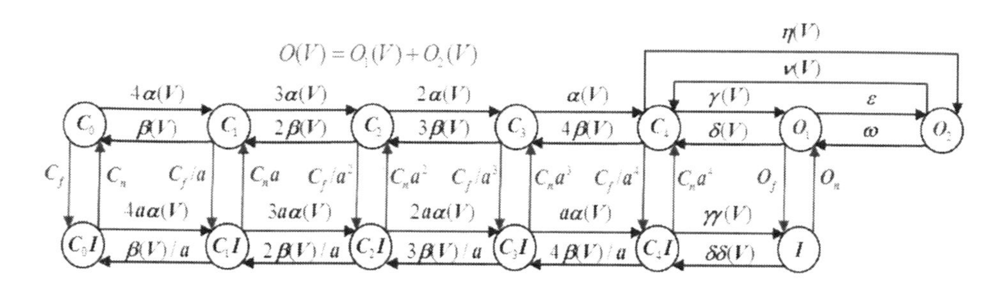

Fig. 5. The 13-variable MDP of the IMW model. The observation function is now $O = O1 + O_2$, and the transition rates of the h-type gate are constants. However, they depend on the number of open m-type gates through a. The transition rates are defined in Table 1.

rates of the h-type gate are constant. However, these rates indirectly depend on V through the number of open-closed m gates (encoded as powers of a).

Moreover, taking the path C_0, C_1, C_1I, C_0I is mathematically equivalent to taking a voltage dependent h-transition C_0, C_0I. The longer the paths, the less one can distinguish between the HH-type and the IMW-type transition. Note also that two states O_1 and O_2 are now observable instead of one, and some bookkeeping was also added.

Definition 1. *Consider the 13-state model for sodium-channel dynamics shown in Fig. 5. Let p_j denote the jth state occupancy probability from the vector* $\mathbf{p} = (C_0, C_1, C_2, C_3, C_4, O_1, O_2, C_0I, C_1I, C_2I, C_3I, C_4I, I)$. *The* ***dynamics of the model*** *M_I is described by the following system of differential equations :*

$$\frac{dp_j}{dt} = \sum_{i \neq j} k_{ij}(V)p_i - \sum_{i \neq j} k_{ji}(V)p_j \qquad i, j = 1, \ldots, 13 \tag{3}$$

where V is the transmembrane potential and $k_{ij}(V)$ is the transition rate from the ith to the jth state as defined in Table 1. This system can be re-written as:

$$\frac{dp_j}{dt} = A(V).\mathbf{p}, \tag{4}$$

where $A(V)$ is a 13×13 matrix with $A_{j,i}(V) = k_{ij}(V)\ i \neq j$, $A_{j,j}(V) = -\sum_{i \neq j} k_{ji}$. The linear system M_{I^v} is obtained from M_I by fixing $V = v$ in Eq. 4.

Table 1. Rates of the 13-state CT-MDP M_I shown in Fig. 5. $c = 8.513 \times 10^9$. Values instantiated from Table 6 of [20] at temperature T = 310K.

rate	function	rate	function	rate	function
$\alpha(V)$	$c.e^{-19.6759+0.0113V}$	$\delta\delta(V)$	$c.e^{-38.4839-0.1440V}$	ϵ	0.0227
$\beta(V)$	$c.e^{-26.2321-0.0901V}$	$\gamma\gamma(V)$	$c.e^{-21.9493+0.0301V}$	ω	1.0890
$\gamma(V)$	$c.e^{-16.5359+0.1097V}$	$\eta(V)$	$c.e^{-19.6729+0.0843V}$	c_n	0.7470
$\delta(V)$	$c.e^{-27.0926-0.0615V}$	$O_n(V)$	$c.e^{-20.6726+0.0114V}$	c_f	0.2261
$\nu(V)$	$c.e^{-26.3585-0.0678V}$	$O_f(V)$	$c.e^{-39.7449+0.0027V}$	a	1.4004

3 Abstraction of Sodium Channel Dynamics

We construct an HH-type DEM M_H that can be substituted for M_I within the IMW cardiac-cell model. We perform the following abstractions in this process:

- We reduce the number of activating subunits to 3 and use a single inactivating subunit. This results in abstracting away the I, C_3I, C_4I, C_3 and C_4I, states of 13-state CTMDP in Figure 5.
- We coalesce the two open states into a single open state O.
- We abstract away the conditional dependence between activating and inactivating subunits of the 13-state model M_I. This is done by abstracting away the scaling factor a.
- With the above abstractions, M_I reduces to the 8-state CTMDP. The 8-state abstraction then reduces to the 2-state DEM M_H model due to the invariant manifold reduction.

Our approach to obtaining the 2-state HH-type abstraction M_H from the 13-state physiological model M_I is summarized in Fig. 6 and described next.

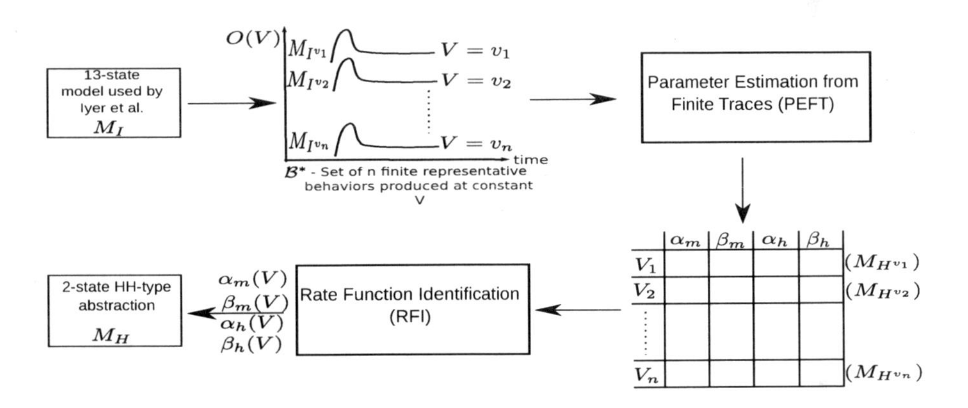

Fig. 6. Abstraction process for sodium channel dynamics

1. **Generating Representative Finite Traces of M_I**
The IMW model was simulated in FORTRAN for a single cell with an integration time step of 10^{-4} ms. Multiple M_{I^v} systems were simulated for the values of V observed during the FORTRAN simulation. The linear system M_{I^v} was simulated in MATLAB using the *ODE45* solver [32]. The integration time step for these simulations was 10^{-2} ms. The simulations ran till the steady state was reached. The initial condition for all the simulations were taken to be the initial condition specified in Table 4 of [20]. The motivation for these initial conditions lies in the voltage-clamp experiments performed in [19]. In these experiments, the voltage was initially maintained at the resting potential, with the neuron conductance also being in the resting state. The voltage was suddenly increased to a specified value and the evolution of conductance was observed till steady state.

The simulations resulted in a set $\mathcal{B}^*$, of finite-length representative behaviors (traces). Each member $\mathcal{B}^*(v)$ is the trajectory of the simulation of M_I^v.

2. **Parameter Estimation from Finite Traces (PEFT)**
This routine takes $\mathcal{B}^*$ as the input and at each of the voltage values v, estimates the parameters of M_{H^v}, the two-state HH model (M_H) at $V = v$. For each voltage v, the following optimization problem was solved to estimate the parameters α_m^v, β_m^v, α_h^v and β_h^v of M_{H^v}:

$$\begin{aligned} &\text{minimize} \sum_{t=0}^{t_S^v} [O^v(t) - m^v(t)^3 h^v(t)]^2 \\ &\text{subject to:} \quad \alpha_m^v, \beta_m^v, \alpha_h^v, \beta_h^v \geq 0 \end{aligned} \tag{5}$$

where
- t is the discrete-time step,
- t_S^v is the number of discrete-time steps taken by M_{I^v} to reach steady state (M_{I^v} was simulated in MATLAB till steady-state),
- $O^v(t) = O_1^v(t) + O_2^v(t)$ is the sum of the occupancy probabilities of states O_1 and O_2 in the trajectory $\mathcal{B}^*(v)$ and
- $m^v(t)$, $h^v(t)$ define a trajectory of M_H^v:

$$\begin{aligned} m^v(t) &= \frac{\alpha_m^v}{\alpha_m^v + \beta_m^v} + \left(m^v(0) - \frac{\alpha_m^v}{\alpha_m^v + \beta_m^v} \right) \exp\left(-\left(\alpha_m^v + \beta_m^v\right) t\right) \\ h^v(t) &= \frac{\alpha_h^v}{\alpha_h^v + \beta_h^v} + \left(h^v(0) - \frac{\alpha_h^v}{\alpha_h^v + \beta_h^v} \right) \exp\left(-\left(\alpha_h^v + \beta_h^v\right) t\right) \end{aligned} \tag{6}$$

where $m^v(0)$ and $h^v(0)$ denote the initial conditions.

We used MATLAB's constrained-optimization solver *FMINCON* [33] for Eq. (5). Details of the active-set optimization algorithm implemented in the function can be found in [30]. Three aspects of our implementation deserve further elaboration:

- **Choosing** $m^v(0)$ **and** $h^v(0)$ - In [19], the authors choose the initial conditions for all the voltages such that the inactivating gating variable h is high and the activating gating variable m is low. We use the same convention but ensure that the initial conductance (observation) $m^v(0)^3 h^v(0) = O^{V_{res}}$, where $O^{V_{res}}$ is the conductance $O_1 + O_2$ of M_I at the resting potential V_{res}. Specifically, we use $m^v(0) = 0.0026$ and $h^v(0) = 0.95$ for all v.
- **Providing seed-values** - For each voltage-value v, FMINCON needs seed values of α_m^v, β_m^v, α_h^v and β_h^v to start optimizing over the parameter space. We implemented a local search strategy for this purpose. The parameters estimated at v_i were used as seed-values for v_{i+1}. For the resting potential, when $i = 1$, the seed values were taken by evaluating Eq. (16)-(18) of [37] at $V = -90.66mV$ (the resting potential).
- **Local minima** - The solver is guaranteed to provide parameter values that locally minimize the objective function. FMINCON was run multiple times until the objective function was minimized to a value below

a pre-defined threshold. The terminal values of an iteration were perturbed and used as seed-values for the next iteration. A maximum of 100 iterations were performed.

PEFT resulted in a table of parameters θ, again indexed by voltage, i.e. θ^v contained the parameters of M_{H^v}.

3. **Rate-Function Identification (RFI)**
RFI combines the parameters θ^v of M_{H^v} and outputs the parameter functions of M_H which are functions of V.This is done by identifying appropriate forms for the parameter functions $\alpha_m(V)$, $\beta_m(V)$, $\alpha_h(V)$ and $\beta_h(V)$ and then using MATLAB's curve-fitting toolbox [31] to estimate the parameters of the chosen form.

$$\alpha_m(V) = -0.6 + \frac{16.31}{1 + exp(-0.05(V + 19.67))} \tag{7}$$

$$\alpha_h(V) = \begin{cases} 0.07 + \frac{0.11}{1+exp(0.2495(V+53.01))} & V \leq -32.00 \\ 0.07 - \frac{0.06}{1+exp(-0.07(V-6.73))} & V > -32.00 \end{cases} \tag{8}$$

$$\beta_h(V) = -4.8 + \frac{145.1}{1 + exp(-0.013(V - 179))} \tag{9}$$

$$\beta_m(V) = \begin{cases} 9.92 - \frac{4.575}{1+exp(-73.73(V+63.78))} & V \leq -60.28 \\ 2.32 + \frac{2.512}{1+exp(0.2173(V+50.69))} & -60.28 < V \leq -33.04 \\ 2.26 + \frac{1.63}{1+exp(-0.2(V+20.72))} & -33.04 < V \leq -1.823 \\ -2.57 + \frac{6.73}{1+exp(0.07(V-40.23))} & V > -1.823 \end{cases} \tag{10}$$

Empirical Validation of the Reduced Model M_H

The 13-state model M_I was substituted by M_H in the IMW model. The modified IMW model was simulated in FORTRAN. This modified model used M_H to produce the sodium current I_{Na}. Both supra and sub-threshold stimuli, lasting for 0.5ms, were used to excite the cardiac cell. S1 and S2 denote supra-threshold stimuli of -100 pA/pF and -120 pA/pF respectively. S3 and S4 denote sub-threshold stimuli of -10 pA/pF and -20 pA/pF.

The results plotted in Fig. 7 show the behavioral equivalence of M_H and M_I. The model retains both normal and anomalous cell-level behaviors on replacing the 13-state sodium-channel component with the 2-state abstraction within the complete cell model.2

4 Approximate Bisimulation Equivalence of M_I and M_H

We use PEFT and RFI to obtain M_H, the two-state HH-type abstraction of the 13-state model for sodium-channel dynamics M_I. We formalize the discrete-time equivalence of M_H and M_I using approximate bisimulation [15].

The approximate bisimulation relation between the state-spaces of the systems can be utilized for gaining physiological insights from formal analysis. Analysis

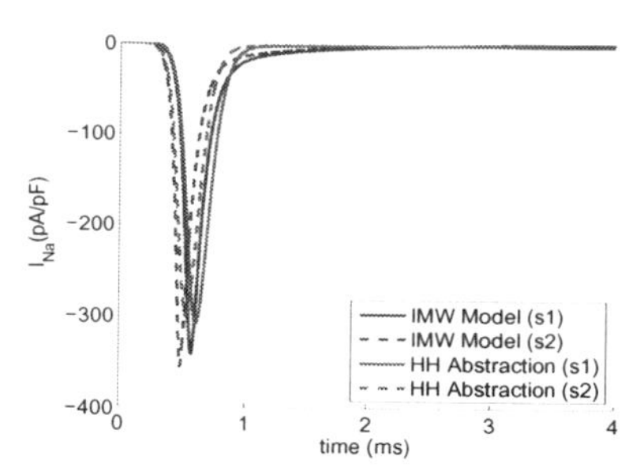

(a) Comparison of I_{Na} during the upstroke phase.

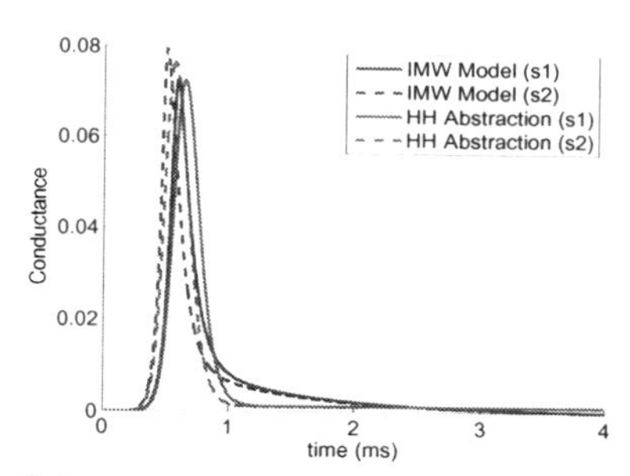

(b) Comparison of conductances $(O_1 + O_2)$ of the 13-state model M_I and (m^3h) of the 2-state abstraction M_H during the upstroke phase.

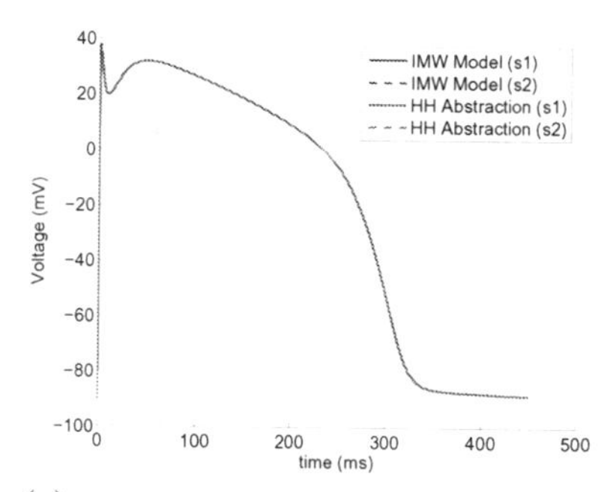

(c) Comparison of AP produced by the original IMW model and the modified version for supra-threshold stimuli.

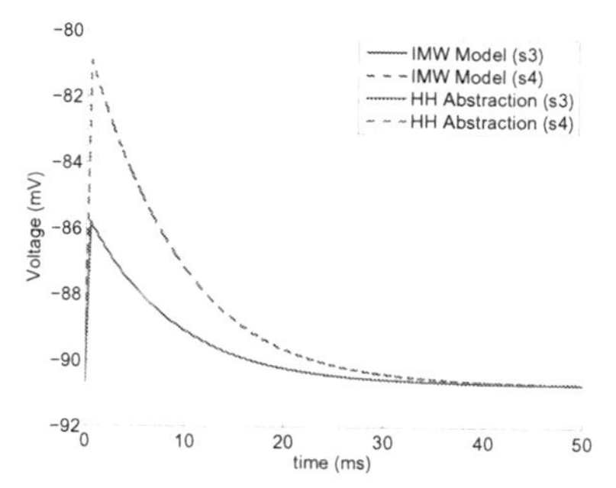

(d) Comparison of AP produced by the original IMW model and the modified version for sub-threshold stimuli.

Fig. 7. Comparison of M_I and M_H when used for I_{Na} in the IMW model. We do not show the currents and conductances for sub-threshold stimuli as they are negligible. Mean L2 errors over the duration of an AP for all stimuli: Conductance: 3.2×10^{-5}, Current: 0.1249 pA/pF, V: 0.12mV.

can be done on the abstract model M_H and the results can be interpreted in the state-space of the physiological model M_I.

In [15], Pappas et al. define approximate bisimulation equivalence of Labeled Transition Systems (LTS), a generic modeling framework. We cast the models M_H and M_I as LTSs and prove approximate bisimulation equivalence of their discrete time versions. First we will establish stability properties of M_{I^v}. We use V_{res} and V_{max} to denote the resting potential and maximum potential attained at the end of the upstroke (UP) phase.

Definition 2. *A $m \times m$ square matrix M is called a closed compartmental matrix if the the following two properties are satisfied:*

1. *$M_{ij} \geq 0$ for $i \neq j$ - Non-diagonal entries are non-negative.*
2. *$\sum_{j=1}^{n} M_{ji} = 0, \quad 1 \leq i \leq m$ - sum of the entries in each column is 0.*

Lemma 1. *Let A^v be the constant matrix obtained by fixing $V = v$ in Eq. (4), where $v \in [V_{res}, V_{max}]$. A^v is a closed compartmental matrix for all $v \in [V_{res}, V_{max}]$.*

Proof. The first condition in Lemma 1 is met by construction.

For every column i, for $i \neq j$, A_{ji} is to the outgoing transition rate from state i to state j: $k_{ji}(V)$. The diagonal entry in the ith column is the negated sum of all these outgoing rates, which satisfies the second condition. □

Lemma 2. *The matrix A^v, obtained by fixing $V = v$, is irreducible for all possible voltage values $v \in [V_{res}, V_{max}]$.*

Proof. A graph-theoretic proof can be made by first inducing a graph from the matrix A^v. Let $G^v(N, E)$ be the graph such that there is a node in the graph for each of the 13 states in the stochastic model in Fig. 5 and an edge $(n_i, n_j) \in E$ if and only if $A^v_{ij} \neq 0$.

Proving that G^v remains connected at all values of V, amounts to proving irreducibility of A^v. This is indeed true because of the exponential functions in Table 1. The graph G^v remains connected for all values $v \in [V_{res}, V_{max}]$. □

Theorem 1. *The model M_{I^v} has a stable equilibrium for $v \in [V_{res}, V_{max}]$.*

Proof. It follows from Proposition 4 in [21]. The prerequisites for the result are:

1. The matrix A^v must be a closed compartmental matrix.
2. The entries in A^v must be constant.
3. The matrix A^v must be irreducible.

The first condition was proved in lemma 1. The second condition holds because the rates in Table 1 are either constants or functions of V (which is fixed). We proved the third prerequisite in Lemma 2.

Proposition 4 in [21] proves that the real part of all eigenvalues of A^v is non-positive. This guarantees stability of the equilibrium. □

Theorem 1 guarantees the existence of t_S, the time taken to reach a stable steady state for $V = v$ by M_{I^v}. We proceed to cast M_I, M_H, M_{I^v} and M_{H^v} as LTSs.

Definition 3. *The LTS corresponding to M_I is the sextuple $\mathcal{I} = (X_I, \mathcal{V}, \rightarrow_I, X^0_I, \Pi_I, \langle\langle . \rangle\rangle_I)$:*

- *$X_I \subseteq \mathbb{R}^{13}$ is the* ***set of states*** *denoting the occupancy probabilities from the vector* $\mathbf{p}$ *in Def. 1.*
- *$\mathcal{V}$ is a family of curves (signals) of the form $[t_0, t_0 + APD] \rightarrow \mathbb{R}$ denoting* ***inputs*** *to the LTS. The lower limit t_0 is the time at which the AP commences and APD is the Action Potential Duration. $\mathcal{V}$ represents different temporal patterns by which the transmembrane potential V can be applied (fed back) to M_I, guaranteeing a solution to it. They are dictated by Eq. 1.*
- *$\rightarrow_I \subseteq X_I \times \mathcal{V} \times X_I$ is the* ***transition relation*** *that captures the dynamics of M_I such that $(\mathbf{x}_I, v, \mathbf{x}'_I) \in \rightarrow_I$, written as $\mathbf{x}_I \xrightarrow{v}_I \mathbf{x}'_I$, holds when there exist $\mathcal{V} \ni v : [0, \tau] \rightarrow \mathbb{R}$ and $\xi : [0, \tau] \rightarrow \mathbb{R}^{13}$ satisfying Eq. 4 with $\xi(0) = \mathbf{x_I}$ and $\xi(\tau) = \mathbf{x_I}'$. The time taken to transit from $\mathbf{x_I}$ to $\mathbf{x_I}'$ is τ.*

- $X_I^0 \subseteq \mathbb{R}^{13}$, *a singleton consisting of the* ***initial condition*** *for* M_I, *is specified in Table 4 of [20] and acts as the initial state for* I.
- $\Pi_I \subseteq \mathbb{R}$, *the set of* ***outputs***, *denotes the observable values of* M_I, *i.e. all possible values of* $O(V) = O_1(V)+O_2(V)$, *the sum of occupancy probabilities of states* O_1 *and* O_2.
- $\langle\langle . \rangle\rangle_I : \mathbb{R}^{13} \to \mathbb{R}$ *is the* ***output map***, *which given a state* $\mathbf{x}_I \in X_I$, *maps it to its corresponding output* $\pi_6(\mathbf{x}_I) + \pi_7(\mathbf{x}_I)$[1], *the sum of* O_1 *and* O_2.

Definition 4. *The LTS corresponding to* M_H *is the sextuple* $\mathcal{H} = (X_H, \mathcal{V}, \to_H, X_H^0, \Pi_H, \langle\langle . \rangle\rangle_H)$:

- $X_H \subseteq \mathbb{R}^2$ *is the* ***set of states*** *denoting the values of* m *and* h *in* M_H.
- $\mathcal{V}$, *the* ***input set*** *is the same as in Def. 3. The curves* $v \in \mathcal{V}$ *guarantee a solution to* M_H.
- $\to_H \subseteq X_H \times \mathcal{V} \times X_H$ *is the* ***transition relation*** *that captures the dynamics of* M_H *such that* $(\mathbf{x}_H, v, \mathbf{x}'_H) \in \to_H$, *written as* $\mathbf{x}_H \xrightarrow{v}_H \mathbf{x}'_H$, *holds when there exists curves* $\mathcal{V} \ni v : [0,\tau] \to \mathbb{R}$ *and* $\psi : [0,\tau] \to \mathbb{R}^2$ *satisfying* M_H, *with* $\psi(0) = \mathbf{x}_H$ *and* $\psi(\tau) = \mathbf{x}'_H$.
- $X_H^0 \subseteq \mathbb{R}^2$ *is a singleton consisting of the* ***initial condition*** *identified by PEFT for* $M_{HV_{rest}}$.
- $\Pi_H \subseteq \mathbb{R}$ *is the set of* ***outputs*** *of the LTS denoting the observables from* M_H. *As* I_{Na} *current depends on the conductance* m^3h *of* M_H, *the set* Π_H *contains all possible values of* m^3h.
- $\langle\langle . \rangle\rangle_H : \mathbb{R}^2 \to \mathbb{R}$ *is the* ***output map***, *which given a state* $\mathbf{x}_H \in X_H$, *maps it to its corresponding output* $(\pi_1(\mathbf{x}_H))^3\pi_2(\mathbf{x}_H)$, *the conductance* m^3h.

Definition 5. *The LTS corresponding to* M_{I^v} *is the sextuple* $\mathcal{I}^v = (X_{I^v}, T, \to_{I^v}, X_{I^v}^0, \Pi_{I^v}, \langle\langle . \rangle\rangle_{I^v})$. *The states* X_{I^v}, *outputs* Π_{I^v} *and output map* $\langle\langle . \rangle\rangle_{I^v}$ *are the same as in Def. 3.*

- $T \subseteq \mathbb{R}_{\geq 0}$ *is the* ***input***, *denoting time.*
- $\to_{I^v}$ *is the* ***transition relation*** *such that* $\mathbf{x}_{I^v} \xrightarrow{t}_{I^v} \mathbf{x}'_{I^v}$ *holds if there exists a solution* ξ^v *to* M_{I^v} *satisfying* $\xi^v(0) = \mathbf{x}_{I^v}$ *and* $\xi^v(t) = \mathbf{x}'_{I^v}$.
- $X_{I^v}^0$ *denotes the* ***initial condition*** *used in step 1 of the three-step procedure in Section 3.*

Definition 6. *The LTS corresponding to* M_{H^v} *is the sextuple* $\mathcal{H}^v = (X_{H^v}, T, \to_{H^v}, X_{H^v}^0, \Pi_{H^v}, \langle\langle . \rangle\rangle_{H^v})$. *The states* X_{H^v}, *outputs* Π_{H^v} *and output map* $\langle\langle . \rangle\rangle_{H^v}$ *are the same as in Def. 4. The input set* T *is the same as in Def. 5.*

- $\to_{H^v}$ *is the* ***transition relation*** *such that* $\mathbf{x}_{H^v} \xrightarrow{t}_{H^v} \mathbf{x}'_{H^v}$ *holds if there exists a solution* ψ^v *to* M_H^v *satisfying* $\psi^v(0) = \mathbf{x}_{H^v}$ *and* $\psi^v(t) = \mathbf{x}'_{H^v}$.
- $X_{H^v}^0$ *is the* ***initial condition*** *determined by PEFT in Sec 3 for* $V = v$.

Definition 7. *The two LTSs* $T_1(Q_1, \Sigma, \to_1, Q_1^0, \Pi, \langle\langle . \rangle\rangle_1)$ *and* $T_2(Q_2, \Sigma, \to_2, Q_2^0, \Pi, \langle\langle . \rangle\rangle_2)$ *are approximately bisimilar, with precision* δ, *denoted as* $T_1 \cong_\delta T_2$, *if there exists a relation* $B_\delta \subseteq Q_1 \times Q_2$ *such that:*

[1] $\pi_j(x)$ is the projection function that projects the jth element from the vector x.

1. *For every* $q_1 \in Q_1^0$*, there exists a* $q_2 \in Q_2^0$ *such that* $(q_1, q_2) \in B_\delta$ *and conversely.*
2. *For every* $(q_1, q_2) \in B_\delta$, $d_\Pi(\langle\langle q_1\rangle\rangle_1, \langle\langle q_2\rangle\rangle_2) \leq \delta$*, where* d_Π *is some distance metric defined on the output set* Π *shared by the two LTS.*
3. *For every* $(q_1, q_2) \in B_\delta$*:*
   (a) $q_1 \xrightarrow{\sigma}_1 q_1'$, $\sigma \in \Sigma$*, implies the existence of* $q_2 \xrightarrow{\sigma}_2 q_2'$ *such that* $(q_1', q_2') \in B_\delta$*.*
   (b) $q_2 \xrightarrow{\sigma}_2 q_2'$, $\sigma \in \Sigma$*, implies the existence of* $q_1 \xrightarrow{\sigma}_1 q_1'$ *such that* $(q_1', q_2') \in B_\delta$*.*

The relation B_δ *is called the approximate bisimulation relation.*

In the case of deterministic systems, such as $\mathcal{I}$, $\mathcal{H}$, $\mathcal{I}^v$ and $\mathcal{H}^v$, proving two LTSs approximately bisimilar is equivalent to proving that the distance between the unique trajectories (behaviors) of the systems is bounded. Next, we state a simple lemma relating finite-length trajectories of two Linear Autonomous Dynamical Systems (LADS)[2], whose proof follows from the uniqueness and continuity of the trajectories.

Lemma 3. *Consider two LADSs* $\{\dot{\mathbf{x}}_1 = M_1.\mathbf{x}_1,\ \mathbf{x}_1(0) = \mathbf{x}_1^0\}$ *and* $\{\dot{\mathbf{x}}_2 = M_2.\mathbf{x}_2,\ \mathbf{x}_2(0) = \mathbf{x}_2^0\}$ *where* $\mathbf{x}_1, \mathbf{x}_2, \mathbf{x}_1^0, \mathbf{x}_2^0 \in \mathbb{R}^n$ *and* M_1 *and* M_2 *are* $n \times n$ *matrices. Let* $\mathbf{x}_1(t)$ *and* $\mathbf{x}_2(t)$ *be the respective solution trajectories. Let* $I_1[t_1, t_2]$ *and* $I_2[t_2, t_3]$ *be two time intervals of arbitrary lengths such that:*

- $|\mathbf{x}_1(t) - \mathbf{x}_2(t)| \leq \delta$ *for* $t \in I_1$*, and*
- $|\mathbf{x}_1(t) - \mathbf{x}_2(t)| \leq \delta$ *for* $t \in I_2$*,*

where $|.|$ *denotes the L2 norm. Then* $|\mathbf{x}_1(t) - \mathbf{x}_2(t)| \leq \delta$ *for* $t \in I_{12}[t_1, t_3]$*.*

Definition 8. *The LTSs* $\mathcal{I}_d$, $\mathcal{H}_d$, $\mathcal{I}_d^v$ *and* $\mathcal{H}_d^v$ *denote discrete time equivalents of the LTSs* $\mathcal{I}$, $\mathcal{H}$, $\mathcal{I}^v$ *and* $\mathcal{H}^v$ *respectively such that:*

- *The input curves* v *for* $\mathcal{I}_d$ *and* $\mathcal{H}_d$ *are discrete time signals of voltage of the the form* $[v_1, v_2, \ldots, v_i, \ldots]$*, where* v_i *is the voltage at the* i^{th} *time step. The inputs to* $\mathcal{I}_d^v$ *and* $\mathcal{H}_d^v$ *are integral multiples of the time step.*
- *The transition relations of the LTSs respect the transitions of the corresponding continuous time ones, except that the dynamics are now defined in discrete time. Chapter 11 of [27] provides details about converting continuous time models to discrete time versions via techniques like sample and hold.*

Note: Discrete time arguments can be justified because the LADS resulting at constant voltages are band-limited as they attain steady state in finite time for all voltages (see Theorem 1). For such systems, the Sampling theorem [27] guarantees the existence of a Digital to Analog Converter (DAC) that can recover the continuous time behaviors from discrete time samples, if a small-enough discretization of time is used. This sampling frequency is determined by the maximum frequency component in the continuous-time behaviors. Theorem 1 ensures that the maximum frequency component of the trajectories is bounded for all voltages.

[2] See Lecture 9 of [3] for a formal definition of LADS.

Theorem 2. *The PEFT procedure can ensure that $\mathcal{I}_d^v \cong_{\delta^v} \mathcal{H}_d^v$ for any $v \in [V_{res}, V_{max}]$. The precision δ^v is the maximum L2 error incurred by the optimizer while solving Eq. (5).*

Proof The approximate bisimulation relation $B_{\delta^v} \subseteq X_{I^v} \times X_{H^v}$ can be constructed as follows.

1. The initial condition in $\mathbf{x}_{I^v}^0 \in X_{I^v}^0$ is paired with the initial condition $\mathbf{x}_{H^v}^0 \in X_{H^v}^0$.
2. Consider a state $\mathbf{x}_{I^v} \in X_{I^v}$ such that $\mathbf{x}_{I^v}^0 \xrightarrow{t}_{I^v} \mathbf{x}_{I^v}$, $t \in T$. Also say $\mathbf{x}_{H^v} \in X_{H^v}$ such that $\mathbf{x}_{H^v}^0 \xrightarrow{t}_{H^v} \mathbf{x}_{H^v}$. Then, $(\mathbf{x}_{I^v}, \mathbf{x}_{H^v}) \in B_{\delta^v}$. The existence of states $\mathbf{x}_{I^v}$ and $\mathbf{x}_{H^v}$ satisfying the conditions is guaranteed due to uniqueness and existence of solutions to LADS.

The relation B_{δ^v} is a valid approximate bisimulation relation. Condition 1 of Def. 7 is satisfied by construction. Suppose we have $(\mathbf{x}_{I^v}, \mathbf{x}_{H^v}) \in B_{\delta^v}$, $\mathbf{x}_{I^v} \xrightarrow{t'}_{I^v} \mathbf{x}'_{I^v}$, and $\mathbf{x}_{H^v} \xrightarrow{t'}_{H^v} \mathbf{x}'_{H^v}$, then we have $\mathbf{x}_{I^v}^0 \xrightarrow{t+t'}_{I^v} \mathbf{x}'_{I^v}$ and $\mathbf{x}_{H^v}^0 \xrightarrow{t+t'}_{H^v} \mathbf{x}'_{H^v}$, due to the uniqueness of the trajectories, where t is the time required to transit from $\mathbf{x}_{I^v}^0$ to $\mathbf{x}_{I^v}$ and from $\mathbf{x}_{H^v}^0$ to $\mathbf{x}_{H^v}$ in $\mathcal{I}_d^v$ and $\mathcal{H}_d^v$ respectively. This ensures that $(\mathbf{x}'_{I^v}, \mathbf{x}'_{H^v}) \in B_{\delta^v}$, thus satisfying condition 2 of Def. 7. Condition 3 is satisfied due to Lemma 3, which also holds for discrete time trajectories. □

We now define perturbed LADS. Then we outline the approximate bisimilarity of $\mathcal{I}_d$ and $\mathcal{H}_d$.

Definition 9. *Consider an LADS $\{\dot{\mathbf{x}} = M.\mathbf{x},\ \mathbf{x}(0) = \mathbf{x}^0\}$, where $\mathbf{x}, \mathbf{x}^0 \in \mathbb{R}^n$, M is a $n \times n$ matrix and $\mathbf{x}(0)$ is the initial condition. An ϵ-perturbation of the LADS is obtained by perturbing any of the entries in M or $\mathbf{x}(0)$ by at-most $\epsilon \in \mathbb{R}$.*

Theorem 3. *The three-step abstraction process explained in Section 3 ensures that $\mathcal{H}_d \cong_\delta \mathcal{I}_d$ with $\delta \leq 7.58 \times 10^{-4}$.*

Proof sketch: In discrete time, the evolution of M_H (M_I) can be modeled as a series of one-step evolutions of M_{H^v} (M_{I^v}) i.e. when the input signal is of the form $[v_1, \ldots, v_i, v_{i+1} \ldots]$, at the ith step, the LADS $M_{H^{v_i}}$ ($M_{I^{v_i}}$) evolves for one time step, followed by $M_{H^{v_{i+1}}}$ ($M_{I^{v_{i+1}}}$) and so on. This idea is also illustrated in Fig. 4(b).

For some voltage $V = v$, the distance between the trajectories of M_{I^v} and M_{H^v} can be bound in terms of the trajectories of $M_{I^{v*}}$ and $M_{H^{v*}}$, where v^* is a voltage that was processed by PEFT and M_{I^v} is a minimal perturbation of $M_{I^{v*}}$. At the ith step, the perturbation is the least for $M_{I^{v_i^*}}$ among all the voltages that were processed by PEFT. We first bound the corresponding perturbation of $M_{H^{v_i*}}$, ϵ, and then use a similar approach for $M_{I^{v_i^*}}$
$\epsilon = max(\epsilon_1, \epsilon_2)$,where

$$\epsilon_1 = \max_{1 \leq j \leq n} [max\{|\alpha_m(v_j) - \alpha_m(v_{j+1})|, |\beta_m(v_j) - \beta_m(v_{j+1})|, |\alpha_h(v_j) - \alpha_h(v_{j+1})|, |\beta_h(v_j) - \beta_h(v_{j+1})|\}],$$

$$\epsilon_2 = max[|\alpha_m(v_\Delta) - \alpha_m(v_{\Delta+1})|, |\beta_m(v_\Delta) - \beta_m(v_{\Delta+1})|, |\alpha_h(v_\Delta) - \alpha_h(v_{\Delta+1})|, |\beta_h(v_\Delta) - \beta_h(v_{\Delta+1})|]$$

and $\Delta = \underset{1 \leq j \leq n}{\text{argmax}}[\frac{|v_j - v_{j+1}|}{2}]$

The limit n is the total number of voltages processed by PEFT. The term ϵ_1 accounts for sharp changes in the rate functions $\alpha_m(V)$, $\alpha_h(V)$, $\beta_m(V)$, $\beta_h(V)$ and ϵ_2 accounts for sparsity in the voltages processed by PEFT. Given the input signal v, the ith step v_i may be at most Δ mV away from a voltage processed by PEFT.

At the ith step, let $M_{H^{v_i}}$ be an ϵ-perturbation of $M_{H^{v_i^*}}$ and at the $(i+1)^{th}$ step, let $M_{H^{v_{i+1}}}$ be an ϵ-perturbation of $M_{H^{v_{i+1}^*}}$. We can always ensure that $v_i^* \neq v_{i+1}^*$. This can be done by first bounding the time-scale, which determines the maximum change in V that can occur over one time step, $(|v_i - v_{i+1}|)$. Once we know the least value of $|v_i - v_{i+1}|$, we can perform the PEFT procedure for voltages that satisfy $\Delta \leq |v_i - v_{i+1}|$. Thus, we can ensure that at the ith step, the perturbed-system $M_{H^{v_i}}$ $(M_{I^{v_i}})$ diverges from $M_{H^{v_i^*}}$ $(M_{I^{v_i^*}})$ for at most one time step.

We first bound the one-step divergence between the trajectories of $M_{H^{v_i}}$ and $M_{H^{v_i^*}}$. We calculate the sensitivity of the variable m to an ϵ change in the parameters and the initial conditions below.

$$\begin{aligned}
\dot{m}^{v_i^*} &= \alpha_m^{v_i^*}(1-m) + \beta_m^{v_i^*} m \\
m^{v_i^*}[1] &= m_0^{v_i^*} + [\alpha_m^{v_i^*}(1 - m_0^{v_i^*}) + \beta_m^{v_i^*} m_0^{v_i^*}] \quad \text{(one time step)} \\
m^{v_i}[1] &= m_0^{v_i^*} + \epsilon + [(\alpha_m^{v_i^*} + \epsilon)(1 - m_0^{v_i^*}) + (\beta_m^{v_i^*} + \epsilon) m_0^{v_i^*}] \text{(perturbed)} \\
|m^{v_i^*}[1] - m^{v_i}[1]| &= |\epsilon[1 + (1 - 2m - \alpha_m^{v_i} - 2\epsilon - \beta_m^{v_i})]| \quad \text{(divergence)} \\
&\leq |2\epsilon|
\end{aligned}$$

The divergence is maximized when $m = 0$ and the transition rates $\alpha, \beta = 0$. Thus given an initial separation of ϵ, the trajectories diverge by at most 2ϵ in one time-step. The same calculation can be repeated independently for h.

Theorem 2 dictates that the trajectories of $M_{I^{v_i^*}}$ and $M_{H^{v_i^*}}$ may not diverge beyond $\delta^{v_i^*}$. This is implied by their approximate bisimulation equivalence.

Using a similar approach as taken for $M_{H^{v_i}}$, we now bound the divergence of trajectories of $M_{I^{v_i}}$ from $M_{I^{v_i^*}}$. At $V = v_i$, the maximum possible perturbation μ of $M_{I^{v_i}}$ from $M_{I^{v_i^*}}$, where v_i^* is the nearest voltage processed by PEFT, can be bound as was done for ϵ, by considering the rate functions of M_I. The solution trajectory of $M_{I^{v_i^*}}$ is given by the matrix exponential $e^{A(v_i^*)t}$, where A is the matrix in Eq.(4). An arbitrary voltage v_i in the input-signal presents a μ-perturbation of the entries in $A(v_i^*)$. The evolution of $M_{I^{v_i}}$ is then approximated by the corresponding perturbation of $e^{A(v_i^*)t}$.

The matrix exponential is determined by the eigenvalues of $A(v_i^*)$. Bauer-Fike theorem [2] bounds the spectral perturbation caused due to a perturbation of the original matrix. It ensures that the eigenvalues of $A(v_i)$ are μ-perturbations

of the eigenvalues of $A(v_i^*)$. Thus, the maximum divergence[3] of $M_I^{v_i}$ from $M_I^{v_i^*}$ in one time-step is at most e^{μ}.
Thus, $\delta \leq 16\epsilon^4 + \underset{1\leq i\leq n}{\text{argmax}}[\delta^{v_i^*}] + e^{\mu}$, sum of the following quantities:

- Maximum divergence of $M_{H^{v_i}}$ from $M_{H^{v_i^*}}$ over one time-step: $16\epsilon^4$. This is due to the conductance being m^3h. We bound the divergence of m and h individually at 2ϵ.
- Maximum divergence of any $M_{H^{v_i^*}}$ from $M_{I^{v_i^*}}$ over all n voltages processed during PEFT: $\underset{1\leq i\leq n}{\text{argmax}}[\delta^{v_i^*}]$. This was estimated to be 2.79×10^{-4}.
- Maximum divergence of $M_{I^{v_i}}$ from $M_{I^{v_i^*}}$ over one time step: e^{μ}. □

5 Related Work

Singular perturbation [24,34] and invariant manifold reduction [6,16] are two popular approaches to reducing multi-scale state-space models of chemical reaction kinetics [7,17,38]. The quasi steady state assumption is central to singular perturbation techniques used in [38]. The derivative of fast variables, which evolve on relatively short time scales, is approximated to be zero, resulting in model reduction. Despite being successful for chemical kinetics models, such techniques are not well-suited for Markovian ion channel models. The former involves a constant rate matrix A that renders the system linear, where as in our Markovian models, the rate matrix A is a function of the transmembrane voltage V. The voltage V is itself dependent on the evolution of the Markovian model and this circular dependency causes the overall model to be nonlinear.

Reduction of Markovian ion-channel models, which is the central topic of this paper, has been explored in [41,42]. The focus is on reducing the simulation time, rather than obtaining a formal reduction. In [39] Smith et al. reduce a stochastic model for the sodium-potassium pump by lumping the states of their model. In [10], Fink et al. use mixed formulations of an HH-type model and a Markovian model to reduce the number of state variables for the calcium current. In this paper, we provide a systematic reduction of the sodium channel. Conventional approaches like [25] use behavioral equivalence to validate the reduced models. Approximate bisimulation, used in this paper, formalize equivalence in a compositional setting and also help in insightful analysis.

6 Conclusions and Future Work

We constructed a two-state Hodgkin-Huxley-type model M_H that can replace the 13-state CT-MDP M_I for sodium-channel dynamics, within the IMW model for ventricular myocytes. The open state of M_I being the only observable was an underlying assumption in the reduction. It should be noted that this is not very restrictive. Any observable state occupancy probability can be handled by

[3] A tighter bound can be found, as was done for $M_{H^{v_i}}$, by projecting the error onto the O_1 and O_2 dimensions.

modifying Eq. (5). Currently we map the open state probabilities of M_I and the 8-state CT-MDP in Fig. 3 to each other. Once such a mapping is established between any two states of the two models, Eq. (5) can then be modified to fit the trajectories of the states of M_I that one is interested in. The invariant manifold of the m and h is related to all the 8 states.

The reduction was formalized by proving the abstract and the concrete models to be approximately bisimilar. This notion of system equivalence can be used for compositional reasoning. When $\mathcal{H}$ is appropriately composed with the rest of the larger whole-cell IMW model, approximate bisimulation guarantees that the newly composed-system retains the properties of the original system. The original system can be modeled as an appropriate composition of $\mathcal{I}$ and rest of the IMW model. In the future, further complicated non-deterministic models will be explored and reduced. Tighter bounds will also be pursued for the precision of the bisimulation relation. We then plan to use the *towers of abstraction* constructed from the strategy outlined in the paper, for insightful analysis of cardiac models.

References

1. Bartocci, E., Cherry, E., Glimm, J., Grosu, R., Smolka, S.A., Fenton, F.: Toward real-time simulation of cardiac dynamics. In: Proceedings of the 9th International Conference on Computational Methods in Systems Biology, CMSB 2011, pp. 103–112. ACM (2011)
2. Bauer, F.L., Fike, C.T.: Norms and Exclusion Theorems. Numerische Matematik (1960)
3. Boyd, S.: EE 263: Introduction to Linear Dynamical Systems, lecture notes. In: Stanford Engineering Everywhere, SEE (2010)
4. Bueno-Orovio, A., Cherry, E.M., Fenton, F.H.: Minimal model for human ventricular action potentials in tissue. J. of Theor. Biology 253(3), 544–560 (2008)
5. Cherry, E.M., Fenton, F.H.: Visualization of spiral and scroll waves in simulated and experimental cardiac tissue. New Journal of Physics 10, 125016 (2008)
6. Chiavazzo, E., Gorban, A.N., Karlin, I.V.: Comparisons of invariant manifolds for model reduction in chemical kinetics. Comm. Comp. Phys. 2, 964–992 (2007)
7. Epstein, I.R., Pojman, J.A.: An Introduction to Nonlinear Chemical Dynamics. Oxford University Press, London (1998)
8. Fenton, F., Karma, A.: Vortex dynamics in three-dimensional continuous myocardium with fiber rotation: Filament instability and fibrillation. Chaos 8(1), 20–47 (1998)
9. Fenton, F.H., Cherry, E.M.: Models of cardiac cell. Scholarpedia 3, 1868 (2008)
10. Fink, M., Noble, D.: Markov models for ion channels: Versatility versus identifiability and speed. Philosophical Transactions of the Royal Society A: Mathematical, Physical and Engineering Sciences 367(1896), 2161–2179 (2009)
11. Fisher, J., Piterman, N., Vardi, M.Y.: The Only Way Is Up. In: Butler, M., Schulte, W. (eds.) FM 2011. LNCS, vol. 6664, pp. 3–11. Springer, Heidelberg (2011)
12. Girard, A.: Controller synthesis for safety and reachability via approximate bisimulation. Automatica 48, 947–953 (2012)
13. Girard, A., Pappas, G.J.: Approximate bisimulations for nonlinear dynamical systems. In: Proc. of CDC 2005, The 44th Int. Conf. on Decision and Control, Seville, Spain. IEEE (December 2005)

14. Girard, A., Pappas, G.J.: Approximate bisimulation relations for constrained linear systems. Automatica 43, 1307–1317 (2007)
15. Girard, A., Pappas, G.J.: Approximation metrics for discrete and continuous systems. IEEE Transactions on Automatic Control 52(5), 782–798 (2007)
16. Gorban, A.N., Karlin, I.V.: Method of invariant manifold for chemical kinetics. Chem. Eng. Sci. 58, 4751–4768 (2003)
17. Gorban, A.N., Kazantzis, N., Kevrekidis, I.G., Ottinger, H.C., Theodoropoulos, C.: Model reduction and coarse-graining approaches for multiscale phenomena. Springer (2006)
18. Grosu, R., Batt, G., Fenton, F.H., Glimm, J., Le Guernic, C., Smolka, S.A., Bartocci, E.: From Cardiac Cells to Genetic Regulatory Networks. In: Gopalakrishnan, G., Qadeer, S. (eds.) CAV 2011. LNCS, vol. 6806, pp. 396–411. Springer, Heidelberg (2011)
19. Hodgkin, A.L., Huxley, A.F.: A quantitative description of membrane current and its application to conduction and excitation in nerve. Journal of Physiology 117, 500–544 (1952)
20. Iyer, V., Mazhari, R., Winslow, R.L.: A computational model of the human left-ventricular epicardial myocytes. Biophysical Journal 87(3), 1507–1525 (2004)
21. Jahnke, T., Huisinga, W.: Solving the chemical master equation for monomolecular reaction systems analytically. Journal of Mathematical Biology 54, 1–26 (2007)
22. Fisher, J., Harel, D., Henzinger, T.A.: Biology as reactivity. Communications of the ACM 54(10), 72–82 (2011)
23. Keener, J.: Invariant manifold reductions for markovian ion channel dynamics. Journal of Mathematical Biology 58(3), 447–457 (2009)
24. Kevorkian, J., Cole, J.D.: Multiple Scale and Singular Perturbation Methods. Springer (1996)
25. Kienker, P.: Equivalence of aggregated markov models of ion-channel gating. Proceedings of the Royal Society of London. B. Biological Sciences 236(1284), 269–309 (1989)
26. Kuo, C.-C., Bean, B.P.: Na channels must deactivate to recover from inactivation. Neuron 12, 819–829 (1994)
27. Lee, E., Varaiya, P.: Structure and Interpretation of Signals and Systems. Pearson Education (2003)
28. Irvine, L.A., Saleet Jafri, M., Winslow, R.L.: Cardiac sodium channel markov model with tempretature dependence and recovery from inactivation. Biophysical Journal 76, 1868–1885 (1999)
29. Luo, C.H., Rudy, Y.: A dynamic model of the cardiac ventricular action potential. I. Simulations of ionic currents and concentration changes. Circulation Research 74(6), 1071–1096 (1994)
30. MATLAB. Choosing a solver, `http://www.mathworks.com/help/toolbox/optim`
31. MATLAB. Curve fitting toolbox, `http://www.mathworks.com/products/curvefitting`
32. MATLAB. Nonlinear numerical methods, `http://www.mathworks.com/help/techdoc/ref/f16-5872.html`
33. MATLAB. Optimization toolbox, `http://www.mathworks.com/help/toolbox/optim`
34. Murray, J.D.: Mathematical Biology. Springer (1990)
35. Myers, C.J.: Engineering Genetic Circuits. CRC Press (2010)
36. National Science Foundation (NSF). Computational Modeling and Analysis of Complex Systems (CMACS), `http://cmacs.cmu.edu`

37. Noble, D.: A modification of the Hodgkin-Huxley equations applicable to purkinje fibre action and pace-maker potentials. J. Physiol. 160, 317–352 (1962)
38. Radulescu, O., Gorban, A.N., Zinovyev, A., Lilienbaum, A.: Robust simplifications of multiscale biochemical networks. BMC Systems Biology 2(1), 86 (2008)
39. Smith, N., Crampin, E.: Development of models of active ion transport for whole-cell modelling: Cardiac sodium–potassium pump as a case study. Progress in Biophysics and Molecular Biology 85(2-3), 387–405 (2004), Modelling Cellular and Tissue Function
40. ten Tusscher, K.H., Noble, D., Noble, P.J., Panfilov, A.V.: A model for human ventricular tissue. American Journal of Physiology 286, H1573–H1589 (2004)
41. Wang, C., Beyerlein, P., Pospisil, H., Krause, A., Nugent, C., Dubitzk, W.: An efficient method for modeling kinetic behavior of channel proteins in cardiomyocytes. IEEE/ACM Trans. on Computational Biology and Bioinformatics 9(1), 40–51 (2012)
42. Whiteley, J.P.: Model reduction using a posteriori analysis. Mathematical Biosciences 225(1), 44–52 (2010)

Efficient Handling of Large Signalling-Regulatory Networks by Focusing on Their Core Control

Aurélien Naldi[1], Pedro T. Monteiro[2], and Claudine Chaouiya[2]

[1] Center for Integrative Genomics, Fac. of Biology and Medicine, Univ. of Lausanne, Switzerland
[2] IGC, Instituto Gulbenkian de Ciência, Oeiras, Portugal

Abstract. Considering the logical (Boolean or multi-valued) asynchronous framework, we delineate a reduction strategy for large signalling and regulatory networks. Consequently, focusing on the core network that drives the whole dynamics, we can check which attractors are reachable from given initial conditions, under fixed or varying environmental conditions.

More specifically, the dynamics of logical models are represented by (asynchronous) state transition graphs that grow exponentially with the number of model components. We introduce adequate reduction methods (preserving reachability of the attractors) and proceed with model-checking approaches.

Input nodes (that generally represent receptors) and output nodes (that constitute readouts of network behaviours) are each specifically processed to reduce the state space. The proposed approach is made available within GINsim, our software dedicated to the definition and analysis of logical models. The new GINsim functionalities consist in a proper reduction of output components, as well as the corresponding symbolic encoding of logical models for the NuSMV model checker. This encoding also includes a reduction over input components (transferring their values from states to transitions labels). Finally, we demonstrate the interest of the proposed methods through their application to a published large scale model of the signalling pathway involved in T cell activation.

Keywords: Qualitative modelling, Logical modelling, Model checking, Regulatory networks, Signalling networks.

1 Introduction

As ever larger signalling and regulatory maps are being identified, there is a growing need for efficient computational means to analyse the behaviours induced by these networks. Among the numerous existing modelling approaches (see *e.g.*, reviews [4,18]), the logical framework provides a convenient way to convey current qualitative knowledge and proved useful to study a significant number of published models. Here, we rely on the formalism initially proposed by R. Thomas

D. Gilbert and M. Heiner (Eds.): CMSB 2012, LNCS 7605, pp. 288–306, 2012.

and co-workers [20], where discrete (Boolean or multi-valued) dynamics are represented as (asynchronous) state transition graphs (STG). For such models, the number of states grows exponentially with the number of regulatory components. We propose to tackle this combinatorial explosion, by applying adequate reduction methods and by proceeding with model-checking approaches.

This manuscript focuses on signalling-regulatory networks that encompass large numbers of input components (denoting external stimuli) and output components (used as readouts of network behaviours). In concrete biological networks, input components vary, often being externally controlled (*e.g.*, light availability, presence of nutrients, heat shock, etc.). It is thus relevant to consider that input components freely vary (*i.e.*, are under no specific control) and to slightly extend the current definition of logical regulatory graphs. Then, the STG encompasses transitions between all the states that share the same values for internal components, denoting the sole changes of the input components.

For these signalling-regulatory networks, our rationale is to reduce the complexity of the corresponding models, while ensuring the full preservation of the properties of interest. These relate to asymptotical behaviours embodied in terminal strongly components of the state transition graphs (referred to as attractors) as well as the reachability of those attractors from given initial conditions. Moreover, we analyse the possible switches between attractors, upon variations of the input components.

The reduction method presented in [11] possibly leads to the loss of some trajectories. Here, we show that when applied to output cascades, this reduction has no impact on reachability properties. However, it cannot be applied to input cascades if attractor reachability is to be preserved. Hence, we propose another strategy to lessen the size of the state space, by transferring the values of input nodes to transition labels, thus reducing the state space by at least 2^n (for n input Boolean nodes). Furthermore, we discuss the nature of stable states in these labelled state transition graphs, in the case of varying input components. Finally, we resort to model-checking to analyse these complex dynamics.

We demonstrate the potential of this approach on the large scale Boolean model that accounts for T cell activation as defined by Saez-Rodriguez *et al.* [16].

In [1], the authors introduce a "decimation algorithm", which amounts to removing variables that have no impact on the long-term behaviour. While similar to ours, their method is valid for deterministic Boolean networks (with synchronous updates). Indeed, for such models, the reduction method presented in [11] has clearly different impacts on the dynamics. Considering asynchronous Boolean dynamics, Saadatpour *et al.* recently proposed a reduction strategy for large signalling transduction networks, relying on the fact that input cascades stabilise under constant input conditions [15]. Here, we aim at going further, first by ensuring that all the asynchronous dynamics is preserved (reducing only output cascades), second by considering varying input conditions.

The paper is organised as follows. First, we recall the basics of the logical formalism and of the model reduction method in Section 2. Section 3 presents the method that allows us to focus on the core network, reducing the output cascades and projecting the dynamics over the input components. Implementation aspects are discussed in Section 4. The proposed method is then applied to further analyse a published model of the signalling pathway involved in T cell activation [16]. The paper ends with some conclusions and prospects.

2 Background

In this section, we recapitulate (and extend) essential definitions concerning the logical formalism. Then, we recall the basics of the model reduction as defined in [11].

Definition 1. A logical signalling-regulatory graph (LSRG) $\mathcal{R} = (\mathcal{G}, \mathcal{K})$ *is defined by:*

- $\mathcal{G}$ *a set of* n *components partitioned into three subsets:* $\mathcal{I} = \{g_j\}_{j=1\ldots n_i}$, *the set of input components,* $\mathcal{P} = \{g_j\}_{j=(n_i+1)\ldots(n_i+n_p)}$, *the set of proper components, and* $\mathcal{O} = \{g_j\}_{j=(n_i+n_p+1)\ldots n}$, *the set of output components.*
- *Discrete variables denoting the* levels *of the components:* $\forall_{g_i} v_i \in \mathcal{D}_i = \{0 \ldots M_i\}$. *Then* $v = (v_i)_{g_i \in \mathcal{G}}$ *is a* state *and* $\mathcal{S} = \prod_{g_i \in \mathcal{G}} \mathcal{D}_i$ *is the* state space.
- *Logical functions defining the behaviours of the components:*
  - *For* $g_i \in \mathcal{P} \cup \mathcal{O}$, $\mathcal{K}_i$ *is a multi-valued function that specifies the (unique) target value* $\mathcal{K}_i(v)$ *of* g_i, *given the current state* v: $\mathcal{K}_i : \mathcal{S} \to D_i$.
  - *Input components* $g_i \in \mathcal{I}$ *are either set to constant values in their domains* ($\mathcal{K}_i : \mathcal{S} \to D_i$):

$$\forall v \in \mathcal{S}, \mathcal{K}_i(v) = v_i,$$

or freely vary ($\mathcal{K}_i : \mathcal{S} \to D_i \cup D_i^2$):

$$\forall v \in \mathcal{S}, \begin{cases} \textit{if } v_i = 0, & \mathcal{K}_i(v) = v_i + 1, \\ \textit{if } v_i = M_i, & \mathcal{K}_i(v) = v_i - 1, \\ \textit{if } M_i > 1 \textit{ and } 0 < v_i < M_i, & \mathcal{K}_i(v) = (v_i + 1, v_i - 1). \end{cases}$$

This definition slightly extends the classical definition of Logical Regulatory Graphs as introduced in *e.g.*, [11], since it specifically distinguishes input components, which account for environmental conditions and are either strictly controlled (*i.e.*, kept constant to their current values) or not (*i.e.*, freely vary).

The partition of $\mathcal{G}$, the set of regulatory components, will be useful in what follows. Note however that it could be partially derived from the logical functions $\mathcal{K}$. Indeed, given a proper or output component $g_i \in \mathcal{P} \cup \mathcal{O}$, one can define the set of nodes that influence g_i, denoted $Reg(g_i)$: $\forall g_k \in \mathcal{G}$, $g_k \in Reg(g_i)$ iff $\exists v \in \mathcal{S}, \mathcal{K}_i(v) \neq \mathcal{K}_i(v')$, where $v_k = v'_k \pm 1$ and $v_j = v'_j$, $\forall j \neq k$. If $g_k \in Reg(g_i)$, then there is an interaction from g_k to g_i and its sign can also be deduced from the logical function $\mathcal{K}_i$ (see *e.g.*, [11] for further detail). Moreover, we have

$$\begin{aligned} & g \in \mathcal{O} \Leftrightarrow \nexists g' \in \mathcal{G}, g \in Reg(g'), \\ & g \in \mathcal{P} \Leftrightarrow Reg(g) \neq \emptyset \text{ and } \exists g' \in \mathcal{G}, g \in Reg(g'), \\ & g \in \mathcal{I} \Leftrightarrow g \notin \mathcal{P} \cup \mathcal{O}. \end{aligned}$$

2.1 State Transition Graphs Representing LSRGs Dynamics

The behaviours of LSRGs are represented as State Transition Graphs, defined below. For proper and output components, we denote $\Delta_i(v)$ the "direction" of the update of g_i in state v:

$$\Delta_i(x) = \begin{cases} 0 & \text{if } K_i(v) = v_i, \\ \frac{|K_i(v)-v_i|}{K_i(v)-v_i} & \text{otherwise.} \end{cases}$$

Definition 2. *Given a LSRG $\mathcal{R} = (\mathcal{G}, \mathcal{K})$, its full, asynchronous State Transition Graph (*STG*) is a graph $\mathcal{E}(\mathcal{R}) = (\mathcal{S}, \mathcal{T})$ where:*

- *the nodes are the states in $\mathcal{S}$,*
- *the arcs denote transitions between states,*
  - *transitions over proper and output components are such that:*

$$(v, w) \in \mathcal{T} \subset \mathcal{S}^2 \iff \exists g_i \in \mathcal{P} \cup \mathcal{O} \text{ s.t. } \mathcal{K}_i(v) \neq v_i,\ w_i = v_i + \Delta_i(v) \text{ and } w_j = v_j\ \forall j \neq i,$$

  - *transitions over a varying input component g_i are as follows (depending on v_i and M_i, there is a transition increasing v_i, a transition decreasing v_i or both):*

$$\begin{cases} (v, w) \in \mathcal{T}, \text{ with } w_i = v_i + 1,\ w_j = v_j\ \forall j \neq i, \iff v_i < M_i, \\ (v, w) \in \mathcal{T}, \text{ with } w_i = v_i - 1,\ w_j = v_j\ \forall j \neq i \iff v_i > 0. \end{cases}$$

We denote $\mathcal{E}_{cste}(\mathcal{R})$ (or simply $\mathcal{E}_{cste}$), the STG where input components are kept constant, and $\mathcal{E}_{var}$ the STG where input components freely vary. The STG $\mathcal{E}_{cste}$ is made of at least as many disconnected sub-graphs as the number of fixed input combinations (see Section 3.3 and Figure 1). In $\mathcal{E}_{var}$ these sub-graphs are connected through transitions over varying inputs, connecting all states that differ only in their values of input components.

Note that one can consider a sub-graph of the full STG, by defining initial state(s). This graph can be constructed by visiting all successors of the initial state(s) and proceeding with the exploration until no new state is encountered.

The main relevant properties to be analysed relate to LSRGs asymptotical behaviours that are called attractors. In STGs, they correspond to terminal Strongly Connected Components (SCCs). When input components are maintained constant, we have:

- *Stable states* (*i.e.*, terminal SCCs reduced to a unique state);
- *Complex attractors* (*i.e.*, terminal SCCs encompassing at least two states). Within these complex attractors, we can further distinguish *elementary terminal cycles*, in which all states have a unique outgoing transition.

In $\mathcal{E}_{var}$, there is no stable state (as defined above) and we need to revisit the definition of these attractors (see Figure 1 and Section 3.3).

Beside the identification of attractors, it is often important to check reachability properties, *e.g.*, which attractors are reachable from an initial condition, what are the properties of all trajectories leading to those reachable attractors, etc.

As already mentioned, we face a combinatorial explosion of the number of states that hampers efficient analysis of STGs. The consideration of priority classes, based on well-founded assumptions, amounts to choose between concurrent transitions and thus constitutes a convenient way of reducing the size of a STG (see [5]). Another approach, yet related, consists in reducing the size of the model (the number of its components). The method is briefly described below, a full description being available in [11].

2.2 Model Reduction

The reduction method presented in [11] consists in iteratively taking components off the model, which is adequately modified. The intuitive idea is to transfer the role of the reduced component to its regulators. Importantly, reduction of auto-regulated components is not allowed, to ensure that essential dynamical properties are preserved. Indeed, regulatory circuits are known to be at the origin of multi-stability (for positive circuits) and of stable oscillations (for negative circuits) [19]. In the same way, with our extended definition of logical models, where input components may freely vary, we will not allow reduction of these input nodes.

More precisely, taking a (non-autoregulated) component g_r off a LSRG $\mathcal{R} = (\mathcal{G}, \mathcal{K})$ leads to a new LSRG $\mathcal{R}^r = (\mathcal{G}^r, \mathcal{K}^r)$ with a reduced state space denoted $\mathcal{S}^r = \prod_{g_i \in \mathcal{G}^r = \mathcal{G} \setminus \{g_r\}} \mathcal{D}_i$. It is useful to define:

- The projection $\pi^r : \mathcal{S} \to \mathcal{S}^r$ such that $\forall v \in \mathcal{S}$, $\forall g_i \neq g_r$, $\pi^r(v)_i = v_i$;
- The "retrieval" function $s^r : \mathcal{S}^r \to \mathcal{S}$ such that $\forall x \in \mathcal{S}^r$, $\forall g_i \neq g_r$, $s^r(x)_i = x_i$ and $s^r(x)_r = \mathcal{K}_r(v)$, for $v \in \mathcal{S}$ such that $\pi^r(v) = x$. We say that $s^r(x)$ is the *representative state* of the equivalence class $[s^r(x)]_{\sim r}$ containing the states that are projected on $x \in \mathcal{S}^r$ (all these states differ solely in their values for the component g_r).

Furthermore, the reduction of g_r consists in modifying the logical functions (more precisely, the functions of those components g_i such that $g_r \in Reg(g_i)$):

$$\forall x \in \mathcal{S}^r,\ \mathcal{K}_i^r(x) = \mathcal{K}_i(s^r(x)).$$

Note that excluding auto-regulated and input components as candidates for reduction ensures the existence and uniqueness of the representative state. We recapitulate a number of properties concerning the dynamical behaviour of a reduced LSRG (details and proofs can be found in [11]). Let consider $\mathcal{R} = (\mathcal{G}, \mathcal{K})$ a LSRG and $\mathcal{R}^r = (\mathcal{G}^r, \mathcal{K}^r)$ its reduced version (taking off g^r). Let denote $\mathcal{E} = (\mathcal{S}, \mathcal{T})$ and $\mathcal{E}^r = (\mathcal{S}^r, \mathcal{T}^r)$ their STGs. We have,

1. $\forall u, v \in \mathcal{S}$, if u is a representative state ($u_r = \mathcal{K}_r(u)$), then: $(u, v) \in \mathcal{T} \Rightarrow (\pi^r(u), \pi^r(v)) \in \mathcal{T}^r$;

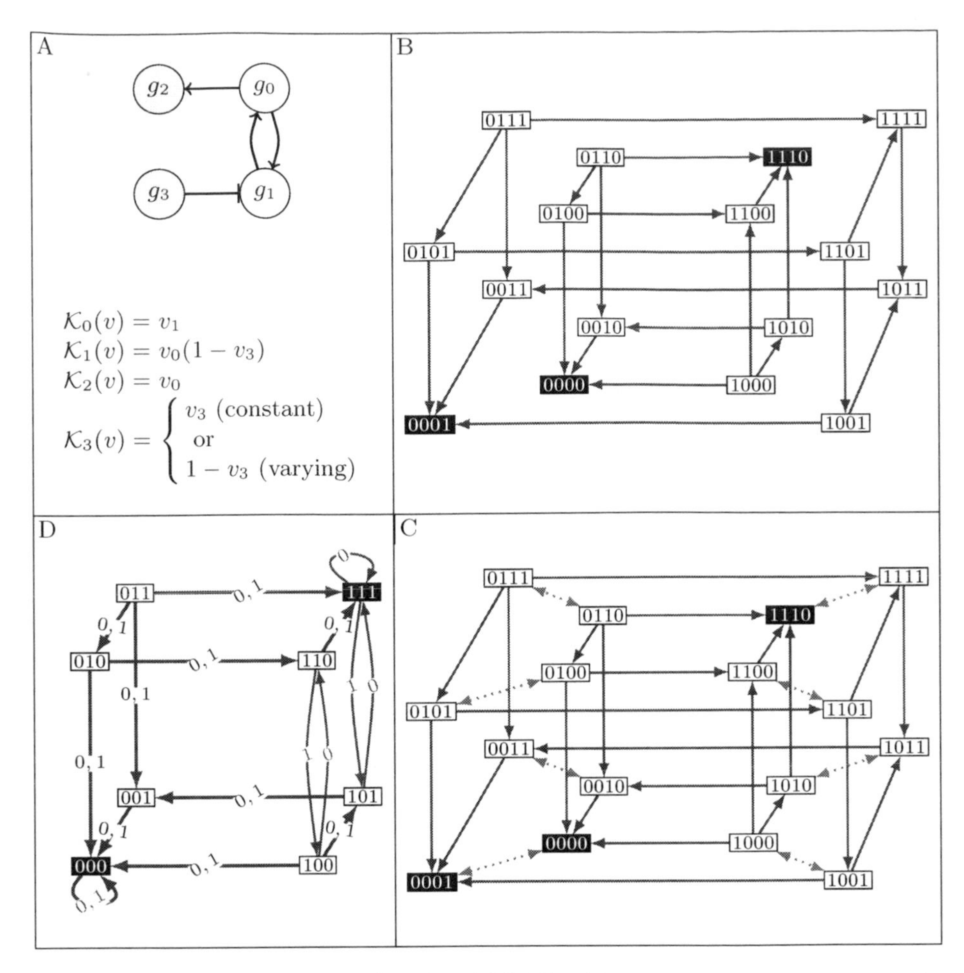

Fig. 1. (A) Simple example of a (Boolean) LSRG, with two proper components, one input and one output, and the associated logical functions. **(B)** The corresponding (full) STG $\mathcal{E}_{cste}$, when the input component g_3 is kept constant (states denote the values of g_0, g_1, g_2 and g_3 in this order). There are 3 stable states. **(C)** The (full) STG $\mathcal{E}_{var}$, considering a free variation of g_3. **(D)** The labelled STG $\mathcal{E}^{|\{g_3\}}$, resulting from the projection over g_3 and the labelling of the transitions with its values (see Section 3.3).

2. a transition $(u, v) \in \mathcal{T}$ is not preserved (*i.e.*, $(\pi^r(u), \pi^r(v)) \notin \mathcal{T}^r$ and $\pi^r(u) \neq \pi^r(v)$) iff the following conditions are fulfilled:
   - u is not a representative state ($u_r \neq \mathcal{K}_r(u)$),
   - $\exists i \neq r$, $v_i \neq u_i$, *i.e.*, u and v differ on their values for a component g_i, which is not g_r, hence $(u, v) \in \mathcal{T}$ is a transition over g_i,
   - and $\Delta_i(u) \neq \Delta_i(s^r(\pi^r(u)))$ (the updating call on g_i in state u is not preserved in the representative state $s^r(\pi^r(u))$);
3. Stable states in $\mathcal{E}$ are conserved in $\mathcal{E}^r$: u stable in $\mathcal{E}$ implies that u is a representative state and $\pi^r(u)$ is stable. Moreover, if z is stable in $\mathcal{E}^r$, then $s^r(z)$ is stable in $\mathcal{E}$;

4. If $(u_1 \dots u_p)$ is a elementary terminal cycle in $\mathcal{E}$ then $(\pi^r(u_1)\dots\pi^r(u_p))$ is a elementary terminal cycle in $\mathcal{E}^r$;
5. If $C \in \mathcal{S}$ is a complex attractor in $\mathcal{E}$, $\pi^r(C)$ contains at least one complex attractor in $\mathcal{E}^r$.

Summarising, stable states and elementary attractive cycles are preserved by the reduction (they only occur for constant input components). Complex attractors may appear as the result of an SCC disconnection provoked by the loss of transitions. Note that, all transitions over varying input components are preserved in $\mathcal{T}^r$ (since, for any state $v \in \mathcal{S}$, they equally exist in $\mathcal{T}$ for all states in the equivalence class $[v]_{\sim r}$).

3 Focusing on Core Networks in LSRGs

In this section, based on topological considerations, we define three set of components, each playing a distinct role in the dynamics of a LSRG. We then describe the reduction of output cascades and projection over input components as relevant means to reduce the size of the dynamics, yet keeping all its properties.

3.1 Splitting the Set of Components into Three Relevant Subsets

Given an LSRG $\mathcal{R} = (\mathcal{G}, \mathcal{K})$, its set of components is defined as the union of the set of inputs $\mathcal{I}$, the set of output $\mathcal{O}$ and the set of proper components $\mathcal{P}$. Here, still on topological considerations, we define another partition of $\mathcal{G}$ in three sets that play different role in the emergence of the dynamical properties.

The *set of input and pseudo-input components*, denoted $\widetilde{\mathcal{I}}$ is recursively defined as follows:

- $\mathcal{I} \subset \widetilde{\mathcal{I}}$ (all input components are in $\widetilde{\mathcal{I}}$);
- $\forall g_i \in \mathcal{G}$, if $Reg(g_i) \subset \widetilde{\mathcal{I}}$ then $g_i \in \widetilde{\mathcal{I}}$ (if all regulators of g_i are inputs or pseudo-inputs, then g_i is a pseudo-input).

Similarly, the *set of output and pseudo-output components*, denoted $\widetilde{\mathcal{O}}$, is defined as follows:

- $\mathcal{O} \subset \widetilde{\mathcal{O}}$ (all output components are in $\widetilde{\mathcal{O}}$);
- $\forall g_i \in \mathcal{G}$, if $\forall g_k \in \mathcal{G}$ s.t. $g_i \in Reg(g_k)$ we have $g_k \in \widetilde{\mathcal{O}}$, then $g_i \in \widetilde{\mathcal{O}}$ (if all targets of g_i are outputs or pseudo-outputs, then g_i is a pseudo-output).

Finally, *Core*, the *set of core components* of a LSRG is defined as the set of components that are neither in $\widetilde{\mathcal{I}}$ nor in $\widetilde{\mathcal{O}}$:

$$Core = \mathcal{G} \setminus (\widetilde{\mathcal{I}} \cup \widetilde{\mathcal{O}})$$

When input components (elements of $\mathcal{I}$) are maintained constant, for any attractor made up of a set of states A, we have: $\forall g_i \in \widetilde{\mathcal{I}}$, $\forall v, v' \in A$, $v_i = v'_i$ (pseudo-input components are stable).

Moreover, while input and pseudo-input components transmit external stimuli to the core components, these drive the dynamics of output and pseudo-output components. We also refer to the sets $\widetilde{\mathcal{O}}$ as output cascades, and $\widetilde{\mathcal{I}}$ as input cascades.

3.2 Reduction of Output Cascades

Since an output has no effect on other components, output components can be removed from a LSRG, with no impact on the behaviour. This is formalised by the following property.

Property 1. Let $\mathcal{R} = (\mathcal{G}, \mathcal{K})$ a LSRG, $g_r \in \mathcal{O}$ an output component of $\mathcal{R}$ and $\mathcal{E} = (\mathcal{S}, \mathcal{T})$ the associated STG. Then $\mathcal{E}^r = (\mathcal{S}^r, \mathcal{T}^r)$ the STG of the LSRG $\mathcal{R}^r$ resulting from the reduction of g_r, verifies:

$$\forall u, v \in \mathcal{S},\ (u, v) \in \mathcal{T} \Longrightarrow (\pi^r(u), \pi^r(v)) \in \mathcal{T}^r \text{ or } \pi^r(u) = \pi^r(v), \tag{1}$$

$$\forall x, y \in \mathcal{S}^r,\ (x, y) \in \mathcal{T}^r \Longrightarrow (s^r(x), s^r(y)) \in \mathcal{T}. \tag{2}$$

Proof. We only prove the first point that corresponds to the preservation of all the transitions, the proof of Eq. 2 can be found in [11], lemma 1. Let us consider $(u, v) \in \mathcal{T}$, then,

- if transition (u, v) involves g_r (the reduced component), u and v are in the same equivalence class $[u]_{\sim r} = [v]_{\sim r}$ (they only differ in their values of g_r), therefore their projection is equal: $\pi^r(u) = \pi^r(v)$;
- if transition (u, v) involves $g_i \in \mathcal{I}$, an input component, which freely varies, then, this transition is obviously preserved: $(\pi^r(u), \pi^r(v)) \in \mathcal{T}$;
- if transition (u, v) involves $g_i \in \mathcal{P} \cup \mathcal{O}$, a proper or an output component (different from g_r), then $v_i = \mathcal{K}_i(u) + \Delta_i(u)$ and, because g_r is an output component, we have also $\mathcal{K}_i^r(\pi^r(u)) = \mathcal{K}_i(u)$, hence this transition is preserved: $(\pi^r(u), \pi^r(v)) \in \mathcal{T}$.

As a consequence, attractors are fully preserved in $\mathcal{E}^r$, including complex ones, and more than that, reachability of these attractors is conserved. This follows from the fact that a path in $\mathcal{E}$ is preserved if the reduction preserves all its transitions (see [11]). We say that the reduction of an output component is *lossless*.

As mentioned before, output components often serve as readouts of a model. Hence it is important to retrieve their values (typically in a stable state or in a complex attractor). This is easily done because the behaviour of an output component is fully described by the sole representative states in the original STG. Hence retrieving the behaviours of output components only requires to store their logical functions (see Section 4).

Since the reduction of an output component is lossless, it is obviously also the case for the reduction of several output components.

Following the reduction of an output component, some of its former regulators become output components. These are the pseudo-outputs (that only regulate outputs or other pseudo-outputs). As such pseudo-outputs become outputs after reduction of their targets, they can be reduced as well, still preserving the dynamical behaviour. Therefore, the reduction of the whole set $\widetilde{\mathcal{O}}$ of output and pseudo-output components does not affect the dynamics.

To be able to recover the values of the components in $\widetilde{\mathcal{O}}$, we need to keep trace of the reduction of pseudo-outputs, redefining the logical functions of the targets of previously reduced components (see Section 4).

3.3 Input Components

Regulatory functions of proper and output components define their behaviours, depending on the current state of their regulators. An input component has no regulator and its function is thus assumed to be either constant or to freely vary as specified in Definition 1. For a LSRG with constant input values, its STG is composed by a set of disconnected graphs, one for each combination of input values (see Figure 1, panel B). Given an initial value of all the input variables, the behaviour is thus restricted to a sub-graph, easing the analysis of large systems. However, when input components freely vary (*i.e.*, are under no specific control), the STG encompasses transitions between all the states that have the same values of the internal components, denoting the sole changes of the input components. Considering that states are characterised by proper and output components values, the whole behaviour can be represented by a STG, with transitions labelled by the values of the input variables, yielding a compacted, labelled STG (see Figure 1). Below, we define such a projection over the input variables.

Definition 3. *Given* $\mathcal{R} = (\mathcal{G} = \mathcal{I} \cup \mathcal{P} \cup \mathcal{O}, \mathcal{K})$*, a LSRG and* $\mathcal{E} = (\mathcal{S}, \mathcal{T})$ *its STG. The corresponding* labelled STG $\mathcal{E}^{|\mathcal{I}} = (\mathcal{S}^{|\mathcal{I}}, \mathcal{T}^{|\mathcal{I}})$ *is defined as follows:*

- $\mathcal{S}^{|\mathcal{I}} = \prod_{g_i \in \mathcal{P} \cup \mathcal{O}} D_i$,
- $\forall v^{|\mathcal{I}}, w^{|\mathcal{I}} \in \mathcal{S}^{|\mathcal{I}}$, $(v^{|\mathcal{I}}, L, w^{|\mathcal{I}}) \in \mathcal{T}^{|\mathcal{I}}$ *iff* $\exists v, w \in \mathcal{S}$ *with* $(v, w) \in \mathcal{T}$ *such that* $\forall g_i \in \mathcal{P} \cup \mathcal{O}, v_i^{|\mathcal{I}} = v_i$ *and* $w_i^{|\mathcal{I}} = w_i$*, with L the label of this transition defined as the set of all the values of the input components for which this transition is observed in E:*

$$L = \{u \in \Pi_{g_i \in \mathcal{I}} D_i \ s.t. \ \forall g_i \in \mathcal{I}, \ v_i = u_i (= w_i)\} .$$

When a LSRG has a significant number of input components, this representation may presents a true gain in the number of states ($\Pi_{g_i \in \mathcal{G} \setminus \mathcal{I}} |D_i|$ instead of $\Pi_{g_i \in \mathcal{G}} |D_i|$), still keeping all the information regarding the input components. Such a graph structure combining labels on both states and transitions is already used by the formal verification community, and is called a Kripke Transition System (KTS) [7].

By definition, stable states in a STG have no output transitions (see Figure 1, panel B). However, when using model checking techniques, states of the system to be verified must give rise to at least one transition. Notably, a self-loop must be added to each stable state when translating the system into a KTS (*e.g.*, the implemented export to NuSMV). This is particularly useful to represent labelled STGs (Definiton 3).

Definition 4. *Given a labelled STG* $E^{|\mathcal{I}} = (S^{|\mathcal{I}}, T^{|\mathcal{I}})$*, a state* $v \in S^{|\mathcal{I}}$ *is:*

a strong stable state *iff* $\forall w \in S^{|\mathcal{I}}, \forall L \in \Pi_{g_i \in \mathcal{I}} D_i, w \neq v \Rightarrow (v, L, w) \notin T^{|\mathcal{I}}$,
a weak stable state *iff* $\forall w \in S^{|\mathcal{I}}, \exists L \in \Pi_{g_i \in \mathcal{I}} D_i, w \neq v \Rightarrow (v, L, w) \notin T^{|\mathcal{I}}$.

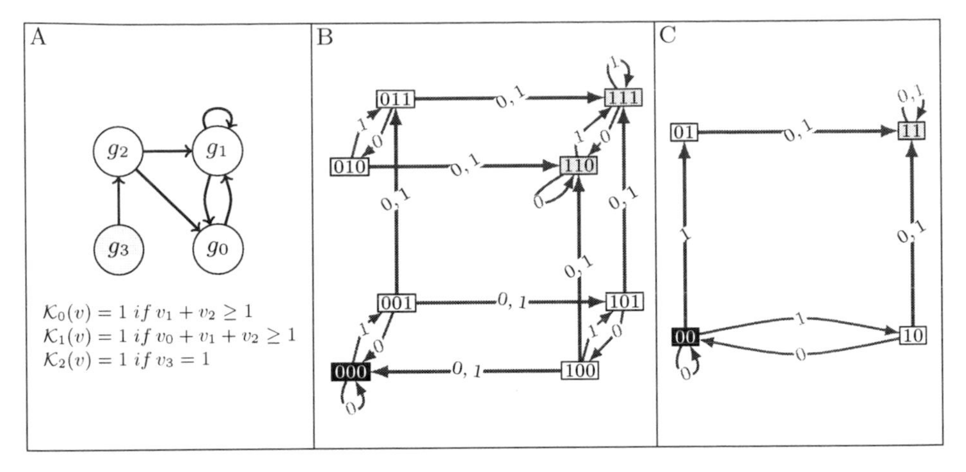

Fig. 2. (A) Simple example of a (Boolean) LSRG, with two proper components, an input and a pseudo-input, and the associated logical functions. Here $\widetilde{\mathcal{I}} = \{g_3, g_2\}$. **(B)** The labelled STG $\mathcal{E}^{|\{g_3\}}$, resulting from the projection over the input component g_3 and the labelling of the transitions with its values. The states 000, 110 and 111 are weak stable states. The last two states constitute a strong stable core ensemble (see text). **(C)** The labelled STG of the model where g_2 has been reduced. Here, the previous core ensemble {110, 111} can be recovered from the strong stable state 11.

These definitions are sufficient for fixed input components and also cover all the cases observed in our toy example (Figure 1, panel D), and in the T cell activation model described in Section 5. They are however not sufficient in other situations, where signalling cascades do vary upon input variations and the "core" network remains stable. This is illustrated in Figure 2.

It is thus necessary to better classify stable patterns, and identify what we would call stable core patterns (see states 110 and 111 in Figure 2). To assess pattern stability over input component variations, we rely on the behaviour of the components in *Core*, the set of components that belong to the core network (see Section 3.1).

Let $Stable \subset \mathcal{S}$ be the set of stable states for constant input components: $Stable = \{v \in \mathcal{S}, \mathcal{K}_i(v) = v_i, \forall i \in \mathcal{G}\}$. It is worth noting that GINsim provides a very efficient algorithm to determine this set [9,12]. Then, for all $v \in Stable$, we define $Core(v) = \{v' \in Stable, \forall g_i \in Core, v'_i = v_i\}$, the set of stable states that have the same values for the core components. Then, for varying inputs, we classify the stable states as follows,

- if $|Core(v)| = 1$, then v is a *weak stable state* (there is a unique input configuration for which this state is stable);
- if $\forall v' \in Core(v), \forall g_i \in \mathcal{P}, v_i = v'_i$, then v is a *strong stable state* (since all these stable states only differ in their input component values);
- otherwise ($|Core(v)| > 1$ and $\exists v' \in Core(v), \exists g_i \in \mathcal{P} \setminus Core, v_i \neq v'_i$), v defines what we could call a *stable core ensemble.* Then, similarly to the stable states, we could define *strong stable core ensembles* (such that $|Core(v)|$

equals the number of configurations of the input values) and *weak stable core ensembles*.

Note that if $v' \in Core(v) \subset Stable$, states v and v' necessarily share the same values on their output (and pseudo-output) components.

Another method to assess the stability of patterns upon input variations, would consist in reducing the input cascades (iteratively all the pseudo-input components in $\widetilde{\mathcal{I}} \setminus \mathcal{I}$). Then, the strong stable states and core ensembles of the original model are recovered from the (strong) stable states of this reduced model (see Figure 2, panel C). Similarly, weak stable states and core ensembles are recovered from the weak stable states of the reduced model. However, it is important to recall here that this reduction, although it conserves the number of stable states, could modify their reachability. This method is rather similar to that described by Saadatpour *et al.* who, for constant input values, consider the components that will reach stable values and propose to reduce them [15].

Although such considerations on stability upon input variations could also apply to complex attractors, we leave this extension for future work.

4 Implementation

The software GINsim is dedicated to the definition and analysis of logical models [9]. It provides, among a variety of functionalities, the reduction method presented in Section 2 [11]. Here, we briefly describe implementation aspects of the methodology presented in this paper. First, the reduction of the output cascades is implicitly made in GINsim. The model is then exported to be verified using the model-checker NuSMV. A new stable release of GINsim is expected in the near future. Meanwhile, a beta version of the tool with these new functionalities is available, along with supplementary material, at `http://compbio.igc.gulbenkian.pt/nmd/node/46`.

4.1 Output Nodes Manipulation in GINsim

GINsim [9] has been extended to automatically annotate output nodes based on the structure of the LSRG. We have added an internal method that identifies the pseudo-outputs and turns them into output components. For this, we apply the reduction method to remove references to pseudo-outputs from the logical functions of their targets. This trick ensures that all outputs can be defined as depending only on core components, their values can thus be computed independently. This is supported by the argument below.

Considering g_i an output and g_j a pseudo-output regulating g_i ($g_j \in Reg(g_i)$). $\mathcal{K}_j$, the logical function of g_j, is not modified, whereas $\mathcal{K}_i$, the logical function of g_i, is replaced by $\mathcal{K}_i^j$ its new function obtained by the reduction of g_j.

Note that, for the time being, the LSRG obtained through this manipulation is used only for the NuSMV export, where outputs are defined as macros, not characterising a state.

4.2 NuSMV Model Encoding and Verification

In [8], we have implemented an export functionality in the context of GINsim, making possible the use of the NuSMV symbolic model-checker [2]. This NuSMV model description enabled the consideration of the following updating policies: synchronous, asynchronous or priority classes. This export also permitted to distinguish between input and non-input components, enabling the reduction of the state space over input components (described in section 3.3).

Here, this GINsim export functionality is extended with the capability to distinguish between proper and output components (Section 3.2), containing only the logical rules governing the proper components. The input valuations are still projected over the transitions where they are valid, and the output valuations are computed as macros depending on the state (defined only by the proper components), permitting a reduction of the state space proportional to the number of (pseudo-)output components. Pseudo-outputs are tackled through the previously discussed method: from the NuSMV export point of view, they are treated just like outputs.

As mentioned earlier, to be able to represent information both on states (proper components) and on transitions (input components), we consider a graph structure known in the formal verification community as Kripke Transition System (KTS) [7]. The main version of NuSMV supports the verification of temporal logic properties over KTSs, but only when these properties do not make a reference to input components, *i.e.*, do not impose restrictions on the transitions of the KTS. This version only permit us to verify properties considering varying inputs, with no query on their values.

In order to perform verifications over KTSs with properties imposing restrictions on transitions, we consider a particular NuSMV extension, denoted NuSMV-ARCTL-TLACE [6]. This extension supports the ARCTL temporal logic [13], an extension of CTL [3], which adds action-restricted operators through an additional argument allowing the specification of restrictions on transitions over KTSs. Through the use of the ARCTL temporal logic, it is then possible to perform verifications over reduced models of signalling-regulatory networks, considering restrictions on their input components. These restrictions thus allow the verification of a property with specific (combinations of) values of the input components, or a chaining of CTL and ARCTL operators to verify reachability properties over KTS paths with unrestricted and specific input values, respectively.

A CTL temporal logic operator accepts as argument a set of restrictions over non-input components, *e.g.*, `EF(var1 & !var2)`, exploring each transition independently of the value of the input components. On the other hand, an ARCTL temporal logic operator accepts an additional argument defining a restriction over input components, *e.g.*, `EAF(inp4 & !inp7)(var1 & !var2)`, exploring only the transitions satisfying `inp4 & !inp7` (without imposing restrictions on other input components that might exist).

5 Application to the Model of T Cell Activation

T cells (or T lymphocytes) are immune cells reacting to the presence of specific antigens in the organism and playing a key role in the selection of the immune response. The T cell family is divided into several subfamilies, each with a specific role. Their antigen specificity arises from a randomly-generated membrane receptor: the T cell receptor (TCR). New T cells are continuously generated, then undergo a selection mechanism to avoid self-immunity before circulating in the organism. In the absence of their specific antigen, these cells will enter apoptosis after a few days. However, when it encounters its specific antigen, a T cell is activated and elicits the corresponding immune response. This activation is triggered by binding of the antigen on their T cell Receptor (TCR). Cells are kept alive as long as they receive TCR stimulation. The TCR activation pathway is thus a crucial part of the initiation and maintenance of a specific immune response.

The TCR recognises specific antigens (peptides) associated to the Major Histocompatibility Complex (MHC) on the surface of Antigen Presenting Cells. This recognition also involves TCR co-receptor (CD4 for T helper cells, CD8 for cytotoxic T cells) and is accompanied by CD28 co-stimulation. Saez-Rodriguez *et al.* [17] proposed a comprehensive logical model of the TCR activation pathway, taking into account the CD4 co-receptor and the action of the CD28 co-stimulatory molecule. This TCR activation model encompasses 94 components, including 35 proper components (see Figure 3).

The dynamical analysis performed in [17] focuses on short-term effects thus studying the states reached by signal propagation. This is done by impeding the feedback loops to function through the definition of slow events. Here, we are interested in the more complex behaviours that arise when feedback loops are taken into account. Firstly, we have defined a GINsim version of this model and confirmed that we obtain the same stable states in absence of the feedback loops, both in the wild-type condition and after applying model perturbations.

For the full model (with feedback loops), we observe that stable states only exist in the absence of TCR stimulation (with or without CD4 and CD28). Indeed, the TCR signal triggers oscillations, embodied in complex attractors. Note that such oscillations will be transient by nature as the TCR signal is not a stable stimulation. In what follows, we start by identifying the attractors (stable states in the absence of TCR signal and oscillations when TCR is present) using GINsim capabilities. We then resort to model checking to gain further insights into long term behaviours upon changes of the input signals; in particular, switches from one attractor to another.

5.1 Attractor Identification

To study the reachability of attractors by model-checking, we first need to identify these attractors, which is easy for stable states but more challenging for complex attractors as shown hereafter.

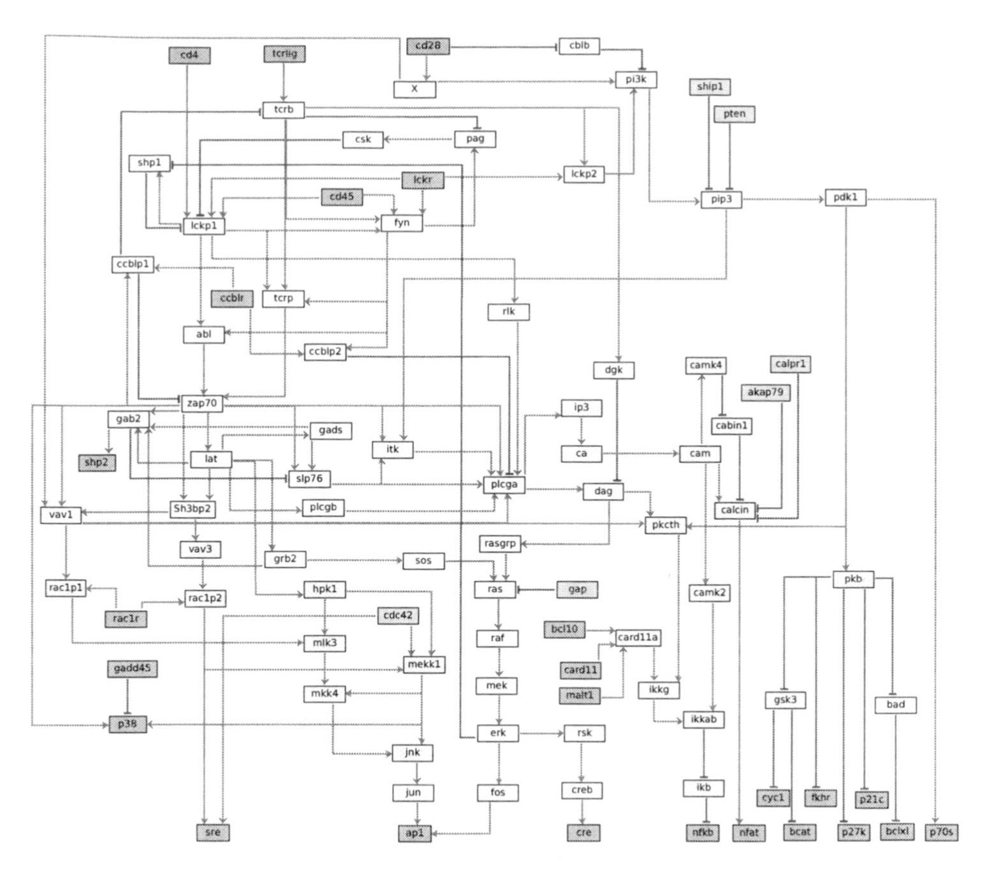

Fig. 3. TCR activation model proposed in [17]. This model encompasses 94 components: 3 inputs (blue), 14 fixed components (green for active and grey for inactive), and 14 outputs (orange), 28 pseudo-outputs (yellow), 35 proper components (white). Green arrows denote activations, while red T-arrows denote inhibitions.

Using the stable state search tool available in GINsim [12], we have identified all the stable states and, using the STG construction, we have checked their reachability from the initial state defined in [16]. This initial state comprises all components set to zero, except the three repositories `lckr`, `ccblr`, `rac1r` and five fixed components `cd45`, `gadd45`, `bcl10`, `card11` and `malt1`.

We have then identified the complex attractors by performing simulations in the presence of the TCR ligand. In this case, the computation of the STG is not tractable on the full model, even after the reduction of the output cascades. We have thus considered a further reduced model in which 10 internal components have been manually selected for reduction (aiming for a minimal impact of this reduction on the dynamical behaviour, we selected: `cblb`, `cblbp1`, `cblbp2`, `sos`, `lckp1`, `lckp2`, `mek`, `raf`, `ras`, and `X`). We recall that this reduction of core components ensures that no attractor can be lost, but may impede their reachability. We thus use this technic to identify complex attractors before using model-checking on the full model to assess their reachability. Using this simplified model, we could identify its complex attractors. Starting from states within these sets of states, we could further refine the descriptions of these attractors for the full model (without output cascades, though).

Table 1 illustrates the identified attractors. We named these attractors `SSxxx` or `SCCxxx` to represent weak stable states or strongly connected components, respectively. Additionally, we specified the name of each attractor according to the (set of) input combination(s) for which each attractor is stable, following the order `cd28`, `cd4` and `tcrlig`.

The set of states composing each complex attractor is covered by a schema, or pattern (see Table 1). For each complex attractor, we specify an ARCTL property and use the NuSMV-ARCTL-TLACE model-checker to verify that, for each of these patterns, no state covered by the pattern enables a transition leaving that set of states. The following property exemplifies the case of the pattern `SCC001` corresponding to complex attractor that arises in the absence of both `cd28` and `cd4` and the presence of `tcrlig`: `INIT SCC001; SPEC EAF(!cd28 & !cd4 & tcrlig)(!SCC001)`. All the equivalent properties defined for each complex attractor (see Supplementary files) returned `false`, ensuring that each pattern indeed captures the terminal SCC, containing at most some states belonging to its (strict) basin of attraction. The patterns describing the attractors given in Table 1, show that the weak stable state `SS0*0`, where `cd28` is absent, is part of the complex attractor `SCC001`, which in turn is part of the complex attractor `SCC011`. The analogous occurs when `cd28` is present in the case of the remaining attractors.

In the attractor summary of Table 1, considering the wild-type condition, we can observe that the `tcrlig` input variable is responsible for the existence (resp. absence) of oscillations, whenever it is present (resp. absent). This is confirmed by a circuit analysis, which provides the conditions under which the existing negative circuits are functional (see [12,14,19]). Here, the main functional negative circuit, `zap70 ccbblp1`, depends on the presence of `tcrp`, which in turn depends on the presence of `tcrb`, which is directly controlled by the value of the `tcrlig` input variable. The other negative circuit, `shp1 lckp1`, depends on the absence of `csk`, therefore on the presence of `tcrb`, but also directly depends on the `cd4` input variable.

While the other input components alone are not capable to trigger oscillations, we can note that in absence of `tcrlig`, `cd4` has no effect on the stable state identity, while `cd28` changes the activity of some proper and output components. In particular, `cd28` activates `pkb`, which triggers anti-apoptotic signals. This role of `cd28` is maintained in the presence of `tcrlig`, while `cd4` increases the number of oscillating components, dramatically increasing the number of states in the attractor.

If we consider more closely the activation pattern of some crucial output components, we can observe that `ap1`, `nfat` and `nfkb` share a pattern: their activation requires both `tcrlig` and one of the other inputs (`cd4` or `cd28`). This result is consistent with the fact that the TCR activation is crucial for the activity (through `nfat`) and survival (through `nfkb` and `pkb` in presence of `cd28`) of T cells.

Table 1. List of patterns for the wild-type, for the single mutant Δ`fyn` and double mutant Δ`fyn-lckr`. Two weak stable states (`SS0*0` and `SS1*0`) are shared by all conditions: wild-type, single Δ`fyn` mutant and double Δ`fyn-lckr` mutant. Light and dark grey cells, highlight the values for the proper and output variables, which depend on `cd28` and `cd4` input variables, respectively. The `tcrlig` input variable discriminates between the weak stable states and the complex attractors. Some proper and output variables are omitted for sake of space, their values being easily deduced from those listed in the table.

Conditions		Input			Internal																											Output				
		cd28	cd4	tcrlig	abl	ccblb	ccblp1	ccblp2	csk	dag	dgk	erk	fyn	gab2	itk	lat	lckp1	lckp2	mek	pag	pi3k	pip3	plcga	ras	shp1	slp76	tcrb	tcrp	vav1	X	zap70	ap1	jnk	nfkb	nfat	pkb
All	SS0*0	0	*	0	0	1	0	0	1	0	0	0	0	0	0	0	0	0	0	1	0	0	0	0	0	0	0	0	0	0	0	0	0	0	0	0
	SS1*0	1	*	0	0	0	0	0	1	0	0	0	0	0	0	0	0	0	0	1	1	1	0	0	0	0	0	0	1	1	0	0	1	0	0	1
Wild-type	SCC001	0	0	1	*	1	*	*	*	0	*	0	*	*	0	*	0	*	0	*	0	0	0	0	0	*	*	*	*	0	*	0	*	0	0	0
	SCC011	0	1	1	*	1	*	*	*	*	*	*	*	*	0	*	*	*	*	*	0	0	*	*	*	*	*	*	*	0	*	*	*	*	*	0
	SCC101	1	0	1	*	0	*	*	*	*	*	*	*	*	*	*	0	*	*	*	1	1	*	*	0	*	*	*	1	1	*	*	*	*	*	1
	SCC111	1	1	1	*	0	*	*	*	*	*	*	*	*	*	*	*	*	*	*	1	1	*	*	*	*	*	*	1	1	*	*	*	*	*	1
ΔFyn	SS001	0	0	1	0	1	0	0	0	0	1	0	0	0	0	0	0	1	0	0	0	0	0	0	0	0	1	0	0	0	0	0	0	0	0	0
	SS101	1	0	1	0	0	0	0	0	0	1	0	0	0	0	0	0	1	0	0	1	1	0	0	0	0	1	0	1	1	0	0	1	0	0	1
	SCC011	0	1	1	*	1	*	0	*	*	*	*	0	*	0	*	0	*	*	*	0	0	*	*	*	*	*	*	*	0	*	*	*	0	*	0
	SCC111	1	1	1	*	0	*	0	*	*	*	*	0	*	*	*	*	*	*	*	1	1	*	*	*	*	*	*	1	1	*	*	1	*	*	1
ΔFyn ΔLck	SS0*1	0	*	1	0	1	0	0	0	0	1	0	0	0	0	0	0	0	0	0	0	0	0	0	0	0	1	0	0	0	0	0	0	0	0	0
	SS1*1	1	*	1	0	0	0	0	0	0	1	0	0	0	0	0	0	0	0	0	1	1	0	0	0	0	1	0	1	1	0	0	1	0	0	1

We further performed the attractor search for a given set of perturbations and we observed that some of the complex attractors are replaced by stable states in the Δ`fyn` single mutant, as well as in the Δ`fyn-lckr` double mutant (see Table 1). In the Δ`fyn` single mutant, we observe that most of the proper components that underwent oscillations in the wild-type condition are now fixed at zero in the stable states `SS001` and `SS101`, while a small subset of them becomes tightly dependent on the presence of `tcrlig` (`dgk`, `lckp2` and `tcrb`). However, the remaining complex attractors `SCC011` and `SCC111` differ from the wild-type condition by having `fyn` and `ccblp2` not expressed, and `lckp1`, `jnk` and `nfat` dependent on the presence of `cd28`. The Δ`fyn-lckr` double mutant, additionally prevents the expression of `lckp2`, making the attractors `SS001` and `SS101` no longer dependent on the absence of `cd4`. It is worth remembering that Saez-Rodriguez *et al.* [17] performed the mutant simulations focusing only on "slow" events, thus breaking the feedback loops, which allows the activation of `pkb`. In this paper, considering the full model, we observe that `pkb` becomes tightly dependent on the presence of `cd28` due to the influence of `cblb` upon `pi3k`.

5.2 Reachability Analysis

After the identification and characterisation of the attractors, we have analysed their reachability, assessing which input conditions permit to reach (or leave) each attractor.

This is done by first encoding the patterns given in Table 1 in the NuSMV model description. The characterisation of the complex attractors considers only a restriction on the fixed variables (described by the corresponding pattern). Then, for each of the attractors, we specify a set of ARCTL temporal logic reachability properties, testing the existence of a path from each stable/complex attractor to every other attractor, for all the combinations of (varying or fixed) input components. These combinations of input components are obtained by fixing some of them using ARCTL temporal operators, possibly leaving the others to freely vary (see Supplementary files).

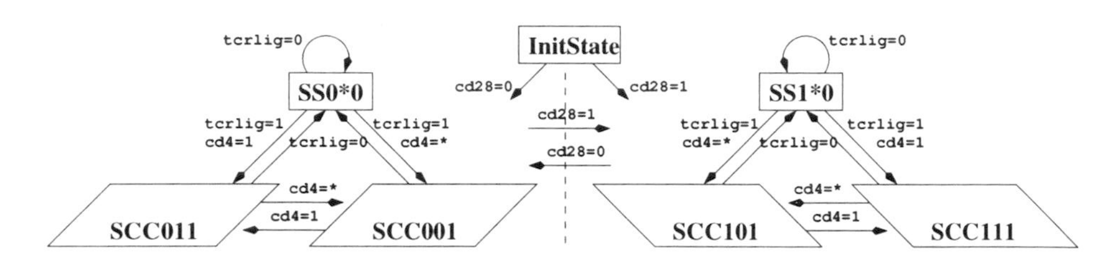

Fig. 4. State space characterization, of the necessary input conditions to switch between the stable and complex attractors specified in Table 1, with respect to the input variables valuation. It is worth noting that the `SS0*0` stable state is included in the `SCC001` complex attractor, and this is itself included in the `SCC011` complex attractor. The transitions between these attractors are then dependent on value restrictions of the `cd4` and `tcrlig` input variables. The analogy is valid for the `SS1*0`.

Figure 4 presents the verification results, indicating the necessary input conditions to switch between attractors. First, we confirm that the input component `cd28` divides the state space in half, setting the dynamics to focus on one group of attractors or the other. These are mirroring each other, where each is composed by a stable state and two complex attractors. Like previously mentioned, the presence of `tcrlig` controls the exit from a stable state towards its corresponding complex attractor and vice-versa. Finally, within each group of complex attractors, the presence (resp. absence) of the `cd4` input variable allows (resp. restricts) the dynamics to evolve to a larger (resp. smaller) set of states, `SCC011` or `SCC111` (resp `SCC001` or `SCC101`).

6 Conclusions and Prospects

The analysis of qualitative models of large signalling-regulatory networks is hampered by a combinatorial explosion of their state spaces. This is particularly true when properties of interest relate to reachability that often requires extensive search of the state transition graphs. Here, we propose to lessen this problem by a specific handling of input and output components. For the input components, their values are taken into account by proper labels on the transitions. This leads to a significant reduction when the model encompasses a large number of input components (as it is the case, for instance, in the model accounting for T cell differentiation defined in [10], which includes 13 inputs for a total number of 65 components). Furthermore, we used the reduction method as defined in [11] to get rid of output components and of what we called pseudo-output components. We prove that this reduction is lossless, in the sense that it preserves all the attractors and their reachability.

We aim at using the methodology presented here to revisit the T cell differentiation model [10]. In particular, we can now systematically analyse the impacts of input variations on attractor switches (accounting for a possible plasticity of the T cells), as well as mutant conditions.

In the near future, we will make the reduction of output cascades fully functional in GINsim. More precisely, upon user request, output cascades will be reduced and made implicit, STGs will be computed disregarding the corresponding variables, which values will be possibly recovered for given set of states (*e.g.*, in attractors).

In Section 3.3, we have discussed the determination of stable patterns, including in the case of varying input components. When inputs freely vary, pseudo-inputs also vary, but these variations may not affect the stability of the core network. We thus introduced the notions of (strong or weak) stable states and stable core ensembles. While GINsim implements an efficient algorithm to identify all stable states (for constant input components), we still need to delineate a method to determine the complex attractors. Indeed, for large models (as it is the case for the TCR model revisited in Section 5), the full characterisation of all the complex attractors is often difficult or even intractable. We then could extend the concepts of strong and weak stability to these complex attractors.

Ackowledgements. We acknowledge the support from the Portuguese FCT (PM post-doctoral fellowship SFRH/BPD/75124/2010 and project grant PTDC/EIACCO/099229/2008). We are grateful to D. Thieffry and E. Remy for inspiring discussions. The first version of the TCR model in GINsim was defined by D. Thieffry. Finally, we thank E. Mohr for her help to interpret the results concerning the TCR model.

References

1. Bilke, S., Sjunnesson, F.: Stability of the Kauffman model. Phys. Rev. E 65(016129) (2001)
2. Cimatti, A., Clarke, E., Giunchiglia, E., Giunchiglia, F., Pistore, M., Roveri, M., Sebastiani, R., Tacchella, A.: NuSMV 2: An OpenSource Tool for Symbolic Model Checking. In: Brinksma, E., Larsen, K.G. (eds.) CAV 2002. LNCS, vol. 2404, pp. 359–364. Springer, Heidelberg (2002)
3. Clarke, E.M., Emerson, E.A., Sistla, A.P.: Automatic verification of finite-state concurrent systems using temporal logic specifications. ACM Trans. Program. Lang. Syst. 8, 244–263 (1986)
4. de Jong, H.: Modeling and simulation of genetic regulatory systems: a literature review. J. Comput Biol. 1(9), 67–103 (2002)

5. Fauré, A., Naldi, A., Chaouiya, C., Thieffry, D.: Dynamical analysis of a generic boolean model for the control of the mammalian cell cycle. Bioinformatics 22(14), 124–131 (2006)
6. Lomuscio, A., Pecheur, C., Raimondi, F.: Automatic verification of knowledge and time with nusmv. In: Veloso, M.M. (ed.) Proc. 20th Intl. Joint Conf. on Artificial Intelligence (IJCAI 2007), pp. 1384–1389. Morgan Kaufmann Publishers Inc. (2007)
7. Müller-Olm, M., Schmidt, D., Steffen, B.: Model-Checking: A Tutorial Introduction. In: Cortesi, A., Filé, G. (eds.) SAS 1999. LNCS, vol. 1694, pp. 330–354. Springer, Heidelberg (1999)
8. Monteiro, P.T., Chaouiya, C.: Efficient Verification for Logical Models of Regulatory Networks. In: Rocha, M.P., Luscombe, N., Fdez-Riverola, F., Rodríguez, J.M.C. (eds.) 6th International Conference on PACBB. AISC, vol. 154, pp. 259–267. Springer, Heidelberg (2012)
9. Naldi, A., Berenguier, D., Fauré, A., Lopez, F., Thieffry, D., Chaouiya, C.: Logical modelling of regulatory networks with GINsim 2.3. Biosystems 97(2), 134–139 (2009)
10. Naldi, A., Carneiro, J., Chaouiya, C., Thieffry, D.: Diversity and plasticity of th cell types predicted from regulatory network modelling. PLoS Comput. Biol. 6(9) (2010)
11. Naldi, A., Remy, E., Thieffry, D., Chaouiya, C.: Dynamically consistent reduction of logical regulatory graphs. Theor. Comput. Sci. 412, 2207–2218 (2011)
12. Naldi, A., Thieffry, D., Chaouiya, C.: Decision Diagrams for the Representation and Analysis of Logical Models of Genetic Networks. In: Calder, M., Gilmore, S. (eds.) CMSB 2007. LNCS (LNBI), vol. 4695, pp. 233–247. Springer, Heidelberg (2007)
13. Pecheur, C., Raimondi, F.: Symbolic Model Checking of Logics with Actions. In: Edelkamp, S., Lomuscio, A. (eds.) MoChArt IV. LNCS (LNAI), vol. 4428, pp. 113–128. Springer, Heidelberg (2007)
14. Remy, E., Ruet, P.: From minimal signed circuits to the dynamics of boolean regulatory networks. Bioinformatics 24(16), i220–i226 (2008)
15. Saadatpour, A., Albert, I., Albert, R.: Attractor analysis of asynchronous boolean models of signal transduction networks. J. Theor. Biol. 266(4), 641–656 (2010)
16. Saez-Rodriguez, J., Simeoni, L., Lindquist, J., Hemenway, R., Bommhardt, U., Arndt, B., Haus, U.-U., Weismantel, R., Gilles, E., Klamt, S., Schraven, B.: A logical model provides insights into T cell receptor signaling. PLoS Comput. Biol. 3(8), e163 (2007)
17. Saez-Rodriguez, J., Simeoni, L., Lindquist, J.A., Hemenway, R., Bommhardt, U., Arndt, B., Haus, U.-U., Weismantel, R., Gilles, E.D., Klamt, S., Schraven, B.: A logical model provides insights into T cell receptor signaling. PLoS Comput. Biol. 3(8), e163 (2007)
18. Schlitt, T., Brazma, A.: Current approaches to gene regulatory network modelling. BMC Bioinformatics 8(suppl. 6), S9 (2007)
19. Thieffry, D.: Dynamical roles of biological regulatory circuits. Brief. Bioinform. 8(4), 220–225 (2007)
20. Thomas, R.: Regulatory networks seen as asynchronous automata: A logical description. J. Theor. Biol. 153, 1–23 (1991)

Sequence Dependent Properties of the Expression Dynamics of Genes and Gene Networks

Ilya Potapov[1], Jarno Mäkelä[1], Olli Yli-Harja[1,2], and Andre Ribeiro[1]

[1] Computational Systems Biology Research Group, Department of Signal Processing, Tampere University of Technology, P.O. Box 527, FIN-33101 Finland
[2] Institute for Systems Biology, 1441N 34th St, Seattle, WA, 98103-8904, USA

Abstract. The sequence of a gene determines the protein sequence and structure, but to some extent also the kinetics of protein production. Namely, the DNA and the codon sequence affect the kinetics of transcription and translation elongation, respectively. Here, using a stochastic model of transcription and translation at the nucleotide and codon levels, we investigate the effects of the codon sequence on the dynamics of single gene expression and of a genetic switch. We find that the ribosome binding site region sequence affects mean expression rates. In the genetic toggle switch, the sequence is shown to affect the switching frequency.

Keywords: gene expression, codon sequence, dynamics, genetic switch.

1 Introduction

The dynamics of gene expression has an essential effect on the prokaryotic phenotype. Given that these organisms have a short life time and most genes are transcribed only a few times in a cell's lifetime [1], the moment when genes are transcribed, the mean frequency of transcription and the variability in the intervals between transcription events are relevant in determining how a cell behaves and responds, for example, to environmental changes.

The kinetics of gene expression is subject to multiple regulatory mechanisms at various stages. Some of these mechanisms and their effects have been studied by measurements and models. For example, several studies characterized how the kinetics of transcription initiation affects both mean and fluctuations in RNA and protein numbers [2, 3]. The effects of other sequence dependent events and mechanisms on gene expression dynamics still require much study. One example of such a mechanism is transcriptional pausing, which has been suggested to be an important regulator of transcription in both prokaryotes and eukaryotes [4–6].

There are several sequence dependent events in translation elongation whose effects on the kinetics of protein production, and possible regulatory roles, still require much study. In [7] different mutants of the LacZ gene were used to study the kinetics of translation of individual codons. The sequences were designed so as to

D. Gilbert and M. Heiner (Eds.): CMSB 2012, LNCS 7605, pp. 307–321, 2012.

enhance queue formation and traffic in elongation. One sequence corresponded to the wild-type lacZ while the other two differed in that a region of slow-to-translate codons was inserted. The speed of protein chain elongation was measured by subjecting the cells to a pulse of radioactive methionines and measuring the level of radioactivity in cells of each population, every 10 s after the pulse. Each strand contained 23 methionines, spread out unevenly on the DNA sequence, causing the incorporation curve to be non-linear. The study revealed that the translation elongation speed of these strands differed, as the speed of incorporation of an amino acid depends on which synonymous codon is coding for it.

These studies allowed the development of realistic kinetic models of transcription at the nucleotide level [8, 9] and translation at the codon level [10, 11]. In [10] it was shown that measurements of sequence dependent translation rates of synonymous codons cannot be modeled by neither deterministic nor uniform stochastic models. For proper mimicking of the kinetics, the models need to include explicit translation elongation. Another study [11] showed that the degree of coupling between fluctuations in RNA and protein numbers is sequence dependent and that events in elongation, such as sequence-specific transcriptional pauses, affects the temporal levels of proteins.

Here, using a detailed stochastic model of transcription and translation in bacteria [11], we investigate how the codon sequence of a gene affects the dynamics of expression of proteins. The codon sequences are randomly generated according the natural codons frequency of occurrence in *E. coli*.

We first study how the protein numbers differ between models with randomly generated codon sequences and the underlying causes. Next, we compare the kinetics of the model with measured mean expression levels of genes differing only in synonymous codon usage [12]. After, we introduce sequences of codons at the start and at the end of the RNA sequence, known as 'slow ramps', as these are found in many genes in *E. coli* [13]. We compare the kinetics of protein production of these sequences with those of 'null models' with uniform codons translation efficiencies. Finally, we investigate whether codons translation efficiencies affect the kinetics of genetic circuits, namely, genetic switches [14].

2 Methods

In prokaryotes, gene expression comprises several events that cannot be modeled as common bimolecular chemical reaction since they take a relevant time to be completed once initiated and the chemical species involved exist in very small numbers in the cell. Thus, to model the dynamics of gene expression we use the delayed Stochastic Simulation Algorithm (delayed SSA) [9], which while with stochastic kinetics, it allows delaying the release of products of a reaction in the system following the reactive event. For that, it stores products in a waitlist and release them after enough time has elapsed. The delayed SSA is based on the SSA, a method for the exact numerical simulation of well-stirred chemically reacting systems [15]. In both methods, each chemical species is a variable of integer value and time advances in discrete steps. Each time a reaction occurs, the number of molecules of the species involved is updated according to the reaction's formula. All simulations throughout the work were performed using the simulator for stochastic gene networks "SGNSim" [16].

The model of gene expression used [11] is at the nucleotide and codon level. By modeling transcription and translation elongation at this level it allows starting translation as soon as the ribosome binding site region of the RNA is formed, which was found to be necessary to accurately account for the fluctuations in RNA and protein numbers in bacterial gene expression [11]. Further, it allows including events at each elongation step, such as transcriptional pauses in transcription elongation and back-translocation in translation elongation.

The reactions modeling transcription and the stochastic rate constants are shown in Table 1. The model of transcription at the nucleotide level [8] includes transcription initiation (1), promoter occupancy time (τ_{oc} in (1)), and promoter clearance (2) [2]. It follows elongation, which includes nucleotide activation (3) [17] and stepwise elongation (4) [17]. At the end of this process, a complete RNA molecule and the RNA polymerase (Rp) are released in the system (12) [18].

Several events compete with elongation at each nucleotide, such as transcriptional pauses (5) [19]. These can end spontaneously (5) or by collisions with preceding Rp's (6) [20]. Collisions can also induce pauses (7). There are ubiquitous arrests and release from arrest (8) [8], misincorporation and editing (9) [19], premature terminations (10) [21], and pyrophosphorolysis (11) [22]. The model accounts for the nucleotides occupied by an Rp on the DNA strand. Finally, two RNA polymerases can never occupy simultaneously the same nucleotide. Reaction (13) models RNA degradation [23].

The reactions modeling translation and the stochastic rate constants are shown in Table 2. The delayed stochastic model of translation at the codon level includes initiation (14) [10], and ribonucleotide activation (15) [24] followed by stepwise translocation (16–18) [10]. Reactions competing with translocation are back-translocation (19) [25], ribosome drop-off (20) [26], and transtranslation (21) [27]. It ends with elongation completion [10] followed by protein folding (22) [28]. The model accounts for the ribonucleotides occupied by a ribosome when on the RNA. Also accounted for are codon-specific translation rates [7], implying that each codon has a specific translocation rate. While, in general, this rate is either k_{trA}, k_{trB} or k_{trC}, in some models we introduce other rates so as to model the slow ramps. Finally, proteins undergo degradation (23) [28]. Finally, the time for a ribosome to move from the Ribosome Binding Site (RBS) region to the start codon [10, 29] is accounted in the kinetic rate of the translation initiation chemical rate constant k_{tl}.

Ribosomes on the mRNA strand occupy 31 nucleotides, covering ΔR (=15) nucleotides upstream and ΔR nucleotides downstream from the nucleotide being processed. This region cannot be occupied by other ribosomes. Initiation proceeds as follows. A ribosome binds to the first nucleotide of the growing mRNA strand, occupying ΔR nucleotides downstream. The first nucleotide only becomes available for the next ribosome to bind when the previous ribosome

Table 1. Chemical reactions, rate constants (in s^{-1}), and delays (in s) used to model transcription. Pro — promoter, Rp — RNA polymerase, U — unoccupied nucleotide and O — nucleotide occupied by Rp, A — activated nucleotide (after processing by Rp), n denotes the number of a nucleotide in the sequence under process. $(2\Delta_P + 1)$ — range of nucleotides that Rp occupies, $\Delta_P = 12$ (range occupied upstream/downstream from the nucleotide being processed).

1. Initiation and promoter complex formation	$\mathrm{Pro} + \mathrm{Rp} \xrightarrow{k_{init}} \mathrm{Rp} \cdot \mathrm{Pro}(\tau_{oc})$	$k_{init} = 0.0245$, $\tau_{oc} = 40 \pm 4$
2. Promoter clearance	$\mathrm{Rp} \cdot \mathrm{Pro} + \mathrm{U}_{[1,(\Delta_P+1)]} \xrightarrow{k_m} \mathrm{O}_1 + \mathrm{Pro}$	$k_m = 150$
3. Activation	$\mathrm{O}_n \xrightarrow{k_a} \mathrm{A}_n$	$k_a = 150, n > 10$; $k_a = 30, n \leq 10$
4. Elongation	$\mathrm{A}_n + \mathrm{U}_{n+\Delta_P+1} \xrightarrow{k_m} \mathrm{O}_{n+1} + \mathrm{U}_{n-\Delta_P} + \mathrm{U}^R_{n-\Delta_P}$	$k_m = 150$
5. Pausing	$\mathrm{O}_n \underset{1/\tau_p}{\overset{k_p}{\rightleftharpoons}} \mathrm{O}_{n_p}$	$k_p = 0.55$, $\tau_p = 3$
6. Pause release by collision	$\mathrm{O}_{n_p} + \mathrm{A}_{n-2\Delta_P-1} \xrightarrow{0.8k_m} \mathrm{O}_n + \mathrm{A}_{n-2\Delta_P-1}$	$k_m = 150$
7. Pause by collision	$\mathrm{O}_n + \mathrm{A}_{n-2\Delta_P-1} \xrightarrow{0.2k_m} \mathrm{O}_{n_p} + \mathrm{A}_{n-2\Delta_P-1}$	$k_m = 150$
8. Arrests	$\mathrm{O}_n \underset{1/\tau_{ar}}{\overset{k_{ar}}{\rightleftharpoons}} \mathrm{O}_{n_{ar}}$	$k_{ar} = 0.000278$, $\tau_{ar} = 100$
9. Editing	$\mathrm{O}_n \underset{1/\tau_{ed}}{\overset{k_{ed}}{\rightleftharpoons}} \mathrm{O}_{n_{corr}}$	$k_{ed} = 0.00875$, $\tau_{ed} = 5$
10. Premature termination	$\mathrm{O}_n \xrightarrow{k_{pre}} \mathrm{Rp} + \mathrm{U}_{[(n-\Delta_P),(n+\Delta_P)]}$	$k_{pre} = 0.00019$
11. Pyrophosphorolysis	$\mathrm{O}_n + \mathrm{U}_{n-\Delta_P-1} + \mathrm{U}^R_{n-\Delta_P-1} \xrightarrow{k_{pyr}} \mathrm{O}_{n-1} + \mathrm{U}_{n+\Delta_P-1}$	$k_{pyr} = 0.75$
12. Completion	$\mathrm{A}_{last} \xrightarrow{k_f} \mathrm{Rp} + \mathrm{U}_{[\mathrm{last}-\Delta_P,\mathrm{last}]} + \mathrm{mRNA}$	$k_f = 2$
13. mRNA degradation	$\mathrm{mRNA} \xrightarrow{k_{dr}} \emptyset$	$k_{dr} = 0.025$

moves by at least $2\Delta R + 2 = 32$ nucleotides, thus revealing the first $\Delta R + 1 = 16$ nucleotides of the RNA needed for the next ribosome to bind. This length is referred to as RBS region of the RNA.

Trans-translation corresponds to the release of the ribosome from the RNA template after stalling, which can occur for several reasons, such as incorporation of an incorrect codon, premature mRNA degradation, or frame-shifting [27, 30]. In the model, stalling followed by trans-translation can occur spontaneously with a given probability at any codon via (21). When occurring, the mRNA strand is degraded and all translating ribosomes are released. Estimates from the observation of expression activity in *E. coli* suggest that, on average, 0.4% of translation reactions are terminated by trans-translation [27], meaning that the probability of occurrence of this event at each nucleotide depends on the length

Table 2. Chemical reactions, rate constants (in s^{-1}), and delays (in s) used to model translation. Rib — ribosome, $[Rib^R]$ — number of translating ribosomes on mRNA strand, P — complete protein, U^R — unoccupied nucleotide and O^R — nucleotide occupied by Rib, A^R — activated nucleotide (after processing by Rib), n denotes the number of a nucleotide in the sequence. $(2\Delta_R + 1)$ — range of ribonucleotides that ribosome occupies, $\Delta_R = 15$ (range occupied upstream/downstream from the nucleotide being processed).

14. Initiation	$Rib + U^R_{[1,\Delta_R+1]} \xrightarrow{k_{tl}} O^R_1 + Rib^R$	$k_{tl} = 0.53$
15. Activation	$O^R_n \xrightarrow{k_{tr\{A,B,C\}}} A^R_n$	Codon dependent: $k_{trA} = 35$; $k_{trB} = 8$; $k_{trC} = 4.5$
16–18. Stepwise translocation	$A^R_{n-3} + U^R_{[n+\Delta_R-3,n+\Delta_R-1]} \xrightarrow{k_{tm}} O^R_{n-2}$ $O^R_{n-2} \xrightarrow{k_{tm}} O^R_{n-1}$ $O^R_{n-1} \xrightarrow{k_{tm}} O^R_n + U^R_{[n-\Delta_R-2,n-\Delta_R]}$	$k_{tm} = 1000$
19. Back-translocation	$O^R_n + U^R_{[n-\Delta_R-2,n-\Delta_R]} \xrightarrow{k_{bt}} A^R_{n-3} + U^R_{[n+\Delta_R-3,n+\Delta_R-1]}$	$k_{bt} = 1.5$
20. Drop-off	$O^R_n \xrightarrow{k_{drop}} Rib + U^R_{[n-\Delta_R,n+\Delta_R]}$	$k_{drop} = 1.14 \cdot 10^{-4}$
21. Trans-translation	$mRNA \xrightarrow{k_{tt}} [Rib^R] \times Rib$	$k_{tt} = 0.000159$
22. Elongation completion and protein folding	$A^R_{last} \xrightarrow{k_{tlf}} Rib + U^R_{[last-\Delta_R,last]} + P(\tau_{fold})$	$k_{tlf} = 2$, $\tau_{fold} = 420 \pm 100$
23. Protein degradation	$P \xrightarrow{k_{dp}} \emptyset$	$k_{dp} = 0.0029$

of the gene. Since we only model sequences 1000 nucleotides long, the trans-translation reaction is always set to the same value, defined by k_{tt} (Table 2).

We generate codon sequences randomly, according to the known statistical frequency of each codon (extracted from NCBI GenBank, Dec. 1st, 2011) [31]. Slow ramps at the start or end of a sequence are set to be 35 codons long [13]. The codons activation rates (k_{tr} in (15)) change linearly with the codon position in the ramp, with k_{tr} values ranging from 1 to 35 s^{-1}, varying by 1 s^{-1} per codon.

In addition to the reactions from Table 1 and 2 for the genetic toggle switch there are reactions modeling the interactions between the repressor proteins and the promoters' regions to which they bind to:

$$\textbf{24. } Pro_i + P_j \underset{k_{unrep}}{\overset{k_{rep}}{\rightleftharpoons}} Pro_i \cdot P_j \,,$$

where $i, j = 1, 2$ and $i \neq j$.

Reaction (24) models the binding (k_{rep}) of a repressor protein to a promoter region and its unbinding (k_{unrep}). The expected time that the promoters are available for transcription can be regulated by these rates. For this study the default values are: $k_{rep} = 1\,s^{-1}$, and $k_{unrep} = 0.1\,s^{-1}$.

To characterize the dynamics of a switch we quantify, from time series of RNA and protein numbers, the stability of the switch, that is, the robustness of its noisy attractors to fluctuations in protein and RNA numbers. For a wide range of parameter values, the delayed stochastic switch has two noisy attractors [32, 33]. We assess the stability from the number of switches between the noisy attractors in the time series. We define a switch as an event that satisfies the following conditions: the number of one of the proteins becomes, from one moment to the next, bigger than the number of the other protein, and there is no other switch event within the next 1000 s (so that a single switch between noisy attractors is not counted multiple times). Given these conditions, the stability, S, is:

$$S = \frac{T}{n+1}, \tag{1}$$

where T is the total time of simulation and n is number of switches during T. Thus, S is the average time between two consecutive switches.

3 Results and Discussion

3.1 Kinetics of Gene Expression for Randomly Generated Codon Sequences

Based on the frequency of occurrence of codons in *E. coli* [31] we generated 30 random sequences, each 1000 nucleotides long. The sequences differ only in the frequency of occurrence of its codons. We performed 200 simulations of the kinetics of gene expression of each sequence, each 10^5 s long with a sampling frequency of 1 s^{-1}. Reactions and stochastic rate constants are shown in Table 1 and Table 2. Since simulations are initialized without mRNA or proteins, in all calculations below we disregard an initial transient of $2{\cdot}10^3$ s, found sufficient to allow mRNA and proteins to reach numbers near-equilibrium.

First, we computed the mean mRNA (E(RNA)) and protein (E(P)) numbers in each simulation. Distributions of the mRNA mean numbers at near equilibrium are show in Fig. 1(left) for the 30 sequences as well as for a single sequence, chosen at random from the 30 sequences. This allows comparing the variability between simulations due to the stochasticity in the kinetics. Similarly, we show the same distributions for mean protein numbers in Fig. 1(right), to compare the variability in protein numbers between simulations due to stochasticity and differing codon sequences.

Figure 1 (left) shows that there is no significant difference in mRNA kinetics for different codon sequences, as expected. However, the kinetics of protein production in Fig. 1 (right), differs between the 30 sequences causing the overall distribution to be much wider than the distribution of the individual sequences (Fig. 1 (right)). Such differences arise solely from the differing rates of protein elongation and thus overall production, as the kinetics of transcription and RNA

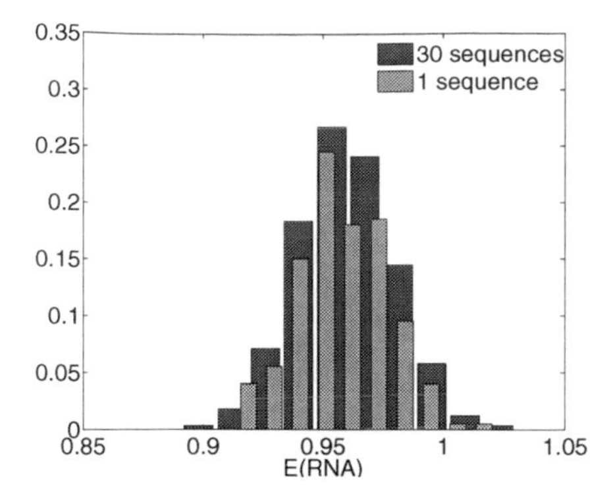

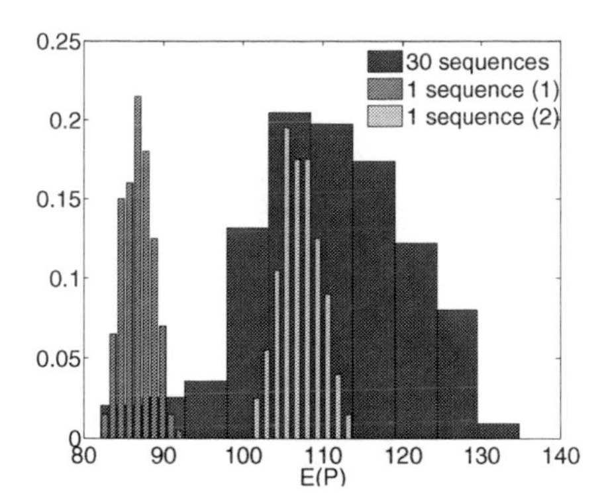

Fig. 1. (Left) Distribution of mean mRNA numbers (E(RNA)) and (Right) mean protein numbers (E(P)) at near-equilibrium from 200 simulations of individuals sequences. Also shown are the distributions resulting from 30 sequences, each of which simulated 200 times.

and proteins degradation do not differ between models. The different proteins production rate is due to differing codon sequences. These affect the mean translation elongation time, in particular, how long ribosomes remain, on average, in the RBS region, which affects the mean of the intervals between initiations of translation events.

To demonstrate this, in Fig. 2 (left) we show the mean rate of translation events per time unit versus the mean protein numbers for each simulation, for two sequences (arbitrarily named sequences 1 and 2). There is a clear positive correlation between the two quantities, showing that the codon sequence affects the rate of translation initiation events. Having a faster rate of translation initiation events leads to a higher number of premature terminations during translation (ribosome drop-offs) (Fig. 2 (right)). However, this is not sufficient to cancel out the effect of having different rates of initiation on the mean protein numbers.

We next analyze the fluctuations in protein numbers. From the formula for the stationary variance in protein abundance [34], one expects that differences in mean protein numbers cause the stationary variances to differ as well due to the low-copy noise term ($(\langle p \rangle^{-1})$) and because the kinetics of RNA production does not differ between sequences. In Fig. 3 (top) we show the CV^2 (square of the coefficient of variation) in protein numbers at near-equilibrium for the 30 sequences, averaged over all simulations for each sequence. Also shown are the distributions of protein numbers from the sequence that exhibited the lowest CV^2 (Fig. 3, bottom left) and from the sequence that exhibited the highest CV^2 (Fig. 3, bottom right), out of the 30 sequences.

The differences in CV^2 in protein numbers between sequences are of the order of 5%. Note that this quantity is also affected by the stochasticity in protein degradation which tends to diminish differences arising from the kinetics of production. The relevance of these differences is, as noted later, context dependent. While, at the single gene level they are likely not very significant, at a gene

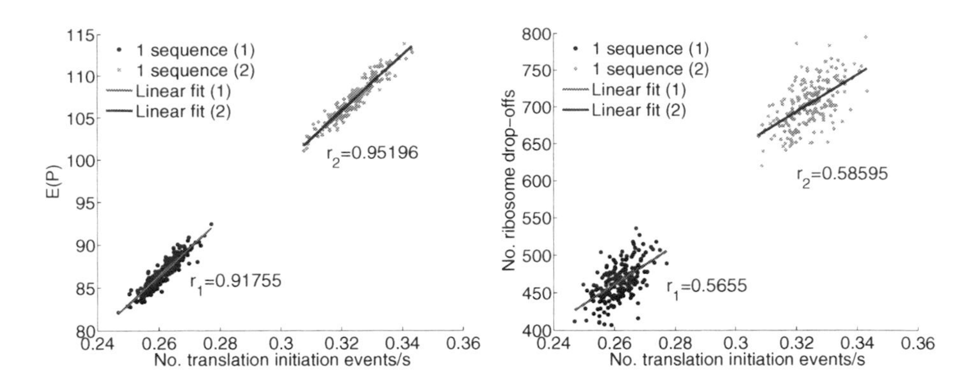

Fig. 2. Left: mean number of protein E(P) is positively correlated (with correlation coefficients r_i) with number of translation events. Right: number of ribosome drop-offs is positively correlated with number of translation events. Data shown for the same two codon sequences, denoted as 1 and 2.

network level they may be of significance, e.g., in allowing to cross thresholds in protein numbers due to fluctuations.

The above results are derived from simulations using codon sequences randomly generated. However, genes' sequences are under selection and thus have codon sequences that are far from random. Recent DNA analysis appears to support this hypothesis, providing evidence for the existence of slow ramps in specific regions of the mRNA sequences, relative to the RBS region [13]. Next, we study the kinetics of expression of RNA sequences with such ramps.

3.2 Kinetics of Gene Expression in the Presence of Ramps of Slow Codons

Recent studies revealed universally conserved profiles of RNA translation efficiencies in different classes of organisms including bacteria [13]. In *E. coli*, many mRNA sequences are such that the first 15–35 codons are, on average, translated with lower efficiency than the rest of the sequence. Some genes also have slow ramps at the end of the sequence. In general, these ramps exhibit an approximately linear increase/decrease in translation speed from beginning to end when at the start/end of the RNA sequence [13].

In this section, we study the gene expression kinetics of such sequences. We introduce linearly increasing codon rates for codons in positions 1 to 35 (counting forward from the translation start site) or linearly decreasing codon rates for codons in positions -35 to -1 (counting backwards from the stop codon). The rest of the sequence is generated randomly as previously. We also generated three sequences with uniform codon activation rates (models "A", "B" and "C"), corresponding to the three groups of codon activation rates (k_{trA}, k_{trB}, k_{trC} in Table 2). These are used as "null models" that we compare with those with slow ramps. We simulate each of the five models (the three null models and the two

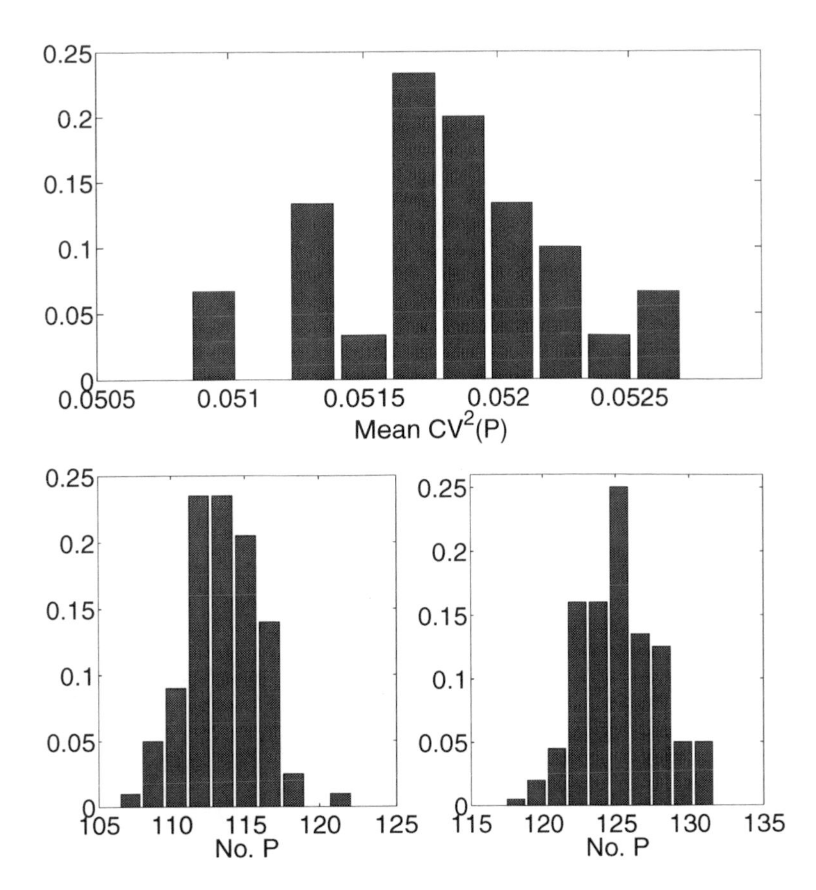

Fig. 3. Distribution of mean $CV^2(P)$ of protein numbers for the 30 different codon sequences (top) and distributions of protein numbers(P) for the models that exhibited the lowest (bottom left) and the highest (bottom right) values of $CV^2(P)$. All distributions are obtained from the complete time series except for an initial transient of $2{\cdot}10^3$ s since the networks are initialized without mRNA and proteins.

models with a slow ramp at the start and at the end) 200 times, each simulation lasting 10^5 s with 1 s^{-1} sampling frequency. Results are shown in Fig. 4.

From Fig. 4, the model with a slow ramp at the start of the sequence exhibits the smallest E(P), due to having the slowest rate of protein production, along with null model A. It is of interest to note these two models barely differ in kinetics, supporting the hypothesis that the mean rate of protein production is largely controlled by the sequence of the RBS region of the RNA. The model with a slow ramp at the end, along with null model B, have an E(P) distribution similar to the E(P) distributions from randomly generated codon sequences (see Fig. 1 (right)). From this, we conclude that adding slow ramps in the end of the sequences does affect significantly the mean rate of protein production.

Interestingly, while the ramps affect E(P), they do not affect the degree of fluctuations in protein numbers (Fig. 5). From Fig. 5, the distributions of values

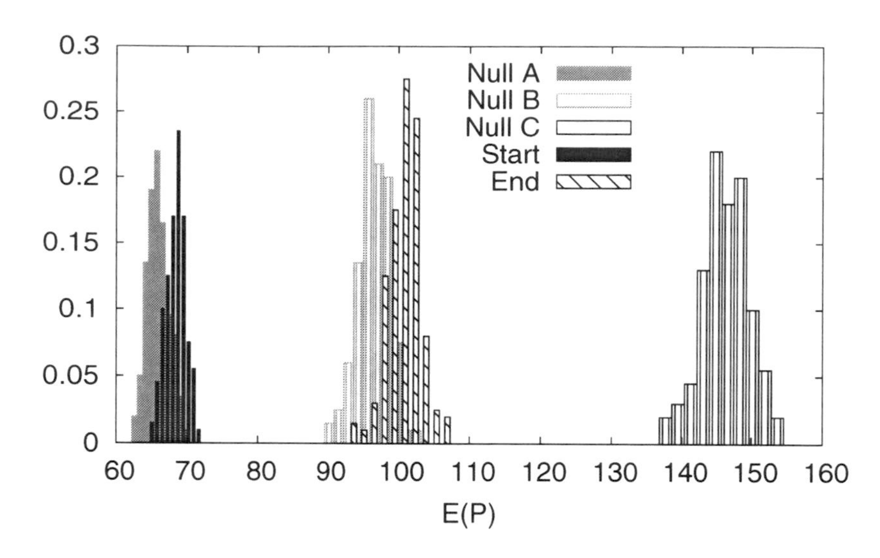

Fig. 4. Distributions of mean protein numbers, E(P), at near-equilibrium of the two models with a slow ramp at the start ("Start") and the end ("End") and of the three null models with the uniform codons sequences. Null model A (slowest codon activation rate, 4.5 s^{-1}/codon for all codons), null model B (8 s^{-1}/codon for all codons) and, null model C (fastest activation rate, 35 s^{-1}/codon for all codons).

of CV^2 of the sequences with a ramp at the start and end are identical (from the same simulations used for Fig. 4). These distributions are also identical to those of the null models (data not shown).

3.3 Effects of Ramps of Slow Codons on Bistable Genetic Circuits

We next study the effect of ramps in the kinetics of a genetic toggle switch (TS) [14], which consists of two genes, each expressing a protein that represses the expression of the other gene. Due to the interactions that define the structure of a genetic switch, usually one protein level is "high" while the other is "low". In the context of stochastic genetic circuits, regions of the state space where the network remains in for long periods of time, yet can leave due to fluctuations in the molecule numbers, are usually referred to as "noisy attractors" [35]. The term is used instead of the usual concept of attractor, since stochastic systems do not have attractors. 2-gene switches usually have two noisy attractors [35].

All model genetic toggle switches in this study have genes that are 1000 nucleotides long each. All models are simulated 200 times for either 10^5 or 10^6 s, sampled every second, and are described by reactions 1–24 and corresponding kinetic parameters from Sec. 2.

First, we build 10 random switches, having the same codon sequences for both genes. In order to assess variability of stability of the switches, obtained by randomly varying the codon composition of genes belonging to the switch,

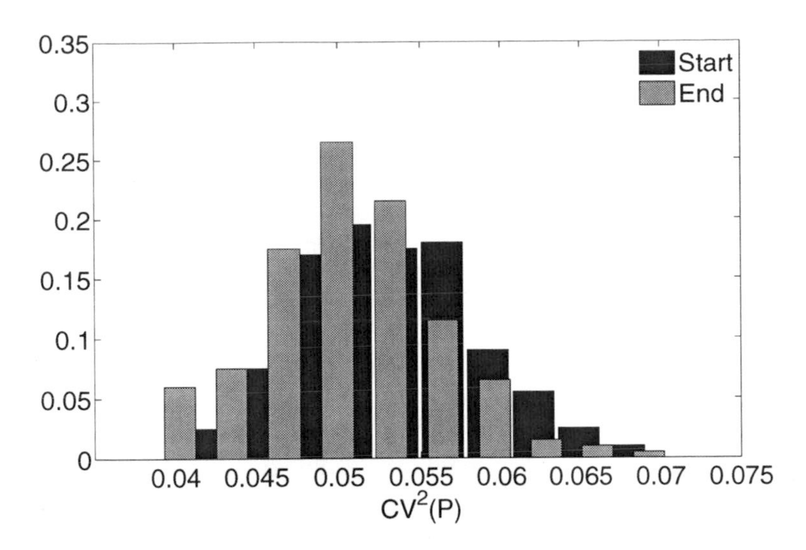

Fig. 5. Distribution of mean CV^2 of protein numbers of the models with a slow ramp at the start and at the end of the RNA sequence

we simulate those 10 switches for 10^5 s and compute distributions of stability according to (1). The results are merged from all switches simulated and shown in Fig. 6 (left). The figure also shows the contribution of a single toggle switch to the overall distribution. Mean and standard deviation for the total distribution in Fig. 6 (left) are 15820 and 8304 s, respectively, giving the total variation of stability possible due to randomly varied codon sequence of genes within the switch.

Next, we build null model TS, where the codon sequence is formed from codons from one of the three classes of codons (A, B or C) according to their translation rate (see reaction 15 in Table 2). Thus, we have three null model switches with uniform codon rates distribution along the sequence. We simulated three models 200 times for 10^5 s and computed stability. Resulting distributions are shown in Fig. 6 (right). One can see that the bigger translation rate of the null model, the

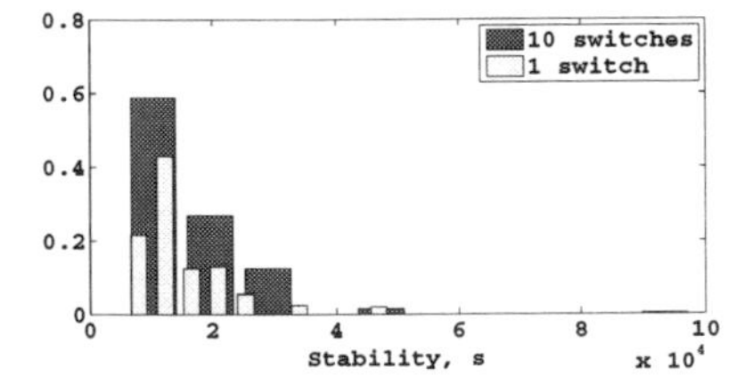

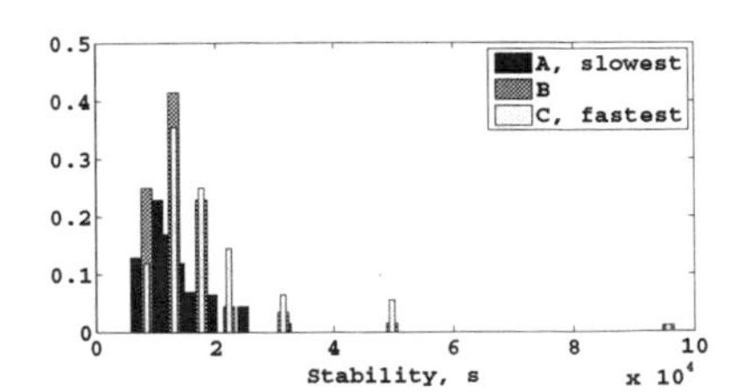

Fig. 6. Total stability of 10 toggle switches with randomly generated codon sequences (both genes of a switch have the same sequence); also shown is example of one of those distributions (left). Stability of three null model toggle switches with uniform codon translation rates (right).

bigger the tail of the distribution towards larger values of stability and, thus, the bigger mean stability. The bigger rate of translation of a codon provides less time that a ribosome spends on the codon and thus the faster the production of the outcoming protein is. Increasing production rate increases the mean protein numbers and, consequently, stability increases.

Next, we compare the kinetics of other three model switches. One switch has random codon sequences, generated as previously. Another switch has slow ramps at the start of the sequences of both genes. The last model switch has slow ramps at the end of the sequences of both genes. In all cases, both genes are 1000 nucleotides long.

We simulated the dynamics of each of the three model switches 200 times, each simulation 10^6 s long, sampled every second. From the time series of protein numbers we calculated the stability of the noisy attractors given by (1).

Compared to the model switch without ramps, the slow ramp at the end of the sequences does not alter S significantly due to the weak effects on E(P) numbers. However, the ramp at the start of the sequences causes S to decrease by ~25%. Thus, codon sequences can significantly affect the kinetics of small genetic circuits and the location of a slow ramp, relative to the RBS region, is relevant to the degree of change in the dynamics (Fig. 7).

Finally, we simulated a model switch such that only one gene has a slow ramp, located at the start of the sequence (here named 'biased switch'). Again, 200 simulations were performed. This model exhibits a stability increase of ~400% in comparison to the model without ramps. As such, we conclude that ramps can bias the switch's behavior and this bias has a strong effect in the ability of the switch to 'hold state'.

In Fig. 8 we show examples of the kinetics of protein numbers of two model switches. In the left panel of Fig. 8 is shown the kinetics of the model switch without ramps, while in the right panel is shown the kinetics of the model switch where one of the two genes has a ramp, placed at the start of the RNA sequence. Note the bias in the choice of noisy attractor in the latter model.

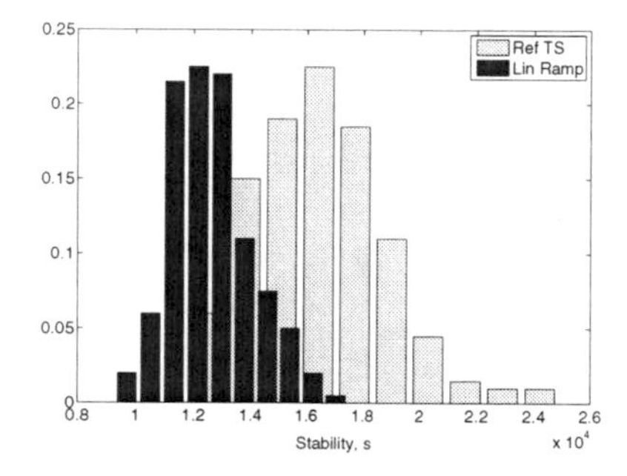

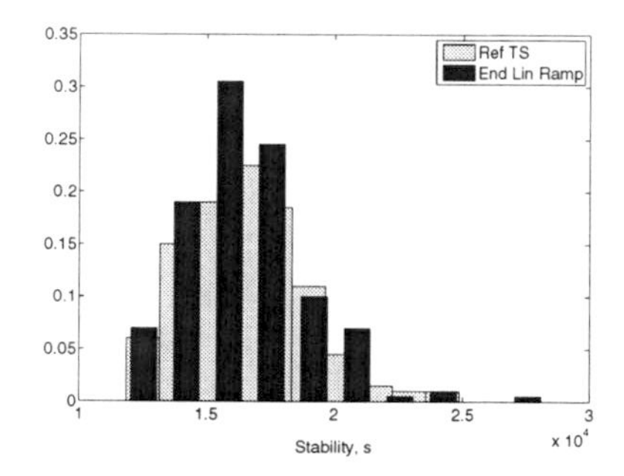

Fig. 7. (Left) Slow ramp in the beginning of each sequence of the toggle switch ('Lin Ramp') decreases stability as compared to the switch with randomly generated codon sequence ('Ref TS'). (Right) Slow ramp in the end of each sequence of the toggle switch ('End Lin Ramp') does not alter significantly stability as compared to the switch with randomly generated codon sequence.

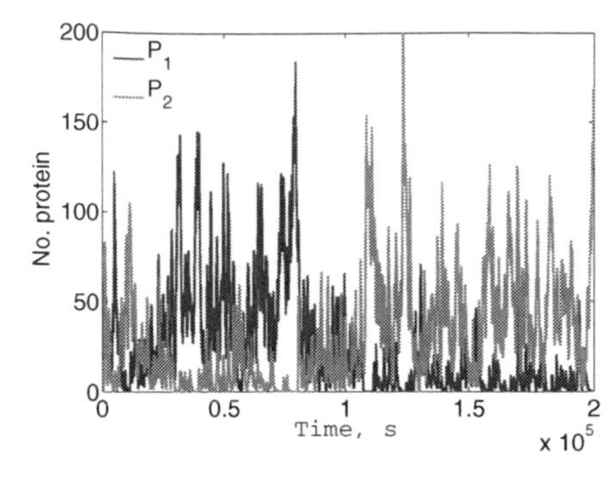

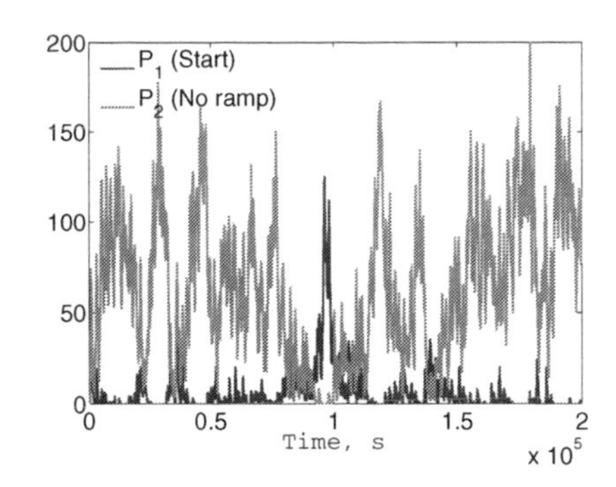

Fig. 8. Examples of the kinetics of protein numbers of the model toggle switch. Left: model toggle switch with codon sequences randomly generated. Right: model toggle switch with one gene (gene 1) having a slow ramp at the start of the RNA sequence

4 Conclusions

Using a model of transcription and translation at nucleotide and codon levels with realistic parameters values extracted from measurements in *E. coli* we studied how the codon sequence affects the kinetics of protein production of individual genes and the dynamics of genetic toggle switches.

From the studies of the kinetics of single gene expression of genes with codon sequences generated randomly according to the natural probabilities of occurrence in *E. coli*, we found that the codon sequence affects the mean protein numbers in near-equilibrium significantly, due to affecting the rate of translation initiation (by changing how long ribosomes remain at the RBS region). Interestingly, the noise in protein numbers, as measured by square of coefficient of variation (CV^2), does not change significantly with the codon sequence, indicating that altering the codon bias may allow to engineer more strongly expressing genes with weaker fluctuations in protein numbers.

Slow ramps were found to influence the kinetics of both single genes as well as genetic switches. Relevantly, their effect likely depends on their location relative to the RBS region, and both our kinetics studies, as well as sequence analysis studies [13], suggest that selection may have shaped these ramps. The strongest effects are attained by placing a slow ramp at the start of the sequence. In [13] it was hypothesized that slow ramps can reduce ribosomal jams by averaging the spacing between ribosomes on the mRNA strand. Our results support the hypothesis of reduction of ribosomal jams, and further indicate that this is achieved mostly by reducing the rate of translation initiation, rather than by averaging the spacing between ribosomes (although this also occurs). At the level of small genetic circuits we found evidence that slow ramps have tangible effects in the kinetics of bistable networks. Again, effects are stronger when slow ramps are placed at the start of the RNA sequences.

In conclusion, our results suggest that the codon sequences affect significantly the dynamics of single gene expression and genetic circuits of prokaryotes. These results may be of relevance in synthetic biology, when engineering genetic circuits for specific purposes, and in computational biology, when modeling genetic circuits. Finally, they may assist in a better understanding of the evolutionary pressures that genomes are subject to.

Acknowledgments. Work was supported by Academy of Finland (ASR), CIMO (IP), and FiDiPro program of Finnish Funding Agency for Technology and Innovation (JM, OYH, and ASR). The funders had no role in study design, data collection and analysis, decision to publish, or preparation of the manuscript.

References

1. Taniguchi, Y., Choi, P., Li, G., Chen, H., Babu, M., Hearn, J., Emili, A., Xie, X.: Quantifying *E. coli* proteome and transcriptome with single-molecule sensitivity in single cells. Science 329(5991), 533–538 (2010)
2. McClure, W.: Rate-limiting steps in RNA chain initiation. PNAS 77, 5634–5638 (1980)
3. Arkin, A., Ross, J., McAdams, H.: Stochastic kinetic analysis of developmental pathway bifurcation in phage λ-infected *Escherichia coli* cells. Genetics 149(4), 1633–1648 (1998)
4. Herbert, K., Porta, A.L., Wong, B., Mooney, R., Neuman, K., Landick, R., Block, S.: Sequence-resolved detection of pausing by single RNA polymerase molecules. Cell 125(6), 1083–1094 (2006)
5. Landick, R.: The regulatory roles and mechanism of transcriptional pausing. Biochem. Soc. Trans. 34(pt. 6), 1062–1066 (2006)
6. Bender, T., Thompson, C., Kuehl, W.: Differential expression of c-myb mRNA in murine B lymphomas by a block to transcription elongation. Science 237(4821), 1473–1476 (1987)
7. Sørensen, M., Pedersen, S.: Absolute in vivo translation rates of individual codons in *Escherichia coli*: The two glutamic acid codons gaa and gag are translated with a threefold difference in rate. Journal of Molecular Biology 222(2), 265–280 (1991)
8. Ribeiro, A., Smolander, O., Rajala, T., Häkkinen, A., Yli-Harja, O.: Delayed stochastic model of transcription at the single nucleotide level. J. Comp. Biol. 16(4), 539–553 (2009)
9. Roussel, M., Zhu, R.: Validation of an algorithm for delay stochastic simulation of transcription and translation in prokaryotic gene expression. Phys. Biol. 3, 274–284 (2006)
10. Mitarai, N., Sneppen, K., Pedersen, S.: Ribosome collisions and translation efficiency: Optimization by codon usage and mRNA destabilization. Journal of Molecular Biology 382, 236–245 (2008)
11. Mäkelä, J., Lloyd-Price, J., Yli-Harja, O., Ribeiro, A.: Stochastic sequence-level model of coupled transcription and translation in prokaryotes. BMC Bioinformatics 12(1), 121 (2011)
12. Welch, M., Govindarajan, S., Ness, J., Villalobos, A., Gurney, A., Minshull, J., Gustafsson, C.: Design parameters to control synthetic gene expression in Escherichia coli. PLoS ONE 4(9), e7002 (2009)
13. Tuller, T., Carmi, A., Vestsigian, K., Navon, S., Dorfan, Y., Zaborske, J., Pan, T., Dahan, O., Furman, I., Pilpel, Y.: An evolutionarily conserved mechanism for controlling the efficiency of protein translation. Cell 141(2), 344–354 (2010)

14. Gardner, T., Cantor, C., Collins, J.: Construction of a genetic toggle switch in *Escherichia coli*. Nature 403(6767), 339–342 (2000)
15. Gillespie, D.: Exact stochastic simulation of coupled chemical reactions. J. Phys. Chem. 81, 2340–2361 (1977)
16. Ribeiro, A.S., Lloyd-Price, J.: SGN Sim, a stochastic genetic networks simulator. Bioinformatics 23, 777–779 (2007)
17. Proshkin, S., Rahmouni, A., Mironov, A., Nudler, E.: Cooperation between translating ribosomes and RNA polymerase in transcription elongation. Science 328(5977), 504–508 (2010)
18. Greive, S., Weitzel, S., Goodarzi, J., Main, L., Pasman, Z., von Hippel, P.: Monitoring RNA transcription in real time by using surface plasmon resonance. Proc. Natl. Acad. Sci. USA 105(9), 3315–3320 (2008)
19. Greive, S., von Hippel, P.: Thinking quantitatively about transcriptional regulation. Nat. Rev. Mol. Cell. Biol. 6, 221–232 (2005)
20. Epshtein, V., Nudler, E.: Cooperation between RNA polymerase molecules in transcription elongation. Science 300(5620), 801–805 (2003)
21. Lewin, B.: Genes IX. Jones and Bartlett Publishers, USA (2008)
22. Erie, D., Hajiseyedjavadi, O., Young, M., von Hippel, P.: Multiple RNA polymerase conformations and GreA: control of the fidelity of transcription. Science 262(5135), 867–873 (1993)
23. Yu, J., Xiao, J., Ren, X., Lao, K., Xie, S.: Probing gene expression in live cells, one protein molecule at a time. Science 311, 1600–1603 (2006)
24. Wen, J., Lancaster, L., Hodges, C., Zeri, A., Yoshimura, S.H., Noller, H.F., Bustamante, C., Tinoco, I.: Following translation by single ribosomes one codon at a time. Nature 452(7187), 598–603 (2008)
25. Shoji, S., Walker, S., Fredrick, K.: Ribosomal translocation: one step closer to the molecular mechanism. ACS Chem. Biol. 4(2), 93–107 (2009)
26. Jørgensen, F., Kurland, C.: Processivity errors of gene expression in *Escherichia coli*. Journal of Molecular Biology 215(4), 511–521 (1990)
27. Moore, S., Sauer, R.: Ribosome rescue: tmRNA tagging activity and capacity in *Escherichia coli*. Mol. Microbiol. 58(2), 456–466 (2005)
28. Cormack, B., Valdivia, R., Falkow, S.: Facs-optimized mutants of the green fluorescent protein (GFP). Gene. 173, 33–38 (1996)
29. Ringquist, S., Shinedling, S., Barrick, D., Green, L., Binkley, J., Stormo, G., Gold, L.: Translation initiation in *Escherichia coli*: sequences within the ribosome-binding site. Molecular Microbiology 6(9), 1219–1229 (1992)
30. Keiler, K.: Biology of trans-translation. Annu. Rev. Microbiol. 62, 133–151 (2008)
31. Benson, D., KarschMizrachi, I., Lipman, D., Ostell, J., Wheeler, D.: Genbank: update. Nucleic Acids Research 32(suppl. 1), D23–D26 (2004)
32. Ribeiro, A.: Dynamics of a two-dimensional model of cell tissues with coupled stochastic gene networks. Phys. Rev. E 76(5), 051915 (2007)
33. Ribeiro, A., Dai, X., Yli-Harja, O.: Variability of the distribution of differentiation pathway choices regulated by a multipotent delayed stochastic switch. J. Theor. Biol. 260(1), 66–76 (2009)
34. Pedraza, J., Paulsson, J.: Effects of molecular memory and bursting on fluctuations in gene expression. Science 319(5861), 339–343 (2008)
35. Ribeiro, A., Kauffman, S.: Noisy attractors and ergodic sets in models of gene regulatory networks. Journal of Theoretical Biology 247(4), 743–755 (2007)

Simulating Insulin Infusion Pump Risks by *In-Silico* Modeling of the Insulin-Glucose Regulatory System*

Sriram Sankaranarayanan[1] and Georgios Fainekos[2]

[1] University of Colorado, Boulder, CO.
firstname.lastname@colorado.edu
[2] Arizona State University, Tempe, AZ.
fainekos@asu.edu

Abstract. We present a case study on the use of robustness-guided and statistical model checking approaches for simulating risks due to insulin infusion pump usage by diabetic patients. Insulin infusion pumps allow for a continuous delivery of insulin with varying rates and delivery profiles to help patients self-regulate their blood glucose levels. However, the use of infusion pumps and continuous glucose monitors can pose risks to the patient including chronically elevated blood glucose levels (hyperglycemia) or dangerously low glucose levels (hypoglycemia).

In this paper, we use mathematical models of the basic insulin-glucose regulatory system in a diabetic patient, insulin infusion pumps, and the user's interaction with these pumps defined by commonly used insulin infusion strategies for maintaining normal glucose levels. These strategies include common guidelines taught to patients by physicians and certified diabetes educators and have been implemented in commercially available insulin bolus calculators. Furthermore, we model the failures in the devices themselves along with common errors in the usage of the pump. We compose these models together and analyze them using two related techniques: (a) robustness guided state-space search to explore worst-case scenarios and (b) statistical model checking techniques to assess the probabilities of hyper- and hypoglycemia risks. Our technique can be used to identify the worst-case effects of the combination of many different kinds of failures and place high confidence bounds on their probabilities.

1 Introduction

The goal of this paper is to combine physiological models of the insulin-glucose regulatory system in diabetic patients with medical device models of infusion pumps and continuous glucose meters to perform *in silico* risk assessments. Modern treatments for type-1 and 2 diabetes mellitus require frequent, periodic monitoring of blood glucose levels and the subcutaneous delivery of artificial insulin. Developments in medical device technologies have enabled software-controlled insulin infusion pumps that can deliver precise amounts of insulin in user programmable patterns. Likewise, advances in sensor technologies have enabled continuous glucose monitors (CGM) that can be

* This work was funded by National Science Foundation (NSF) grants under award numbers CPS-1035845, CNS-1016994 and CNS-1017074. All opinions expressed here are those of the authors and not necessarily of the NSF.

D. Gilbert and M. Heiner (Eds.): CMSB 2012, LNCS 7605, pp. 322–341, 2012.

used to sense the concentration of glucose subcutaneously. These technologies have enabled the development of automatic and manual control strategies, vastly improving the ability of patients to achieve normal glycemic control [12,19,37].

However, the use of these devices and the control strategies are prone to hazards arising from device, software and usage errors. A comprehensive list of these hazards has been compiled by Zhang et al. [43]. These risks primarily arise due to factors such as (a) failures in insulin pumps due to software errors, occlusions, and pump failures; (b) calibration and dropout errors in the glucose monitors; and (c) usage errors including discrepancies between planned and actual meals, incorrect insulin-to-carbs ratios, sensitivity factors, and basal insulin levels. These hazards can expose the patient to significant levels of hypoglycemia (low levels of blood glucose) or hyperglycemia (high levels of glucose), each of which leads to dangerous complications including loss of consciousness for hypoglycemia and ketacidosis for significant hyperglycemia. The long term consequences of elevated post-prandial glucose levels include kidney damage (nephropathy) and eye damage (retinopathy). Given the severity of these risks, a careful study of the various kinds of faults involved in the infusion process and the associated risks is of great importance.

In this paper, we create mathematical models of the overall infusion process by modeling the components involved in an infusion. Our model incorporates physiological models of the gluco-regulatory system [22,11,7,36,9,27,29], models of the various devices involved, the user's infusion strategy [34] and some of the possible faults that can arise during the infusion process [43]. We specify metric temporal logic (MTL) properties [25] for the executions of this model which include (a) absence of hypoglycemia ($G(t) \geq 3$ mmol/L), (b) absence of significant hyperglycemia, ($G(t) \leq 20$ mmol/L), and (c) settling of the blood glucose level to a normal range 3 hours after a meal ($\forall\ t, t \geq 150 \;\Rightarrow\; G(t) \in [4, 10]$ mmol/L). Unfortunately, the resulting models are nonlinear and include discrete mode switches due to the user and device models. Existing symbolic verification tools are inadequate for exhaustively exploring the behaviors of the overall model. Therefore, we adapt two recent approaches based on simulations:

Robustness-Guided Model Checking: We use robustness-guided sampling, assuming that faults are non-deterministic [28,4] to explore the possible worst case scenarios involving a combination of faults. Robustness guided sampling is based on the idea of providing real-valued robustness semantics to formulas in metric temporal logic [30,16,17]. The robustness of a trace w.r.t a given specification can be used as an objective function for a global stochastic optimization approach that seeks to minimize the robustness to falsify a given temporal property. This is a suitable approach in cases where the model is infinite state and non-linear. Such models are generally not amenable to existing symbolic verification techniques. In this paper, we employ this scheme to search for combinations of faults that can cause severe hyperglycemia, hypoglycemia and delayed return to normal glycemic levels following a meal.

Statistical Model Checking: We use statistical model checking by associating probabilities with faults to quantify the risk of hyper- and hypoglycemia with some confidence interval bounds [42,10,44]. Statistical model checking (SMC) repeatedly simulates a stochastic system while evaluating probabilistic temporal logic queries with high confidence. SMC approaches allows us to place bounds on the probability that a formula holds for a given stochastic system.

The models described in this paper are integrate inside the Matlab Simulink/Stateflow(tm) modeling environment. We use our tool S-Taliro which incorporates robustness-guided state-space exploration using many different global optimization engines including Monte-Carlo search [28], Ant-colony optimization [3], genetic algorithms and cross-entropy sampling [32][1]. Recently, we have extended S-Taliro to support Bayesian Statistical Model Checking (SMC) [23,44]. Using S-Taliro, we examine numerous fault scenarios involving a combination of faults to analyze the worst-case scenarios arising from these situations and to quantify the risk of hyper or hypoglycemia, assuming some prior probabilities for the various faults.

To our knowledge, the use of statistical and robustness-guided model checking to analyze infusion risks in the insulin infusion pump setting is novel. Previously, there have been attempts at quantifying risks involved in model-predictive controllers (MPC) for overnight glycemic control using a simulation environment [40,20]. These simulations derive a risk score for hypoglycemia risk using numerous simulations. However, no confidence intervals are derived for the risk scores. The use of SMC in this paper provides a more systematic and potentially less computationally expensive approach. Jha et al. employ statistical model checking to discover parameters for a PI controller for managing insulin infusion pumps using the Bergman minimal model [24].

Assessing risks in infusion pumps has received increasing attention recently [43,5,39,38]. Our previous work considered the effects of infusion risks in a hospital setting using drug infusion pumps and linear phramacokinetic models [33]. Therein, we were able to employ bounded-model checking techniques for linear hybrid automata to drive the worst-case search. Currently, the state-of-the-art in symbolic verification techniques are inapplicable to our model which involves non-linear dynamics with switching.

2 Overview

We provide a brief overview of the problem and the proposed solutions.

Insulin Infusion Scenario: Consider a commonly occurring scenario of planning a meal for a patient suffering from type-1 (insulin dependent) diabetes. The patient uses an insulin infusion pump to deliver an appropriate bolus dosage of insulin [2] before the meal commences. The planning process requires the patient to decide on the following parameters:

1. The insulin bolus amount to be infused through an insulin infusion pump,
2. The timing of the bolus relative to the planned meal time,
3. The width of the bolus,
4. The timings and amounts of any planned corrective dosages to accommodate higher than normal post-prandial blood glucose levels.

[1] S-Taliro can be downloaded for free from `https://sites.google.com/a/asu.edu/s-taliro/`

[2] A bolus dosage is a fixed amount of a drug that is delivered over a relatively short period of time to achieve a near-term effect.

Typically, patients suffering from diabetes undergo training by physicians, certified diabetes educators and numerous books on the topic to arrive at suitable strategies for planning meal infusions [34]. A typical calculation that is often automated by an insulin bolus calculator involves the steps detailed below using the planned meal data:

- Divide the amount of carbohydrates in the meal by a personal *insulin-to-carbs ratio* to obtain an appropriate bolus,
- Decide on the timing, shape and width of the bolus based on the Glycemic Index (GI) of the planned meal and the blood glucose reading measured prior to the meal,
- Decide on a correction bolus a few hours after the meal by measuring blood glucose levels and dividing it by a personal *insulin sensitivity factor*.

There are many rules of thumb for deciding upon an appropriate insulin-to-carbs ratio or an insulin sensitivity factor. Often, the patients are required to carefully monitor and adjust these ratios until they can achieve good glycemic control. However, there are numerous risks involved in a typical infusion that can lead to elevated blood glucose levels (hyperglycemia) or very low levels (hypoglycemia). A few commonly occurring faults are summarized below (Zhang et al. provide an exhaustive list of hazards [43]):

1. Software errors in the insulin pump, affecting its ability to deliver insulin of the specified amount and at the specified rates.
2. Calibration errors in the glucose monitors, whose readings are used to compute the correction bolus.
3. Mismatches between the *planned meal* used in the bolus calculations and the *actual meal* consumed.
4. Incorrect timing of the insulin dosage.
5. Incorrect usage of insulin infusion pump (eg., entering a wrong dosage, incorrect bolus shape, unit errors).
6. Failures due to occlusions or pump hardware faults.

The systematic study of the effects of the faults on the overall infusion process is necessary to find and remedy common causes that may result in significant hyperglycemia (elevated blood glucose levels) causing dangerous conditions such as ketacidosis, or hypoglycemia (low blood glucose levels) that may lead to a loss of consciousness or a dangerous coma in the worst case. For instance, it is natural to ask questions such as (a) what are the worst-case effects of a particular single fault or a combination of faults? (b) given probabilities of individual faults, what is the overall probability of a severe hypoglycemia?

While it is possible to predict the qualitative effects of a single fault in isolation, the combined effect of multiple faults are often be hard to predict *quantitatively*. Naturally, a clinical study with real patients using the pump with various controls is the gold standard for providing answers to some of the questions above. However, such studies are expensive, requiring a large set of participants since some of the faults occur infrequently.

An emerging line of research consists of modeling the various components involved in the infusion: the infusion pump, the user's meal planning strategies, physiological models of the insulin-glucose regulation, the glucometer incorporating models of the various faults that may occur [40,27,29]. Such models can then be analyzed for finding worst-cases and their likelihood. This can often point the way towards improving the process to make it safer for patients.

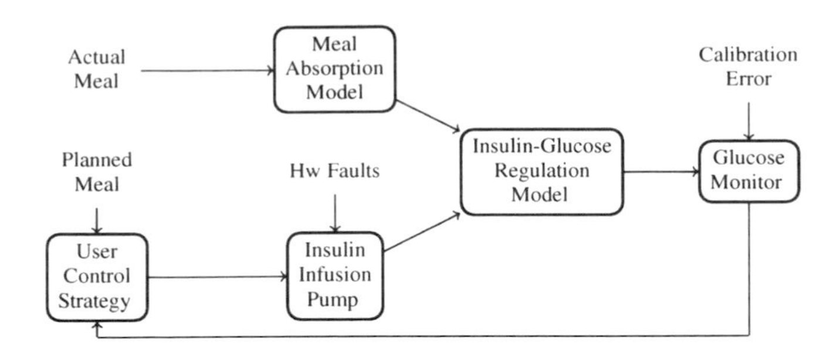

Fig. 1. Key components in modeling meal insulin infusion pump usage scenario

Modeling and Simulation: Figure 1 shows the basic components that are modeled in this scenario and the interactions between these models. Our approach integrates models of the insulin-glucose regulatory system [22,40], meal absorption models [41], a minimal infusion pump model, glucose meter models [40,15] and a model of the patient's usage of the pump to cover meals. The latter model is based on an understanding of common pump usage recommendations by physicians and certified diabetes educators [34]. These sub-models are integrated to yield a Matlab Simulink(tm) model.

Figure 2 shows a set of possible blood glucose and insulin levels over time obtained by running simulations with randomly chosen glucose monitor calibration errors and discrepancies between planned and actual meals. The shaded region shows acceptable limits for glucose levels. We note that presence of errors and faults have the effect of potentially causing hyperglycemia as well as hypoglycemia. However, repeated simulations do not suffice to explore worst-case scenarios. If simulations are performed uniformly at random, the number of simulations required to uncover these scenarios is often prohibitively large. The analysis techniques used here explore the worst case outcomes using state-space exploration guided by trace robustness [28,4] and estimate the probability of hypo- and hyperglycemia, given the probabilities for the individual machine faults and user errors [42,23].

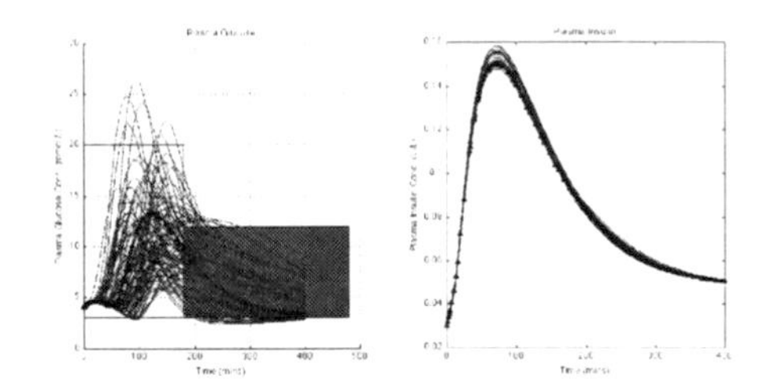

Fig. 2. Variation of plasma glucose (left) and plasma insulin (right) concentrations with varying calibration errors and infusion timings. The reference trajectories with no infusion faults is highlighted.

3 Background

We first provide some brief background on diabetes mellitus and its treatment using intensive insulin therapy. More information on topics related to diabetes can be obtained from clinical textbooks on this topic [35].

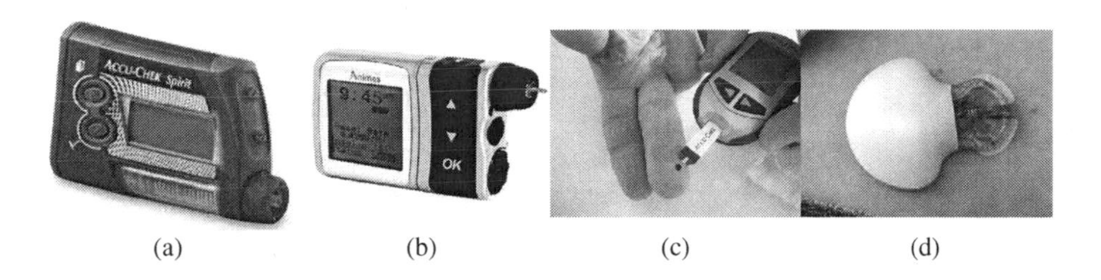

(a) (b) (c) (d)

Fig. 3. (a,b) Commercial insulin infusion pump models, (c) blood glucose monitor and (d) continuous sub-cutaneous glucose monitors

Insulin-Glucose Regulatory System. Diabetes mellitus is the generic name for a class of diseases where critical parts of the natural glyco-control system fail. Type-1 diabetes results from the loss of pancreatic insulin secretion due to to auto-immune destruction of insulin producing β cells in the pancreas. Likewise, type-2 diabetes results from insulin sensitivity, wherein damage to insulin receptors in the cells makes the action of insulin weaker, resulting in the inability of the pancreas to keep up with the demand.

Diabetes is a commonly occurring ailment in the developed world as well as the developing world. A common treatment for chronic diabetes involves the external delivery of artificial insulin (or insulin analogs) directly through a syringe, or sub-cutaneously through an insulin infusion pump. The everyday delivery of insulin is controlled by the patient with advance knowledge of their activities such as diet and exercise. Furthermore, diabetic patients are required to monitor their blood glucose levels intermittently through "finger stick" blood glucose monitors, or recently by continuous glucose monitors (CGMs) that provide a continuous reading of the subcutaneous glucose levels.

Insulin Infusion Pump and Continuous Glucose Monitors. We will now review some of the basics of insulin infusion pumps and blood glucose monitors. The monograph by Chee and Fernando contains further details [9].

An insulin infusion pump is a device that delivers insulin at a programmable rate over time. Insulin infusion pumps have been shown to deliver insulin accurately even when the requested rate of delivery is very small. This allows the pump to deliver insulin continuously throughout the day at a basal rate to counteract the endogenous production of glucose in the body. Furthermore, pumps also allow for various bolus doses of insulin to be infused before or just after meals to limit the occurrence of hyperglycemia following a meal. The shape, width and amount of the bolus can be fine tuned according to the planned meal. Starchy foods such as rice have a high glycemic index, requiring rapid infusion of (short-acting) insulin while fat and protein-rich foods have a lower glycemic index, requiring an infusion with a spread out peak (eg., square-wave bolus).

A continuous glucose monitor (CGM) provides frequent estimates of the blood glucose level by sensing the amount of glucose subcutaneously in the interstitial fluid. Currently available CGM devices provide readings that can be quite accurate. Furthermore, these devices can communicate wirelessly with a computer or an insulin pump to transmit readings directly. CGMs have proven useful in providing feedback to diabetic patients and their physicians to improve the patient's ability to achieve normoglycemia.

Table 1. Commonly used mathematical models of insulin-glucose regulation

Model Name	Type	Vars	Remarks
Ackerman	Affine	2	Two compartment linear model [2,1]
Bergman	Nonlinear	3	2 insulin + 1 glucose compartment [7,6]
Cobelli	Nonlinear	~ 11	Comprehensive model including glucagon submodel and renal function model [11,13]
Sorensen	Nonlinear	~ 19	Comprehensive physiological model with compartments for brain, vascular, kidney, renal and peripheral systems [36].
Hovorka	Nonlinear	~ 11	Comprehensive model incorporating endogenous glucose production and renal flitration [22,21,40].

4 Modeling Insulin Infusion

In this section, we present the overall model developed for insulin infusion. Figure 1 shows the overall model for the insulin infusion setup. The setup consists of an insulin infusion pump used to deliver insulin to the patient to counteract the effect of a meal. The meal itself is modeled based on the meal time, duration and its "glycemic factor" that dictates the time from the start of the meal to the peak in blood glucose. The gut absorption of the meal is modeled using a simple linear two compartment model proposed originally by Worthington [41]. Finally, the model incorporates a user model that attempts to capture the user's insulin infusion and correction dosages. An ideal user model is first formulated and calibrated based on the best practices advocated in many guide books that are used by patients using infusion pumps (Cf. [34], for instance). We then attempt to model various user mistakes such as discrepancies between the planned meal and the actual meal, calibration errors in glucose readings, mis-timing of the correction dosages, miscalibration of basal insulin levels, insulin-to-carbs ratio and insulin sensitivity factors.

We will now describe the construction of each of the sub-models in detail.

4.1 Glucoregulatory Models

There have been numerous attempts to derive mathematical models of the regulation of glucose by insulin in diabetic patients. In this paper we employ the Hovorka model [22,21,40] originally proposed by Hovorka and co-workers. We note that there are many other models that are widely used. A few of the notable models are summarized in Table 1. We refer the reader to many comprehensive surveys on this topic including Cobelli et al. [12], Hovorka [19] and the monograph by Chee and Fernando [9]. Comparing predictions obtained by various models of the insulin-glucose regulatory system in our risk assessment framework is an important future work.

The Hovorka model refers to a modeling approach that has been used to model the regulatory system based on tracer studies during a standard intravenous glucose tolerance test [22]. The test measures the amount of insulin and the time needed to restore normal plasma glucose concentrations after the direct infusion of an unit of glucose under fasting conditions. Data from this test were fitted to a model that considers the

$$\begin{aligned}
\frac{dQ_1(t)}{dt} &= -\left[\frac{F_{01}^c}{Q_1(t)} + x_1(t)\right] Q_1(t) + k_{12}Q_2(t) - F_R(t) + \mathrm{EGP}_0(1 - x_3(t)) + U_g(t)\\
\frac{dQ_2(t)}{dt} &= x_1(t)Q_1(t) - [k_{12} + x_2(t)]Q_2(t)\\
\frac{dS_1(t)}{dt} &= U_I(t) - \frac{S_1(t)}{t_{max,I}}\\
\frac{dS_2(t)}{dt} &= \frac{1}{t_{max,I}}(S_1(t) - S_2(t))\\
\frac{dI(t)}{dt} &= \frac{S_2(t)}{t_{max,I}V_I} - k_e I(t)\\
\frac{dx_j(t)}{dt} &= k_{a,j}x_j(t) + k_{b,j}I(t),\ j = 1, 2, 3\\
G(t) &= \frac{Q_1(t)}{V_g}
\end{aligned}$$

$$F_{01}^c(t) = \begin{cases} F_{01}, & \text{if } G \geq 4.5mmol/L \\ \frac{F_{01}G(t)}{4.5}, & \text{otherwise} \end{cases}$$

$$F_R(t) = \begin{cases} 0.003(G(t) - 9)V_G, & \text{if } G \geq 9mmol/L \\ 0, & \text{otherwise} \end{cases}$$

Fig. 4. Hovorka's model for insulin-glucose regulatory system. See [9] for an explanation and comparison with other models. Inputs to the model are $U_I(t)$ the rate of insulin infusion and $U_g(t)$, the rate of plasma glucose infusion. The output is $G(t)$ the blood glucose concentration.

various factors affecting glucose concentration: its uptake by cells, its endogenous production, renal clearance and production due to meal absorption. The complete ODE (with details omitted) is summarized in Figure 4. A detailed explanation is available elsewhere [21,40,9].

The parameter values for a group of "virtual patients" are summarized by Wilinska et al. [40]. These parameter sets capture the observed intra- and inter subject variations seen in real-life patient studies. The Hovorka model has been the basis of a model-predictive controller that has been designed to automatically regulate overnight insulin levels in diabetic through an insulin infusion pump and subcutaneous measurements of glucose concentrations through continuous glucose monitors [21]. The controller has been extensively simulated in-silico to estimate the risk of hypoglycemia [40] and recently has been tested successfully in clinical trials [20].

Meal Sub-Model. The meal sub-model is part of the overall glucoregulatory model described by Hovorka et al. to model the rate of absorption of the meal into the bloodstream by the digestive system. We use a two compartment model

$$\frac{dG_1(t)}{dt} = -\frac{G_1(t)}{t_{max,G}} + B \cdot U_D(t) \text{ and } \frac{dG_2(t)}{dt} = \frac{1}{t_{max,G}}(G_1(t) - G_2(t)) .$$

Here $G_1(t), G_2(t)$ model the amounts of glucose in the two hypothetical compartments, B refers to the bio-availability of the meal (taken to be 0.8 in our simulations), $U_D(t)$ refers to the meal input in terms of millimoles of glucose ingested at time t, and $t_{max,G}$ refers to the time to peak glucose absorption rate. In general, $t_{max,G}$ is a function of the meal glycemic index, wherein meals with high glycemic indices such as starch cause the glucose absorption to peak relatively quickly, while meals with lower glycemic indices such as protein and fat rich meals result in relatively flatter peaks that appear slowly. Throughout our simulation, we will use $t_{max,G}$ as being synonymous with the glycemic index of the chief carbohydrate source in the consumed meal. The input $U_g(t)$ of glucose to the bloodstream resulting from the meal is given by $U_g(t) = \frac{G_2(t)}{t_{max,G}}$.

Recently, Dalla Man et al. consider non-linear models of gut absorption wherein the rate constant of glucose absorption from the gut is itself dependent on the amount of glucose present. This model is shown to fit tracer meal data better than the Worthington model [14,26].

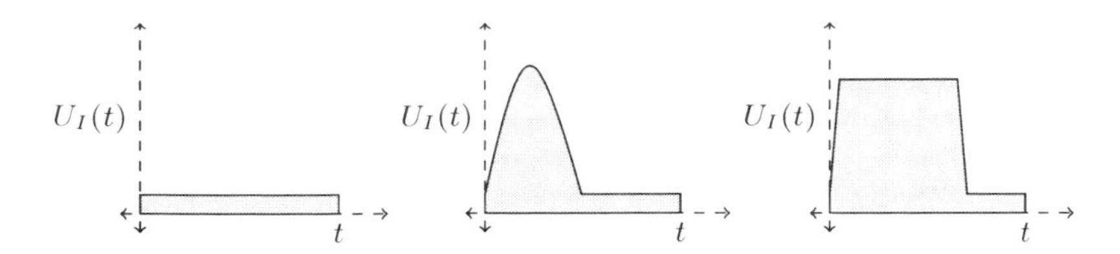

Fig. 5. Schematics for insulin delivery profiles supported by most insulin infusion pump models: (left) basal, (middle) spike bolus and (right) square wave bolus

Glucose Monitors. The glucose monitor model periodically samples the output of the insulin-glucose regulatory model to simulate readings of the subcutaneous glucose. We assume that the value read by the glucose monitor is subject to a systematic calibration error. Calibration errors in continuous glucose monitors (CGMs) have been studied by Wilinska et al. (ibid.), Castle and Ward [8] and Cobelli and co-workers [15]. CGMs need periodic re-calibration using traditional "finger stick" blood glucose readings. It is conceivable, however, that the user may often delay this process leading to significant calibration error wherein the reading may be off by as much as 40-50%. Our setup models a fixed calibration error parameter that can be set at the start of the simulation. The assumed calibration error for each simulation can occur in either direction. Apart from calibration errors, "dropouts" have been commonly reported wherein the reading from the CGM is attenuated for brief stretches of time. The simulation of "dropouts" due to physical sensor errors is not currently considered in our setup.

Insulin Pump Sub-model. The pump model is itself quite simple: it supports (a) basal delivery of insulin at a fixed rate and upon receiving a command, it provides a spike bolus dosage of a given amount, shape and width. Figure 5 schematically presents the basic modes supported by infusion pumps. Our model has "hardwired" bolus profiles representing a unit bolus amount over a unit time, in the form of lookup tables that summarize fractions of requested amounts against sub intervals. Given a particular amount and time, the values from the lookup table are scaled appropriately and the insulin inputs $U_I(t)$ to the gluco-regulatory models are set.

Insulin pump faults include inaccurate doses delivered due to hardware or software errors, stoppages due to occlusions or pump failure. These can also be modeled during the infusion by using input parameters that specify the times and durations of the various faulty situations.

4.2 User Infusion Control Strategy Model

Typically, the infusion pump is used to deliver a continuous flow of short-acting insulin through the day, and intermittent bolus infusions to cover the glucose level increase after a meal. Correction doses of insulin are administered to correct for higher than normal levels immediately before a meal and/or a few hours after a meal. This section is based on information available from diabetes education websites and books. We refer the reader to the book by Scheiner for more information [34].

Basal Insulin Requirement. Basal insulin refers to a constant flow of insulin delivered by a pump all day to compensate for endogenous glucose production. The basal insulin level I_B requires periodic calibration by the user based on glucose levels observed during the night 3-4 hours after dinner. An appropriately calibrated basal level I_B ensures that the increase in blood glucose level during an extended period of fasting (eg., during the night well after dinner) is as small as possible.

A typical recommendation for basal insulin for "moderately active" adult with type-1 diabetes is $I_B(U/day) = 0.4 \times$ weight in kilograms. Starting from this rule of thumb, the basal levels are adjusted with feedback obtained by frequent monitoring of blood glucose levels to fine tune a basal insulin requirement.

Calculating Pre-Prandial Insulin Bolus. In order to adjust for the increase in post-prandial glucose levels, a bolus dose of insulin is delivered through the pump before the meal. The amount and width of a bolus infusion can be directly programmed by the user or calculated by the pump using the planned meal parameters as inputs. The inputs to the calculation include (1) grams of carbohydrates (CHO), (2) glycemic index (GI) of the major CHO source in the food, and (3) personal ratio for insulin-to-carbohydrates.

The bolus size is calculated as follows:

$$\text{pre-meal bolus amt.}(U) = \text{amt. of CHO}(gms) \;\times\; \text{insulin-to-carbs ratio}(U/gm)\,.$$

Additionally, based on the current blood glucose reading a correction bolus may also be required. The correction bolus uses the formula

$$\text{correction bolus}(U) = \frac{G(t) - G_{\text{desired}}}{\text{sensitivity}}\,,\ \text{if}\ G(t) > G_{\text{desired}}\,.$$

Here $G(t)$ refers to the current blood glucose reading, G_{desired} refers to the desired level, and sensitivity is a parameter that is discovered by calibration during the initial period of pump usage by the patient.

There are many different "rules of thumb" for arriving at an initial estimate of the basal insulin requirement, the insulin-to-carb ratio and the sensitivity factor. A starting guess at the insulin-to-carbs ratio is given by $\frac{850}{weight(kgs)}$. An initial sensitivity factor is obtained using the formula $\frac{\text{total daily insulin reqd. (mmol/L)}}{94}$. Often starting from these values, the patient is asked to carefully adjust these values over a period of weeks to achieve robust control of their blood sugar levels.

Timing the Bolus: Another key parameter is the timing of the bolus relative to the meal time. The recommended time for the pre-meal bolus depends on the *glycemic index* (GI) of the meal. For instance, high glycemic index meals (starches such as rice, potatoes, white bread) require the bolus infusion $30 - 40$ minutes pre-meal while lower glycemic index foods require an infusion that starts with the meal. Another parameter is the shape and the width of the infusion. Typically starch heavy meals are covered by a spike bolus while a low GI meal rich in fat and proteins is covered by a square wave bolus. Insulin infusion pumps incorporate infusion modes that can support these bolus shapes including combinations for meals that combine various food types.

Overall infusion control model: The infusion control model incorporates a program that calculates the pre-meal bolus requirements using the insulin-to-carbs ratio and the correction insulin using the sensitivity factors. These factors are "calibrated" for the insulin-glucose regulation model using the robustness guided state-space exploration technique described in Section 5. The timing and shape of the bolus are determined by classifying the planned meal GI into three categories high, medium and low [34]. The models developed for this paper and the analysis results will be made available for download as part of the S-Taliro tool.

5 Robustness Guided Search

In this section, we present the basic concepts used in the analyses of models with respect to metric temporal logic properties [4]. We present the details of our analysis methodology at a high level. More details are available from our prior work which deals with the problem of using robustness guided state-space exploration to find falsifying traces for MTL specifications of non-linear hybrid systems [28,17].

5.1 Metric Temporal Properties and Trace Robustness

Metric Temporal Logic (MTL) is a formalism to specify temporal properties of continuous time signals [25]. Table 2 summarizes the syntax and semantics of MTL formulae. MTL formulae can be used to succinctly express key properties of desirable post-prandial blood glucose levels. Let $t = 0$ model the start of a meal and we assume $t = 400$ to be end of the simulation period being considered. $G(t)$ is a signal modeling the blood glucose concentration in terms of mmol/L at time t. Table 3 shows the three properties of interest to us along with their descriptions.

Our goal is to find executions of the overall infusion process model that falsify at least one of the properties in Table 3. Here each execution trace corresponds to a different values of planned vs. actual meal data, and calibration error. However, there are potentially infinitely many executions for various values of the input parameters and the models for the insulin-glucose regulatory system are non-linear. Therefore, we use robustness metrics over execution traces to define an objective function over traces to guide us in the search for a falsifying input.

Table 2. Metric Temporal Logic (MTL) Operators and their formal semantics at time $t = t_0$. $\sigma : [0, T] \mapsto \mathbb{R}^n$ refers to a continuous time signal, $\mathcal{I}$ refers to a real time interval, AP refers to a set of atomic proposition symbols, $\mathcal{O}$ maps each atomic proposition to a subset of $\mathbb{R}^n$.

Formula φ	Semantics $(\sigma, t_0, \mathcal{O}) \models \varphi$	Remarks
$\top$	*true*	Tautology
$p \in$ AP	$\sigma(t_0) \in \mathcal{O}(p)$	Atomic Proposition holds.
$\varphi_1 \wedge \varphi_2$	$(\sigma, t_0, \mathcal{O}) \models \varphi_1 \ \wedge \ (\sigma, t_0, \mathcal{O}) \models \varphi_2$	Conjunction
$\varphi_1 \vee \varphi_2$	$(\sigma, t_0, \mathcal{O}) \models \varphi_1 \ \vee \ (\sigma, t_0, \mathcal{O}) \models \varphi_2$	Disjunction
$\neg \varphi$	$(\sigma, t_0, \mathcal{O}) \not\models \varphi$	Negation
$\Box_{\mathcal{I}} \varphi$	$(\forall t \in \mathcal{I})((t_0 + t < T) \Rightarrow (\sigma, t_0 + t, \mathcal{O}) \models \varphi)$	φ is Invariant in $\mathcal{I}$
$\Diamond_{\mathcal{I}} \varphi$	$(\exists t \in \mathcal{I})((t_0 + t < T) \wedge (\sigma, t_0 + t, \mathcal{O}) \models \varphi)$	φ eventually holds in $\mathcal{I}$
$\varphi_1 \mathcal{U}_{\mathcal{I}} \varphi_2$	$(\exists t \in \mathcal{I})((t_0 + t < T) \wedge (\sigma, t_0 + t, \mathcal{O}) \models \varphi_2 \wedge$ $(\forall t' \in [0, t)) \, (\sigma, t_0 + t', \mathcal{O}) \models \varphi_1)$	φ_1 until φ_2

Table 3. MTL specifications for normal post-prandial glycemic control

No Hypoglycemia φ_{hypo}	$\Box_{[0,400]}(G \geq 3)$
No significant hyperglycemia φ_{hyper}	$\Box_{[0,400]}(G \leq 20)$
Glucose levels settle after digestion φ_{settle}	$\Box_{[200,400]}(G \in [3, 10])$

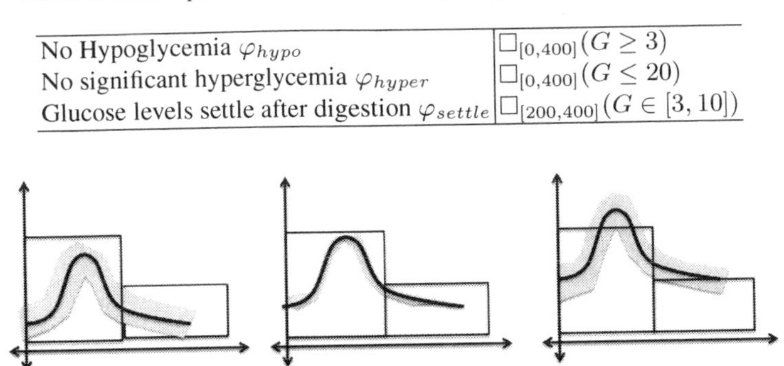

Fig. 6. Illustration of robustness of time trajectories. The trajectory is required to lie inside the union of the two rectangles. A "cylindrification" around each trajectory is shown such that the any trace in cylindrification has same outcome w.r.t trace as the original trajectory. Leftmost trajectory satisfies property with strictly positive robustness due to larger cylindrification radius, middle trajectory satisfies but with a small robustness value while rightmost trajectory violates property with negative robustness.

5.2 Trace Robustness

The robustness of signals obtained by simulating hybrid systems is a useful concept that generalizes the standard true/false interpretation of MTL formulae to real valued semantics. Informally, robustness provides a measure of how far away a given trace is from satisfying or violating a property. Real-valued semantics for temporal specifications were considered by Rizk et al. for applications in systems biology [30] and independently by Fainekos and Pappas for testing control systems [16,17]. Figure 6 illustrates the main idea behind the robustness value of a trace σ w.r.t a MTL formula φ. Informally, the robustness value ε indicates the size of the smallest cylinder that can be drawn around σ so that any other trace σ' contained inside this cylinder also has the same valuation for the property as φ. I.e, σ' satisfies φ iff σ does.

Formally, the robustness of a trace σ w.r.t a formula φ, denoted $R(\sigma, \varphi)$, is a real number such that

1. If $R(\sigma, \varphi) > 0$ then $\sigma \models \varphi$. Likewise, if $R(\sigma, \varphi) < 0$ then $\sigma \not\models \varphi$.
2. If $R(\sigma, \varphi) = \varepsilon$, then any trace that lies inside a cylinder of radius $|\varepsilon|$ defined around σ will also have the same outcome for the property φ as σ.

Details on the systematic calculation of robustness values from a sampled continuous-time trace σ and a bounded-time MTL formula φ are available elsewhere [17]. The approach to falsification of a property φ given a model $\mathcal{M}$ is to minimize the objective $R(\sigma, \varphi)$ over all traces σ of the model $\mathcal{M}$. As noted in our previous work [28], this optimization problem is non-convex for most systems and furthermore, the objective $R(\sigma, \varphi)$ cannot be written down in a closed form. However, it can be evaluated for a given trace σ. Our previous works have explored the use of various global optimization techniques such as Monte-Carlo simulation using Simulated Annealing [28],

Table 4. Infusion faults and assumed ranges for minimal robustness search

Parameter	Ideal	Actual	Remarks
Meal Time (mins)	40	[0, 80]	discrepancy between planned (t=40) and actual meal time.
Meal Carbs (gms)	250	[150,350]	discrepancy between planned and actual carbs ingested.
Meal $t_{g,max}$ (mins)	40	[20,80]	meal start time to peak glucose absorption (planned vs. actual GI).
Meal Duration (mins)	30	[10,50]	meal duration planned vs. actual.
Correction Bolus Time	200	[100,300]	time when correction bolus is administered.
Calibration Error	0	[-0.3,0.3]	CGM calibration error.

Ant-Colony Optimization [3], Genetic Algorithms and more recently the Cross-Entropy Method [32]. These techniques have been implemented in a Matlab toolbox called S-Taliro [4], which supports state-space exploration of Simulink/Stateflow models with MTL specifications.

6 Worst-Case Scenario Search

We fix a scenario consisting of a planned meal at time $t = 40$ minutes after the start of the simulation, a planned duration of 30 minutes, and consisting of 200 grams of CHO (~ 850 calories) with $t_{g,max} = 40$ minutes, indicating a high GI meal (eg., bread, rice or pasta). Our goal here is to explore the risks to the patient arising from these faults in the infusion process. We note that our general framework allows us to explore other meal scenarios and a different set of faults, as well. The overall methodology for this exploration involves the following steps: (a) Formulating MTL properties to falsify (Cf. Table 3); (b) Calibrating the model, so that the MTL properties are satisfied robustly under normal, fault-free situations; (c) Setting up various combinations of faults (Cf. Table 4); (d) Using S-Taliro tool to search for falsifications of the properties in Table 3 in the presence of infusion faults; and (e) Repeating the analysis with various faults disabled to understand the minimal set of faults that are responsible for a given scenario.

Even though our model allows for transient infusion faults, the study performed in this section, does not include such faults. This is primarily due to the lack of available data on the frequency and timings of pump failures.

Model Calibration: First, we calibrate the patient-specific parameters involved in deciding the size of the infusion including the basal insulin I_B, the insulin-to-carbs ratio and the sensitivity factor. After fixing a planned meal as described previously and disabling all faults, we perform this calibration by using S-Taliro to search for parameter values that *maximize* (usually, falsification involves a minimization) the robustness of the correctness property $\psi : \varphi_{hypo} \wedge \varphi_{hyper} \wedge \varphi_{settle}$, as defined in Table 3. By maximizing the robustness, we are, in effect, searching for a trace that (a) satisfies the property ψ, i.e, achieves

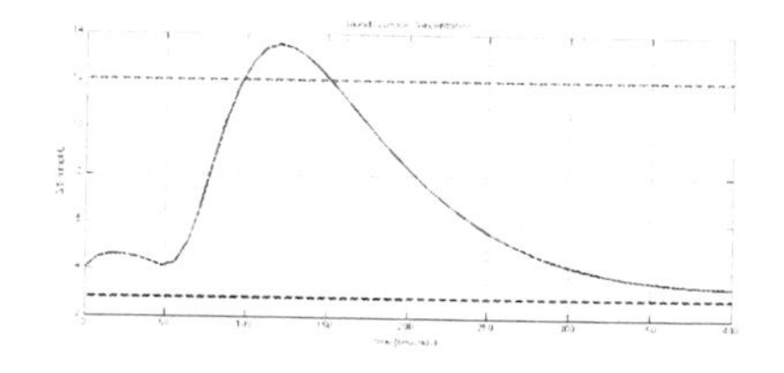

Fig. 7. Post-meal glucose on "calibration parameters" found using S-Taliro

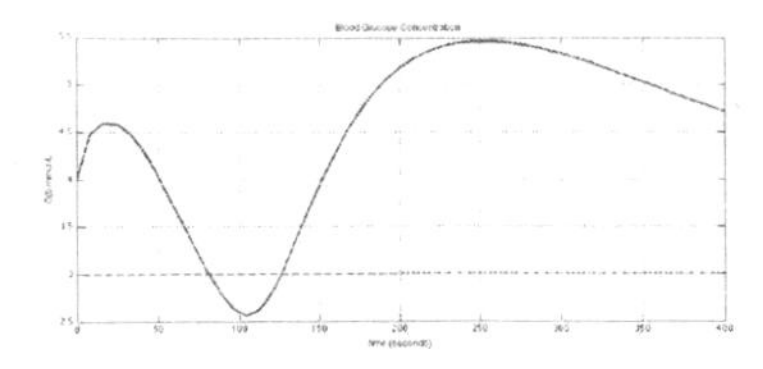

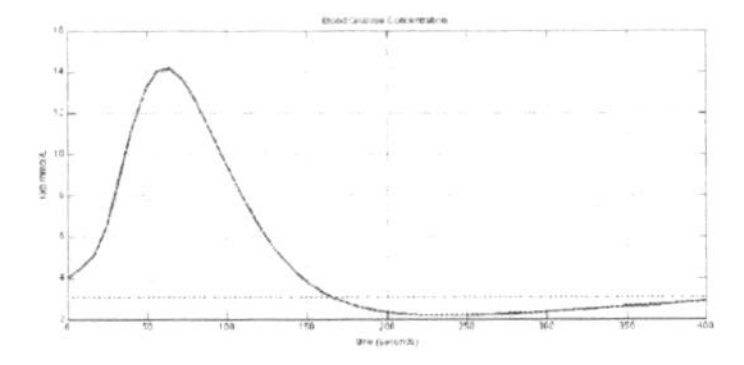

Fig. 8. Two minimal robustness scenarios for potential hypoglycemia risk. (left) meal too late/infusion too early and (right) meal too early/infusion too late.

ideal glycemic control and (b) does so robustly with nearby traces also satisfying the property. The parameter values corresponding to the maximal robustness are chosen as the ideal parameters that achieve the best overall control for a long enough time horizon. Figure 7 shows the output of a fault-free execution using the calibration results to control the infusion.

The analysis using S-Taliro identified four dangerous scenarios, all characterized by discrepancies between the meal times. The assumed sensor calibration errors of up to 30% in either direction were found to have very little effect. Furthermore, analysis performed here was repeated for many different "optimal" values of basal insulin, insulin-to-carbs ratio, and sensitivity factor found by running the model calibration procedure. Each repetition yields the same qualitative results described in the paper, but with slightly varying robustness values for $G(t)$ (upto 20% variation seen). We will now discuss these scenarios in detail.

6.1 Hypoglycemia Scenarios

We first study the effect of faults to falsify the property $\varphi_{hypo} : \Box G \geq 3$. We performed the search inside S-Taliro using our model implemented in Simulink/Stateflow (tm), the calibrated values for the user's strategy reported previously and the meal timings drawn non-deterministically from the ranges specified in Table 4. The optimization was run using the stochastic optimization algorithms based on Simulated Annealing (SA) and Cross Entropy (CE) roughly twenty five times, each time using different random seeds to produce different minimal robustness scenarios. Each run required approximately $3 - 10$ minutes with up to 1000 simulations per run.

Each optimization run discovered property violations that expose potential hypoglycemia. The minimal values of G were around $G = 2.3$. Examining these violations, we found that each scenario falls in one of two distinct categories. Figure 8 shows the blood glucose outputs for both scenarios.

Potential Hypoglycemia Scenario-1: By disabling various faults in turn, we identified three sufficient faults for this scenario:

1. The planned meal time ($t = 40$) is significantly earlier than actual meal time ($t \sim 70$). Alternatively, the insulin bolus is delivered too early.
2. The planned GI ($t_{max,G} = 40$) is lower than its actual GI ($t_{max,G} \sim 20$).
3. The actual amount of CHO ingested in this scenario is *less than* the planned amount of CHO.

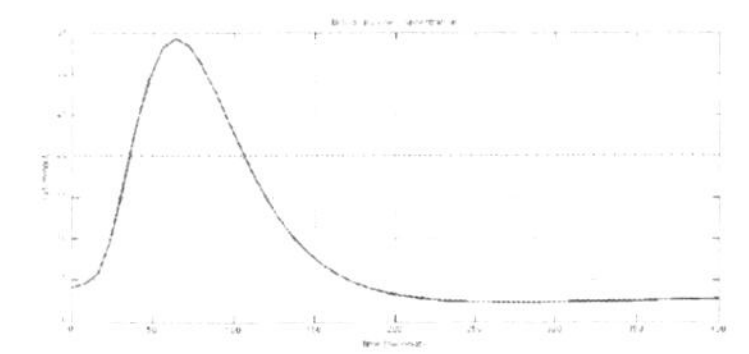

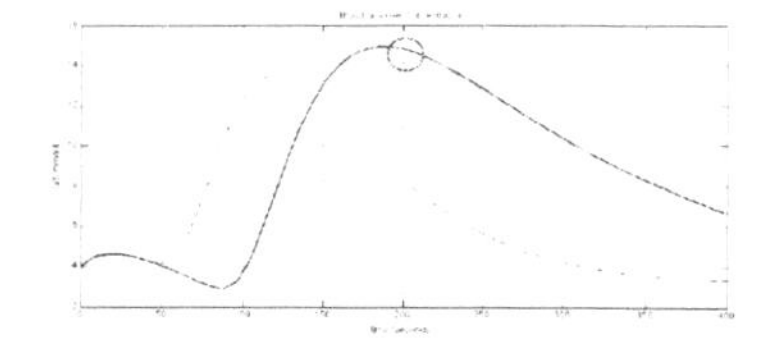

Fig. 9. Glucose concentration for significant hyperglycemia (left) and failure to settle (right). The fault-free output is shown as a dashed line.

Additionally, the actual meal durations were slightly larger than planned.

Potential Hypoglycemia Scenario-2: This scenario is characterized by the following combination of faults:

1. The planned meal time ($t = 40$) is significantly *later* than the actual meal time ($t \sim 5$). Alternatively, the bolus is delivered at or after the start of a high GI meal.
2. The meal's planned GI is lower than the meal's actual GI.
3. The actual amount of CHO ingested is lower than the planned CHO.

The actual meal durations were slightly less than the planned duration.

6.2 Scenario Analysis for Significant Hyperglycemia

Next, we consider scenarios for significant hyperglycemia $G \geq 25$ that can lead to dangerous conditions such as ketacidosis. Unlike hypoglycemia, this property is found to be easier to falsify, requiring fewer iterations to find falsifying inputs. The minimal robustness scenarios found by S-Taliro depend chiefly on two faults: (a) discrepancy between planned meal GI ($t_{max,G} = 40$) and actual meal GI $t_{max,G} \sim 20$ and (b) discrepancy between planned meal CHO ($= 200$) and actual meal CHO (~ 300). Other faults have a minor impact on the maximum value of the blood glucose level $G(t)$. In such a scenario, the bolus of insulin is supplied too late and is insufficient to "cover" the meal.

6.3 Scenario Analysis for Failure to Settle

Finally, we consider minimal robustness scenarios w.r.t failure to settle. Figure 9 (right) shows the output glucose level for this scenario. The scenario depends on a combination of two faults: (a) actual meal time significantly later than the planned time, and (b) the actual CHO is higher than the planned CHO. In this scenario, the peak value of the insulin precedes the peak gut absorption of glucose. Notice the initial dip in $G(t)$ much like the first hypoglycemia scenario followed by a delayed rise in $G(t)$ that fails to settle even after 6 hours.

6.4 Evaluation

We have thus far derived some situations that can cause significant hypoglycemia, hyperglycemia and failure to settle. Can the predictions made by our model be tested? The gold standard evaluation would be to conduct clinical studies of patients to determine if the violations observed can be borne in real life. While we are planning to conduct extensive patient studies with collaborators from medical sciences as part of our ongoing work in this area, such studies require time and significant effort to carry out.

Preliminary evidence is available from web logs maintained by many diabetic patients to check if any of the situations reported by us are also confirmed by diabetic patients. We surveyed many such weblogs and some of the scenarios such as scenario-1 for hypoglycemia and the scenario for failure to settle seem to be well known [3]. However, barring a few exceptions incident reports by patients focused on the effects and not the root causes.

7 Statistical Model Checking

We have, thus far, used robustness guided state-space explorations to explore the extreme, worst-case scenarios that may happen in the infusion. However, it is equally interesting to find out the probability that the infusion may result in a hypoglycemia, hyperglycemia or a failure to settle (Cf. Table 3). In order to do so, a simple approach is to assume a probability distribution for each of the faults described in Table 4. We assume that the faults are uniformly distributed within their intervals and independent of each other. We then simulate the model by sampling faults from this distribution and find out the fraction of executions that violate each property. However, if the probabilities of these bad outcomes are tiny to begin with, the number of executions required become prohibitively large. Recent advances in *Statistical Model-Checking* (SMC) have given rise to techniques that can estimate these probabilities efficiently while running as few simulations as possible. SMC was originally formulated by Younes and Simmons [42]. A promising extension involving the use of Bayesian reasoning, has been used to applied to problems in systems biology [23] and control systems verification [44]. Rather than estimating the probabilities of hypoglycemia or hyperglycemia empirically, these techniques use repeated simulations to bound the required probability inside an interval of given half-width δ with a given confidence c. In addition to worst-case search, we have extended S-Taliro to support Bayesian SMC. Existing techniques implemented inside S-Taliro such as the Cross-Entropy Method can be directly used to find a suitable prior distribution to reduce the number of simulations required.

Table 5 shows the probability estimates for significant hypoglycemia with various confidence levels and intervals around the probability. We find that hypoglycemia ($G < 3$) has roughly 30% chance of occurrence, while significant hypoglycemia ($G < 2.7$) has an estimated 7% chance of occurrence with very high confidence. The probabilities for hyperglycemia are also presented in Table 5. Once again, we estimate a 2.2% chance of significant hyperglycemia $G > 25$ and 0.3% chance of finding $G > 35$ (which exposes the patient to dangerous ketacidosis).

The findings of this section are qualitatively borne out by an informal analysis of incidents reported by diabetic patients and some clinical studies [18]. The majority of

[3] For a discussion of meal timing from a patient's perspective, see `http://thethirstthatchangedmylife.blogspot.com/2010/09/loading-on-carbs.html`

Table 5. Results of SMC for estimating probability of avoiding hypoglycemia (left) and hyperglycemia (right). Each entry shows the posterior probability estimate $\hat{p}$ such that the probability lies within $[\hat{p} - \delta, \hat{p} + \delta]$ with confidence (coverage) indicated by c.

	$\delta = 0.05$			$\delta = 0.01$				$\delta = 0.05$			$\delta = 0.01$		
Coverage (c)	.95	.99	.999	.95	.99	.999	Coverage(c)	.95	.99	.999	.95	.99	.999
$\Box G \geq 3$	0.75	0.7	0.74	0.73	0.72	0.72	$\Box G \leq 25$	0.97	0.97	0.96	0.97	0.97	0.98
$\Box G \geq 2.8$	0.87	0.88	0.88	0.89	0.88	0.88	$\Box G \leq 30$	0.97	0.98	0.98	0.99	0.99	0.99
$\Box G \geq 2.7$	0.91	0.92	0.93	0.93	0.93	0.93	$\Box G \leq 35$	0.97	0.98	0.99	0.99	0.99	0.99

the infusion faults due to discrepancies between planned and actual meals result in hypoglycemia. Hyperglycemia risks commonly occur due to silent pump failures that were not modeled in this study.

8 Threats to Validity

In this section, we discuss some of the threats to validity and address remedial steps taken to ensure that the results in this work are applicable to real-life situations.

With any result involving *in silico* simulations, there is a risk that we are observing modeling quirks that are not reflective of what happens in reality. However, the models used here have been extensively evaluated against studies on real patients [22,20], providing evidence for their validity.

Another concern is that we assume that the food ingested has a single carbohydrate source with fixed (high) GI. While this can be a good approximation in some cases (eg., a meal consisting mostly of pasta or a CHO heavy drink), such meals are not advised for diabetic patients. Insulin pumps provide combination boluses to offset for different types of foods with varying GIs. We plan to investigate these effects as part of our ongoing research. Mixed meal simulations have been considered in the past by Della Man et al. [14,27]. The effects of exercise are also a factor. However, modeling physical activity and its effect on the blood glucose regulation is an active area of research with few established models [31].

A shortcoming of assigning probabilities on the occurrence of faults is that there is no available mathematical evidence that the distribution of planned meal times vs. actual meal times are uniformly distributed in the interval of interest. Another assumption is that of independence of the various faults. It is conceivable that a larger discrepancy between planned mealtimes and actual mealtimes indicates a larger discrepancy between the planned and actual meal CHO or GI. Building fault models based on observations of insulin infusion pump usage of real patients are critical to construct these models. The probabilities reported by SMC are likely to change if we included more types of faults such as pump failures in our study.

9 Conclusions

We have presented an *in-silico* evaluation of the risks involved in the infusion process. Our approach has been two-fold: (a) using robustness-guided model checking to search for potential worst-case scenarios. Here, we report on some scenarios causing hypoglycemia, hyperglycemia and failure of the blood glucose to settle to normoglycemia. Some of our scenarios are borne out by patient reports reported online. (b) We use Statistical Model Checking to place small bounds on the risks with very high confidence.

Our future direction is to consider *individualized risk studies*. Here, we seek to develop models and risk analysis fitted to individual patients. This can yield lifestyle analysis tools that can help advise patients on the best pump calibration parameters to maintain normal glucose levels.

Acknowledgments. We gratefully acknowledge helpful discussions with Dr. Phillip Dzwonczyk, NY State Veterans Home, Oxford, NY and Prof. Clayton Lewis, University of Colorado, Boulder. We thank the anonymous reviewers for their comments.

References

1. Ackerman, E., Gatewood, L., Rosevear, J., Molnar, G.: Blood glucose regulation and diabetes. In: Heinmets, F. (ed.) Concepts and Models of Biomathematics, pp. 131–156. Marcel Dekker (1969)
2. Ackerman, E., Rosevear, J., McGuckin, W.: A mathematical model of the insulin-glucose tolerance test. Physics in Medicine and Biology 9, 202–213 (1964)
3. Annapureddy, Y.S.R., Fainekos, G.E.: Ant colonies for temporal logic falsification of hybrid systems. In: Proceedings of the 36th Annual Conference of IEEE Industrial Electronics, pp. 91–96 (2010)
4. Annpureddy, Y., Liu, C., Fainekos, G., Sankaranarayanan, S.: S-TALIRO: A Tool for Temporal Logic Falsification for Hybrid Systems. In: Abdulla, P.A., Leino, K.R.M. (eds.) TACAS 2011. LNCS, vol. 6605, pp. 254–257. Springer, Heidelberg (2011)
5. Arney, D.E., Jetley, R., Jones, P., Lee, I., Ray, A., Sokolsky, O., Zhang, Y.: Generic infusion pump hazard analysis and safety requirements: Version 1.0, CIS Technical Report, University of Pennsylvania (2009), `http://repository.upenn.edu/cis_reports/893` (accessed May 2011)
6. Bergman, R.N.: Minimal model: Perspective from 2005. Hormone Research, 8–15 (2005)
7. Bergman, R.N., Urquhart, J.: The pilot gland approach to the study of insulin secretory dynamics. Recent Progress in Hormone Research 27, 583–605 (1971)
8. Castle, J., Ward, K.: Amperometric glucose sensors: Sources of error and potential benefit of redundancy. J. Diabetes Sci. and Tech. 4(1) (January 2010)
9. Chee, F., Fernando, T.: Closed-Loop Control of Blood Glucose. Springer (2007)
10. Clarke, E., Donzé, A., Legay, A.: Statistical Model Checking of Mixed-Analog Circuits with an Application to a Third Order $\Delta - \Sigma$ Modulator. In: Chockler, H., Hu, A.J. (eds.) HVC 2008. LNCS, vol. 5394, pp. 149–163. Springer, Heidelberg (2009)
11. Cobelli, C., Federspil, G., Pacini, G., Salvan, A., Scandellari, C.: An integrated mathematical model of the dynamics of blood glucose and its hormonal control. Mathematical Biosciences 58, 27–60 (1982)
12. Cobelli, C., Man, C.D., Sparacino, G., Magni, L., Nicolao, G.D., Kovatchev, B.P.: Diabetes: Models, signals and control (methodological review). IEEE Reviews in Biomedical Engineering 2, 54–95 (2009)
13. Cobelli, C., Mari, A.: Control of diabetes with artificial systems for insulin delivery — algorithm independent limitations revealed by a modeling study. IEEE Trans. on Biomed. Engg. BME-32(10) (October 1985)

14. Dalla Man, C., Rizza, R.A., Cobelli, C.: Meal simulation model of the glucose-insulin system. IEEE Transactions on Biomedical Engineering 1(10), 1740–1749 (2006)
15. Facchinetti, A., Sparacino, G., Cobelli, C.: Modeling the error of continuous glucose monitoring sensor data: Critical aspects discussed through simulation studies. J. Diabetes Sci. and Tech. 4(1) (January 2010)
16. Fainekos, G., Pappas, G.J.: Robustness of temporal logic specifications for continuous-time signals. Theoretical Computer Science 410, 4262–4291 (2009)
17. Fainekos, G.E.: Robustness of Temporal Logic Specifications. PhD thesis, Department of Computer and Information Science, University of Pennsylvania (2008)
18. Fox, L., Buckloh, L., Smith, S.D., Wysocki, T., Mauras, N.: A randomized controlled trial of insulin pump therapy in young children with type 1 diabetes. Diabetes Care, 28(6) (June 2005)
19. Hovorka, R.: Continuous glucose monitoring and closed-loop systems. Diabetic Medicine 23(1), 1–12 (2005)
20. Hovorka, R., Allen, J.M., Elleri, D., Chassin, L.J., Harris, J., Xing, D., Kollman, C., Hovorka, T., Larsen, A.M., Nodale, M., Palma, A.D., Wilinska, M., Acerini, C., Dunger, D.: Manual closed-loop delivery in children and adoloscents with type 1 diabetes: a phase 2 randomised crossover trial. Lancet 375, 743–751 (2010)
21. Hovorka, R., Canonico, V., Chassin, L., Haueter, U., Massi-Benedetti, M., Frederici, M., Pieber, T., Shaller, H., Schaupp, L., Vering, T., Wilinska, M.: Nonlinear model predictive control of glucose concentration in subjects with type 1 diabetes. Physiological Measurement 25, 905–920 (2004)
22. Hovorka, R., Shojaee-Moradie, F., Carroll, P., Chassin, L., Gowrie, I., Jackson, N., Tudor, R., Umpleby, A., Hones, R.: Partitioning glucose distribution/transport, disposal and endogenous production during IVGTT. Am. J. Physiol. Endocrinol. Metab. 282, 992–1007 (2002)
23. Jha, S.K., Clarke, E.M., Langmead, C.J., Legay, A., Platzer, A., Zuliani, P.: A Bayesian Approach to Model Checking Biological Systems. In: Degano, P., Gorrieri, R. (eds.) CMSB 2009. LNCS, vol. 5688, pp. 218–234. Springer, Heidelberg (2009)
24. Jha, S.K., Datta, R., Langmead, C., Jha, S., Sassano, E.: Synthesis of insulin pump controllers from safety specifications using bayesian model validation. In: Proceedings of 10th Asia Pacific Bioinformatics Conference, APBC (2012)
25. Koymans, R.: Specifying real-time properties with metric temporal logic. Real-Time Systems 2(4), 255–299 (1990)
26. Man, C., Camilleri, M., Cobelli, C.: A system model of oral glucose absorption: Validation on gold standard data. IEEE Transactions on Biomedical Engineering 53(12), 2472–2478 (2006)
27. Man, C.D., Raimondo, D.M., Rizza, R.A., Cobelli, C.: GIM, simulation software of meal glucose-insulin model. J. Diabetes Sci. and Tech. 1(3) (May 2007)
28. Nghiem, T., Sankaranarayanan, S., Fainekos, G.E., Ivančić, F., Gupta, A., Pappas, G.J.: Monte-carlo techniques for falsification of temporal properties of non-linear hybrid systems. In: Hybrid Systems: Computation and Control, pp. 211–220. ACM Press (2010)
29. Patek, S., Bequette, B., Breton, M., Buckingham, B., Dassau, E., Doyle III, F., Lum, J., Magni, L., Zisser, H.: In silico preclinical trials: methodology and engineering guide to closed-loop control in type 1 diabetes mellitus. J. Diabetes Sci. Technol. 3(2), 269–282 (2009)
30. Rizk, A., Batt, G., Fages, F., Soliman, S.: On a Continuous Degree of Satisfaction of Temporal Logic Formulae with Applications to Systems Biology. In: Heiner, M., Uhrmacher, A.M. (eds.) CMSB 2008. LNCS (LNBI), vol. 5307, pp. 251–268. Springer, Heidelberg (2008)
31. Roy, A., Parker, R.: Dynamic modeling of exercise effects on plasma glucose and insulin levels. J. Diabetes Sci. and Tech. 1(3), 338–347 (2007)

32. Sankaranarayanan, S., Fainekos, G.E.: Falsification of temporal properties of hybrid systems using the cross-entropy method. In: HSCC, pp. 125–134. ACM (2012)
33. Sankaranarayanan, S., Homaei, H., Lewis, C.: Model-Based Dependability Analysis of Programmable Drug Infusion Pumps. In: Fahrenberg, U., Tripakis, S. (eds.) FORMATS 2011. LNCS, vol. 6919, pp. 317–334. Springer, Heidelberg (2011)
34. Scheiner, G.: Think like a pancreas: A Practical guide to managing diabetes with insulin. Da Capo Press (2011)
35. Skyler, J.S.: Atlas of Diabetes, 4th edn. Springer Science+Business Media (2012)
36. Sorensen, J.: A Physiological Model of Glucos Metabolism in Man and its use to Design and Access Improved Insulin Therapies for Diabetes. PhD thesis, Massachussetts Inst. of Technology. MIT (1985)
37. Teixeira, R.E., Malin, S.: The next generation of artificial pancreas control algorithms. J. Diabetes Sci. and Tech. 2, 105–112 (2008)
38. Thimbleby, H.: Ignorance of interaction programming is killing people. ACM Interactions, 52–57 (2008)
39. Thimbleby, H.: Is it a dangerous prescription? BCS Interfaces 84, 5–10 (2010)
40. Wilinska, M., Chassin, L., Acerini, C.L., Allen, J.M., Dunber, D., Hovorka, R.: Simulation environment to evaluate closed-loop insulin delivery systems in type 1 diabetes. J. Diabetes Science and Technology 4 (January 2010)
41. Worthington, D.: Minimal model of food absorption in the gut. Medical Informatics 22(1), 35–45 (1997)
42. Younes, H.L.S., Simmons, R.G.: Statistical probabilitistic model checking with a focus on time-bounded properties. Information & Computation 204(9), 1368–1409 (2006)
43. Zhang, Y., Jones, P.L., Jetley, R.: A hazard analysis for a generic insulin infusion pump. J. Diabetes Sci. and Tech. 4(2), 263–282 (2010)
44. Zuliani, P., Platzer, A., Clarke, E.M.: Bayesian statistical model checking with application to simulink/stateflow verification. In: HSCC, pp. 243–252. ACM (2010)

Revisiting the Training of Logic Models of Protein Signaling Networks with ASP

Santiago Videla[1,2,⋆], Carito Guziolowski[3,4,⋆], Federica Eduati[5,6], Sven Thiele[2,1,7], Niels Grabe[3], Julio Saez-Rodriguez[6,⋆⋆], and Anne Siegel[1,2,⋆⋆]

[1] CNRS, UMR 6074 IRISA, Campus de Beaulieu, 35042 Rennes cedex, France
[2] INRIA, Centre Rennes-Bretagne-Atlantique, Projet Dyliss, Campus de Beaulieu, 35042 Rennes cedex, France
[3] University Heidelberg, National Center for Tumor Diseases, TIGA Center
[4] École Centrale de Nantes, IRCCyN, UMR CNRS 6597, 1 rue de la Noë, 44321 Nantes cedex 3, France
[5] Department of Information Engineering, University of Padova, Padova, 31050, Italy
[6] European Bioinformatics Institute (EMBL-EBI) Wellcome Trust Genome Campus, Cambridge CB10 1SD, UK
[7] University of Potsdam, Institute for Computer Science, Germany
saezrodriguez@ebi.ac.uk, anne.siegel@irisa.fr

Abstract. A fundamental question in systems biology is the construction and training to data of mathematical models. Logic formalisms have become very popular to model signaling networks because their simplicity allows us to model large systems encompassing hundreds of proteins. An approach to train (Boolean) logic models to high-throughput phosphoproteomics data was recently introduced and solved using optimization heuristics based on stochastic methods. Here we demonstrate how this problem can be solved using Answer Set Programming (ASP), a declarative problem solving paradigm, in which a problem is encoded as a logical program such that its answer sets represent solutions to the problem. ASP has significant improvements over heuristic methods in terms of efficiency and scalability, it guarantees global optimality of solutions as well as provides a complete set of solutions. We illustrate the application of ASP with *in silico* cases based on realistic networks and data.

Keywords: Logic modeling, answer set programming, protein signaling networks.

1 Introduction

Cells perceive extracellular information via receptors that trigger signaling pathways that transmit this information and process it. Among other effects, these pathways regulate gene expression (transcriptional regulation), thereby defining the response of the cell to the information sensed in its environment. Over

⋆ These authors contributed equally to this work.
⋆⋆ Corresponding author.

D. Gilbert and M. Heiner (Eds.): CMSB 2012, LNCS 7605, pp. 342–361, 2012.

decades of biological research we have gathered large amount of information about these pathways. Nowadays, there exist public repositories such as Pathways Commons [1] and Pathways Interaction Database [2] that contain curated regulatory knowledge, from which signed and oriented graphs can be automatically retrieved [3,4]. These signed-oriented graphs represent molecular interactions inside the cell at the levels of signal transduction and (to a lower extent) of transcriptional regulation. Their edges describe causal events, which in the case of signal transduction are related to the molecular events triggered by cellular receptors. These networks are derived from vast generic knowledge concerning different cell types and they represent a useful starting point to generate predictive models for cellular events.

Phospho-proteomics assays [5] are a recent form of high-throughput or 'omic' data. They measure the level of phosphorylation (correlated with protein-activity) of up to hundreds of proteins at the same moment in a particular biological system [6]. Most cellular key processes, including proliferation, migration, and cell cycle, are ultimatelly controlled by these protein-activity modifications. Thus, measurement of phosphorylation of key proteins under appropriate conditions (experimental designs), such as stimulating or perturbing the system in different ways, can provide useful insights of cellular control.

Computational methods to infer and analyze signaling networks from high-throughput phospho-proteomics data are less mature than for transcriptional data, which has been available for much longer time [6]. In particular, the infererence of gene regulatory networks from transcriptomics data is now an established field (see [7,8] for a review). In comparison to transcriptomics, data is harder to obtain in (phospho) proteomics, but prior knowledge about the networks is much more abundant, and available in public resources as mentioned above.

An approach to integrate the prior knowledge existing in databases with the specific insight provided by phospho-proteomics data was recently introduced and implemented in the tool CellNOpt (CellNetOptimizer; www.cellnopt.org) [9]. CellNOpt uses stochastic optimization algorithms (in particular, a genetic algorithm), to find the Boolean logic model compatible that can best describe the existing data. While CellNOpt has proved able to train networks of realistic size, it suffers from the lack of guarantee of optimum intrinsic of stochastic search methods. Furthermore, it scales poorly since the search space (and thus the computational time) increases exponentially with the network size.

In this paper, we propose a novel method to solve the optimization problem posed in [9] that overcomes its limitations. Our approach trains generic networks based on experimental measures equally as CellNOpt in order to obtain a complete set of global optimal networks specific to the experimental data. This family of optimal networks could be regarded as an explanatory model that is specific to a particular cell type and condition; from these models it should be possible to derive new, more accurate biological insights. To illustrate our approach we used a generic Prior Knowledge Network (PKN) related to signaling events upon stimulation of cellular receptors in hepatocytes, and trained this network with *in silico* simulated phospho-proteomics data. This network was used as a benchmark for network inference in the context of the DREAM (Dialogues for Reverse Engineering Assessment of Methods; www.the-dream-project.org) Predictive Signaling Network Challenge [10].

The proposed solution encodes the optimization problem in Answer Set Programming (ASP) [11]. ASP is a declarative problem solving paradigm from the field of logic programming. Distributed under the GNU General Public Licence, it offers highly efficient inference engines based on Boolean constraint solving technology [12]. ASP allows for solving search problems from the complexity class NP and with the use of disjunctive logic programs from the class Σ_2^P. Moreover, modern ASP tools allow handling complex preferences and multi-criteria optimization, guaranteeing the global optimum by reasoning over the complete solution space.

Our results show significant improvements, concerning computation time and completeness in the search of optimal models, in comparison with CellNOpt. We note that similar features can be obtained by formulation of the problem as an integer linear optimization problem [13]. The perspectives of this work go towards the exploration of the complete space of optimal solutions in order to identify properties such as the robustness of optimal models, and relate them to the quality of the obtained predictions.

2 Formalization

The biological problem that we tackle in this work is essentially a combinatorial optimization problem over the possible logic models representing a given PKN. In this section, first we introduce the graphical representation of logic models by giving a simple example that motivates our formalization. Then, we give a formal definition for the inputs of the problem, we formally define a Protein Signaling Logic Model (PSLM) and we show how predictions are made for a given model. Finally, we define an objective function used for the optimization over the space of possible logic models.

2.1 Motivation

The functional relationships of biological networks, such as PKNs, cannot be captured using only a graph [15,9]. If, for example, two proteins (nodes) A and B have a positive effect on a third one C (encoded in a graph as $A \rightarrow C$, and $B \rightarrow C$), is not clear if either A or B can active C, or if both are required (logic OR and AND gate, respectively). To represent such complex (logical) relations between nodes and offer a formal representation of cellular networks, hypergraphs can be used. Since hypergraphs were already described and used to represent logic models of protein signaling networks in [15,16,9,17,14], here we adopt the same formalism and we simply give an example to introduce this representation. For more details, we refer the reader to the cited literature.

Example 1. Given the toy PKN described in Fig. 1(a), an arbitrary compatible logic model is given by the following set of formulas $\{d = (a \wedge b) \vee \neg c; e = c; f = d \wedge e\}$. Moreover, a representation of this logic model is given in Fig. 1(b) as a signed and directed hypergraph. Note that each conjunction clause gives place to a different hyperedge having as its source all the present literals in the clause.

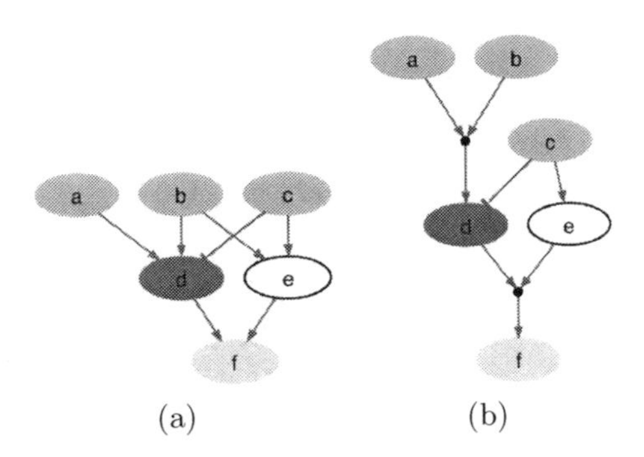

Fig. 1. Hypergraph representation of Logic Models. The green and red edges correspond to activations and inhibitions, respectively. Green nodes represent ligands that can be experimentally stimulated. Red nodes represent those species that can be inhibited by using a drug. Blue nodes represent those species that can be measured by using an antibody. White nodes are neither measured, nor manipulated. **(a)** A toy PKN as a directed and signed graph. **(b)** An arbitrary Logic Model compatible with the PKN shown in (a). Black filled circles represent AND gates, whereas multiple edges arriving to one node model OR gates.

Informally, the training of logic models to phospho-proteomics data consist in finding the hypergraph(s) compatible with a given PKN that best explains the data (using some criteria). Even though a graphical representation is quite intuitive and has been widely used in the literature, it is not the most appropriate way to give a formal and clear formulation of this problem. Thus, in what follows we give a formalization based on propositional logic and in the rest of this work, whenever convenient, we will refer interchangeably to Protein Signaling Logic Models as hypergraphs and to conjunctive clauses as hyperedges.

2.2 Problem Inputs

We identify three inputs to the problem: a Prior Knowledge Network (PKN), a set of experimental conditions or perturbations and for each of them, a set of experimental observations or measurements. For the sake of simplicity, in this work we have considered only PKNs with no feedback loops. They account for the main mechanisms of transmission of information in signaling pathways, but do not include feedback mechanisms that are typically responsible for the switching off of signals once the transmission has taken place [9,6]. In what follows, we give a mathematical definition for each of these inputs.

Definition 1 (Prior Knowledge Network). *A PKN is a signed, acyclic, and directed graph (V, E, σ) with $E \subseteq V \times V$ the set of directed edges, $\sigma \subseteq E \times \{1, -1\}$ the signs of the edges , and $V = S \cup K \cup R \cup U$, the set of vertices where S are the stimulus (inputs), K are the inhibitors (knock-outs), R are the readouts (outputs) and U are neither measured, nor manipulated. Moreover, the subsets S, K, R, U are all mutually disjoint except for K and R.*

Note that in the previous definition, σ is defined as a relation and not as a function since it could be the case where both signs are present between two vertices. This is even more likely to happen when a PKN, either extracted from the literature or from one of the mentioned databases, is compressed as described in [9] in order to remove most of the nodes that are neither measured, nor manipulated during the experiments. Also note that the subset of nodes K correspond to those proteins (e.g. kinases) that can be forced to be inactive (inhibited) by various experimental tools such as small-molecule drugs, antibodies or RNAi.

Definition 2 (Experimental condition). *Given a PKN (V, E, σ) an experimental condition over (V, E, σ) is a function $\varepsilon : S \cup K \subseteq V \to \{0, 1\}$ such that if $v \in S$, then $\varepsilon(v) = 1$ (resp. 0) means that the stimuli v is present (resp. absent), while if $v \in K$, then $\varepsilon(v) = 1$ (resp. 0) means that the inhibitor for v is absent and therefore v is not inhibited (resp. the inhibitor for v is present and therefore v is inhibited).*

Definition 3 (Experimental observation). *Given a PKN (V, E, σ) and an experimental condition ε over (V, E, σ), an experimental observation under ε is a function $\theta : R(\varepsilon) \subseteq R \to \{0, 1\}$ such that $R(\varepsilon)$ denotes the set of observed readouts under ε and if $v \in R(\varepsilon)$, $\theta(v) = 1$ (resp. 0) means that the readout v is present (resp. absent) under ε.*

Since the phospho-proteomics data used here represents an average across a population of cells, each of which may contain a different number of proteins in active or inactive (1 or 0) state, the values are continuous. Thus, we have to discretize the experimental data somehow in order to fit the previous definition. A simple but yet effective approach is to use a threshold $t = 0.5$ such that values greater than t are set to 1, while values lower that t are set to 0. Other approaches could also be used but, since in this paper we work with discrete *in silico* data, we left this discussion for a future work. Indeed, this paper focuses on comparing the performance of training and formal approaches to the optimization problem, for which *in silico* datasets appear more relevant.

2.3 Protein Signaling Logic Models

Here we state the combinatorial problem as a Constraint Satisfaction Problem (CSP) in order to have a clear and formal definition of a Protein Signaling Logic Model (PSLM) as a solution to this problem. Recall that a CSP is defined by a set of variables X, a domain of values D, and a set of constraints or properties to be satisfied. A solution to the problem is a function $e : X \to D$ that satisfies all constraints [18].

Next, we define two properties that we use later as the constraints of the CSP formulation. The first property defines for a given PKN, the conditions that must be satisfied by a logical formula in order to define the truth value of any node. For example, if we look the Fig. 1 is quite clear that the hypergraph in (b) is not just some arbitrary hypergraph, but instead is strongly related to the graph in (a). This relation is captured by the following definition.

Definition 4 (PKN evidence property). *Given a PKN (V, E, σ) and $v \in V$, a logical formula φ in Disjunctive Normal Form (DNF) has an evidence in (V, E, σ) with respect to v if and only if for every propositional variable w that occurs positively (resp. negatively) in φ, it exists an edge $(w, v) \in E$ and $((w, v), 1) \in \sigma$ (resp. $((w, v), -1) \in \sigma$).*

The second property identifies those logical formulas in DNF for which exist some equivalent but simpler formula. For example, for two literals X and Y, it is easy to see that $X \vee (X \wedge Y) \equiv X$. In such case we say that $X \vee (X \wedge Y)$ is redundant since, as we will see later, we are interested in minimizing the complexity of the logic models. This concept was previously introduced in [9] as a way to reduce the search space of all possible logical formulas.

Definition 5 (Redundancy property). *Given a logical formula φ in DNF, with $\varphi = \bigvee_{j \geq 1} c_j$ where each c_j is a conjunction clause, φ is a redundant formula if and only if for some $k, l \geq 1$ with $k \neq l$ and some logical conjunction r it holds that $c_k = c_l \wedge r$.*

Now, based on the general form of a CSP given above, and the properties defined in (4) and (5), we define a PSLM as follows.

Definition 6 (Protein Signaling Logic Model). *Given a PKN (V, E, σ) with $V \backslash S = \{v_1, \ldots, v_m\}$ for some $m \geq 1$, let the set of variables $X = \{\psi_{v_1}, \ldots, \psi_{v_m}\}$ and the domain of values D given by all the formulas in DNF having V as the set of propositional variables. Then, a function $e : X \rightarrow D$ defines a compatible Protein Signaling Logic Model $\mathcal{B}$ if it holds for $i = 1, \ldots, m$ that $e(\psi_{v_i})$ has an evidence in (V, E, σ) with respect to v_i. Moreover, if it also holds for $i = 1, \ldots, m$ that $e(\psi_{v_i})$ is not redundant, then we say that $\mathcal{B}$ is a non redundant logic model.*

Example 2. Given the PKN in Fig. 1(a) the function that defines the logic model in Fig. 1(b) is given by:

$$e(\psi_v) = \begin{cases} (a \wedge b) \vee \neg c & \text{if } v = d \\ c & \text{if } v = e \\ (d \wedge e) & \text{if } v = f \end{cases}$$

for $\psi_v \in \{\psi_d, \psi_e, \psi_f\}$. Note that in every case, each formula satisfies both properties: PKN evidence (Definition 4) and Non-redundancy (Definition 5).

2.4 Predictive Logic Model

A given Protein Signaling Logic Model (PSLM) describes only the static structure of a Boolean network. Even though Boolean networks are either synchronous or asynchronous, in any case the set of Logical Steady States (LSSs) is the same. Therefore, and since we focus on a Logical Steady State Analysis (LSSA) which offers a number of applications for studying functional aspects in cellular interactions networks [15], choosing between synchronous and asynchronous is not relevant in this work. Moreover, we do not need to compute all possible LSSs, but only the one that can be reached from a given initial state. Note that the existence of a unique LSS is guaranteed by the assumption of no feedbacks loops in the given PKN. Next, we describe how we compute this LSS in terms of satisfiability of a particular logical formula. This is based on the formalization given by [19] for a related problem named IFFSAT.

Let (V, E, σ) a PKN and $\mathcal{B}$ a compatible PSLM. Recall that $\mathcal{B}$ is defined by a function from every non-stimuli in (V, E, σ) to a DNF formula satisfying the PKN evidence property defined in Definition 4.

First, we define a logical formula $\mathcal{R}$ representing the regulation of every non-stimuli or non-inhibitor node in (V, E, σ).

$$\mathcal{R} = \bigwedge_{v \in V \setminus (S \cup K)} (\mathcal{B}(v) \iff v) \tag{1}$$

Note that we use $(\mathcal{B}(v) \iff v)$ instead of just $(\mathcal{B}(v) \Rightarrow v)$ to enforce that every activation must have a "cause" within the model. Next, we define two logical formulas $\mathcal{S}$ and $\mathcal{K}$ in order to fix the values of *stimulus* and the values or regulations of *inhibitors* in (V, E, σ) under a given experimental condition ε.

$$\mathcal{S} = \bigwedge_{v \in S} \begin{cases} v & \text{if } \varepsilon(v) = 1 \\ \neg v & \text{if } \varepsilon(v) = 0 \end{cases} \qquad \mathcal{K} = \bigwedge_{v \in K} \begin{cases} \mathcal{B}(v) \iff v & \text{if } \varepsilon(v) = 1 \\ \neg v & \text{if } \varepsilon(v) = 0 \end{cases} \tag{2}$$

Thereafter, we look for the truth assignment such that $\mathcal{R} \wedge \mathcal{S} \wedge \mathcal{K}$ evaluates to *true* and which represents the LSS of the network for the given initial conditions. The biological meaning behind this concept is that the input (*stimuli*) signals are propagated through the network by using the faster reactions and after some time, the state of each protein will not change in the future. Thus, we say that the network is stabilized or that it has reached an steady state [15]. Finally, we can define the model prediction under a given experimental condition as follows.

Definition 7 (Model prediction). *Given a PKN (V, E, σ), a PSLM $\mathcal{B}$ compatible with (V, E, σ) and a experimental condition ε over (V, E, σ), the prediction made by $\mathcal{B}$ under the experimental condition ε is given by the truth assignment $\rho : V \rightarrow \{0, 1\}$ such that $\mathcal{R} \wedge \mathcal{S} \wedge \mathcal{K}$ evaluates to true.*

Note that without the assumption of no feedbacks loops in the given PKN, the existence of multiple steady states or cycle attractors should be considered. Then, in order to guarantee that ρ is well defined, new constraints should be added to the CSP instance defined in (6), but this is left as a future work.

Example 3. Let $\varepsilon : \{a, b, c, d\} \rightarrow \{0, 1\}$ an experimental condition over the PKN given in Fig. 1(a) defined by:

$$\varepsilon(v) = \begin{cases} 1 & \text{if } v \in \{a, b, c\} \\ 0 & \text{if } v = d \end{cases}$$

That is, a, b, c are stimulated while d is inhibited. Then, the prediction made by the PSLM given in Fig. 1(b) under ε, is given by the truth assignment such that the formula

$$((d \wedge e) \iff f) \wedge (c \iff e) \wedge a \wedge b \wedge c \wedge \neg d$$

evaluates to *true*. Thus, e is assigned to 1 and f is assigned to 0.

2.5 Objective Function

Given all the PSLMs compatible with a given PKN, our goal is to define an objective function in order to capture under different experimental conditions, the matching between the corresponding experimental observations and model predictions. To this end, we adopt and reformulate the objective function proposed in [9] in terms of our formalization. The objective function represents a balance between fitness of model to experimental data and model size using a free parameter chosen to maximize the predictive power of the model. Of course, other objective functions can be defined in the future, but here we focus on a comparison against one of the state of the art approaches and thus, we choose to use the same objective function.

Before going further, we define the size of a model as follows.

Definition 8 (Size of Protein Signaling Logic Models). *Given a PKN (V, E, σ) and a PSLM $\mathcal{B}$ compatible with (V, E, σ), the size of $\mathcal{B}$ is given by $|\mathcal{B}| = \sum_{v \in V \setminus S} |\mathcal{B}(v)|$ where $|\mathcal{B}(v)|$ denotes the canonical length of logical formulas.*

Example 4. If we consider the PSLM given in Fig. 1(b), its size is given by: $|(a \wedge b) \vee \neg c| + |c| + |(d \wedge e)| = 3 + 1 + 2 = 6$. This can be seen also as the size of the hypergraph, where each hyperedge is weighted by the number of source nodes and the size of the hypergraph is the sum of all weights.

Finally, we define the combinatorial optimization problem of learning PSLMs from experimental observations under several experimental conditions as follows.

Definition 9 (Learning Protein Signaling Logic Models). *Given a PKN (V, E, σ), n experimental conditions $\varepsilon_1, \ldots, \varepsilon_n$ and n experimental observations $\theta_1, \ldots, \theta_n$ with each θ_i defined under ε_i, for a given PSLM $\mathcal{B}$ compatible with (V, E, σ), and n model predictions $\rho_1, \ldots, \rho_n$ over $\mathcal{B}$ with each ρ_i defined under ε_i, we want to minimize*

$$\Theta(\mathcal{B}) = \Theta_f(\mathcal{B}) + \alpha \times \Theta_s(\mathcal{B}) \tag{3}$$

where $\Theta_f(\mathcal{B}) = \frac{1}{n_o} \times \sum_{i=1}^{n} \sum_{v \in R(\varepsilon_i)} (\theta_i(v) - \rho_i(v))^2$ such that n_o, the total number of output measures, is given by $\sum_{i=1}^{n} |R(\varepsilon_i)|$ and $\Theta_s(\mathcal{B}) = \frac{1}{b} \times |\mathcal{B}|$ such that b is the size of the union of all PSLMs compatible with (V, E, σ).

3 Methods

In this section we describe the methods used to perform a comparison between our ASP-based approach and the one presented in [9]. First we provide the ASP implementation that we used to run the experiments and then, we describe the method proposed to systematically generate *in silico* study cases based on realistic networks and data.

3.1 ASP Implementation

Our goal is to provide an ASP solution for learning PSLMs from experimental observations under several experimental conditions (Definition 9). Here we provide a logic program representation of the problem described in Section 2 in the input language of the ASP grounder *gringo* [20]. After describing the format of any input instance, we show how we generate non-redundant candidate solutions having an evidence in the given PKN, then we describe how model predictions are made and finally, we show the minimization of the objective function.

Input Instance. We start by describing the input instance for the PKN given by (V, E, σ), the experimental conditions $\mathcal{E} = \varepsilon_1, \ldots, \varepsilon_n$ and the experimental observations $\mathcal{O} = \theta_1, \ldots, \theta_n$.

$$
\begin{aligned}
\mathcal{G}((V,E,\sigma),\mathcal{E},\mathcal{O}) =& \{vertex(v) \mid v \in V\} \\
&\cup \{edge(u,v,s) \mid (u,v) \in E, ((u,v),s) \in \sigma\} \\
&\cup \{exp(i,v,s) \mid \varepsilon_i(v) = s, v \in S \cup K\} \\
&\cup \{obs(i,v,s) \mid \theta_i(v) = s, v \in R(\varepsilon_i)\} \\
&\cup \{nexp(n)\} \\
&\cup \{stimuli(v) \mid v \in S\} \\
&\cup \{inhibitor(v) \mid v \in K\} \\
&\cup \{readout(v) \mid v \in R\}
\end{aligned}
\tag{4}
$$

Candidate solutions. We follow a common methodology in ASP known as "guess and check" where using non-deterministic constructs, we "guess" candidate solutions and then, using integrity constraints we "check" and eliminate invalid candidates. Since we are interested only on those logical formulas having an evidence in (V, E, σ), first we generate all the possible conjunction clauses having such evidence by computing for every $v \in V$ all the possible subsets between the predecessors of v. This is done by the following rules.

$$
\begin{aligned}
subset(U,S,null,1,V) &\leftarrow edge(U,V,S). \\
subset(U,S_U,subset(W,S_W,T),N+1,V) &\leftarrow subset(U,S_U,null,1,V), \\
&\quad\; subset(W,S_W,T,N,V), \\
&\quad\; vertex(U) < vertex(W).
\end{aligned}
\tag{5}
$$

The idea is to start with the singleton subsets containing only a single predecessor, and to create a bigger subset by recursively extending a singleton subset with any other subset until a fix point is reached. The first rule defines all the singleton subsets related to V. We represent the subsets here as linked lists where U, the first argument in the predicate $subset/5$, represents the head of the list (5 is the arity of the predicate). The second argument represents the sign of the edge from U to V. The third argument represents the tail of the linked list ($null$ in case of a singleton). The fourth argument represents the subset cardinality, and the last argument keeps track of the target vertex. The head is here used as a identifier such that we can order all subsets. The second rule recursively extends a singular subset identified by head argument U with any subset identified by W as long as $U < W$. We exploit the order between the predicates $vertex/1$ to avoid different permutations of the same subsets.

The following rules define the inclusion relationship between these subsets of predecessors.

$$\begin{aligned} in(U,S,subset(U,S,T)) &\leftarrow subset(U,S,T,N,V). \\ in(W,S_W,subset(U,S_U,T)) &\leftarrow in(W,S_W,T), \\ &\qquad subset(U,S_U,T,N,V). \end{aligned} \tag{6}$$

The first rule declares that every subset contains its "head" element. The second rule declares that if W is included in T, and if there is another subset having T as its "tail", then W is also included in it.

Since each subset generated by the rules in (5) represents a possible conjunction clause, we can generate all possible logical formulas in DNF by considering each subset as either present or absent.

$$\begin{aligned} \{clause(subset(U,S,T),N,V)\} &\leftarrow subset(U,S,T,N,V). \\ &\leftarrow clause(C_1,N,V), clause(C_2,M,V), \\ &\qquad C_1 \neq C_2, in(U,S,C_2) : in(U,S,C_1). \end{aligned} \tag{7}$$

The first rule is a choice rule that declares the non-deterministic generation of predicates $clause/3$ from a subset. A clause represents the conjunction of all the elements included in the subset. The second rule declares an integrity constraint to avoid the generation of redundant logical formulas by using the predicates generated in (6).

Model predictions. Next, we show the representation for the input signals propagation. For each experiment, first the truth values for stimuli and inhibited nodes are fixed and then, truth values are propagated to all nodes by exploiting the fact that in order to assign *true* to any node, it is enough that one conjunction clause over it evaluates to *true*.

$$\begin{aligned} fixed(E,V) &\leftarrow nexp(N), E = 1..N, stimuli(V). \\ fixed(E,V) &\leftarrow inhibitor(V), exp(E,V,0). \\ active(E,V) &\leftarrow exp(E,V,1), stimuli(V). \\ inactive(E,V) &\leftarrow exp(E,V,0). \end{aligned} \tag{8}$$

The first and second rules simply declare which nodes have fixed truth values because they are either an input node, or an inhibited node in a particular experiment. Thereafter, the third and fourth rules declare the truth assignments that are given by the experimental condition.

The following rules model the signal propagation in every experiment.

$$\begin{aligned} active(E,V) \leftarrow\ & nexp(N), E = 1 \ldots N, \\ & clause(C,M,V), not\ fixed(E,V), \\ & active(E,U) : in(U,1,C), \\ & inactive(E,U) : in(U,-1,C). \\ inactive(E,V) \leftarrow\ & vertex(V), nexp(N), E = 1 \ldots N, \\ & not\ fixed(E,V), not\ active(E,V). \end{aligned} \tag{9}$$

The first rule declares that for each experiment, if there is at least one conjunction clause having all its positive literals assigned to *true* and all its negated literals assigned to *false*, then the complete clause evaluates to *true*. While the second rule declares that every node that is not assigned to *true*, it is assigned by default to *false*.

Optimization. Finally, we show the declaration of the objective function. In Section 2 we defined the objective function Θ, but since ASP can only minimize integer functions, we transformed Θ into Θ_{int} trying to lose as less information as possible. To this end, if we assume that the free parameter $\alpha = \frac{N}{D}$ for some $N, D \in \mathbb{N}$, multiplying Θ by $N \times \frac{1}{\alpha} \times n_o \times b$ we define Θ_{int} as follows.

$$\begin{aligned} \Theta_{\text{int}}(\mathcal{B}) &= D \times n_o \times b \times \Theta(\mathcal{B}) \\ &= D \times b \times \sum_{i=1}^{n} \sum_{v \in R(\varepsilon_i)} (\theta_i(v) - \rho_i(v))^2 + N \times n_o \times |\mathcal{B}| \end{aligned} \tag{10}$$

This new (integer) objective function is represented as follows in our ASP encoding.

$$\begin{array}{lll} \#const\ npenalty = 1. & & \\ \#const\ dpenalty = 1000. & & \\ penalty\_N(npenalty). & & \\ penalty\_D(dpenalty). & & \\ b(B) & \leftarrow & B = [subset(\_,\_,N,\_) = N]. \\ n_o(E) & \leftarrow & E = [obs(\_,\_,\_)]. \\ mismatch(E,V) & \leftarrow & obs(E,V,0), active(E,V). \\ mismatch(E,V) & \leftarrow & obs(E,V,1), inactive(E,V). \end{array} \tag{11}$$

The rules in (11) declare the predicates that we need to give a representation of Θ_{int}. First, we declare two predicates to represent the free parameter α as a fraction of integers. Then, we use a weighted sum to declare the size of the union of all logic models and we count the number of single experimental observations. The two last rules declare in which cases a model prediction does not match the corresponding output measurement.

Last but not least, we require the minimization of the (integer) objective function Θ_{int} simply by using the *#minimize* directive.

$$\begin{aligned} \#minimize[\ & mismatch(\_,\_) : b(B) : penalty\_D(PD) = B \times PD, \\ & clause(\_,N,\_) : n_o(E) : penalty\_N(PN) = E \times PN \times N]. \end{aligned} \tag{12}$$

3.2 Benchmark Datasets

We wanted to compare the ASP approach with CellNOpt and analyze the scalability of the methods. Also we wanted to determine how the inference of the network is influenced by specific parameters of the problem. For this purpose, we generated meaningful benchmarks that covered a broad range of these influential parameters.

Middle and Large-Scale Benchmark Datasets We constructed a middle (see Fig 2) and a large scale (see Fig 3) optimization problem. Both PKNs were derived from literature and in each case we randomly selected compatible PSLMs or hypergraphs (middle: Fig. 2(b), large: Fig. 3(b)), from which we generated *in silico* datasets under several experimental conditions giving place to different numbers of output measures. The main parameters used to compute the objective function for the optimization are shown in the Table 1.

Table 1. Middle and Large optimization problems

Scale	Nodes	Edges	Compatible hyperedges	Size of union hypergraph	Selected hypergraph size	Experimental conditions	Output measures
Middle	17	34	87	162	20	34	210
Large	30	53	130	247	37	56	840

Large Set of Benchmark Datasets. We relied on a literature derived PKN for growth and inflammatory signaling [10] to derive compatible PSLMs and generate 240 benchmark datasets with *in silico* observation data. Given the literature derived network (V, E, σ) with $V = S \cup K \cup R \cup U$, we created 4 derivative networks $(V_i, E, \sigma), i = 1 \ldots 4$ with $V_i = V$, $V_i = S \cup K_i \cup R_i \cup U_i$, $K_i \subseteq K$, $R_i \subseteq R$, and $U \subseteq U_i$. Each network differing in sets of inhibitors and readouts. For these networks we compressed (bypassed) nodes that are neither measured, nor manipulated during the experiments, which were not affected by any perturbation, lay on terminal branches or in linear cascades, as described in [9], yielding to 4 compressed networks.

To investigate the influence of the size of the networks in the optimization both in terms of computational times and recovered edges, we randomly selected PSLMs of 3 different sizes (20, 25, or 30) for each compressed network. The size of each model was obtained as defined in Definition 8. Then, 5 different PSLMs were generated for each compressed network and for each size, giving a total of 60 different models (20 of each size). We use these 60 models to run simulated experiments and generate *in silico* experimental observations.

Moreover, we wanted to investigate how the amount of experimental observation data influences the network inference. Therefore, we generated 4 datasets $D_1 \ldots D_4$ of experimental observations for each model. The first dataset D_1

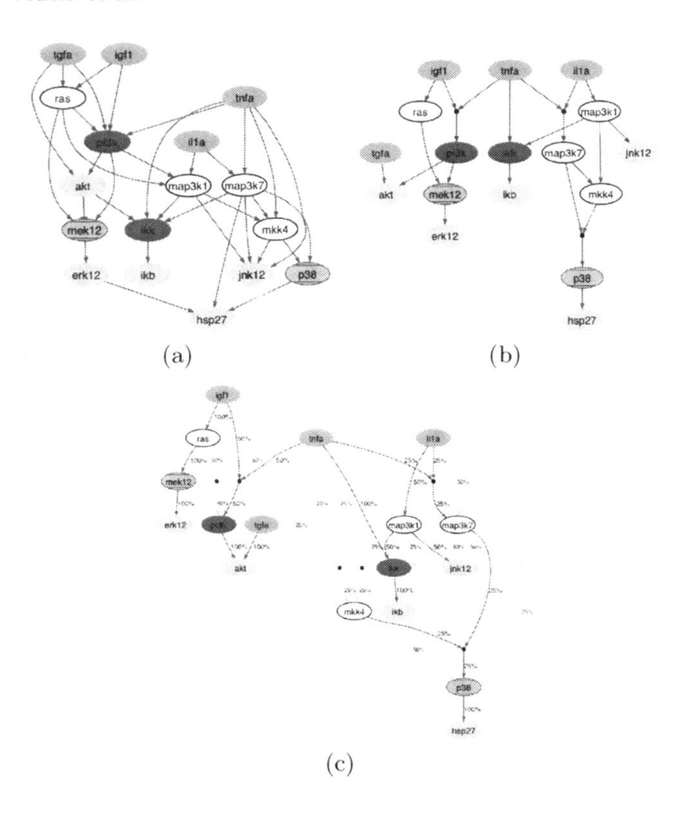

Fig. 2. Input and outputs of a middle-scale optimization problem. (a) A literature-derived Prior Knowledge Network (PKN) of growth and inflammatory signaling. **(b)** An hypergraph which is compatible with the PKN shown in (a). From this model we derive 240 output measures under 34 experimental conditions. **(c)** The ASP optimization enumerated all the minimal PSLMs that predict the *in silico* measures produced in (b) with no mismatches. The union of the 8 optimal models is shown with a specific edge encoding: edges are labeled according to their percentage of occurrence in all the 8 models. The thick green edges correspond to those edges that also appear in the hypergraph used to generate the *in silico* datasets.

contained only experimental observations from single-stimulus/single-inhibitor experiments. The other datasets $D_2 \dots D_4$ contained observations from multiple-stimuli/multiple-inhibitors experiments with 30, 50 and 60 experimental conditions respectively. The larger datasets always include the smaller datasets, such that $D_1 \subset D_2 \subset D_3 \subset D_4$. In total we generated 240 different datasets of 4 different sizes, generated from 60 different models of 3 different sizes. The whole method is illustrated in the Fig. 4.

4 Results and Discussions

First, we focused in finding minimal PSLMs compatible with the given PKN and predicting the generated dataset for the middle (see Fig. 2) and large-scale (see

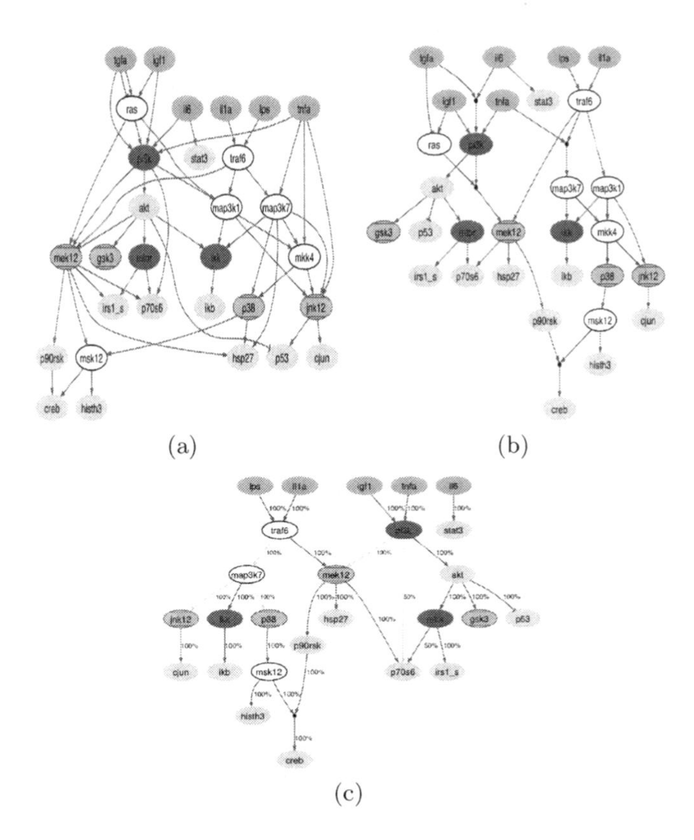

Fig. 3. Large-scale optimization problem. **(a)** A Large literature-derived PKN of growth and inflammatory signaling obtained from [21]. **(b)** An hypergraph which is compatible with the PKN shown in (a). From this model we derive 840 output measures under 56 experimental conditions. **(c)** Union of the two minimal PSLMs predicting the whole dataset with no mismatches.

Fig. 3) benchmarks. Second, general comparisons between our logical approach implemented in ASP and the genetic algorithm implemented in CellNOpt, were performed over the 240 datasets generated as described in Section 3.2.

4.1 Enumeration of Solutions to the Optimization Problems

We used the ASP implementation detailed in Section 3.1 to identify the PSLMs compatible with the middle-scale PKN (respectively the large-scale PKN) having an optimal score with respect to the generated dataset.

Notice that in both cases, by the construction of the datasets, we knew that there exists a compatible PSLM which predicts the whole datasets without mismatches. As a consequence, if $\alpha \in (0,1)$ (see Eq. 3, Eq. 10), the optimization problem is equivalent to find the logic models with perfect fit and minimal size.

In a first run, the ASP implementation allowed us to compute the minimal score of the optimization problem. Afterwards, we run the ASP solver again to

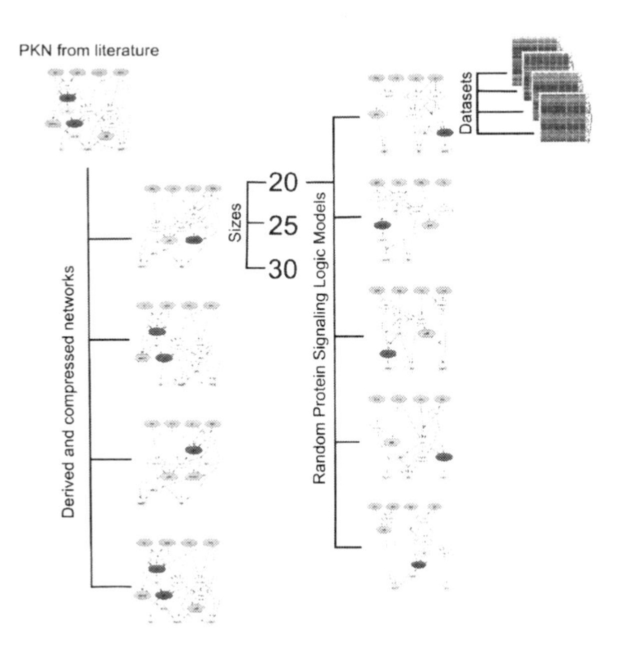

Fig. 4. Pipeline of the generation of the 240 benchmarks

enumerate all the models having a score lower or equal than the minimal score. All together, we obtained a complete enumeration of all minimal models. Below, we show the results obtained using the ASP-based approach to solve the middle and large optimization problems.

- **Middle-scale** The minimal score was computed in 0.06 seconds[1]. The enumeration took 0.03 seconds and found 8 global optimal Boolean models with size equal to 16. The union of the 8 optimal models found is shown in Fig. 2(c).
- **Large-scale** The minimal score was computed in 0.4 seconds. The enumeration took 0.07 seconds and found 2 global optimal Boolean models with size equal to 26. In Fig. 3(c) we show the union of the 2 optimal models found.

Both optimization problems were also run with CellNOpt, based on its genetic algorithm (see Materials and Methods section in [9]) performing generations over a population of 500 models[2].

- **Middle-scale**The optimization was run for 9.2 hours and the best score was reached after 7.2 hours (299 generations). During the optimization, 66 Boolean models with perfect fit were found, with sizes going from 16 to 24. Out of the 66 models, only 2 models were minimal (i.e. with size equal to 16).

[1] All ASP computations were run in a MacBook Pro, Intel Core i7, 2.7 GHz and 4 GB of RAM using Gringo 3.0.3 and Clasp 2.0.5 versions.

[2] All CellNOpt computations were run in a cluster of 542 nodes, each with 32 GB of memory, and a total of 9000 cores using CellNOptR 1.0.0.

- **Large-scale** The optimization problem was run for 27.8 hours and the best score was reached after 24.5 hours (319 generations). During the optimization, 206 models with perfect fit were found, with sizes going from 27 to 36. Note that in this case, CellNOpt did not find any of the minimal models (i.e. with size equal to 26).

Our main conclusion here is as follows: in both cases, due to the use of *in silico* data, models with perfect fit were exhibited by both approaches. The main advantage of the formal approach is to be able to explicitly compute the minimal score, allowing us to enumerate all models with this score in a very short time. Meanwhile, genetic algorithms are not able to exhibit this information and therefore cannot develop strategies to compute all minimal models. At the same time, this leads to the question about the biological relevance of optimal models and if it is possible to discriminate between them. A precise study of the biological pathways selected in each optimal model did not allow us to specifically favor one model according to biological evidences. That is why we choose to show the union of them in each case (Fig. 2(c) and Fig. 3(c)).

Nonetheless, the ASP search was strongly supported by the fact that there exists at least one model with perfect fit. This considerably reduces the optimization problem to the search of compatible models with minimal size by canceling the Θ_f term in Eq. (3). Performing optimizations over real data will induce that there will no more exist models with perfect fit, which may have a strong effect over the performance of our formal approach, while for genetic algorithms performances may probably be less affected by real data, but this will have to be studied. An interesting perspective is therefore to test the efficiency of these approaches in a real case experiment.

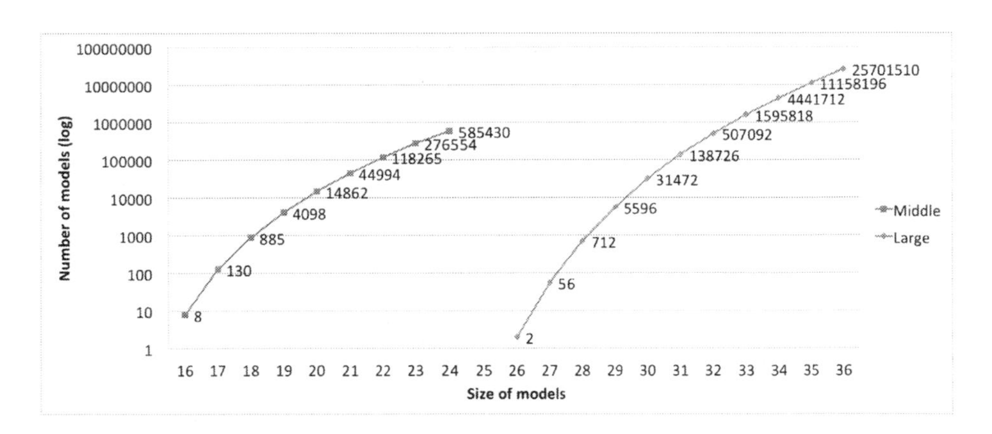

Fig. 5. Number of suboptimal solutions to the middle-scale (red curve) and large-scale (blue curve) optimization problems. Each curve describes the number of models with perfect fit with a given size, where the size ranges from its minimal value to the maximal size of models found by CellNOpt.

4.2 Dependency to the Model Size

To have a first view of the space of solutions, we investigated the role of the model size over the optimization process. Indeed, the optimization criteria moderates the choice of a model of minimal size -according to a parsimonious principle- by a free parameter related to the fitting between observations and predictions. (see Eq. 3). However, as we mentioned above, in all our experiments we known that there exists at least one model which predicts the whole datasets without mismatches and thus, the optimization problem is focused on finding minimal models. Therefore strongly favoring the size of the model. To evaluate the impact of this for the middle and large optimization problems depicted in Fig. 2 and Fig. 3, we used ASP to enumerate all the models with perfect fit having their size less or equal to the size of the models found by CellNOpt. Results are depicted in Fig. 5, providing a first insight on the structure of the space of compatible PSLMs with perfect fit. It appears that the number of compatible PSLMs increases exponentially with the size of the model. Therefore, optimizing over the size criteria appears quite crucial. A prospective issue is to elucidate whether the topology of the space of suboptimal models informs about the biological relevance of minimal models.

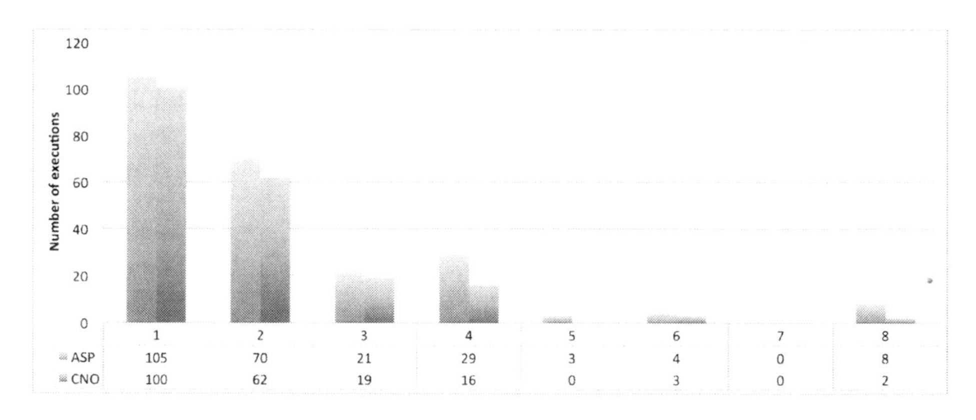

Fig. 6. Executions of ASP and CellNOpt optimizations that found all global optimal models. The total number of runs was of 240 in account of the in-silico data generated. The x-axis represents the number of global optimal models that each problem had. The y-axis, shows the number of executions where ASP and CellNOpt found the total number of global optimums.

4.3 Accuracy of Predictions

The study of the middle and large optimization problems evidenced that genetic algorithms may not find all minimal models. In order to elucidate whether this phenomenon is frequent, we used the 240 benchmark datasets generated with the method described in Section 3.2. In Fig. 6 we show the number of executions of the optimization process for ASP and CellNOpt where both approaches found the complete set of global optimal (minimal) models. Recall that ASP ensures finding the complete set of global optimal models (blue bars in Fig. 6) while this is not the case for CellNOpt. We observed that in 202 executions out of 240 (84%), ASP and CellNOpt both found all the minimal models. This is particularly clear in the 105 executions with a single minimal model, which was found by CellNOpt in 95% of executions. Nonetheless, in the 44 cases with more than 4 optimal models, CellNOpt found all optimal models in only 47% cases. More generally, as the number of minimal solutions to the optimization problem increases, the percentage of minimal solutions identified by CellNOpt decreases.

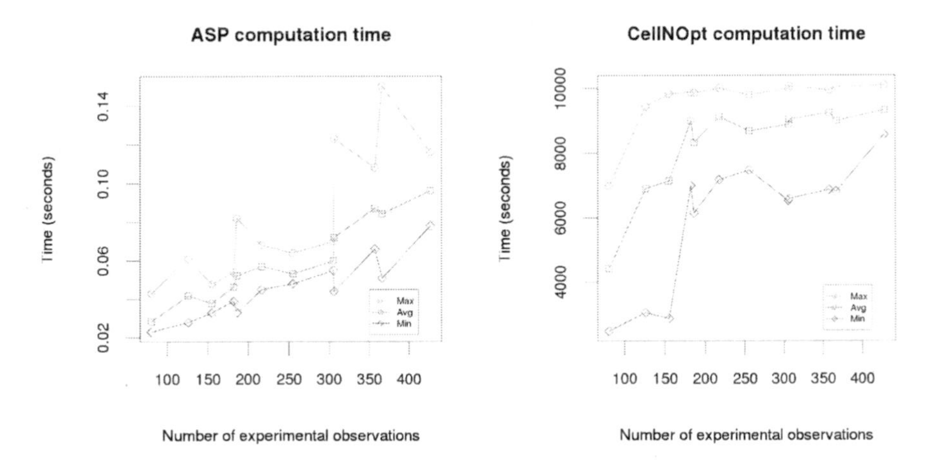

Fig. 7. Computation time of ASP and CellNOpt with respect to the number of experimental observations. The x-axis is the number of experimental observations: sum of number of readouts for each experimental condition. The maximum, average, and minimum computation times are plotted in green, red, and blue respectively.

4.4 Computation Times

In Fig. 7 we plot the computation time evolution for ASP and CellNOpt with respect to the number of experimental observations (i.e. output measures) included in the *in silico* datasets used to run the optimizations. Since we generated multiple datasets which contained the same number of experimental observations, for each optimization related to these multiple datasets we obtained minimum, maximum, and average times. We observe that the ASP computation times are in a range that goes from 0.02 to 0.15 seconds, while CellNOpt computation times to find the best score goes from 43 minutes to 2.7 hours, which was set as the limit running time. We see from these results that ASP outperforms CellNOpt in 5 order of magnitude guaranteeing in all cases global optimality. As discussed in a previous subsection, the main prospective issue is to test the relevance of this conclusion when optimizing with real data instead of *in silico* data.

5 Conclusion

We have proposed a formal encoding of a combinatorial optimization problem related to the inference of Boolean rules describing protein signaling networks. We have used ASP, a declarative problem solving paradigm, to solve this optimization problem and compared its performance against the stochastic method implemented by CellNOpt. Our ASP formulation relies on powerful, state-of-the-art and free open source software [20,12]. As main conclusion, we prove that our ASP-based approach ensures to find all optimal models by reasoning over the complete solution space. Moreover, in the experiments presented in this work, ASP outperforms CellNOpt in up to 5 orders of magnitude.

Our analyses provide concrete illustrations of the potential applications, in our opinion under-explored, of ASP in this field. Recently, Integer Linear Programming (ILP) have been used to solve the same problem that we described here [13]. In principle, ILP solvers can also provide the complete set of optimal solutions but a detailed comparison between ASP and ILP for this particular problem remains to be done.

As discussed within the results section, several prospective issues shall now be investigated. We first have to study the robustness of our results when optimizing over real networks and datasets. Second, we shall develop tools to explore the topology of the space of suboptimal models in order to gain in biological relevance in the inference process and try to elucidate whether this topology informs about the biological relevance of minimal models. Finally, by considering the presence of feedbacks loops in the input PKN and by studying the effect of different discretization approaches, we hope to improve the state of the art in protein signaling network inference and offer a useful tool for biologists.

Acknowledgements. The work of the first author was supported by the project ANR-10-BLANC-0218. The work of the second author was financed by the BMBF MEDSYS 0315401B. The work of the third author was partially supported by the 'Borsa Gini' scholarship, awarded by 'Fondazione Aldo Gini', Padova, Italy.

References

1. Cerami, E.G., Gross, B.E., Demir, E., Rodchenkov, I., Babur, O., Anwar, N., Schultz, N., Bader, G.D., Sander, C.: Pathway Commons, a web resource for biological pathway data. Nucleic Acids Research 39(Database issue), D685–D690 (2011)
2. Schaefer, C.F., Anthony, K., Krupa, S., Buchoff, J., Day, M., Hannay, T., Buetow, K.H.: PID: the Pathway Interaction Database. Nucleic Acids Research 37(Database issue), D674–D679 (2009)
3. Zinovyev, A., Viara, E., Calzone, L., Barillot, E.: BiNoM: a Cytoscape plugin for manipulating and analyzing biological networks. Bioinformatics 24(6), 876–877 (2008)

4. Guziolowski, C., Kittas, A., Dittmann, F., Grabe, N.: Automatic generation of causal networks linking growth factor stimuli to functional cell state changes. FEBS Journal (2012)
5. Palmisano, G., Thingholm, T.E.: Strategies for quantitation of phosphoproteomic data. Expert Review Of Proteomics 7(3), 439–456 (2010)
6. Terfve, C., Saez-Rodriguez, J.: Modeling Signaling Networks Using High-throughput Phospho-proteomics. Advances in Experimental Medicine and Biology 736, 19–57 (2012)
7. Bansal, M., Belcastro, V., Ambesi-Impiombato, A., di Bernardo, D.: How to infer gene networks from expression profiles. Mol. Syst. Biol. 3, 78 (2007)
8. Hecker, M., Lambeck, S., Toepfer, S., Van Someren, E., Guthke, R.: Gene regulatory network inference: data integration in dynamic models-a review. Bio Systems 96(1), 86–103 (2009)
9. Saez-Rodriguez, J., Alexopoulos, L.G., Epperlein, J., Samaga, R., Lauffenburger, D.A., Klamt, S., Sorger, P.K.: Discrete logic modelling as a means to link protein signalling networks with functional analysis of mammalian signal transduction. Molecular Systems Biology 5(331), 331 (2009)
10. Prill, R.J., Saez-Rodriguez, J., Alexopoulos, L.G., Sorger, P.K., Stolovitzky, G.: Crowdsourcing network inference: the DREAM predictive signaling network challenge. Sci. Signal 4(189), mr7 (2011)
11. Baral, C.: Knowledge Representation, Reasoning and Declarative Problem Solving. Cambridge University Press (2003)
12. Gebser, M., Kaufmann, B., Neumann, A., Schaub, T.: Conflict-driven answer set solving, pp. 386–392 (2007)
13. Mitsos, A., Melas, I., Siminelakis, P., Chairakaki, A., Saez-Rodriguez, J., Alexopoulos, L.G.: Identifying Drug Effects via Pathway Alterations using an Integer Linear Programming Optimization Formulation on Phosphoproteomic Data. PLoS Comp. Biol. 5(12), e1000591 (2009)
14. Klamt, S., Haus, U.U., Theis, F.J.: Hypergraphs and Cellular Networks. PLoS Comput. Biol. 5(5), e1000385 (2009)
15. Klamt, S., Saez-Rodriguez, J., Lindquist, J., Simeoni, L., Gilles, E.: A methodology for the structural and functional analysis of signaling and regulatory networks. BMC Bioinformatics 7(1), 56 (2006)
16. Saez-Rodriguez, J., Simeoni, L., Lindquist, J., Hemenway, R., Bommhardt, U., Arndt, B., Haus, U.U., Weismantel, R., Gilles, E., Klamt, S., Schraven, B.: A Logical Model Provides Insights into T Cell Receptor Signaling. PLoS Comput. Biol. 3(8), e163 (2007)
17. Christensen, T.S., Oliveira, A.P., Nielsen, J.: Reconstruction and logical modeling of glucose repression signaling pathways in Saccharomyces cerevisiae. BMC Systems Biology 3, 7 (2009)
18. Tsang, E.: Foundations of constraint satisfaction. Academic Pr. (1993)
19. Haus, U.U., Niermann, K., Truemper, K., Weismantel, R.: Logic integer programming models for signaling networks. J. Comput. Biol. 16(5), 725–743 (2009)
20. Gebser, M., Kaminski, R., Ostrowski, M., Schaub, T., Thiele, S.: On the Input Language of ASP Grounder *Gringo*. In: Erdem, E., Lin, F., Schaub, T. (eds.) LPNMR 2009. LNCS, vol. 5753, pp. 502–508. Springer, Heidelberg (2009)
21. Morris, M.K., Saez-Rodriguez, J., Clarke, D.C., Sorger, P.K., Lauffenburger, D.A.: Training signaling pathway maps to biochemical data with constrained fuzzy logic: quantitative analysis of liver cell responses to inflammatory stimuli. PLoS Comput. Biol. 7(3), e1001099 (2011)

JAK-STAT Signalling as Example for a Database-Supported Modular Modelling Concept

Mary Ann Blätke[1], Anna Dittrich[1], Monika Heiner[2], Fred Schaper[1], and Wolfgang Marwan[1]

[1] Otto-von-Guericke-Universität, Magdeburg, Germany
[2] Brandenburg Technical University, Cottbus, Germany
mary-ann.blaetke@ovgu.de

Abstract. We present a detailed model of the JAK-STAT pathway in IL-6 signaling as non-trivial case study demonstrating a new database-supported modular modeling method. A module is a self-contained and autonomous Petri net, centred around an individual protein. The modelling approach allows to easily generate and modify coherent, executable models composed from a collection of modules and provides numerous options for advanced biomodel engineering.

1 Background

The evolutionary conserved JAK-STAT pathway is one of the major signalling components in most of the eukaryotic cells [4]. JAK-STAT transmits stimuli from the cell membrane to the nucleus and is therefore mainly responsible for the gene regulation to control cell growth, differentiation and death. The dysfunctionality of the JAK-STAT pathway can lead to cancers or immune deficiency syndromes. An outstanding characteristic of the JAK-STAT pathway is the extensive crosstalk among its components (Fig. 1.A). The JAK protein as well as the STAT protein can appear in different isoforms that interact with several cytokine receptors and gene promoters, which leads to a combinatorial problem. It is a challenging task to explore the extensive cross-talk of the regulatory components, which we address by a new modular approach to biomodel engineering.

2 Results and Conclusions

Instead of creating a large monolithic mathematical model, we have developed a bottom-up modular description of the JAK-STAT pathway [1]. Each module as model component represents an individual protein and all its reactions with other interaction partners. In addition, each module contains metadata for documentation purposes; it, thus, corresponds to a wiki-like mini-review. Modules can be linked to each other in arbitrary combination accounting for the combinatorial complexity of regulation (Fig. 1.B). A coherent network is obtained via specific connection interfaces, these are identical shared subnets describing the interaction between to proteins. The characteristic advantage of the approach is that no further modifications are required in order to obtain an executable model. As modelling framework we chose Petri nets, which are intuitively understandable, allow

D. Gilbert and M. Heiner (Eds.): CMSB 2012, LNCS 7605, pp. 362–365, 2012.

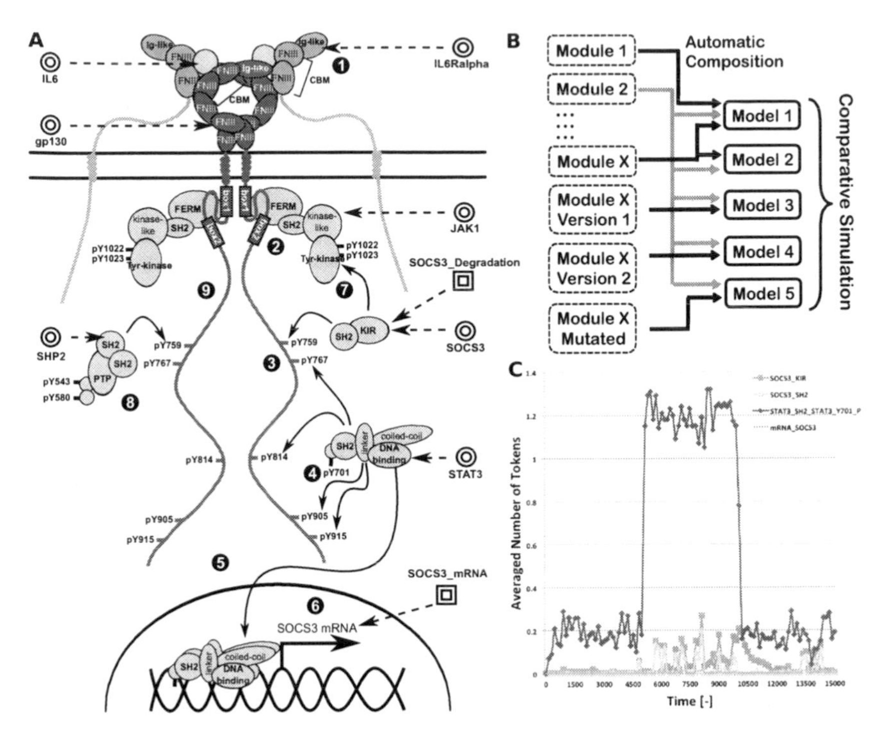

Fig. 1. JAK/STAT pathway in IL-6 signalling. (A) Mechanism of the JAK-STAT pathway in IL-6 signaling consists of the following steps: (1) Binding of interleukin-6 (IL-6) to IL6-receptor α (IL-6Rα) and glycoprotein gp130 and thereby formation of an active receptor complex, (2) activation of the JAK kinase by transphosphorylation, (3) phosphorylation tyrosine residues of gp130 by active JAK, (4) binding of STAT to phosphotyrosines of gp130 and phosphorylation by STAT, (5) dimerization of phosphorylated STATs and translocation to the nucleus, (6) activation of transcription of multiple genes, including SOCS, (7) binding of SOCS to gp130 and thereby inhibition of JAK (negative feedback), (8) dephosphorylation of gp130 by the SHP2 phosphatase, (9) phosphorylation of SHP2 by JAK and thereby inhibition of SHP2 (negative feedback); see [4] for a detailed review. The molecular mechanisms of each involved protein have been modelled in the form of separate Petri nets (modules) indicated by the corresponding coarse places (two nested circles) representing the underlying Petri net. The synthesis and degradation of SOCS3 are modelled within a biosynthesis/degradation module as indicated by the corresponding coarse transition (two nested squares) representing the underlying Petri net. The figure was redrawn from [4]. **(B)** Modules can be reused and recombined in various combinations. The obtained models can be used to test for the effect of alternative or modified reaction mechanisms. **(C)** Here, we employed stochastic simulations to demonstrate the response of the involved components dependent on IL-6 supply. IL-6 is injected only in the second third of the simulation time. The signalling activity increases during stimulation. The system shows basal activity before and after the stimulation.

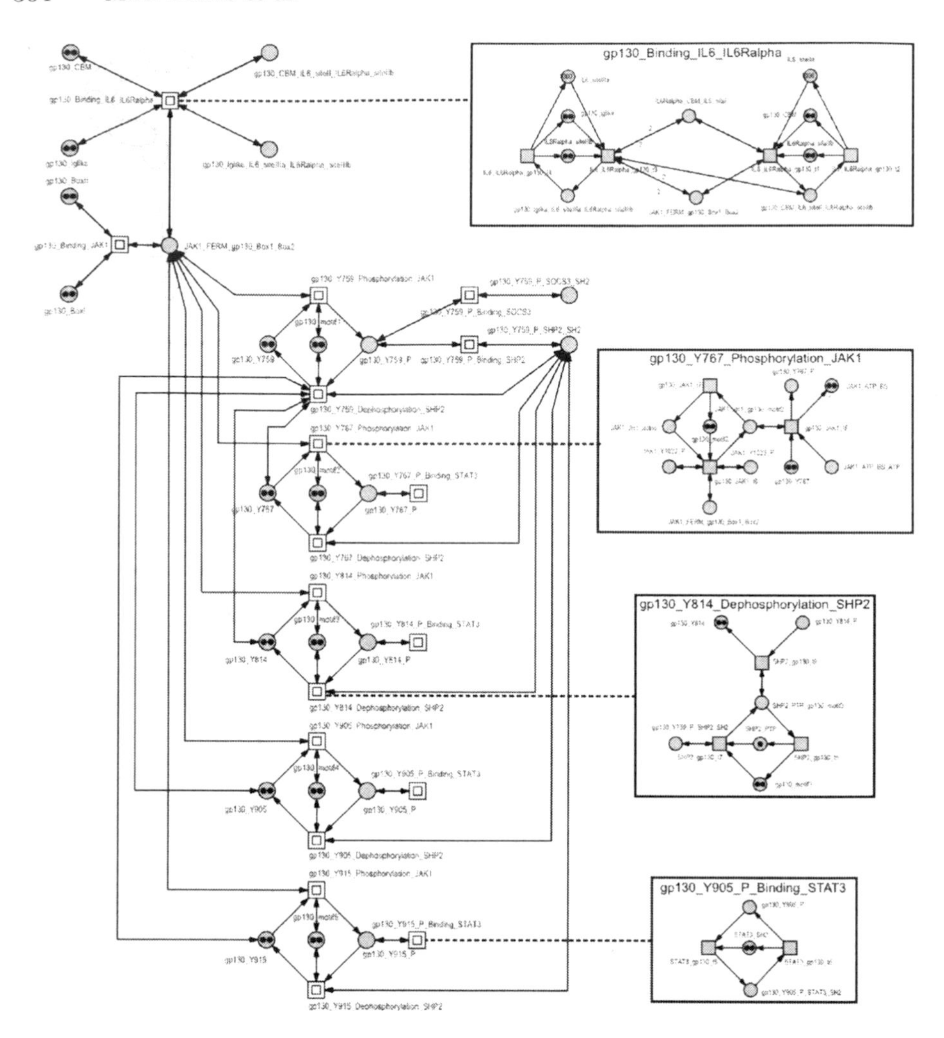

Fig. 2. The gp130 module. The extracellular part of the gp130 glycoprotein consists of the Ig-like domain and a cytokine-binding site, which are responsible for complex formation with IL-6 and IL6-Rα. The intracellular part of gp130 has an interbox 1-2 region, where JAKs are constitutively bound. Downstream of this binding site the gp130 receptor has several important tyrosine residues, which are phosphorylated by activated JAK. The phosphotyrosine pY759 is the specific binding site of SHP2 and SOCS isoforms via their SH2 domain, where STAT isoforms can interact via their SH2 domain with the other phosphotyrosines. In the corresponding Petri net model of gp130, the extracellular binding site of IL-6 and the intracellular binding site of JAK1 to the Box 1 and Box 2 sites are represented as coarse transitions. Three coarse transitions are assigned to each tyrosine residue downstream of Y759 describing the phosphorylation by JAK1, the dephosphorylation by SHP2, and the binding interaction with STAT3 (in the case of Y759 two coarse transitions are used to represent the interaction with SHP2 and SOCS3). The boxed panels on the right side of the Figure exemplify four subnets that are included in the corresponding coarse transitions in the gp130 module.

visual modelling and are executable. Petri nets have been shown to be perfectly suited to describe the inherently concurrent mechanisms in biological systems [3]. For providing a proof-of-principle for our modular modelling concept, we focus on JAK-STAT signalling induced by the IL-6 cytokine with only one isoform of each protein involved (JAK1, STAT3, SOCS3, SHP2, gp130, IL-6R, IL-6). Accordingly we constructed seven protein modules, one mRNA module for SOCS3 biosynthesis and one SOCS3 degradation module. Exemplarily, we show here the module of gp130 (Fig. 2). The networks which can be generated from this set of modules consist of up to 90 places and 100 transitions. The modules are hierarchically designed to obtain a clear graphical representation of the reaction mechanisms. The composed models comprise up to 58 pages with a nesting depth of 4. The modules were created and composed with *Snoopy* [5] and the built-in simulator was used to run stochastic simulations (Fig. 1.C). Structural analysis was performed with *Charlie* [2] to validate the structure of each module (not shown here, see [1]).

The model can be extended to include transcriptional regulation by employing gene modules, mRNA modules, degradation modules, and causal interaction modules. We have established a prototype database with a publically accessible web-interface [1]. The database can manage multiple versions of each module by strict version control. It supports the curation, documentation, and update of individual modules and the subsequent automatic composition of complex models, without requiring mathematical skills. The case study has demonstrated that modular modelling is ideally suited for exploring signalling networks with extensive cross-talk like in JAK-STAT. The supporting database essentially contributes to a powerful, comprehensive and unifying platform for modelling and analysis. The platform can be easily used by wet lab scientists to re-engineer individual modules in order to test the global consequences of alternative reaction mechanisms [2] (Fig. 1.B). The database associates meta-data to the individual modules, and thus is ideally suited for the documentation and validation of alternative reaction mechanisms. In the context of advanced biomodel engineering our modelling framework encourages the automated generation of biologically realistic synthetic and synthetically rewired network models.

References

1. Blätke, M., et al.: JAK/STAT signalling - an executable model assembled from molecule-centred modules demonstrating a module-oriented database concept for systems- and synthetic biology (submitted, 2012), http://arxiv.org/abs/1206.0959v1
2. Franzke, A.: Charlie 2.0 – a multi-threaded Petri net analyzer. Diploma thesis, BTU Cottbus, Dep. of CS (December 2009)
3. Heiner, M., Gilbert, D., Donaldson, R.: Petri Nets for Systems and Synthetic Biology. In: Bernardo, M., Degano, P., Zavattaro, G. (eds.) SFM 2008. LNCS, vol. 5016, pp. 215–264. Springer, Heidelberg (2008)
4. Heinrich, P., et al.: Principles of interleukin (il)-6-type cytokine signalling and its regulation. Biochemical Journal 374, 1–20 (2003)
5. Rohr, C., Marwan, W., Heiner, M.: Snoopy - a unifying Petri net framework to investigate biomolecular networks. Bioinformatics 26(7), 974–975 (2010)

ManyCell: A Multiscale Simulator for Cellular Systems

Joseph O. Dada[1,2] and Pedro Mendes[1,2,3]

[1] Manchester Institute of Biotechnology
[2] School of Computer Science, Oxford Road, The University of Manchester, UK
[3] Virginia Bioinformatics Institute, Blacksburg, VA
{joseph.dada,pedro.mendes}@manchester.ac.uk

Abstract. The emergent properties of multiscale biological systems are driven by the complex interactions of their internal compositions usually organized in hierarchical scales. A common representation takes cells as the basic units which are organized in larger structures: cultures, tissues and organs. Within cells there is also a great deal of organization, both structural (organelles) and biochemical (pathways). A software environment capable of minimizing the computational cost of simulating large-scale multiscale models is required to help understand the functional behaviours of these systems. Here we present ManyCell, a multiscale simulation software environment for efficient simulation of such cellular systems. ManyCell does not only allow the integration and simulation of models from different biological scales, but also combines innovative multiscale methods with distributed computing approaches to accelerate the process of simulating large-scale multiscale agent-based models. Thereby opening up the possibilities of understanding the functional behaviour of cellular systems in an efficient way.

Keywords: multiscale, simulation, modelling, agent-based, ODE, software.

1 Introduction

Biological organisms are complex systems. They are made up of various spatial and temporal scales with complex interactions across various processes and mechanisms at different scales. One of the main aims of multiscale simulation in biology is to be able to simulate entire organs or even entire organisms at various levels of details. Several methods have already been used, some taking a mean-field approach, others representing the entire hierarchy of entities [1]. While the objective of these models is to represent entire organs, they should still be able to simulate all of its cells and their intracellular pathways. However that would imply simulating $10^6 - 10^{12}$ cells. It would be best if the simulation software could somehow not have to carry out such large numbers of calculations. Thus it is important to make use of multiscale nature of the biological systems. The software ManyCell, that is described here, is intended to achieve

D. Gilbert and M. Heiner (Eds.): CMSB 2012, LNCS 7605, pp. 366–369, 2012.

this requirement and implements technical solutions that contribute to numerical efficiency and scalability. The software uses ordinary differential equations (ODEs) to model the internal biochemistry of cells; each cell is modelled as an agent, where its state changes depend on discrete events that are triggered by the ODEs. The system also allows other entities, such as the extracellular medium or other extrinsic factors, to be modelled as agents. We present and illustrate the conceptual design, architecture, implementation, and functionality of ManyCell for multiscale simulation of cellular systems. We illustrate its usage using an exemplary multiscale model of a yeast cell culture. ManyCell is designed to support the integration of the different scales of cellular systems from molecular scale to a tissue/culture scale, and eventually to a whole organism.

2 Design and Implementation

The basic requirements for multiscale simulation software to be able to deal with very large numbers of cells are scalability and time efficiency. ManyCell is time efficient and scalable – it is not limited by the finite resource of a single computer. It includes technologies to specifically address these issues. ManyCell is designed as a component-based system with each component having a well-defined interface for interacting with other components. Fig. 1 (a) depicts the modular architecture of ManyCell. To improve efficiency we have adopted an In Situ Adaptive Tabulation (ISAT) mechanism [2]. To provide scalability the system uses distributed computing such that arbitrary numbers of processing units can be added to share the work. The Workload Manager (WM) component is responsible for distributing the computational intensive workloads. Unlike most agent-based simulators that keep the entire ensemble of agents in memory, we adopted a new solution where a relational database management system (RDBMS) is used to manage the agents. This allows the number of transactions (events triggered by the underlying ODEs) to be extremely large, and as a bonus the simulation results are easily managed and queried. This capability is handled by the Simulation Database (SimDB) component. DataManager component provides a unified interface for accessing the database and XML based files (SBML model and XML multiscale model file).

The agents themselves require simulating ODE-based pathway models, which is carried out by the software COPASI [3] via COPASI Web services (CopasiWS) [4], each instance being able to be run on a different processing unit. Web service technology enables the distribution of the computing intensive workload across many processors. Because of this, ManyCell readily accepts models encoded in SBML [5], such as those in the BioModels database [6], and encodes simulation results in SBRML [7]. The tissue/culture organization rules are encoded in a simple XML format called Multiscale XML model file (MXML). MXML is the main input to ManyCell. It consist of five main sections for describing various aspects of the multiscale model: **bioreactor**, **cell**, **sub-cellular**, **simulation** and **database** sections. The Multi-Agents Simulator (MAS) is the core of ManyCell. It simulates the Cell agents and handles the connection of the discrete events defined in the sub-cellular SBML model to the decisions of individual cells. For example, the firing of a division event during simulation of the ODEs model of a cell signifies the birth of a new daughter cell, which means creation of a new cell agent in MAS. MAS automatically generates serial, parallel or ISAT-based executable code for the multiscale model in the MXML file. A Master agent is responsible for the scheduling, management and synchronization of cell agent activities, while the Medium agent manages the nutrients and stress in the micro-environment. ManyCell comes with both command line and a Web interface.

ManyCell components are implemented in different programming languages. MAS, DataManager, CopasiWS are implemented in Java. The CopasiWS Client is implemented in C++ with the help of gSOAP toolkit [8], ISAT component in C++ with the help of Fortran library developed for algorithm in [2] and COPASI simulation engine (CopasiSE) is a C++ based software. Further details about ManyCell are available from `http://www.comp-bio-sys.org/ManyCell`.

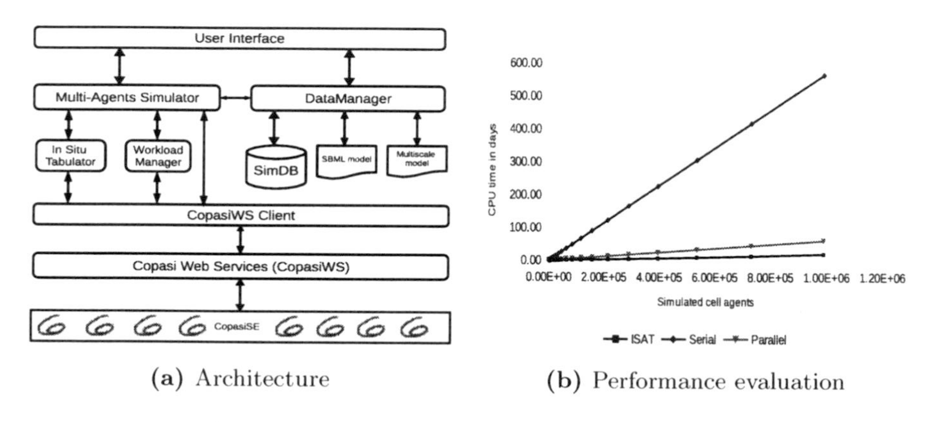

(a) Architecture **(b)** Performance evaluation

Fig. 1. Architecture and performance of ManyCell

3 Results and Discussion

We used proliferating yeast cell cultures as a case study. The Cell-agent encapsulates a number of variables describing the biochemical variables of a yeast biological cell in culture, such as age, cell cycle phase, and concentration of various chemical species. It also performs autonomous calculations of the intracellular dynamics of the biochemical reaction networks [9]. The base model for each cell intracellular reaction network is the one developed by Chen et al. [10], modified here to explictly consume a medium substrate for its growth. Each Cell-agent is equipped with the biochemical network model encoded in SBML format. We defined the multiscale agent-based model in MXML for the case study and performed various simulation experiments on the model. We tested three cases *a)* serial – where only one cell at a time solves its system of ODEs, *b)* parallel – where two or more cells solve their ODEs simultaneously (in different processing units); *c)* ISAT – where the ODE solutions make use of the ISAT scheme. Fig. 1 (b) shows the performance of the three computational approaches. While CPU time for all the approaches increases with number of simulated cell agents, that of serial increases at a faster rate than the parallel and ISAT approaches (note: average serial simulation time for a cell agent is around 48 seconds). ISAT scheme provides some 45- and 5-fold acceleration over the serial and parallel computations respectively. Using only two computers in an intranet environment (one for ODE calculations, the other for the RDBMS) we were able to simulate growth up to one million cells. Thus simulations can easily be scaled up to run at least 10^9 cells in computer farms or cloud computing. The ManyCell architecture allows scaling up the number of computers dedicated to numerics, as well as those dedicated to the data management (RDBMS). This system is thus expected to be able to simulate entire organs by representing all of its cells.

Acknowledgements. This work is funded by the European Commission within the seventh Framework Programme, theme 1 Health through the UNICELLSYS project (Grant 201142).

References

1. Dada, J.O., Mendes, P.: Multi-scale modelling and simulation in systems biology. Integr. Biol. 3(2), 86–96 (2011)
2. Pope, S.: Computation efficient implementation of combustion chemistry using in situ adaptive tabulation. Combustion Theory Modelling 1, 41–63 (1997)
3. Hoops, S., et al.: Copasi—a complex pathway simulator. Bioinformatics 22(24), 3067–3074 (2006)
4. Dada, J.O., Mendes, P.: Design and Architecture of Web Services for Simulation of Biochemical Systems. In: Paton, N.W., Missier, P., Hedeler, C. (eds.) DILS 2009. LNCS, vol. 5647, pp. 182–195. Springer, Heidelberg (2009)
5. Hucka, M., et al.: The systems biology markup language (SBML): a medium for representation and exchange of biochemical network models. Bioinformatics 19(4), 524–531 (2003)
6. Le Novère, N., et al.: Biomodels database: a free, centralized database of curated, published, quantitative kinetic models of biochemical and cellular systems. Nucleic Acids Res. 34, 689–691 (2006)
7. Dada, J.O., et al.: SBRML: a markup language for associating systems biology data with models. Bioinformatics 26(7), 932–938 (2010)
8. van Engelen, R.A., Gallivany, K.A.: The gsoap toolkit for web services and peer-to-peer computing networks. In: 2nd IEEE International Symposium on Cluster Computing and the Grid, pp. 128–135 (2002)
9. Da-Jun, T., Tang, F., Lee, T., Sarda, D., Krishnan, A., Goryachev, A.: Parallel Computing Platform for the Agent-Based Modeling of Multicellular Biological Systems. In: Liew, K.-M., Shen, H., See, S., Cai, W. (eds.) PDCAT 2004. LNCS, vol. 3320, pp. 5–8. Springer, Heidelberg (2004)
10. Chen, K.C., et al.: Integrative analysis of cell cycle control in budding yeast. Mol. Biol. Cell. 15(8), 3841–3862 (2004)

Inferring Reaction Models from ODEs

François Fages, Steven Gay, and Sylvain Soliman

INRIA Paris-Rocquencourt, EPI Contraintes, France

Many models in Systems Biology are described as Ordinary Differential Equations (ODEs), which allows for numerical integration, bifurcation analyses, parameter sensitivity analyses, etc. Before fixing the kinetics and parameter values however, various analyses can be performed on the structure of the model. This approach has rapidly developed in Systems Biology in the last decade, with for instance, the analyses of structural invariants in Petri net representation [4] model reductions by subgraph epimorphims [2], qualitative attractors in logical dynamics or temporal logic properties by analogy to circuit and program verification. These complementary analysis tools do not rely on kinetic information, but on the structure of the model with reactions.

The Systems Biology Markup Language (SBML) of [3] is now a standard for sharing and publishing reaction models. However, since SBML does not enforce any coherence between the structure and the kinetics of a reaction, an ODE model can be transcribed in SBML without reflecting the real structure of the reactions, hereby invalidating all structural analyses.

In this paper we propose a general compatibility condition between the kinetic expression and the structure of a reaction, describe an algorithm for inferring a reaction model from an ODE system, and report on its use for automatically curating the writing in SBML of the models in the repository biomodels.net.

1 Theory of Well-Formed Reaction Kinetics

Let us consider a finite set $\{x_1, \ldots, x_v\}$ of molecular species. A *reaction model R* is a finite set of n reactions, written $R = \{\ e_i\ \texttt{for}\ \ r_i\ /\ m_i\ \texttt{=>}\ \ p_i\ \}_{i=1,\ldots,n}$, where e_i is a mathematical expression over species concentrations, also written x_i by abuse of notation, and symbolic parameters that are supposed positive; r_i, m_i and p_i are multisets of species which represent the reactants, the inhibitors, and the products of the reaction respectively. The species that are both reactants and products in a reaction are called catalysts. Catalysts and inhibitors are called modifiers in SBML. For a multiset r of molecular species, i.e. a function $V \to \mathbb{N}$, we denote by $r(x)$ the multiplicity of x in r, i.e. $r(x) = 0$ if x does not belong to r, and $r(x) \geq 1$ if x belongs to r, which is also written $x \in r$. The empty multiset is written _.

Let us call *non-decomposable* a mathematical expression that is syntactically not an addition nor a subtraction, and that cannot be reduced at top-level by the laws of distributivity of the product and division, and let us say

Definition 1 *A reaction* e `for` $r\ /\ m$ `=>` p *over molecular species* $\{x_1, \ldots, x_v\}$ *is* well-formed *if the following conditions hold:*

D. Gilbert and M. Heiner (Eds.): CMSB 2012, LNCS 7605, pp. 370–373, 2012.

1. *e is a well-defined, nonnegative and partially differentiable mathematical expression for any values* $x_1 \geq 0, \ldots, x_v \geq 0$;
2. $x \in r$ *if and only if* $\partial e/\partial x > 0$ *for some* $x_1 \geq 0, \ldots, x_v \geq 0$;
3. $x \in m$ *if and only if* $\partial e/\partial x < 0$ *for some* $x_1 \geq 0, \ldots, x_v \geq 0$;

The reaction is non-decomposable *if its kinetic expression e is non-decomposable.*

These well-formedness conditions are met by standard kinetics, such as mass action law, Michaelis-Menten, Hill, and negative Hill kinetics. However there are ODE system which cannot be presented by a well-formed non-decomposable reaction model.

Example 1. The ODE system $\dot{x} = -k$ can be associated to the reaction `k for x => _` which is not well-formed, or to `k+1*x for x => _` and `1*x for x => 2*x` which are not non-decomposable, but cannot be associated to a well-formed non-decomposable system.

Example 2. On the other hand, the ODE with symbolic parameters $k1, k2, k3$

$$\dot{pMPF} = k2 * [MPF] * [Wee1] - k1 * [pMPF] * [Cdc25]$$
$$\dot{MPF} = k1 * [pMPF] * [Cdc25] - k2 * [MPF] * [Wee1]$$
$$\dot{Wee1} = k3/(k4 + [Clock]),\ \dot{Cdc25} = 0,\ \dot{Clock} = 0$$

is associated to the well-formed non-decomposable model

```
k1*[pMPF]*[Cdc25] for pMPF + Cdc25 => MPF + Cdc25
k2*[MPF]*[Wee1]   for MPF + Wee1 => pMPF + Wee1
k3/(k4+[Clock])   for _ / Clock => Wee1
```

2 Reaction Model Inference Algorithm

The following algorithm for inferring reactions from ordinary differential equations is based on a syntactical normal form for ODE systems. Unlike the algorithm proposed in [5], our algorithm does not rely on uniqueness properties and always succeeds even when there are no corresponding well-formed reaction models, by inferring possibly non well-formed reactions.

Let us say that an expression is in *additive normal form* if it is of the form $\sum_{s=1}^{t} c_s * f_s$ with c_s numerical coefficients and f_s non-decomposable terms without coefficients. An ODE system is in additive normal form if each equation is in additive normal normal form as follows $\dot{x_i} = \sum_{s=1}^{t} c_{i,s} * f_s,\ 1 \leq i \leq v$ where t is the number of non-decomposable terms in the system.

The idea of our reaction inference algorithm is to normalize the ODE and infer a corresponding reaction model by sorting the terms of the equations and using their coefficients as stoichiometric coefficients. To test the sign of partial derivatives, we content ourselves with an approximate test by comparing the exponents.

Algorithm 2 input: *ODE system O*

1. $O \leftarrow$ *additive-normal-form(O)*

2. $R \leftarrow \emptyset$
3. *for each non-decomposable term* f *of an equation in* O
   (a) *let* $r \leftarrow \_ , p \leftarrow \_ , m \leftarrow \_$
   (b) *for each variable* x *where* f *occurs with coefficient* c *in* $\dot{x}$ *in* O
      i. *if* $c < 0$ *then* $r(x) \leftarrow -c$
      ii. *if* $c > 0$ *then* $p(x) \leftarrow c$
   (c) *for each variable* x *such that* $r(x) = 0$ *and* $\frac{\partial f}{\partial x} > 0$ *for some values*
      i. $r(x) \leftarrow 1$
      ii. $p(x) \leftarrow p(x) + 1$
   (d) *for each variable* x *such that* $\frac{\partial f}{\partial x} < 0$ *for some values*
      i. $m(x) \leftarrow 1$
   (e) $R \leftarrow R \cup \{f$ `for` r / m `=>` $p\}$
4. **output**: *reaction model* R

Proposition 3 *The reaction model inferred by Algorithm 2 from an ODE system in additive normal form* $\dot{x}_i = \sum_{s=1}^{t} c_{i,s} * f_s$ *for* $1 \leq i \leq v$ *is the set of non-decomposable reactions* $\{f_s$ *for* $r_s/m_s \rightarrow p_s\}_{1 \leq s \leq t}$ *where*

$$r_s = \sum_{c_{i,s}<0} (-c_{i,s}) * x_i + \sum_{\{i \mid c_{i,s} \geq 0,\ \frac{\partial f_s}{\partial x_i} > 0\}} x_i \qquad p_s = \sum_{c_{i,s}>0} c_{i,s} * x_i + \sum_{\{i \mid c_{i,s} \geq 0,\ \frac{\partial f_s}{\partial x_i} > 0\}} x_i \qquad m_s = \{x \mid \frac{\partial f_s}{\partial x} < 0\}$$

Algorithm 2 always computes a reaction model with an equivalent associated ODE system but this reaction model may not be well-formed. In particular, step 3b may infer reactions with reactants that do not occur in the kinetic expression. On the other hand, all variables appearing in the kinetics will now appear in the reaction as either catalysts (step 3c), inhibitors or both (step 3d). We have:

Proposition 4 *The ODE system associated to the reaction model inferred from an ODE system* O *is equivalent to* O. *The reaction model inferred from the ODEs associated to a non-decomposable well-formed reaction model is well-formed and non-decomposable.*

For space reasons, we do not describe here the preprocessor that is applied on the ODE system for detecting simplifications by mass conservation linear invariants and inferring hidden molecules, prior to the inference of reactions.

Example 3. The model of Example 2 has one invariant: $pMPF + MPF$ is a constant c. Replacing $pMPF$ by $c - MPF$ yields to the system

$\dot{MPF} = k1 * (c - [MPF]) * [Cdc25] - k2 * [MPF] * [Wee1]$

$\dot{Wee1} = k3/(k4 + [Clock])$ $\qquad \dot{Cdc25} = 0$ $\qquad \dot{Clock} = 0$

From this system, Algorithm 2 would infer the reactions:

```
c*k1*[Cdc25]        for Cdc25 => Cdc25 + MPF
  k1*[Cdc25]*[MPF]   for MPF + Cdc25 => Cdc25
  k2*[MPF]*[Wee1]    for MPF + Wee1 => Wee1
  k3/(k4+[Clock])    for _ / Clock => Wee1
```

However, the preprocessor recognizes the linear invariant and introduces a molecule y for the expression $[y] = c - [MPF]$ which yields in this case to the same reactions as of Example 2 with $y = pMPF$.

3 Evaluation Results on Biomodels.net

Out of the 409 models from the curated branch of the latest version (21) of `biomodels.net`, 340 models have proper kinetic laws. We compare the number of non well formed reaction models before and after the automatic curation obtained by exporting SBML to ODE format and applying our reaction inference algorithm to the ODE. The following table summarizes the improvement by counting the number of models with BIOCHAM warnings: *well-formed*:

- "K not R" denotes the number of models in which the concentration of some compound appears in a kinetic law but it is neither a reactant nor a modifier;
- "R not K" denotes the number of models in which some compound is marked as reactant or modifier but does not appear in its kinetic law.
- "Negative" denotes the number of models where a minus sign appears in the kinetic expression at some place that is not inside an exponent expression.

225 models, i.e. 66% of the 340 reaction models of the original "curated" part of biomodels.net are non well formed and produce some warning. Our algorithm is able to automatically reduce the number of non well-formed models by 58%, from 66% to 28%:

Biomodels.net	**K not R**	**R not K**	**Neg.**	**Any warning**
Originally Curated	165	120	148	225 (66.17%)
Automatically Curated	0	81	39	97 (28.52%)

The algorithm completely removes the "K not R" warnings. For the two other warnings, since the algorithm focuses on *non-decomposable* kinetics, it results in curated models quite close to the original ones, but does not tackle thoroughly the case of reactions with rates independent of some reactant, as in Example 1. For these reasons, 97 over 340 models remain with a non well-formedness warning after automatic curation [1].

References

1. Fages, F., Gay, S., Soliman, S.: Automatic curation of SBML models based on their ODE semantics. Technical Report 8014, Inria Paris-Rocquencourt (July 2012)
2. Gay, S., Soliman, S., Fages, F.: A graphical method for reducing and relating models in systems biology. Bioinformatics 26(18), i575–i581 (2010), special issue ECCB 2010
3. Hucka, M., et al.: The systems biology markup language (SBML): A medium for representation and exchange of biochemical network models. Bioinformatics 19(4), 524–531 (2003), `http://sbml.org/`
4. Reddy, V.N., Mavrovouniotis, M.L., Liebman, M.N.: Petri net representations in metabolic pathways. In: Hunter, L., Searls, D.B., Shavlik, J.W. (eds.) Proceedings of the 1st International Conference on Intelligent Systems for Molecular Biology (ISMB), pp. 328–336. AAAI Press (1993)
5. Soliman, S., Heiner, M.: A unique transformation from ordinary differential equations to reaction networks. PLoS One 5(12), e14284 (2010)

Modelling Trafficking of Proteins within the Mammalian Cell Using Bio-PEPA

Vashti Galpin

CSBE/SynthSys and School of Informatics, University of Edinburgh
Vashti.Galpin@ed.ac.uk

Abstract. Bio-PEPA [2], a process algebra developed from PEPA [5] is used to model a process occurring in mammalian cells whereby the Src oncoprotein is trafficked between different parts of the cell [9]. Src is associated with cell movement and adhesion between cells which is linked to tumour formation [9]. A useful model of the protein's behaviour can provide predictions for new experimental hypotheses which may improve our understanding, and in time, lead to new therapies for cancer. The aim is to assess the suitability of Bio-PEPA for more detailed modelling of Src trafficking than that of a previous simpler Bio-PEPA model [4].

Bio-PEPA has the advantage that it provides many types of analysis, such as ordinary differential equations, stochastic simulation, continuous-time Markov chains, trace investigation and model-checking. These analyses are implemented in or accessible via the Bio-PEPA Eclipse Plug-in [3] (see `www.biopepa.org`).

The syntax of well-defined Bio-PEPA species components with locations [1] is given by $C \stackrel{def}{=} \sum_{i=1}^{n}(\alpha_i, \kappa_i)\ \mathtt{op}_i\ C$ such that $\alpha_i \neq \alpha_j$ for $i \neq j$. Here C is a constant of the form $A@loc$, each α_i is a reaction name, each κ_i is the stoichiometric coefficient for C in reaction α_i, and each $\mathtt{op}_i$ is a prefix operator describing the role of C is reaction α_i. The operator $\downarrow$ indicates a reactant, $\uparrow$ a product, $\oplus$ an activator, $\ominus$ an inhibitor and $\odot$ an arbitrary modifier. Hence a species component has a unique name and describes the reaction capabilities of that species. A Bio-PEPA model then has the syntax $P ::= P \underset{L}{\bowtie} P \mid C(x)$ where x represents the quantity of species C. Species cooperate over reactions in L, or $*$ can be used to denote all shared reactions. A species may only appear once in a well-defined model. A Bio-PEPA system is a tuple containing species components, model and additional information about locations, species, constants and rate functions.

The operational semantics consist of a capability transition relation where labels on the transitions record the reaction name and information about the species taking part in the reaction; and the stochastic relation where this information is transformed into a rate using a function specific to the reaction, giving a transition sytem labelled with reaction names and rates (as in PEPA [5]).

The protein under consideration here is the Src protein, a member of the Src family of proteins. It is a non-receptor protein tyrosine kinase which has two domains to which other molecules can bind [9]. In its inactive form, its conformation prevents access to these domains, whereas in its active form these domains

D. Gilbert and M. Heiner (Eds.): CMSB 2012, LNCS 7605, pp. 374–377, 2012.

are available, and Src can interact with other proteins. Research (described below) has shown that Src is trafficked around the cell in endosomes. These are membrane-bound compartments found in the cytoplasm of the cell. They merge with vesicles which engulf molecules on the inner side of the membrane, and their role is to sort molecules for recycling[1] and degradation. Endosomes move along microfilaments and/or microtubules, so movement is typically in one direction, often in towards the nucleus or out towards the cellular membrane. They tend to vary in contents rather than number or speed.

Usually, Src is found in two locations in the cell: a large amount of inactive Src is located around the nucleus, in the perinuclear region, and a much smaller amount of active Src is located on or near the membrane [7,8]. Sandilands, Frame and others have investigated how the addition of growth factor affects Src activity. After stimulation with fibroblast growth factor (FGF), Src is found in endosomes throughout the cytoplasm. Moreover, there is a gradient of inactive Src to active Src from perinuclear region to membrane and hence Src activation takes place in endosomes [8]. Furthermore, the persistence of active Src at the membrane is inversely related to the quantity of FGF added. When 1 nanogram of FGF is added, large quantities of active Src persist two hours after addition; in contrast, when 50 times this amount is added, the quantity of active Src is already reduced after 30 minutes and has returned to normal levels after 1 hour [7]. The goal for the model is to demonstrate this persistence behaviour.

The first challenge of the modelling is lack of experimental data. There is qualititative data: the gradient of active Src, and quantitative data: the timing of the persistence of active Src after growth factor addition. Additionally, the speed of endosomes can be determined because their movement is directional and has been measured. Research into endosomes has shown that there are both long and short recycling loops [6]. Hence from this, one can estimate how long it should take for an endosome to move through a long loop and a short loop. The second and longer term challenge is making the model concrete enough to be useful in prediction but abstract enough to be tractable.

An initial single combined long recycling loop model was developed but it did not demonstrate the required behaviour. After discussion with experimental biologists, a two loop model was built. In the shorter loop, which is always in operation, some of the active Src at the cell membrane is recycled in endosomes, and the hypothesis is that there will always be active Src about to be delivered to the membrane, ensuring the ongoing presence of active Src at the membrane. The longer recycling loop is only active on stimulation by FGF and involves trafficking of active Src bound to the FGF receptor (FGFR). When endosomes in this loop come close to the perinuclear region, they engulf inactive Src which is then activated during the movement of the endosome outwards. In both loops, it can happen that the endosome contents are degraded rather than recycled. Figure 1 illustrates these concepts. The Bio-PEPA model consists of seven species. Inactive Src in the perinuclear region is available in such large quantities that it can be treated implicitly. The model is moderately abstract;

[1] Here, recycling means to “return to the membrane for re-use”.

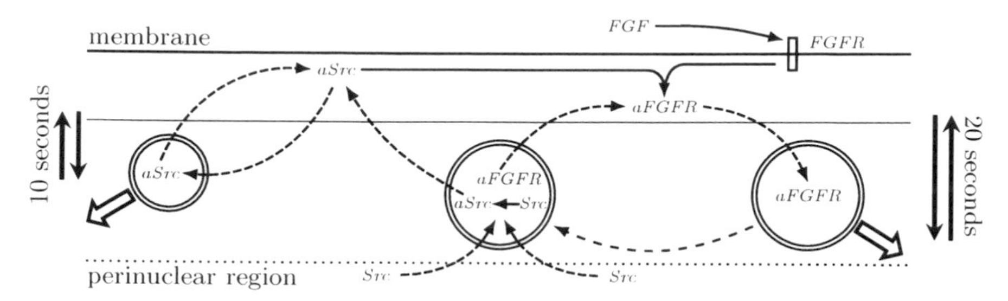

Fig. 1. The two loop model of Src trafficking: dashed arrows denote movement and solid arrows denote reactions. Double circles represent endosomes and the large outlined arrows indicate degradation of endosome contents. The short loop is on the left with the long loop on the right.

active Src at the membrane is a species, as are endosomes containing active Src. This is possible using the stoichiometry coefficient in modelling the interaction of species. Consider the following definition for active Src at the membrane.

$$\begin{aligned} aSrc@membrane \stackrel{def}{=} & (aSrc\_FGFR\_binding, 1) \downarrow aSrc@membrane \\ & + (into\_short\_loop\_endosome, 150) \downarrow aSrc@membrane \\ & + (outof\_short\_loop\_endosome, 150) \uparrow aSrc@membrane \\ & + (outof\_long\_loop\_endosome, 100) \uparrow aSrc@membrane \end{aligned}$$

The binding of active Src and FGFR to form active FGFR is an abstraction of a number of reactions whose detail is not necessary for the modelling but the stoichiometry of 1 indicates that one active Src molecule takes part in the binding (and FGFR has the same stoichiometry). However, for the other reactions which represent amounts of active Src either moving into endosomes or moving out of endosomes, the stoichiometric coefficient represents these quantities. The endosome species taking part in the reaction will have a stoichiometry of one to capture the notion that a single endosome contains multiple Src molecules. The stoichiometry for the *outof_short_loop_endosome* reaction can be made less than that for *into_short_loop_endosome* to describe the loss of active Src within the recycling loop, or alternatively this can be done by removal of endosomes.

Simulation demonstrates a short-lived peak of active Src at the membrane on addition of FGF as shown in Figure 2. This model does not demonstrate the behaviour for smaller amounts of FGF and this requires further exploration of the parameter space. It appears Bio-PEPA provides useful abstraction techniques.

Due to the limited data, this style of modelling can be described as quasi-quantitative and exploratory. However, the model has been useful for discussion with the biologists involved, both for the biological understanding of the modeller, and in raising interesting questions for the biologists about the underlying mechanisms. Ongoing discussions will develop the model further, and new techniques may provide more quantitative data. Recent research about Src in cancerous cells shows that when too much active Src at the membrane could lead to cell death, there is a mechanism to sequester this active Src and hence enable survival of the cell [10]. This shows that there may be multiple mechanisms by which Src is trafficked and suggests scope for a more complex model.

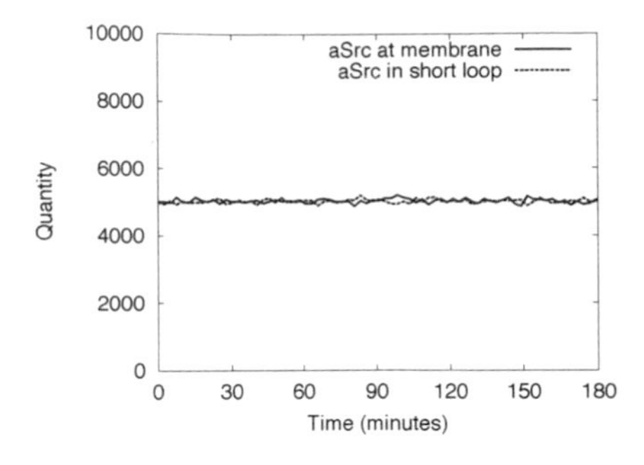

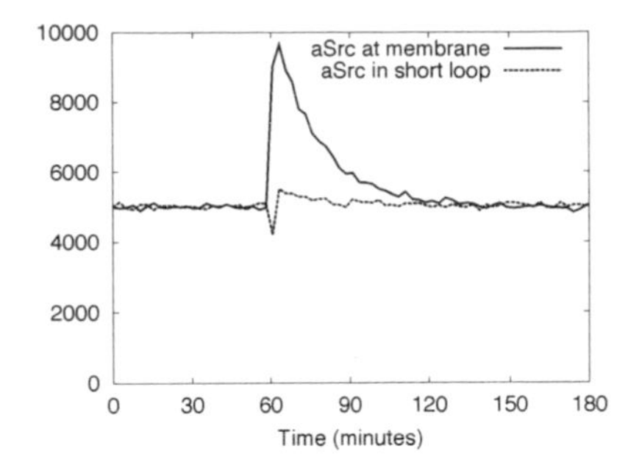

Fig. 2. Stochastic simulations without addition of growth factor (left) and with addition of growth factor at one hour (right). The graphs show the average of 10 simulations.

Acknowledgements. The Centre for Systems Biology at Edinburgh is a Centre for Integrative Systems Biology (CISB) funded by the BBSRC and EPSRC in 2006. The author thanks Jane Hillston for her comments and Margaret Frame and Emma Sandilands of Cancer Research UK in Edinburgh for useful discussions of their research.

References

1. Ciocchetta, F., Guerriero, M.L.: Modelling biological compartments in Bio-PEPA. Electronic Notes in Theoretical Computer Science 227, 77–95 (2009)
2. Ciocchetta, F., Hillston, J.: Bio-PEPA: a framework for the modelling and analysis of biological systems. Theoretical Computer Science 410(33-34), 3065–3084 (2009)
3. Duguid, A., Gilmore, S., Guerriero, M., Hillston, J., Loewe, L.: Design and development of software tools for Bio-PEPA. In: Dunkin, A., Ingalls, R., Yücesan, E., Rossetti, M., Hill, R., Johansson, B. (eds.) Proceedings of the 2009 Winter Simulation Conference, pp. 956–967 (2009)
4. He, X.: A Bio-PEPA Model of Endosomal Trafficking of Src Tyrosine Kinase. MSc dissertation, School of Informatics, University of Edinburgh
5. Hillston, J.: A compositional approach to performance modelling. Cambridge University Press (1996)
6. Jones, M., Caswell, P., Norman, J.: Endocytic recycling pathways: emerging regulators of cell migration. Current Opinion in Cell Biology 18, 549–557 (2006)
7. Sandilands, E., Akbarzadeh, S., Vecchione, A., McEwan, D., Frame, M., Heath, J.: Src kinase modulates the activation, transport and signalling dynamics of fibroblast growth factor receptors. EMBO Reports 8, 1162–1169 (2007)
8. Sandilands, E., Cans, C., Fincham, V., Brunton, V., Mellor, H., Prendergast, G., Norman, J., Superti-Furga, G., Frame, M.: Rhob and actin polymerization coordinate Src activation with endosome-mediated delivery to the membrane. Developmental Cell 7, 855–869 (2004)
9. Sandilands, E., Frame, M.: Endosomal trafficking of Src tyrosine kinase. Trends in Cell Biology 18, 322–329 (2008)
10. Sandilands, E., Serrels, B., McEwan, D., Morton, J., Macagno, J., McLeod, K., Stevens, C., Brunton, V., Langdon, W., Vidal, M., Sansom, O., Dikic, I., Wilkinson, S., Frame, M.: Autophagic targeting of Src promotes cancer cell survival following reduced FAK signalling. Nature Cell Biology 14, 51–60 (2011)

Models of Tet-On System with Epigenetic Effects

Russ Harmer[1], Jean Krivine[1], Élise Laruelle[1], Cédric Lhoussaine[2], Guillaume Madelaine[2,3], and Mirabelle Nebut[2]

[1] PPS (CNRS UMR 7126), University Paris Diderot, France
[2] LIFL (CNRS UMR 8022), University of Lille 1, France
[3] ÉNS Cachan, France

Abstract. We present the first results of ongoing work investigating two models of the artificial inducible promoter Tet-On that include epigenetic regulation. We consider chromatin states and 1D diffusion of transcription factors that reveal, respectively, stochastic noise and a memory effect.

1 Introduction

In gene regulatory systems, transcription factors (TF) usually require activation in order to perform their regulatory function. This generally results from the action of complex signalling pathways, so an investigation of the dynamics of TF activation is important for understanding the underlying gene regulation. Recently, an efficient experimental technique to monitor this dynamics has been proposed, using fluorescent proteins expressed under the control of an inducible promoter by the TF of interest. However, observed fluorescence is not linearly correlated to the TF activation: a delay is induced by fluorescent protein expression and subsequent folding; and fluorescence may persist even after TF deactivation. Huang et al [4] propose a method to reconstruct TF dynamics from observed fluorescent protein dynamics by means of a very simple two-level model of the Tet-On system, an artificial inducible promoter of Green Fluorescent Protein (GFP). The first level (rules (1) to (3) below) models the signal transduction pathway leading to TF activation: this involves the artificial TF $rtTA$, activated by binding with doxycycline Dox_i. Extracellular doxycycline Dox_e is assumed constant and can degrade in the cell. The second level (rules (4) to (9)) models protein synthesis and activation of fluorescence: this includes transcription, translation and GFP activation.

$$Dox_e \xrightarrow{Deff} Dox_e + Dox_i \quad (1)$$

$$Dox_i \xrightarrow{Deff} \emptyset \quad (2)$$

$$rtTA + Dox_i \underset{k_{r2}}{\overset{k_{f2}}{\rightleftharpoons}} rtTA \cdot Dox \quad (3)$$

$$rtTA \cdot Dox \xrightarrow{S'_m} rtTA \cdot Dox + mRNA \quad (4)$$

$$mRNA \xrightarrow{D_m} \emptyset \quad (5)$$

$$mRNA \xrightarrow{S_n} GFP + mRNA \quad (6)$$

$$GFP \xrightarrow{S_f} GFP_a \quad (7)$$

$$GFP \xrightarrow{D_n} \emptyset \quad (8)$$

$$GFP_a \xrightarrow{D_n} \emptyset \quad (9)$$

D. Gilbert and M. Heiner (Eds.): CMSB 2012, LNCS 7605, pp. 378–381, 2012.

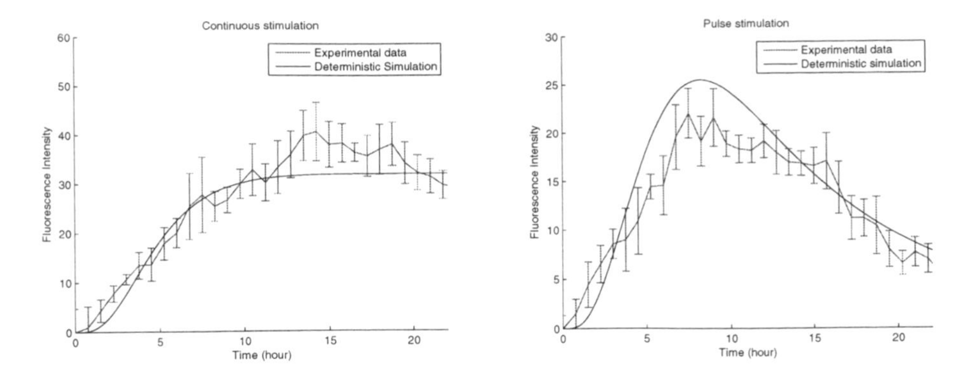

Fig. 1. Continuous (left) and pulse (right) stimulation with doxycycline

Fig. 1 shows that the dynamics predicted by the deterministic model is in good agreement with experimental data in the case of continuous and pulse[1] stimulation with doxycycline. In [4], the authors demonstrate that the dynamics of TF activation is accurately correlated with the dynamics offluorescence.

In this paper, we propose two simple extensions of the basic model proposed in [4] to investigate effects that may result from epigenetic regulation. The first extension deals with *chromatin states*, *i.e.* the (un)availability of the inducible promoter due to chromatin compaction. In the second, we model TF diffusion along DNA, *i.e.* the binding of a TF to non-specific binding sites, followed by its sliding to the operator site.

2 Chromatin States: A Stochastic Noise Effect

The experimental data of [4] appear to be quite noisy, contrary to what is predicted by the original model (Fig. 2 left). Although it is hard to assess quantitatively, we can assume that some of this noise results from intrinsic stochasticity. A recognized likely source of stochasticity in gene regulation is the dynamic alteration of chromatin structure that makes it more or less accessible to the transcriptional machinery. Most of the time, the chromatin is tightly packaged or, roughly speaking, in a *closed* state. This implies that the TF cannot find the promoter to activate transcription; to allow $rtTA \cdot Dox$ to bind with the promoter, the chromatin must be in an *open* state. Depending on the needs of the cell, the chromatin can rapidly switch between these states.

We refine the model given in the introduction by replacing reaction (4) with reactions (10) to (13) below. In words, $rtTA \cdot Dox$ can bind its operator site (rule (10)) when the chromatin is locally open; then, either $rtTA \cdot Dox$ can dissociate from the operator (rule (10)), or transcription can begin (rule (13)). At any time, the chromatin can switch to a closed state (rule (11)); if the TF is bound to the operator when the chromatin closes, then it dissociates (rule (12)).

[1] This means that Dox_e concentration is set to zero after some delay.

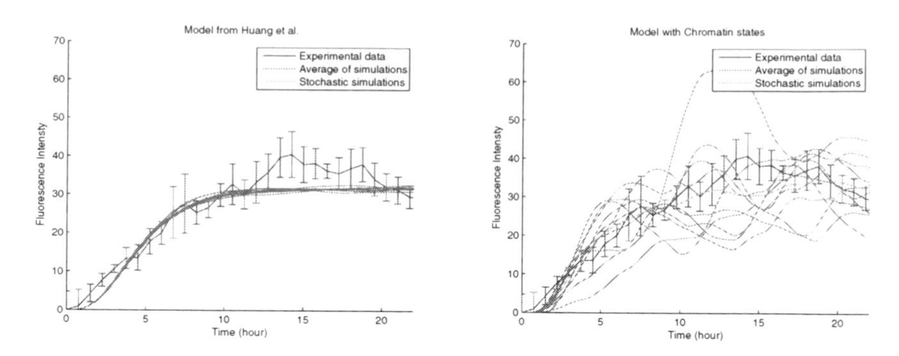

Fig. 2. Comparison of the stochastic noise between the basic model (left) and the model with chromatin states (right)

We take values for k_{rop} and k_{fop} of the same order as those for *TetR* given in [1]. The values of k_{closed} and k_{open} are actually dependent on the promoter used, the position of the gene, the cell type, *etc*, but the values taken here are close to those of promoters found in mammalian cells [6]. Finally, the rate constant of transcription (rule (13)) has been chosen to fit the experimental data.

$$rtTA \cdot Dox + Op_{open} \underset{k_{rop}}{\overset{k_{fop}}{\rightleftharpoons}} Op_{open} \cdot rtTA \cdot Dox \tag{10}$$

$$Op_{closed} \underset{k_{closed}}{\overset{k_{open}}{\rightleftharpoons}} Op_{open} \tag{11}$$

$$Op_{open} \cdot rtTA \cdot Dox \xrightarrow{k_{closed}} Op_{closed} + rtTA \cdot Dox \tag{12}$$

$$Op_{open} \cdot rtTA \cdot Dox \xrightarrow{k_{trans}} Op_{open} \cdot rtTA \cdot Dox + mRNA \tag{13}$$

The average of the stochastic simulations in Fig. 2 is close to the deterministic dynamics given by Huang et al. [4], as expected, but stochastic noise is far more pronounced than in the basic model. The influence of chromatin states could thus provide an explanation for the noise observed in the experimental data.

3 1D Diffusion of TF: A Memory Effect

Another interesting aspect of gene regulation is related to how the TF finds its promoter since it is known that three-dimensional (3D) diffusion is insufficient to explain fast binding [5]. In [3], a diffusion-based model is proposed: once the TF has (3D-) diffused within the nucleus and bound randomly to a non-specific DNA site, it rapidly slides along the DNA (1D diffusion) in a small region. If there is an operator in this region, it has an approximately 50-50 chance to bind it. In the case of binding, transcription can begin; otherwise, the TF either unbinds completely, returning to a search by 3D diffusion, or it jumps to another non-specific binding site in a neighbouring area.

We add this behaviour to the basic model with 14 new reactions available as supplementary material at `www.lifl.fr/~lhoussai/`

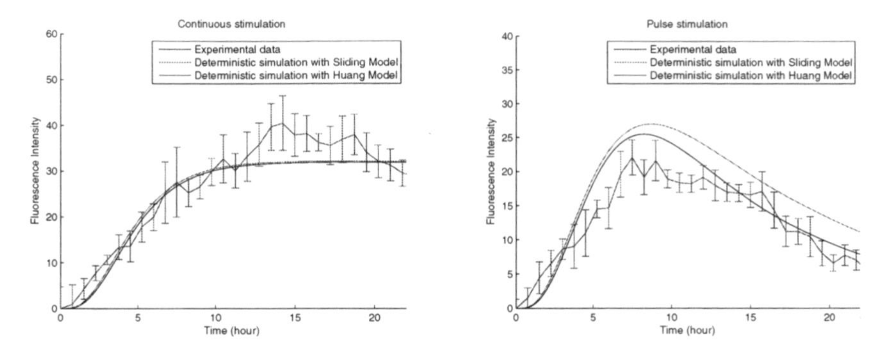

Fig. 3. Comparison between Huang model and Sliding model, with a continuous (left) or pulse (right) stimulation of doxycycline

The transcription rate has again been chosen to fit the experimental data and the other parameters have been taken from [3] and [2]. The results of the deterministic simulations in Fig. 3 are interesting: there is no significant difference in the case of constant stimulation but, in the pulse case, fluorescence decays more slowly than in the basic model which can be interpreted as a sort of "memory effect".

4 Conclusion

We have developed two extensions of the Tet-On model of [4]. The first, adding chromatin states, increases stochastic noise but preserves the average behaviour in accordance with the experimental data. The second, adding a search for the operator by the TF via 1D diffusion, reveals a delayed decay in fluorescence after stimulation has ceased. We are currently investigating this effect in more detail, notably in the case where there are multiple operator sites in close proximity.

References

1. Biliouris, K., Daoutidis, P., Kaznessis, Y.N.: Stochastic simulations of the tetracycline operon. BMC Systems Biology 5 (2011)
2. Bonnet, I.: Mécanismes de diffusion facilitée de l'enzyme de restriction EcoRV (2007)
3. Hammar, P., Leroy, P., Mahmutovic, A., Marklund, E.G., Berg, O.G., Elf, J.: The lac repressor displays facilitated diffusion in living cells. Science 336(6088), 1595–1598 (2012)
4. Huang, Z., Moya, C., Jayaraman, A., Hahn, J.: Using the Tet-On system to develop a procedure for extracting transcription factor activation dynamics. Molecular Biosystems 6(10), 1883–1889 (2010)
5. Riggs, A.D., Bourgeois, S., Cohn, M.: The lac repressor-operator interaction. 3. kinetic studies. Journal of Molecular Biology 53(3), 401–417 (1970)
6. Suter, D.M., Molina, N., Gatfield, D., Schneider, K., Schibler, U., Naef, F.: Mammalian genes are transcribed with widely different bursting kinetics. Science 332(6028), 472–474 (2011)

GeneFuncster: A Web Tool for Gene Functional Enrichment Analysis and Visualisation

Asta Laiho[1,*,**], András Király[2,**], and Attila Gyenesei[1]

[1] Turku Centre for Biotechnology, University of Turku and Åbo Akademi University, Tykistökatu 6A, 20520 Turku, Finland
[2] University of Pannonia, Department of Process Engineering, P.O. Box 158 Veszprém H-8200, Hungary

Abstract. Many freely available tools exist for analysing functional enrichment among short filtered or long unfiltered gene lists. These analyses are typically performed against either Gene Ontologies (GO) or KEGG pathways (Kyoto Encyclopedia of Genes and Genomes) database. The functionality to carry out these various analyses is currently scattered in different tools, many of which are also often very limited regarding result visualization. GeneFuncster is a tool that can analyse the functional enrichment in both the short filtered gene lists and full unfiltered gene lists towards both GO and KEGG and provide a comprehensive result visualisation for both databases. GeneFuncster is a simple to use publicly available web tool accessible at http://bioinfo.utu.fi/GeneFuncster.

Keywords: functional enrichment analysis, pathway analysis, gene expression analysis.

1 Introduction

The technological advance in the field of biotechnology during the last decade has led to the increasing generation of functional genomics data. Especially the well established DNA microarray technology and more recently developed high-throughput short read sequencing technology are producing large data sets that require automated means for analysing and visualising the results by taking the gene functions into account. As a result, functional enrichment analysis has become a standard part of the analysis of high-throughput genomics data sets and many freely available tools with different approaches have been developed during the recent years (see [4] for a good review). These tools vary for example based on the kind of input and organisms supported, the statistical tests used for carrying out the enrichment tests, the selection of databases available to conduct the enrichment analysis against and the way the results are reported and visualised. Many of the tools provide useful and unique features, and thus the researchers typically need to use several tools in order to gain a more complete view on the biological significance behind the list of genes under inspection. For

* Corresponding author.
** Equal contributors.

D. Gilbert and M. Heiner (Eds.): CMSB 2012, LNCS 7605, pp. 382–385, 2012.

example, many of the available tools take a short filtered list of genes as input to be compared to a given background set of genes and then perform a statistical test (typically based on a hypergeometric or binomial model) to detect whether genes belonging to certain functional categorizations appear in the input gene set more often than would be expected by chance alone.

While analysing the functional enrichment among the filtered genes (e.g. most differentially expressed ones) is very useful, the choice of filtering thresholds can have a significant effect on the analysis outcome. When none or only a few of the many related influenced genes are regulated strongly enough to meet the filtering criterion, some important functional categorizations may be missed. As a solution, other tools take the full unfiltered gene list as input and employ threshold free ranking based approaches applying for example a non-parametric Kolmogorov-Smirnov test. These types of methods are efficient in detecting subtle but consistent changes among genes belonging to the same functional category. Thus the two approaches should be regarded as complementary and optimally applied in parallel to gain a complete view of all affected functional categorizations.

While majority of the available tools simply report the results as a table of category terms ranked according to the test p-value, some tools also provide ways to visualise the results in the context of the functional category information. An informative way to present the GO enrichment results is to provide a view on the directed acyclic GO term graph (DAG) in which each gene product may be annotated to one or more terms. Colouring the terms by the enrichment significance allows an easy detection of the clusters of affected closely related GO terms. Similarly, a good way to visualise the KEGG enrichment results is by presenting the pathway maps where the nodes representing genes or gene complexes are coloured. As many of the described useful features are scattered across various tools, there is a clear need for a combined method. In this article, we present GeneFuncster that is able to analyse functional enrichment in both short filtered gene lists and full unfiltered gene lists as well as represent the results by both GO hierarchical graphs and KEGG pathway maps. If fold-change and/or p-value data is provided by the user, gene level bar plots as well as colouring of the KEGG graph gene nodes according to the direction of the regulation becomes available. These kinds of more advanced and extremely useful visualisations are currently missing from most freely available tools.

2 Methods

GeneFuncster takes advantage of several R/Bioconductor packages including topGO, GOstats and gage [1,3]. GeneFuncster is able to run functional enrichment analyses on both short filtered and full unfiltered gene lists. A traditional over-representation analysis is employed to compare a short filtered gene list to a background provided by the user. Alternatively, the list of all known genes of a specific organism can be used as background. With the full unfiltered gene list enrichment analysis the question of how to rank the genes becomes important.

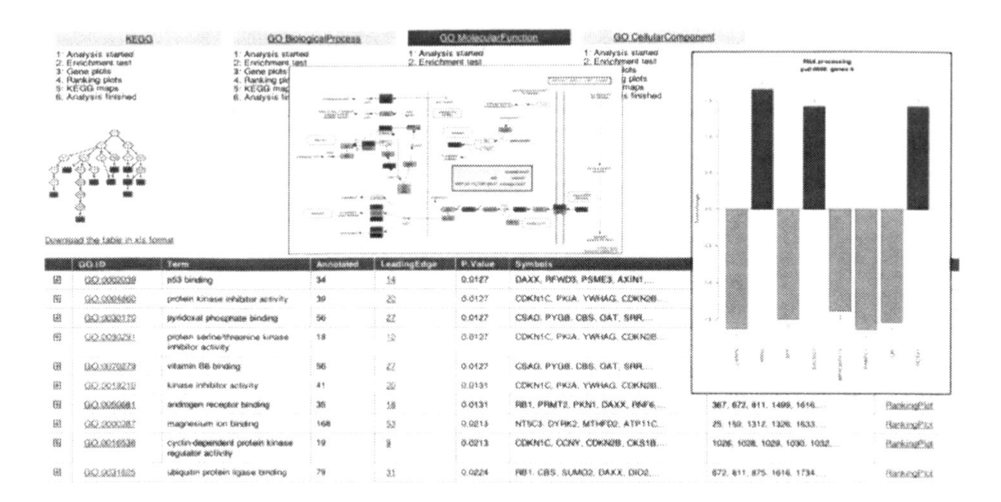

Fig. 1. Overview of various result reports generated by GeneFunscter. Detailed descriptions can be found in the tool web site. The most enriched terms/pathways are listed for KEGG and each main GO category. Result visualisations include coloured KEGG pathway maps, GO term graphs and gene-level plots.

In the context of gene expression profiling data, the goal is to rank the genes according to the strength of evidence for differential gene expression between the sample condition groups. Some tools, like GOrilla, take a list of ranked gene symbols as input while others, like GSEA, start from the matrix of normalized gene expression values across all samples and then perform the statistical analysis between the sample condition groups and rank the genes according to the test statistics. In GeneFuncster, the user may provide a ranked list of gene identifiers, or attach fold-changes and/or p-values and then choose to have the gene list ranked according to either of these. As a unique feature of GeneFuncster, the user may also choose to use the so called average rank method for ranking the genes. This method first ranks the genes separately based on fold-changes and p-values, and then calculates the average ranks based on both of them.

Primary input for GeneFunscter consists of a list of Entrez gene identifiers or gene symbols. The list can be pasted directly to the input form or uploaded from a file with optionally included fold-changes and p-values to be used in visualisations and for allowing the ranking of genes. The user can optionally give a background gene list to be used in the filtered list analysis. There are many parameters available for fine tuning the analysis and result visualisation. Several organisms are currently supported and many others can easily be added when requested.

Results are reported on a summary html page in tables of terms sorted according to the term p-values, separately for all analysed main categories. These tables contain links to official term description pages, GO term graph and KEGG pathway maps where the associated genes are coloured and additional gene level plots taking advantage of the potentially available fold-change and/or p-values for genes. Overview of the various result reports produced with GeneFuncster is shown in Fig. 1.

3 Conclusion

Functional analysis has become a standard tool in elucidating the underlying biology within short unsorted or long sorted lists of genes. Coupled with informative visualisation of the results in a biological context, it has a huge potential in serving the biological research community. GeneFuncster provides functional enrichment analysis with an emphasis especially on the result reporting and visualisation. An earlier version of the tool has been successfully used in several studies including [5].

References

1. Gentleman, R., et al.: Bioconductor: open software development for computational biology and bioinformatics. Genome Biol. 10(5), R80 (2004)
2. Subramanian, A., et al.: GSEA-P: a desktop application for Gene Set Enrichment Analysis. Bioinformatics 23, 3251–3253 (2007)
3. R Development Core Team. R: A Language and Environment for Statistical Computing. R Foundation for Statistical Computing, Vienna, Austria (2008)
4. Huang, D.W., et al.: Bioinformatics enrichment tools: paths toward the comprehensive functional analysis of large gene lists. Nucleic Acids Res. 1(37), 1–13 (2009)
5. Koh, K.P., et al.: Tet1 and Tet2 regulate 5-hydroxymethylcytosine production and cell lineage specification in mouse embryonic stem cells. Cell Stem Cell 8(2), 200–213 (2011)

Effects of Molecular Noise on the Multistability in a Synthetic Genetic Oscillator

Ilya Potapov[1,2] and Evgenii Volkov[3]

[1] Computational Systems Biology Research Group, Dep. of Signal Processing, Tampere University of Technology, Finland
[2] Biophysics Department, Lomonosov Moscow State University, Russia
[3] Department of Theoretical Physics, Lebedev Physical Inst., Moscow, Russia

Abstract. We used 3-genes genetic oscillator as a model of oscillators coupled with quorum sensing, implemented as the production of a diffusive molecule, autoinducer. The autoinducer stimulates expression of the target gene within the oscillator's core, providing a positive feedback. Previous studies suggest that there is a hysteresis in the system between oscillatory (OS) and stationary (SS) dynamical solutions. We question the robustness of these attractors in presence of molecular noise, existing due to small number of molecules in the characteristic processes of gene expression. We showed distributions of return times of OS near and within the hysteresis region. The SS is revealed by the return times duration increase as the system approaches hysteresis. Moreover, the amplitude of stochastic oscillations is larger because of sensitivity of the system to the steady state even outside of the hysteresis. The sensitivity is caused by the stochastic drift in the parameter space.

Keywords: multistability, hysteresis, genetic oscillator.

1 Introduction

Oscillators are common in all contexts of life. For example, genes interact with each other constituting a network [1] which, for a certain structure, may lead to temporal oscillations in protein numbers and, thus, in a whole biochemical regulatory network which is governed by these genes [1].

The ability of living organisms to maintain the period and amplitude of temporal oscillations in presence of molecular noise and environmental fluctuations can be crucial for viability and evolutionary fitness of a single individual as well as a population [2].

We use a model of a synthetic 3-genes oscillator, repressilator [1], with quorum sensing (QS) [3], a mechanism for inter-cellular communication. Each gene in the network inhibits production of a gene next to it, thus, a cyclic structure is formed (Fig. 1). In addition to a ring of three genes, the scheme contains a coupling module implemented as a production of a small diffusive molecule — autoinducer (AI), which is a common agent for QS [3].

The recent study has shown new properties of the model: coexistence (hysteresis) of regular limit cycle (LC) and stable steady state (SSS) in a single cell oscillator [4]. The hysteresis between the LC and the SSS confers a cell the possibility to choose

D. Gilbert and M. Heiner (Eds.): CMSB 2012, LNCS 7605, pp. 386–389, 2012.

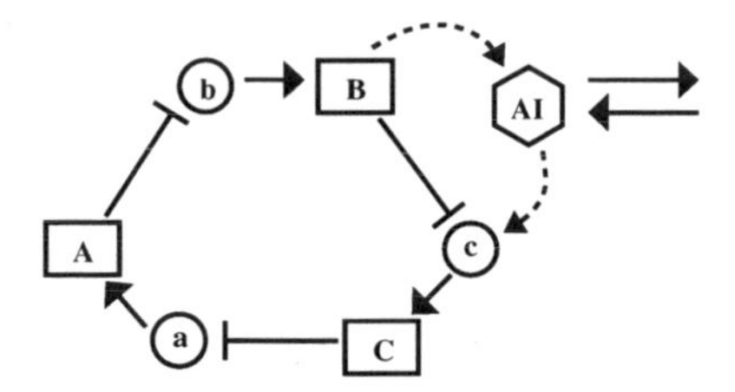

Fig. 1. Scheme of the repressilator with QS. Lowercase and uppercase letters are mRNA and proteins, respectively.

between different responses to external stimuli, for example, additional AI influx. Here we consider effects of noise, occurring due to small number of molecules, on a hysteresis properties of the circuit, since noise may lead to completely new dynamical regimes in a multistable system or destroy existing ones. We show how the dynamical properties of the stochastic system, like period distributions and amplitude of oscillations, change as the system approaches the hysteresis region.

2 Methods

We will use a dimension version (see in [5]) of the dimensionless model presented in [6] to study the stochastic effects on the dynamics of the single cell oscillator [4].

To account for noise due to small numbers of molecules constituting the system we use a standard approach, simulations using Stochastic Simulation Algorithm (SSA) [7]. Linear chemical interactions are modeled as unimolecular reactions. The propensities for the nonlinear reactions are represented by the corresponding nonlinear deterministic functions and computed at each step of the algorithm. We use this model technique due to unknown complex interactions taking place during these reactions.

For each parameter set we perform 100/400 simulations, each 10^5 s long, sampled every second. For the time series we compute the distribution of return times (periods) by taking a Poincaré section in the discrete state space and computing time intervals between moments when trajectory passes the section in one direction. The section is taken so that it is equidistant from maximum and minimum of the deterministic oscillations. If there are fast oscillations in the time series we choose 5000 s to be a minimal possible period: the algorithm sums computed periods until the sum reaches the threshold of 5000 s, then a period value is stored. We found 5000 s to be enough to cut off the fast fluctuations from the time series and not large enough to skew the true period distribution. This analysis is performed on the most abundant in numbers variable B.

3 Results

The dynamics of the deterministic model of the repressilator with QS is characterized by the limit cycle (LC) attractor that corresponds to the temporal oscillations of the system. This stable attractor emerges at the Hopf bifurcation for sufficiently large transcriptional rate [4].

The LC persists in a wide range of the transcriptional rate, but with its increase LC undergoes the infinite period bifurcation (IPB), i.e. the rotation of the representation point at the limit cycle is stopped due to the falling into a fixed point attractor in the phase space and the period of the oscillation goes to infinity. This fixed point attractor corresponds to the stable steady state dynamics of the system and is not related to the emergence of HB. The latter stable steady state (SSS) appears because of the AI influence on the system [4] and in some range of the transcriptional rates there are two stable dynamical behaviors of the system: the oscillations (LC) and the stationary dynamics (SSS). This leads the system to the hysteresis [4]. The study of the hysteresis deserves a special attention because of the new regulatory possibilities of the repressilators with quorum sensing.

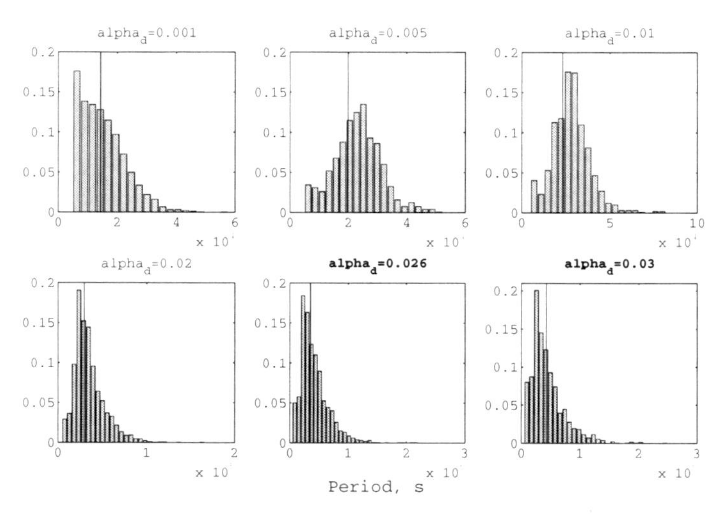

Fig. 2. The period distributions for different values of transcription rate ($\mathrm{alpha_d}$). Two last values are within hysteresis region (shown in bold). Vertical line denotes deterministic period. 100 and 400 simulations were performed for the distributions of the first and second row, respectively.

We perform stochastic simulations of the system to determine to what extent molecular noise affects the hysteresis properties of the system. We use the period distribution analysis (see the Methods) of the stochastic system in the conditions where the hysteresis occurs [4]. We choose 6 values of the transcription rate (here denoted as $\mathrm{alpha_d}$) and compute the distributions of periods for each of them. Results are shown in Fig. 2.

For the smallest value of the transcription rate ($\mathrm{alpha_d} = 0.001$) the peak of the period distribution corresponds to the lowest values of found periods (Fig. 2). This occurs because of the highly intensive intrinsic noise due to small $\mathrm{alpha_d}$. Thus, the LC is smashed by the noise and the dominating fluctuations are mostly captured by the period distribution analysis. As $\mathrm{alpha_d}$ approaches the IPB the LC grows in amplitude and the oscillations become more pronounced, and, additionally, the system becomes increasingly perturbed by the SSS. Namely, the stochasticity causes the drift in the parameter space that in practice creates the hysteresis where there is no hysteresis in the deterministic system. Thus, before the deterministic hysteresis region the peak of the period distribution passes to smaller values as compared to the deterministic period (Fig. 2), which appears because of the fast transitions between two attractors as the SSS becomes stronger in perturbing the system. These fast transitions do not allow the stochastic system to have the same period as that of the deterministic system, which can be clearly seen for much larger period values (data not shown). Namely, due to the high fluctuations' level the system's life time in either of the attractors is significantly shortened.

The amplitude of the stochastic oscillations before the hysteresis region becomes larger as compared to the deterministic case due to the perturbation caused by SSS. This also causes the appearance of the larger periods, which can be seen from the distribution peaks shifted rightwards from the corresponding deterministic periods for some moderate values of $\text{alpha}_\text{d} = \{0.005, 0.01\}$ (Fig. 2), where the LC is not either smashed by the noise or in the hysteresis region determined by the stochastic effects.

4 Conclusions

In this work we questioned the robustness of the multistability of the repressilator with quorum sensing in presence of molecular noise. We have shown that noise highly affects the oscillatory behavior of the repressilator by increasing the amplitude of the oscillations with moderate fluctuations in the period. The stochastic system has been shown to reveal the stable steady state even for the parameters outside of the hysteresis region. We have shown that the straightforward application of the Gillespie algorithm (standard approach in modeling gene expression [2]) to the model indicates that the system is weakly anchored in either of attractors present in the hysteresis region.

References

1. Elowitz, M., Leibler, S.: A synthetic oscillatory network of transcriptional regulators. Nature 403, 335 (2000)
2. McAdams, H.H., Arkin, A.: Stochastic mechanisms in gene expression. Proc. Natl. Acad. Sci. USA 94, 814–819 (1997)
3. Waters, C., Bassler, B.: Quorum sensing: cell-to-cell communication in bacteria. Ann. Rev. Cell Dev. Biol. 21, 319–346 (2005)
4. Potapov, I., Zhurov, B., Volkov, E.: "Quorum sensing" generated multistability and chaos in a synthetic genetic oscillator. Chaos: An Interdisciplinary Journal of Nonlinear Science 22(2), 023117 (2012)
5. `http://www.cs.tut.fi/~potapov/model.pdf`
6. Ullner, E., Koseska, A., Kurths, J., Volkov, E., Kantz, H., García-Ojalvo, J.: Multistability of synthetic genetic networks with repressive cell-to-cell communication. Phys. Rev. E 78 (2008)
7. Gillespie, D.: Exact stochastic simulation of coupled chemical reactions. J. Phys. Chem. 81, 2340–2361 (1977)

Towards an Ontology of Biomodelling

Larisa Soldatova, Qian Gao, and David Gilbert

Department of Information Systems and Computing, Brunel University, Uxbridge, UB8 3PH, London, UK
{Larisa.Soldatova,Qian.Gao,David.Gilbert}@brunel.ac.uk
http://brunel.ac.uk/siscm/disc/

Abstract. We present a core Ontology of Biomodelling (OBM), which formally defines principle entities of modelling of biological systems, and follows a structural approach for the engineering of biochemical network models. OBM is fully interoperable with relevant resources, e.g. GO, SBML, ChEBI, and the recording of biomodelling knowledge with Ontology of Biomedical investigations (OBI) ensures efficient sharing and re-use of information, reproducibility of developed biomodels, retrieval of information regarding tools, methods, tasks, bio-models and their parts. An initial version of OBM is available at disc.brunel.ac.uk/obm.

Keywords: ontology, knowledge representation, systems biology, modelling.

1 Introduction

We propose an Ontology of Biomodelling (OBM) that enables formally defined description of the key information about the design and analysis of biological models, motivated by the need for interoperability and re-usability of scientific knowledge. OBM is an important element in BioModel Engineering, a structured approach for the engineering of biochemical network models [2], which facilitates the design, construction and analysis of computational models of biological systems.

Ontology engineering is a popular solution for integration, interoperability and re-usability of scientific knowledge and is related to several biomodelling ontological resources. Systems Biology Ontology (SBO)[1] is a set of controlled, relational vocabularies of terms commonly used in systems biology, in particular in computational modelling, and informs the development of SBML[2]. The Ontology of Data Mining (OntoDM) enables recording of most essential information about predictive modelling as a type of data mining [8]. The Ontology of Biomedical investigations (OBI) provides semantic descriptors to report the most essential information about scientific investigations carried out in biomedical domains [4].

OBM employs an OBI approach to the reporting of investigations [4], incorporating all the relevant representations from other resources such as OntoDM

[1] http://www.ebi.ac.uk/sbo
[2] http://sbml.org

D. Gilbert and M. Heiner (Eds.): CMSB 2012, LNCS 7605, pp. 390–393, 2012.

and SBO, and is fully interoperable with GO (Gene Ontology), ChEBI (Chemical Entities of Biological Interest), SBML and other biomedical resources. It is designed as a foundation for an environment to support the key steps of the construction and analysis of models of biological systems (see section 2). Such an environment would assist in the selection of methods and tools for the construction and development of a model, searching over available models and their parts, advising appropriate validation methods, and reporting output models in standard formats. An ontology-driven environment would serve as an integration platform for most existing tools.

2 A Workflow of Biomodelling

The following steps are most commonly presented in a typical scenario of biomodelling:

1. Identification of tasks and requirements for a model construction. Construction of a model of a bio-system is a purpose-led process. Such purposes or tasks along with specified requirements could and should be recorded and collected for the benefit of the research community. Some scientific domains already have such dedicated task ontologies [6]. A constructed model can be checked for how well it satisfies the specified requirements. Such analysis can be used for meta-learning to find patterns of what models and design methods are suitable for particular tasks and requirements [1].

2. Modelling a domain of interest. Currently developers of bio-models rely on manual literature searches, interviews with biologists and chemists, analysis of experimental results in order to provide an unambiguous representation of the knowledge about a target biochemical system. Such domain background knowledge relevant for modelling should be represented in a form of domain modelling ontologies and preserved for future re-use. Domain modelling ontologies could be populated by automatic text mining searches that would extract required information from scientific literature, e.g. parameters of a system and recommended values, lists of genes, proteins, and chemical reactions that are associated with a target system; and by facts from already existing resources such as data bases, knowledge bases, and other domain ontologies.

3. Selection of model type and associated construction method. There are different types of models, e.g. static, dynamic – qualitative, quantitative (continuous, stochastic, hybrid), and it is important to select an appropriate for the purpose. OBM (empowered by a task ontology and a methods ontology) enables queries over available methods and models such as "Is there already an existing model A that can be modified to satisfy requirements R?", "Find all methods that are applicable for a task X with requirements R", "What modifications of method A would lead to the satisfaction of requirement R?", "Why is method A not suitable for task B?". Additionally, ready-made 'building blocks' of models could be re-used to construct a new model. Breitling et al. proposed typical building blocks for modelling of cellular signaling models (see [3] for more detail). In a similar way ready building blocks could be collected for other areas of interest, and offered to developers for specified tasks.

4. Selection of tools for model construction, analysis, and validation. There is a great variety of software available for modelling tasks. The Software Ontology (SWO)[3] is a resource for describing software tools, their types, tasks, versions, provenance and data associated. OBM (empowered by a task ontology and SWO) enables such queries over available methods and tools as "List all available tools for a task X and a method M", "Is a tool T well supported and has a large user community?", "Is software S freely available for academic purposes?".

5. Verification of a model. Once a model for a biological system has been created, it needs to be validated in a principled way. Does it produce reasonable predictions of system behavior? What datasets were used to test the model, and what properties do they have? Is the model it safe against deadlock and other system failures? Has it been tested in wet laboratory experiments? OntoDM provides formalized description of various model verification methods, and also of datasets that are used to test models [8], [7]. OBM extends OntoDM descriptors by the description of model checking by wet experimentations.

6. Exploration of a model. Once a model for a biological system has been created and verified, it can be used for simulating system behavior under various conditions. A domain modelling ontology could supply parameters and their values as an input for model simulation. All produced versions of a model should be recorded for further analysis, meta-learning, and re-use. Some tools, e.g. BioNessie[4], allow recording of model versions and simulations runs, but it is still not a common practice in systems biology to record and report this information.

7. Reporting. Currently SBML is widely used for recording and reporting of biomodels. However, it is important to do this not only the final model, but also for all steps in its development, versions, and verification so that scientist could make informative decisions about how to use models. Many parts of models are re-usable and should also be recorded as separate entities. OBM enables the recording and reporting of the key information about the process of the development, analysis, and verification in a machine processable form.

The basic methodology of biomodelling can be extended to meet the challenges of multiscale modelling of complex biological systems (see [5] for more detail).

3 OBM: A Core Ontology of Biomodelling

OBM follows OBI in the representation of a typical scientific workflow [4]. OBM imports from OBI classes that are relevant for the area of modelling of biological systems, e.g. investigator, planned-process, objective, conclusion-textual-entity, and relations between these classes, e.g. has-specified-input, precedes. Additionally, OBM defines biomodelling - specific classes for the representation of the

[3] `http://theswo.sourceforge.net/`

[4] `disc.brunel.ac.uk/bionessie/`

area of modelling of biological systems, e.g. model, model-component, task-identification, model-verification, model-representation, and relations between them, e.g. is-model-of.

We have followed the best practices in ontology engineering in the development of OBM. OBM employs standard upper level classes and relations where possible to ensure full interoperability with key biomedical ontologies and other resources, i.e. ChEBI, GO, OBI. OBM is designed in such a way that it compliments other ontological resources, e.g. SWO for the description of software, OntoDM for the description of predictive modelling, SBO for the description of a model, that are necessary for the efficient recording of the most essential information about biomodelling.

Future Work. The development and application of OBM will have the following next stages: (1) an extension of the coverage of OBM in order to include various biomodelling scenarios, and not only most typical ones; (2) an instantiation of OBM in order to enable search over workflows and their steps (currently OBM provides the conceptual description of biomodelling workflows, i.e. at the class level); (3) support of the development of an ontology-driven environment for biomodelling.

References

1. Brazdil, P., Carrier, G.C., Soares, C., Vilalta, R.: Metalearning. Applications to Data Mining. Springer (2009)
2. Breitling, R., Donaldson, R., Gilbert, D., Heiner, M.: Biomodel engineering - from structure to behavior. Trans. on Comput. Syst. Biol. 12, 1–12 (2010)
3. Breitling, R., Gilbert, D., Heiner, M., Orton, R.: A structured approach for the engineering of biochemical network models, illustrated for signalling pathways. Briefings in Bioinformatics 12 (2008)
4. Brinkman, R.R., Courtot, M., Derom, D., Fostel, J.M., He, Y., Lord, P., Malone, J., Parkinson, H., Peters, B., Rocca-Serra, P., Ruttenberg, A., Sansone, S.-A., Soldatova, L.N., Stoeckert, C.J., Turner, J.A., Zheng, J., The OBI Consortium: Modeling biomedical experimental processes with OBI. J. of Biomedical Semantics 1, 1–12 (2010)
5. Gao, Q., Gilbert, D., Heiner, M., Liu, F., Maccagnola, D., Tree, D.: Multiscale modelling and analysis of planar cell polarity in the drosophila wing. IEEE/ACM Transactions on Computational Biology and Bioinformatics (in press, 2012)
6. Mizoguchi, R., Vanwelkenhuysen, J., Ikeda, M.: Towards very large knowledge bases. Task ontology for reuse of problem solving knowledge. IOS Press (1995)
7. Panov, P., Džeroski, S., Soldatova, L.N.: OntoDM: An ontology of data mining. In: ICDMW 2008: Proceedings of the 2008 IEEE International Conference on Data Mining Workshops, pp. 752–760. IEEE Computer Society (2008)
8. Panov, P., Soldatova, L.N., Džeroski, S.: Towards an Ontology of Data Mining Investigations. In: Gama, J., Costa, V.S., Jorge, A.M., Brazdil, P.B. (eds.) DS 2009. LNCS, vol. 5808, pp. 257–271. Springer, Heidelberg (2009)

Author Index